Biological and Pharmacological Activity of Plant Natural Compounds III

Biological and Pharmacological Activity of Plant Natural Compounds III

Editors

Raffaele Pezzani
Sara Vitalini

MDPI • Basel • Beijing • Wuhan • Barcelona • Belgrade • Manchester • Tokyo • Cluj • Tianjin

Editors

Raffaele Pezzani
Phytotherapy Lab,
Department of Medicine
(DIMED)
University of Padova
Padova
Italy

Sara Vitalini
Department of Biomedical,
Surgical and Dental Sciences
Milan State University
Milano
Italy

Editorial Office
MDPI
St. Alban-Anlage 66
4052 Basel, Switzerland

This is a reprint of articles from the Special Issue published online in the open access journal *Molecules* (ISSN 1420-3049) (available at: www.mdpi.com/journal/molecules/special_issues/NP_III).

For citation purposes, cite each article independently as indicated on the article page online and as indicated below:

LastName, A.A.; LastName, B.B.; LastName, C.C. Article Title. *Journal Name* **Year**, *Volume Number*, Page Range.

ISBN 978-3-0365-7061-7 (Hbk)
ISBN 978-3-0365-7060-0 (PDF)

Cover image courtesy of Raffaele Pezzani

Contents

Preface to "Biological and Pharmacological Activity of Plant Natural Compounds III"

By joining the two previous editions, this Special Issue completes the trilogy of Special Issues entitled "Biological and Pharmacological Activity of Plant Natural Compounds", which deal with the properties and effects of phytonatural products in preclinical models. This third edition has also remained focused on their therapeutic potential, with the aim of increasing and contributing to the expansion of our scientific knowledge on plants in human health. We hope readers will enjoy this Special Issue as much as the previous publications.

Raffaele Pezzani and Sara Vitalini
Editors

molecules

MDPI

Article

Nephroprotective Effect of Diosmin against Cisplatin-Induced Kidney Damage by Modulating IL-1β, IL-6, TNFα and Renal Oxidative Damage

Tarique Anwer [1,*], Saeed Alshahrani [1], Ahmad M. H. Somaili [1], Abdullah H. Khubrani [1], Rayan A. Ahmed [1], Abdulmajeed M. Jali [1], Ayed Alshamrani [1], Hina Rashid [1], Yousra Nomeir [1], Mohammad Khalid [2] and Mohammad Firoz Alam [1]

1 Department of Pharmacology & Toxicology, College of Pharmacy, Jazan University, Jazan 45142, Saudi Arabia
2 Department of Pharmacognosy, College of Pharmacy, Prince Sattam Bin Abdulaziz University, Alkharj 11942, Saudi Arabia
* Correspondence: anwer.tariq@gmail.com; Tel.: +966-565772249

Abstract: Cisplatin (CP) is a platinum compound of the alkylating agent class that is used for the treatment of various types of cancer. However, CP treatments in cancer patients are accountable for nephrotoxicity, as it is a major adverse effect. Hence, this research study was proposed to investigate the nephroprotective effect of diosmin, a flavonoid glycoside of hesperidin derivatives against cisplatin-induced kidney damage. Wistar rats received a single intraperitoneal (i.p) injection of CP (7.5 mg/kg, i.p) to induce nephrotoxicity. The administration of CP significantly ($p < 0.001$) increased the markers of kidney function test (creatinine, blood urea nitrogen, and uric acid) and demonstrated histopathological changes in the kidney of the CP-treated nephrotoxic group. In addition, the CP-treated nephrotoxic group demonstrated a significant ($p < 0.001$) increase in lipid peroxidation (LPO) levels and depleted activities of reduced glutathione (GSH), glutathione peroxidase (GPx), glutathione reductase (GR), superoxide dismutase (SOD) and catalase (CAT).However, diosmin (100 and 200 mg/kg) treatments significantly reduced the elevated levels of kidney function test parameters and restored structural changes in the kidney ($p < 0.001$). The administration of diosmin (100 and 200 mg/kg) significantly ($p < 0.001$) reduced LPO levels, increased GSH content and showed improvements in the activities of GPx, GR, SOD and CAT. The markers of inflammatory cytokines such as IL-1β, IL-6 and TNFα significantly ($p < 0.001$) increased in the CP-treated nephrotoxic group, whereas diosmin (100 and 200 mg/kg) treatments significantly ($p < 0.001$) reduced the elevated levels of these cytokines. The findings of this research demonstrate the nephroprotective effect of diosmin against CP-induced kidney damage. Therefore, we conclude that diosmin may be used as a supplement in the management of nephrotoxicity associated with CP treatments in cancer patients.

Keywords: nephrotoxicity; cisplatin; diosmin; lipid peroxidation; inflammatory cytokines

Citation: Anwer, T.; Alshahrani, S.; Somaili, A.M.H.; Khubrani, A.H.; Ahmed, R.A.; Jali, A.M.; Alshamrani, A.; Rashid, H.; Nomeir, Y.; Khalid, M.; et al. Nephroprotective Effect of Diosmin against Cisplatin-Induced Kidney Damage by Modulating IL-1β, IL-6, TNFα and Renal Oxidative Damage. *Molecules* **2023**, *28*, 1302. https://doi.org/10.3390/molecules28031302

Academic Editors: Raffaele Pezzani and Sara Vitalini

Received: 15 December 2022
Revised: 11 January 2023
Accepted: 13 January 2023
Published: 30 January 2023

1. Introduction

Chemotherapy is one of the major treatment approaches for malignant solid tumors. Cisplatin (CP) is one of the most important anti-cancer drugs that is widely used in combination therapy regimens for the treatment of various malignant diseases such as head and neck, testicular, ovarian, cervical, breast, esophageal, stomach, bladder, small cell and non-small cell lung cancer [1,2]. The undesirable side effect due to CP increases morbidity and reduces the quality of life. Unfortunately, the clinical use of CP is mainly compromised due to its serious adverse effects that include nephrotoxicity, myelosuppression, severe nausea and vomiting, hearing loss and neurotoxicity [3,4]. The nephrotoxicity caused by CP in cancer patients is due to the presence of organic cation transporter-2 on the cell membrane of renal proximal tubular (RPT) cells, which actively accumulate CP and its metabolites [5].

CP-induced nephrotoxicity involves multiple pathways, such as the generation of intracellular reactive oxygen species (ROS) that triggers the cellular oxidative damage of kidney, inflammatory responses and apoptosis, which eventually contribute to renal tubular cell death and kidney dysfunction [6,7]. The important mediators of renal tubular cells damages include ROS-induced free radicals (hydroxyl radicals, hydrogen peroxide and superoxide anions). Previous studies reported that the massive production of ROS after CP treatments by renal tubular cells is possibly due to the disruption of the mitochondrial respiratory chain, the activation of NADPH oxidase and the formation of microsomes via the cytochrome P450 system [8,9]. However, other mechanisms also contribute to the pathogenesis of CP-induced nephrotoxicity, which include apparent inflammatory responses and the activation of mitogen activated protein kinase (MAPK). A recent study also reported that CP-induced nephrotoxicity is associated with the activation of the MAPK pathway that eventually plays an imperative role in the regulation of proliferation, differentiation, and apoptosis [10]. The inflammatory responses after CP treatments are a result of the activation of NF-kB, which further promotes the transcription of TNFα and other inflammatory cytokines such as IL-2, IL-6 and IL-1β [11].Hence, there is an unmet need to find safe and effective drugs that can ameliorate the serious adverse effects associated with the clinical uses of CP.

Diosmin is an important flavonoid glycoside of hesperidin derivatives, and it is present in the peels of citrus fruit (chemical structure is depicted as Figure 1). It introduces important pharmacological activities, such as anti-oxidant, anti-inflammatory, anti-hyperglycemic, anti-mutagenic, anti-rheumatic, anti-allergic activities, etc. [12]. A recent study reported on the protective effect of diosmin against cadmium-induced liver damage [13]. In addition, Ahmed et al. [14] demonstrated that diosmin has nephroprotective potential against alloxan-induced diabetic nephropathy. Therefore, the present study was designed to evaluate the nephroprotective effect of diosmin against cisplatin-induced kidney damage by modulating IL-1β, IL-6, TNFα and renal oxidative damage in rats.

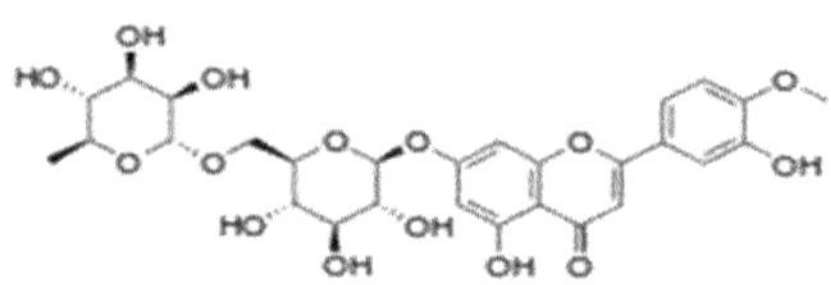

Figure 1. Chemical structure of diosmin.

2. Results

2.1. Diosmin and Markers of Kidney Function Test

The CP group exhibited a significant increase ($p < 0.001$) in the levels of kidney function parameters such as creatinine, BUN and uric acid levels compared to the normal control, as shown in Table 1. The administration of diosmin (100 and 200 mg/kg) demonstrated a significant decrease in the levels of kidney function parameters compared to the CP group ($p < 0.001$).

2.2. Diosmin and Histological Examination of Kidney

A histological examination of the normal and only diosmin-treated group exhibited the normal architecture of kidney without any pathological lesions showing normal glomeruli and no inflammation with normal renal tubules (Figure 2A,E). On the other hand, the CP group showed degenerative changes in the glomerular basement membrane, leukocyte infiltration with marked tubular vacuolization and necrosis (Figure 2B). Treatments with diosmin (100 and 200 mg/kg) attenuated the CP-induced degenerative changes in the kidney (Figure 2C,D). Tubular damage was examined and the tubular epithelial damage in the renal cortex was scored: 0—normal; 4—severe; 2—mild; 1—for minor. Scoring was based on tubular dilation, lesions, degeneration and tubular cell necrosis, etc.

Table 1. Effect of diosmin treatments on creatinine, BUN and uric acid against CP-induced nephrotoxicity in Wistar rats.

Groups	Treatment	Creatinine	BUN	Uric Acid
I	Normal	1.24 ± 0.03	37.30 ± 2.15	2.40 ± 0.06
II	Nephrotoxic (CP, 7.5 mg/kg, i.p)	5.01 ± 0.09 ##	101.50 ± 4.08 ##	6.07 ± 0.10 ##
III	Test group (received diosmin 100 mg/kg, p.o followed by cisplatin 7.5 mg/kg, i.p on 10th day of treatment.	3.95 ± 0.09 **	79.63 ± 2.47 **	4.84 ± 0.09 **
IV	Test group (received diosmin 200 mg/kg, p.o followed by cisplatin 7.5 mg/kg, i.p on 10th day of treatment.	2.67 ± 0.05 **	57.82 ± 2.24 **	3.50 ± 0.06 **
V	Only treated with diosmin (200 mg/kg, p.o)	1.18 ± 0.03	40.33 ± 1.14	2.49 ± 0.04

Group II (nephrotoxic) compared with group I (normal)→ ## $p < 0.001$. Group III and IV (treated with diosmin) compared with group II (nephrotoxic)→ ** $p < 0.001$.

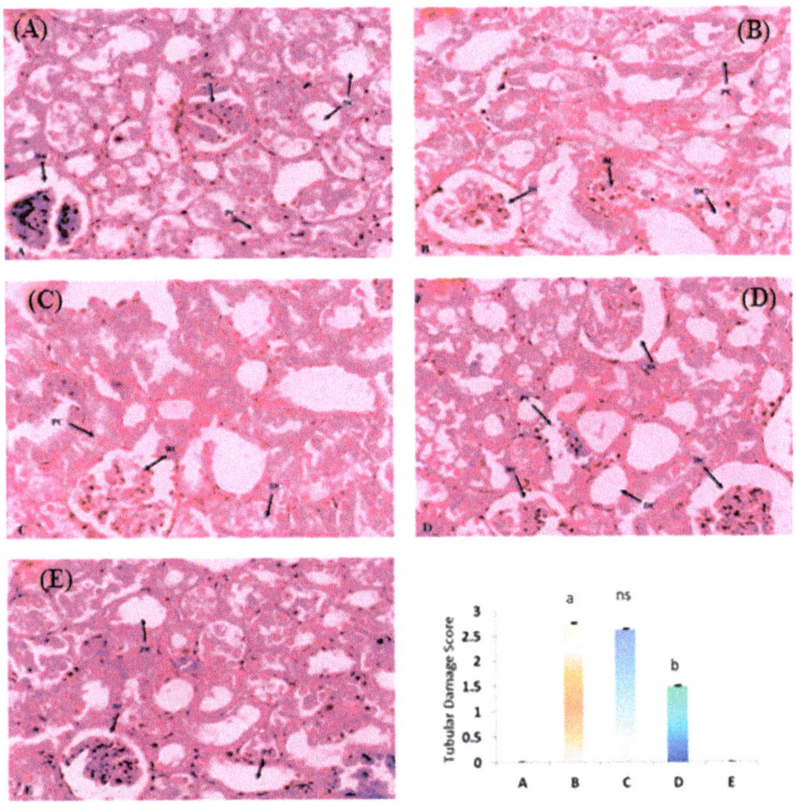

Figure 2. Kidney histology of normal, CP- and diosmin-treated groups. Sections of kidney tissue: (**A**) represents normal control, showing normal glomeruli and normal renal tubules without inflammation and necrosis (Score = 0). (**B**) represents the CP-treated group, showing degenerative changes in the glomeruli and renal tubules with marked tubular vacuolization and necrosis (Score = 3). (**C**) represents diosmin (100 mg/kg) + CP-treated group, showing mild degenerative changes with slight increase in the glomeruli and renal tubules size (Score = 2). (**D**) represents diosmin (200 mg/kg) + CP-treated group, showing a marked increase in the size of glomeruli and renal tubules (Score = 1). (**E**) represents the only diosmin-treated group (200 mg/kg), showing normal glomeruli and renal tubules (Score = 0). [a] $p < 0.001$ vs. NC, [b] $p < 0.001$ vs. CP, [ns] $p > 0.05$ vs. CP.

2.3. Diosmin and LPO Levels

Figure 3 shows the levels of LPO in the kidney tissue homogenate of normal and CP- and diosmin-treated groups. The CP group showed significant ($p < 0.001$) increase in LPO levels compared to the normal control. Diosmin (100 and 200 mg/kg) treatments significantly reduced the high levels of LPO compared to the CP group ($p < 0.001$).

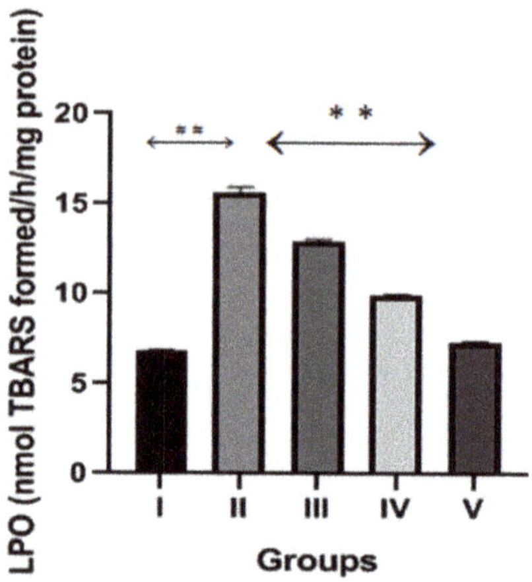

Figure 3. Effect of diosmin treatments on LPO levels against CP-induced nephrotoxicity in Wistar rats. Group II (nephrotoxic) compared with group I (normal)→ ## $p < 0.001$. Group III and IV (treated with diosmin) compared with group II (nephrotoxic)→ ** $p < 0.001$.

2.4. Diosmin and GSH Contents

The CP group demonstrated a significant ($p < 0.001$) decrease in the GSH contents compared to the normal control, as shown in Figure 4. The content of GSH significantly increased after diosmin treatments (100 and 200 mg/kg) when compared to the CP group.

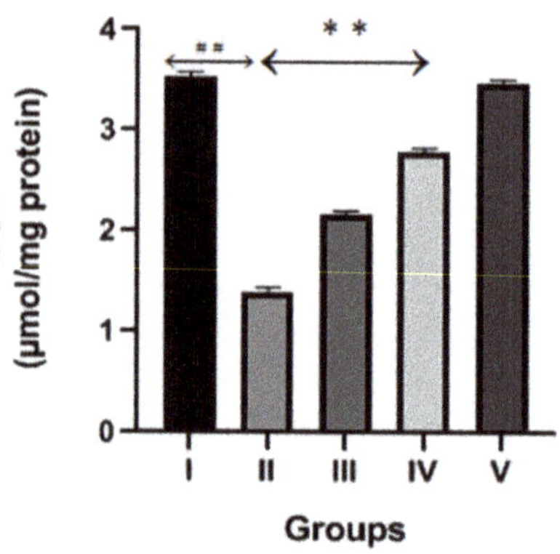

Figure 4. Effect of diosmin treatment on GSH contents against CP-induced nephrotoxicity in Wistar rats. Group II (nephrotoxic) compared with group I (normal)→ ## $p < 0.001$. Group III and IV (treated with diosmin) compared with group II (nephrotoxic)→ ** $p < 0.001$.

2.5. Diosmin and Activities of Antioxidant Enzymes (AOE)

Figure 5 displayed antioxidant enzymes (GPx, GR, SOD and CAT) activities in the kidney tissue homogenate of the normal and CP- and diosmin-treated groups. The activities of antioxidant enzymes were significantly ($p < 0.001$) reduced in the CP group compared to the normal control. However, diosmin (100 and 200 mg/kg) administrations significantly increased the activities of GPx, GR, SOD and CAT enzymes when compared to the CP group ($p < 0.01$–$p < 0.001$).

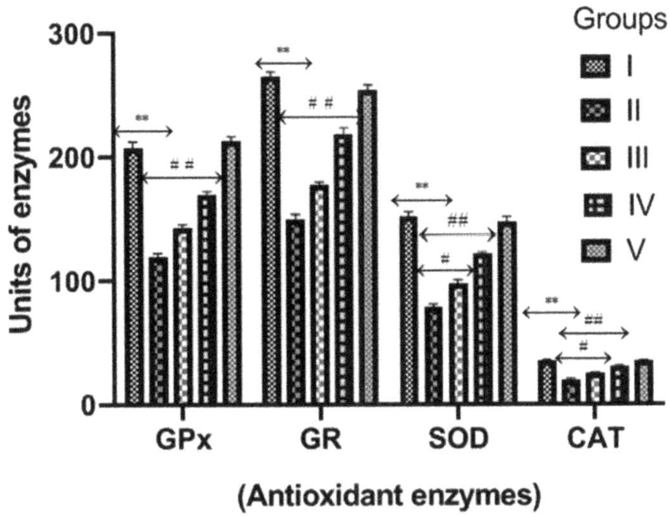

Figure 5. Effect of diosmin on antioxidant enzymes (GPx, GR, SOD, CAT) levels against CP-induced nephrotoxicity in Wistar rats. Group II (nephrotoxic) compared with group I (normal) → ** $p < 0.001$. Group III and IV (treated with diosmin) compared with group II (nephrotoxic) → # $p < 0.01$ → ## $p < 0.001$. Units of GPx and GR: (nmol NADPH oxidized/min/mg protein); SOD: (Units/mg protein); CAT: (nmol H_2O_2 consumed/min/mg protein).

2.6. Diosmin and Inflammatory Markers or Cytokines

The CP group exhibited a significant increase ($p < 0.001$) in inflammatory cytokines markers (TNFα, IL-6 and IL-1β) compared to the normal control, as shown in Figures 6–8. However, diosmin treatments (100 and 200 mg/kg) significantly reversed the elevated levels of these inflammatory cytokines compared to the CP group ($p < 0.001$).

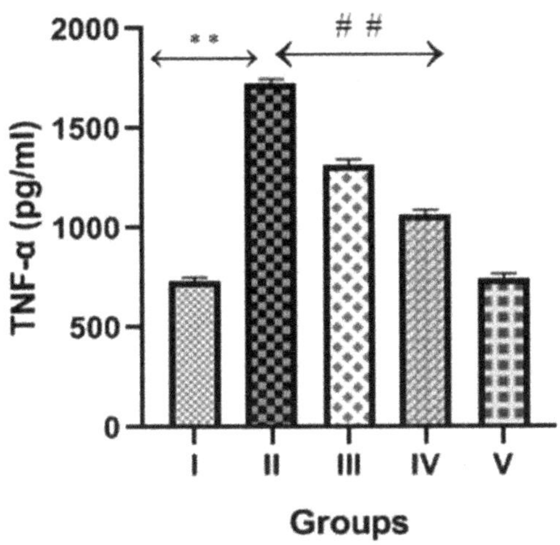

Figure 6. Effect of diosmin on TNFα levels against CP-induced nephrotoxicity in Wistar rats. Group II (nephrotoxic) compared with group I (normal)→ ** $p < 0.001$. Group III and IV (treated with diosmin) compared with group II (nephrotoxic)→ ## $p < 0.001$.

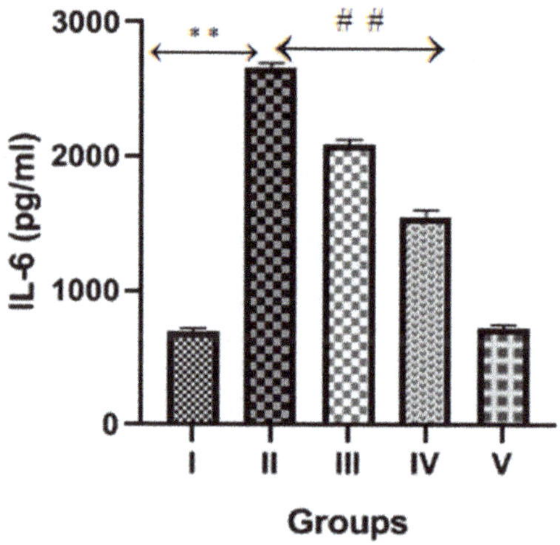

Figure 7. Effect of diosmin on IL-6 levels against CP-induced nephrotoxicity in Wistar rats. Group II (nephrotoxic) compared with group I (normal)→ ** $p < 0.001$. Group III and IV (treated with diosmin) compared with group II (nephrotoxic)→ ## $p < 0.001$.

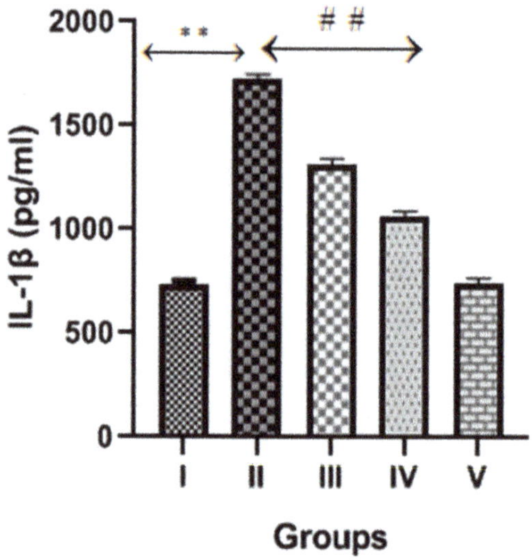

Figure 8. Effect of diosmin on IL-1β levels against CP-induced nephrotoxicity in Wistar rats. Group II (nephrotoxic) compared with group I (normal)→ ** $p < 0.001$. Group III and IV (treated with diosmin) compared with group II (nephrotoxic)→ ## $p < 0.001$.

3. Materials and Methods

3.1. Drugs, Chemicals and Biochemicals

Diosmin and cisplatin (CP) were obtained from Sigma Aldrich Pvt Ltd., St. Louis, MO, USA. Kits for biochemicals parameters such as creatinine, BUN and uric acid were purchased from a UK-based company (Randox) via the official supplier of the country (K.S.A). ELISA kits for cytokines parameters (TNFα, IL-6 and IL-1β) were procured from a USA-based company (MyBioSource) via official supplier of the country.

3.2. Animals (Wistar Rats)

Male Wistar rats within a weight range of 200–220 g were obtained from the animal house of the Medical Research Centre (MRC), Jazan University. All rats were shifted from the MRC animal house to Pharmacy College animal house for acclimatization one week prior to the experiment. The animals were kept under ideal laboratory conditions (temperature 25 ± 2 °C and humidity 45–50%) and provided with standard diets and water during experimental study. This research study was given approval by the standing committee for Scientific Research Ethics of the university (REC41/1-034).

3.3. Experimental Study Design

We used 40 Wistar rats for this study, and they were divided randomly into 5 groups with 8 animals in each.

Group I: normal control (NC) → received normal saline. Group II: CP-treated → received single dose of cisplatin (7.5 mg/kg, i.p) on 10th day of the treatment. Group III: Diosmin (100 mg/kg) + CP → received diosmin (100 mg/kg, p.o) daily for 14 days followed by cisplatin (7.5 mg/kg, i.p) on 10th day of the treatment. Group IV: Diosmin (200 mg/kg) + CP → received diosmin (200 mg/kg, p.o) daily for 14 days followed by cisplatin (7.5 mg/kg, i.p) on 10th day of the treatment. Group V: Only diosmin (200 mg/kg) → received higher dose of diosmin (200 mg/kg) daily for 14 days.

Blood was taken in a vacutainer tube on day 15 of the research study from overnight fasting rats. Serum was separated after the centrifugation of blood at 4000 rpm for 5 min and used for the estimation of creatinine, BUN, uric acid and cytokines such as TNFα, IL-6 and IL-1β. After blood withdrawal, all animals were sacrificed and both kidneys were excised for the assay of lipid peroxidation (LPO), reduced glutathione (GSH), glutathione peroxidase (GPx), glutathione reductase (GR), superoxide dismutase (SOD), catalase (CAT) and histopathological studies.

3.4. Tissue Homogenate Preparation

The homogenate of kidney was prepared in phosphate buffer (pH 7.4) and protease inhibitor (1 μg/mL). It was then centrifuged at 800× *g* for 5 min with a maintained temperature of 4 °C. The supernatant (S1) was separated from the homogenate and used for the assay of LPO and GSH. The remaining homogenate was again centrifuged at 10,500× *g* for 15 min with a maintained temperature of 4 °C to obtain a post-mitochondrial supernatant (PMS). The PMS was further used for the assay of GPx, GR, SOD and CAT.

3.5. Markers of Kidney Function Test

The markers of kidney function tests (Creatinine, BUN and uric acid) were estimated in serum to determine the efficiency of kidneys by using diagnostic kits from Randox, UK. The absorbance of the test sample and standard were recorded, and the values for each parameter were determined according the formula given in the kit.

3.6. Assay of Renal Oxidative Stress Parameters

The protein content in each sample was determined according to the method described by Lowry et al. [15] The assay for LPO was determined according to the method of Islam et al. [16].The assay for GSH was determined according to Jollow et al. [17]. The activities of GPx and GR were determined based on the methodology of Mohandas et al. [18]. The methodology of Claiborne [19] was used to determine the activity of CAT, whereas SOD activities were determined by the methodology of Marklund [20].

3.7. Assay of Inflammatory Markers or Cytokines

The assessments for the inflammatory markers of cytokines such as TNFα, IL-6 and IL-1β were carried out in the serum samples using the simple ELISA sandwich method according to the procedure provided in the kit. Both kidneys were homogenized, centrifuged and supernatants were prepared. The samples were added into a 96-well microplate that is pre-coated with specific antibodies of cytokines (TNFα, IL-6 and IL-1β). Then, the biotinylated detection antibody and horseradish peroxidase (HRP) conjugates were added to each well of the microplate and kept for incubation. The microplate was washed after incubation to remove free components, and the TMB substrate reagent was added. The appearance of yellow was measured by an ELISA plate reader at 450 nm wavelengths. The concentrations of these cytokines in test samples were calculated from the standard curve.

3.8. Histological Examination

Kidney tissue samples were taken out immediately after the animals were sacrificed and washed with ice-cold normal saline. The tissues were then fixed in 10% formalin solution. Later the tissue was cut into appropriate size and embedded in liquid paraffin to solidify and assemble into blocks. These blocks were then used to make sections of 3–5 μm thickness with the help of microtome. The slides of these sections were prepared and stained with hematoxylin and eosin (H&E) for further histological evaluation under a microscope at 40× magnification.

3.9. Statistical Analysis

Analyses of data were carried out using GraphPad Prism software 8 and analyzed by ANOVA followed by Tukey's post hoc test to ascertain statistical significances. The results were compared among each other and presented as ± standard errors mean (SEM). The minimum criterion for the results to be statistically significant was set at a value of $p < 0.05$.

4. Discussion

The clinical use of CP is associated with nephrotoxicity as one of the major reported adverse effects [21]. A preliminary study was conducted in our laboratory by Tripathi et al. [11] to establish the dose of CP (7.5 mg/kg) to induce nephrotoxicity in rats. We also confirmed that CP when given as a single intraperitoneal injection ata dose of 7.5 mg/kg is capable of causing nephrotoxicity. CP and its metabolites are secreted and reabsorbed in the renal tubules, leading to their accumulation in the kidney [22]. Previous studies reported that the concentration of CP and its metabolites in proximal tubular epithelial cells (PETCs) is 4–5 times higher than in the blood [23]. The accumulation of CP and its metabolites is mainly due to the involvement of various transport-mediated uptake processes in renal PTECs. This high concentration of CP and its metabolites causes acute kidney damage possibly due to cellular oxidative damage, inflammation, vascular injury, apoptosis and proximal tubular cell damage [24]. In this context of kidney dysfunction, the process of the glomerular filtration rate (GFR) is compromised, and albumin passes into the urine, leading to its decreased levels during the clinical use of CP. At the same time, creatinine, BUN and uric acid are not filtered properly, resulting in the high levels of these markers of renal functions. Earlier clinical reports also demonstrated that the CP treatment is accountable for the decrease in GFR along with an elevation in serum creatinine, BUN and uric acid [25].

In this study, we also observed a significant increase in creatinine, BUN and uric acid levels after CP treatments, which corroborates with kidney dysfunction. The administration of diosmin at both doses exhibited a significant decrease in creatinine, BUN and uric acid. A recent study reported that pretreatments with diosmin restored the markers of renal function parameters (albumin, creatinine and BUN) in the doxorubicin-treated model of nephrotoxicity in rats [26]. Renal function parameters (albumin, creatinine and BUN) and histological changes in the kidney are considered important markers of kidney injury. The common visible structural changes in the kidney by CP, includes the degeneration of the glomerular basement membrane, tubular vacuolization, tubular epithelial necrosis and leukocyte infiltration. The administration of two doses of diosmin provides protections against CP-induced degenerative changes in the kidney. The outcome of this study clearly demonstrates the protective effect of diosmin against CP-induced nephrotoxicity.

Cellular oxidative stress has been considered as the foremost factor that also participates in the pathogenesis of CP-induced renal dysfunction [27]. CP and its metabolites are retained by the renal tubular cells during their elimination process and stimulate the generation of ROS. These ROSs, which are overproduced after the administration of the CP, impair the oxidant–antioxidant system and damage renal tubular cell structures and functions via the stimulation of lipid peroxidation and the depletion of antioxidant enzymes [28]. Several previous reports documented that LPO is an important marker of cellular oxidative stress that damages cell membrane lipids via the generation of ROS, which subsequently increases malondialdehyde (MDA) levels, a metabolic product of lipid peroxidation [29,30]. The GSH is a well-known endogenous antioxidant, which detoxifies the ROS and protects cellular systems against the deleterious effects of excessively produced lipid peroxidation. The increased production of LPO is possibly due to an excessive production of ROS and the decreased utilization of GSH [31]. Previous studies also documented the significance of oxidative stress markers and ROS in CP-induced renal cell damage and its protection with natural antioxidants [32,33]. In this research study, we also observed significantly high levels of LPO and the decreased contents of GSH in the kidney tissue homogenate of CP group. However, both doses of the diosmin treatment significantly decreased the elevated levels of LPO and increased GSH contents in a dose dependent manner. These observations suggest that diosmin protects renal oxidative damage against CP by scavenging ROS and maintaining GSH contents.

Increased renal oxidative stress and depleted antioxidant enzymes (GSH dependent enzymes, catalase and superoxide dismutase) are accountable for the generation of ROS with a significant increase in lipid peroxidation. SOD is a very important enzyme and is present in ample amounts in all cells to protect them against the renal oxidative injury resulting from ROS. It converts highly toxic superoxide anions ($O_2^{\cdot -}$) to the less toxic hydrogen peroxide (H_2O_2), which is further detoxified by GPx and CAT [34,35]. GPx and CAT protect cells from oxidative damage by converting free H_2O_2 to water by utilizing proton generated from the oxidation of GSH. In order to provide more protons to protect cells from oxidative damage, GSH is oxidized to GSSG; hence, the GSH level starts depleting in cells that undergo oxidative damage. The GSH level is maintained in a cell by GR, which converts GSSG back to GSH. Depleted antioxidant enzymes reduce the clearance of H_2O_2 and facilitate the production of toxic hydroxyl radicals, leading to renal cell oxidative damage [36]. However, CP treatments significantly depleted the activities of antioxidant enzymes, viz., GPx, GR, SOD and CAT. Our results are also consistent with the previously published reports regarding the depleted activities of antioxidant enzymes in the CP group. In the present study, we established that both doses of diosmin augmented the activities of these antioxidant enzymes in the kidney tissue of CP-treated rats, which demonstrates the nephroprotective and antioxidant properties of diosmin.

Other than cellular oxidative stress, inflammatory pathway activations also play vital roles in the pathogenesis of CP-induced nephrotoxicity [6]. When CP and its metabolites accumulate in the kidney, it creates a microenvironment by generating ROS, which further activates transcription factor NF-kB. The activated NF-kB promotes the transcription of TNFα and other inflammatory cytokines (IL-1, IL-2 and IL-6). Several studies proposed that TNFα is the key regulator of inflammatory activities after CP exposure and is responsible for high levels of inflammatory cytokines and chemokines [37]. A previous study reported that diosmin treatments restored the expression of NF-kB and other inflammatory cytokines in the doxorubicin-treated model of nephrotoxicity [26]. In the present study, we observed significantly high levels of TNFα, IL-6 and IL-1β in CP-treated rats, suggesting the involvement of inflammatory cytokines that damage renal tubular cells. Interestingly, the levels of these inflammatory cytokines were reduced after the administration of diosmin, thus demonstrating its protective effect in CP-induced nephrotoxicity.

5. Conclusions

The results obtained from this research study demonstrated the potential protective effect of diosmin against CP-induced nephrotoxicity. This is possibly induced by modulating the kidney function test parameters, IL-1β, IL-6 and TNFα; renal oxidative damage;

and restoring structural changes in the kidney. Hence, we conclude that diosmin may be used as a supplement in the management of nephrotoxicity associated with CP treatments during cancer chemotherapy. Further clinical studies are required to support this research study for its clinical approval.

Author Contributions: Conceptualization, T.A. and S.A.; methodology, T.A., A.M.H.S., A.H.K., H.R., Y.N. and M.F.A.; software, R.A.A., A.M.J. and A.A.; validation, T.A., S.A., R.A.A., A.M.J., M.K. and M.F.A.; formal analysis, T.A., R.A.A. and M.K.; investigation, T.A., A.M.H.S., A.H.K., A.A., H.R., Y.N. and M.F.A.; resources, T.A. and S.A.; data curation, R.A.A., A.M.J. and A.A.; writing—original draft preparation, T.A., S.A. and M.F.A.; writing—review and editing, R.A.A., A.M.J., A.A., H.R., Y.N. and M.K.; visualization, A.M.H.S., A.H.K. and A.A.; supervision, T.A.; project administration, S.A. All authors have read and agreed to the published version of the manuscript.

Funding: The authors extend their appreciation to the Deputyship for Research and Innovation, Ministry of Education in Saudi Arabia for funding this research work via project No.ISP22-17.

Institutional Review Board Statement: The animal study protocol was approved by the standing committee for the Scientific Research and Ethics of Jazan University (REC41/1-034).

Informed Consent Statement: This study is inapplicable as it does not include humans.

Data Availability Statement: The data are incorporated within the article and can be available on request from the corresponding author.

Conflicts of Interest: The authors confirm that they do not have any conflict of interest with this research project.

References

1. Zhang, L.; Gu, Y.; Li, H.; Cao, H.; Liu, B.; Zhang, H.; Shao, F. Daphnetin protects against cisplatin induced nephrotoxicity by inhibiting inflammatory and oxidative stress response. *Int. Immunopharmacol.* **2018**, *65*, 402–407. [CrossRef] [PubMed]
2. Ridzuan, N.R.A.; Rashid, N.A.; Othman, F.; Budin, S.B.; Hussan, F.; Teoh, S.L. Protective role of natural products in cisplatin-induced nephrotoxicity. *Mini Rev. Med. Chem.* **2019**, *19*, 1134–1143. [CrossRef] [PubMed]
3. Yao, X.; Panichpisal, K.; Kurtzman, N.; Nugent, K. Cisplatin nephrotoxicity: A review. *Am. J. Med. Sci.* **2007**, *334*, 115–124. [CrossRef] [PubMed]
4. Hassan, I.; Chibber, S.; Naseem, I. Ameliorative effect of riboflavin on the cisplatin induced nephrotoxicity and hepatotoxicity under photoillumination. *Food Chem. Toxicol.* **2010**, *48*, 2052–2058. [CrossRef]
5. Ciarimboli, G.; Ludwig, T.; Lang, D.; Pavenstädt, H.; Koepsell, H.; Piechota, H.J.; Haier, J.; Jaehde, U.; Zisowsky, J.; Schlatter, E. Cisplatin nephrotoxicity is critically mediated via the human organic cation transporter 2. *Am. J. Pathol.* **2005**, *167*, 1477–1484. [CrossRef]
6. Ozkok, A.; Edelstein, C.L. Pathophysiology of cisplatin-induced acute kidney injury. *BioMed Res. Int.* **2014**, *2014*, 967826. [CrossRef]
7. Yu, X.; Meng, X.; Xu, M.; Zhang, X.; Zhang, Y.; Ding, G.; Huang, S.; Zhang, A.; Jia, Z. Celastrol ameliorates cisplatin nephrotoxicity by inhibiting NF-κB and improving mitochondrial function. *EBioMedicine* **2018**, *36*, 266–280. [CrossRef]
8. Kruidering, M.; Van de Water, B.; de Heer, E.; Mulder, G.J.; Nagelkerke, J.F. Cisplatin-induced nephrotoxicity in porcine proximal tubular cells: Mitochondrial dysfunction by inhibition of complexes I to IV of the respiratory chain. *J. Pharmacol. Exp. Ther.* **1997**, *280*, 638–649.
9. Wang, Y.; Luo, X.; Pan, H.; Huang, W.; Wang, X.; Wen, H.; Shen, K.; Jin, B. Pharmacological inhibition of NADPH oxidase protects against cisplatin induced nephrotoxicity in mice by two step mechanism. *Food Chem. Toxicol.* **2015**, *83*, 251–260. [CrossRef]
10. Cao, X.; Zhang, W.; Moore, P.K.; Bian, J. Protective smell of hydrogen sulfide and polysulfide in cisplatin-induced nephrotoxicity. *Int. J. Mol. Sci.* **2019**, *20*, 313. [CrossRef]
11. Tripathi, P.; Alshahrani, S. Mitigation of IL-1β, IL-6, TNF-α, and markers of apoptosis by ursolic acid against cisplatin-induced oxidative stress and nephrotoxicity in rats. *Hum. Exp. Toxicol.* **2021**, *40*, S397–S405. [CrossRef] [PubMed]
12. Abdel-Daim, M.M.; Khalifa, H.A.; Abushouk, A.I.; Dkhil, M.A.; Al-Quraishy, S.A. Diosmin attenuates methotrexate-induced hepatic, renal and cardiac injury: A biochemical and histopathological study in mice. *Oxidative Med. Cell. Longev.* **2017**, *2017*, 3281670. [CrossRef] [PubMed]
13. Aqir, M.S.; Ersalan, G. The effect of diosmin against liver damage caused by cadmium in rats. *J. Food Biochem.* **2019**, *43*, e12966.
14. Ahmed, S.; Mundhe, N.; Borgohain, M.; Chowdhury, L.; Kwatra, M.; Bolshette, N.; Ahmed, A.; Lahkar, M. Diosmin modulates the NF-kB signal transduction pathways and down-regulation of various oxidative stress markers in alloxan-induced diabetic nephropathy. *Inflammation* **2016**, *39*, 1783–1797. [CrossRef]
15. Lowry, O.H.; Rosenbrough, N.J.; Farr, A.L.; Randall, R.J. Protein measurements with the Folin's reagent. *J. Biol. Chem.* **1951**, *193*, 265–275. [CrossRef]

16. Islam, F.; Zia, S.; Sayeed, I.; Zafar, K.S.; Ahmad, A.S. Selenium-induced alteration of lipids, lipid peroxidation, and thiol group in circadian rhythm centers of rat. *Biol. Trace Element Res.* **2002**, *90*, 203–214. [CrossRef]
17. Jollow, D.J.; Mitchell, J.R.; Zampaglione, N.; Gillette, R. Bromobenzene induced liver necrosis: Protective role of glutathione and evidence for 3, 4–bromobenzene oxide as the hepatotoxic intermediate. *Pharmacology* **1974**, *11*, 151–169. [CrossRef]
18. Mohandas, J.; Marshall, J.J.; Duggin, G.G.; Horvath, J.S.; Tiller, D.J. Low activities of glutathione-related enzymes as factors in the genesis of urinary bladder cancer. *Cancer Res.* **1984**, *44*, 5086–5091.
19. Claiborne, A. Catalase activity. In *Handbook of Methods for Oxygen Radical Research*; Greenwald, R.A., Ed.; CRC Press: Boca Raton, FL, USA, 1985; pp. 283–284.
20. Marklund, S.; Marklund, G. Involvement of the superoxide anion radical in the autoxidation of pyrogallol and a convenient assay for superoxide dismutase. *Eur. J. Biochem.* **1974**, *47*, 469–474. [CrossRef]
21. Lin, M.-T.; Ko, J.-L.; Liu, T.-C.; Chao, P.-T.; Ou, C.-C. Protective effect of D-methionine on body weight loss, anorexia, and nephrotoxicity in cisplatin-induced chronic toxicity in rats. *Integr. Cancer Ther.* **2018**, *17*, 813–824. [CrossRef]
22. Volarevic, V.; Djokovic, B.; Jankovic, M.G.; Harrell, C.R.; Fellabaum, C.; Djonov, V.; Arsenijevic, N. Molecular mechanisms of cisplatin-induced nephrotoxicity: A balance on the knife edge between renoprotection and tumor toxicity. *J. Biomed. Sci.* **2019**, *26*, 25. [CrossRef] [PubMed]
23. Pabla, N.; Dong, Z. Cisplatin nephrotoxicity: Mechanisms and renoprotective strategies. *Kidney Int.* **2008**, *73*, 994–1007. [CrossRef] [PubMed]
24. Miller, R.P.; Tadagavadi, R.K.; Ramesh, G.; Reeves, W.B. Mechanisms of cisplatin nephrotoxicity. *Toxins* **2010**, *2*, 2490–2518. [CrossRef] [PubMed]
25. Abdelrahman, A.M.; Al-Suleimani, Y.; Shalaby, A.; Ashique, M.; Manoj, P.; Al-saadi, H.; Ali, B.H. Effect of levosimendan, a calcium sensitizer, on cisplatin-induced nephrotoxicity in rats. *Toxicol. Rep.* **2019**, *6*, 232–238. [CrossRef]
26. Ali, N.; AlAsmari, A.F.; Imam, F.; Ahmed, M.Z.; Alqahtani, F.; Alharbi, M.; AlSwayyed, M.; AlAsmari, F.; Alasmari, M.; Alshammari, A.; et al. Protective effect of diosmin against doxorubicin-induced nephrotoxicity. *Saudi J. Biol. Sci.* **2021**, *28*, 4375–4383. [CrossRef]
27. Alshahrani, S.; Tripathi, P.; Ashafaq, M.; Sultan, M.H.; Moni, S.S.; Tripathi, R.; Siddiqui, A.H.; Rashid, H.; Malhan, A.M. Role of renin blocker (Aliskiren) on cisplatin induced-nephrotoxicity in rats. *Toxin Rev.* **2020**, *41*, 175–185. [CrossRef]
28. Cetin, R.; Devrim, E.; Kiliçoğlu, B.; Avci, A.; Candir, O.; Durak, I. Cisplatin impairs antioxidant system and causes oxidation in rat kidney tissues: Possible protective roles of natural antioxidant foods. *J. Appl. Toxicol.* **2006**, *26*, 42–46. [CrossRef]
29. Alam, M.F.; Safhi, M.M.; Anwer, T.; Siddiqui, R.; Khan, G.; Moni, S.S. Therapeutic potential of vanillylacetone against CCl4 induced hepatotoxicity by suppressing the serum marker, oxidative stress, inflammatory cytokines and apoptosis in Swiss albino mice. *Exp. Mol. Pathol.* **2018**, *105*, 81–88. [CrossRef]
30. Anwer, T.; Alkarbi, Z.A.; Najmi, A.H.; Alshahrani, S.; Siddiqui, R.; Khan, G.; Alam, M.F. Modulatory effect of zingerone against STZ-nicotinamide induced type-2 diabetes mellitus in rats. *Arch. Physiol. Biochem.* **2021**, *127*, 304–310. [CrossRef]
31. Alam, M.F.; Alshahrani, S.; Alamer, E.A.; Alhazmi, M.A.; Anwer, T.; Khan, G.; Khan, A.; Tanweer, K.T.; Moni, S.S. Nephroprotective effects of 4-4(hydroxyl-3 methoxyphenyl)-2-butane against sodium tellurite induced acute kidney dysfunction by attenuating oxidative stress and inflammatory cytokines in rats. *Arab. J. Chem.* **2022**, *15*, 103857. [CrossRef]
32. Ma, N.; Wei, W.; Fan, X.; Ci, X. Farrerol attenuates cisplatin-induced nephrotoxicity by inhibiting the reactive oxygen species-mediated oxidation, inflammation, and apoptotic signaling pathways. *Front. Physiol.* **2019**, *10*, 1419. [CrossRef] [PubMed]
33. El-Kader, M.A.; Taha, R.I. Comparative nephroprotective effects of curcumin and etoricoxib against cisplatin-induced acute kidney injury in rats. *Acta Histochem.* **2020**, *122*, 151534. [CrossRef] [PubMed]
34. Anwer, T.; Sharma, M.; Pillai, K.K.; Haque, S.E.; Alam, M.M.; Zaman, M.S. Protective effect of bezafibrate on streptozotocin-induced oxidative stress and toxicity in rats. *Toxicology* **2007**, *229*, 165–172. [CrossRef] [PubMed]
35. Alam, M.F.; Khan, G.; Safhi, M.M.; Alshahrani, S.; Siddiqui, R.; Moni, S.S.; Anwer, T. Thymoquinone ameliorates doxorubicin-induced cardiotoxicity in swiss albino mice by modulating oxidative damage and cellular inflammation. *Cardiol. Res. Pract.* **2018**, *2018*, 1483041. [CrossRef] [PubMed]
36. Safhi, M.M.; Qumayri, H.M.; Masmali, A.U.M.; Siddiqui, R.; Alam, M.F.; Khan, G.; Anwer, T. Thymoquinone and fluoxetine alleviate depression via attenuating oxidative damage and inflammatory markers in type-2 diabetic rats. *Arch. Physiol. Biochem.* **2018**, *125*, 150–155. [CrossRef] [PubMed]
37. Ramesh, G.; Reeves, W.B. Cisplatin increases TNF-α mRNA stability in kidney proximal tubule cells. *Ren. Fail.* **2006**, *28*, 583–592. [CrossRef]

molecules 2023, 28, 1111 MDPI

Article

Flavonoids and Phenolic Acids from Aerial Part of *Ajuga integrifolia* (Buch.-Ham. Ex D. Don): Anti-Shigellosis Activity and In Silico Molecular Docking Studies

Fekade Beshah Tessema [1,2], Yilma Hunde Gonfa [1,3], Tilahun Belayneh Asfaw [1,4], Tigist Getachew Tadesse [5], Mesfin Getachew Tadesse [1], Archana Bachheti [6], Devi Prasad Pandey [7], Saikh M. Wabaidur [8], Kholood A. Dahlous [8], Ivan Širić [9], Pankaj Kumar [10], Vinod Kumar [10], Sami Abou Fayssal [11,12] and Rakesh Kumar Bachheti [1,13,*]

1 Department of Industrial Chemistry, Addis Ababa Science and Technology University, Addis Ababa 16417, Ethiopia
2 Department of Chemistry, Faculty of Natural and Computational Science, Woldia University, Woldia 400, Ethiopia
3 Department of Chemistry, Faculty of Natural and Computational Science, Ambo University, Ambo 19, Ethiopia
4 Department of Chemistry, College of Natural and Computational Science, Gondar University, Gondar 196, Ethiopia
5 Bio and Emerging Technology Institute, Health Biotechnology Directorate, Addis Ababa 5954, Ethiopia
6 Department of Environment Science, Graphic Era (Deemed to be University), Dehradun 248002, India
7 Department of Chemistry, Govt Degree College Dehradun Shahar, Suddhowala, Dehradun 248007, India
8 Department of Chemistry, College of Science, King Saud University, Riyadh 11451, Saudi Arabia
9 University of Zagreb, Faculty of Agriculture, Svetosimunska 25, Zagreb 10000, Croatia
10 Agro-Ecology and Pollution Research Laboratory, Department of Zoology and Environmental Science, Gurukula Kangri (Deemed to be University), Haridwar 249404, India
11 Department of Agronomy, Faculty of Agronomy, University of Forestry, 10 Kliment Ohridski Blvd, 1797 Sofia, Bulgaria
12 Department of Plant Production, Faculty of Agriculture, Lebanese University, Beirut 1302, Lebanon
13 Centre of Excellence in Biotechnology and Bioprocess, Addis Ababa Science and Technology University, Addis Ababa 16417, Ethiopia
* Correspondence: rakesh.kumar@aastu.edu.et

Abstract: Shigellosis is one of the major causes of death in children worldwide. Flavonoids and phenolic acids are expected to demonstrate anti-shigellosis activity and anti-diarrheal properties. The aerial part of *A. integrifolia* is commonly used against diarrhea. This study aimed to identify flavonoids and phenolic acids responsible for this therapeutic purpose. Antioxidant activity, total phenol content, and total flavonoid content were determined. The antibacterial activity of the aerial part against *Shigella* spp. was also tested using the agar well diffusion method. HPLC analysis was performed using UHPLC-DAD for different extracts of the aerial part. Autodock Vina in the PyRx platform was used to screen responsible components. Ciprofloxacin was used as a reference drug. An enzyme taking part in pyrimidine biosynthesis was used as a target protein. Molecular docking results were visualized using Discovery Studio and LigPlot1.4.5 software. Antioxidant activity, total phenol content, and total flavonoid content are more significant for the aerial part of *A. integrifolia*. From HPLC analysis, the presence of the flavonoids, quercetin, myricetin, and rutin and the phenolic acids gallic acid, chlorogenic acid, and syringic acid were identified from the aerial part of *A. integrifolia*. Regarding the antibacterial activity, the aerial part shows considerable activity against *Shigella* spp. Binding energies, RMSD and Ki values, interaction type, and distance are considered to identify the components most likely responsible for the therapeutic effects and observed activity. Antioxidant activity, total phenol content, and total flavonoid content of the aerial part are in line with anti-shigellosis activity. The top five components that are most likely potentially responsible for therapeutic purposes and anti-shigellosis activity are chlorogenic acid, rutin, dihydroquercetin, dihydromyricetin, and kaempferol.

Citation: Tessema, F.B.; Gonfa, Y.H.; Asfaw, T.B.; Tadesse, T.G.; Tadesse, M.G.; Bachheti, A.; Pandey, D.P.; Wabaidur, S.M.; Dahlous, K.A.; Širić, I.; et al. Flavonoids and Phenolic Acids from Aerial Part of *Ajuga integrifolia* (Buch.-Ham. Ex D. Don): Anti-Shigellosis Activity and In Silico Molecular Docking Studies. *Molecules* **2023**, *28*, 1111. https://doi.org/10.3390/molecules28031111

Academic Editors: Raffaele Pezzani and Sara Vitalini

Received: 11 December 2022
Revised: 11 January 2023
Accepted: 18 January 2023
Published: 22 January 2023

Keywords: *Ajuga integrifolia*; antioxidant activity; binding energies; flavonoids; phenolic acids; shigellosis

1. Introduction

Ajuga integrifolia Buch.-Ham. Ex D. Don (Syn: *A. remota*; *A. bracteosa*) belonging to the family Lamiaceae is known by the names *Armagussa*, *Etse Libawit*, *Medhanit*, and *Anamuro* [1]. It is an evergreen, clump-forming flowering species of the genus Ajuga. The more than 300 species of the genus *Ajuga* have many variations [2]. Many biological activities were reported for the *Ajuga* spp., specifically for *A. integrifolia* and its synonyms. These include antioxidant and anti-inflammatory [3], antidiabetic [4], antibacterial, diuretic, stimulant, astringent, rheumatism, febrifuge, blood purification [5], and anticancer effects [6]. These activities are achieved in addition to the biological functions of phytochemicals in protecting plant species against pathogens and herbivores and acting as stress-protecting agents [7].

Diarrhea is a major cause of morbidity and mortality for children in developing countries, including Ethiopia [8,9]. It is among the ten major diseases identified and reported to kill human beings [10]. Shigellosis is a leading cause of dysentery and accounts for 5–10% of diarrheal diseases globally. Shigellosis is characterized by inflammation and ulcer formation on the large intestine, with signs such as fever and stomach pain [11]. It is caused by enteroinvasive *E. coli* (EIEC) and other species of *Shigella*, namely *S. dysenteriae*, *S. flexneri*, *S. boydii*, and *S. sonnei* [12]. *Shigella* spp. are easily differentiated from E. coli, but difficult to differentiate among the species due to similar biochemical traits and modes of invasion [13]. *Shigella* spp. invade the gut-lining epithelium and cause shigellosis [14]. The prevalence of *Shigella* spp. in Ethiopia was reported to be 6.6% and resistant to ampicillin (83.1%), amoxicillin (84.1%), erythromycin (86.5%), and multi-drugs (83.2%). Ciprofloxacin (8.9%) and norfloxacin (8.2%) were reported to have low resistance patterns [15].

Shigella spp. infections with increased levels of antimicrobial resistance in children are major public health problems in developing countries [16]. Poor sanitation practices and limited access to clean water are the main reasons for the spread of the disease through the fecal-oral route [17]. *Shigella* spp. is developing resistance to oral antibiotics [18–22]. *Shigella* enters the human body with contaminated food and water as part of its life cycle. It develops an acid-resistant system (glutamate decarboxylase system) which grants the ability to withstand the stomach's acidic environment [17,23]. Passing through the stomach, the pathogen reaches the intestine and starts invasion. Type III Secretion System (T3SS) mediates *Shigella* invasion and leads to infection [17].

The proliferation of many pathogens, including *Shigella* spp., uses the pyrimidine biosynthesis pathway. *N*-carbamoyl-l-aspartate reversible interconversion to 4,5-dihydroorotate is catalyzed by dihydroorotase (DHO). This makes it a therapeutic target for inhibiting bacterial growth [24]. It is a crucial candidate for evaluating the response to antimicrobial resistance because there is considerable variation between bacterial and mammalian DHOs. Ethnobotanical studies indicate that the aerial part of *A. integrifolia* and its synonyms are used for hypertension [25–28], antimalarial and insecticidal activities [29–31], diuretic activity [32], tonsillitis [33], epilepsy [34], breast cancer [35], wound healing [36], abdominal pain and anthelmintic [37–39], and diarrhea [25,40–44]. Individual components have not been identified so far for anti-diarrhea; activity. Flavonoids and polyphenolics are known to have such activity [45,46].

Flavonoids have been discovered in vitro to be potent antibacterial agents against a variety of pathogens, which is not surprising since flavonoids are known to be produced by plants in response to microbial infection. There have been reports of the antibacterial properties of flavonoid-rich plant extracts from various species [47]. Apigenin, galangin, flavone and flavonol glycosides, isoflavones, flavanones, and chalcones are only a few flavonoids that have been proven to have strong antibacterial activity [48]. Rather than focusing on a single site of action, antibacterial flavonoids may have many cellular targets.

Their ability to form complexes with proteins by non-specific forces including hydrogen bonds and hydrophobic effects, as well as by forming covalent bonds, is one of their molecular actions. Their capacity to inactivate microbial adhesins, enzymes, cell membrane transport proteins, etc., may therefore be related to their antibacterial activities. Microbial membranes may also be damaged by lipophilic flavonoids [49].

Despite the diversity of natural inhibitors, it is interesting to notice that many of them are phenolic in origin. Many researchers have thought of appropriate skeletons based on the architectures of natural compounds and developed fresh synthetic inhibitors. The previous analysis of tabulated compounds and the inhibitory outcomes point to the flavonol moiety as a potential scaffold, a key structural component, and a starting point for the continued development of novel inhibitors among all the natural and synthesized flavonoid derivatives [50]. Looking for potential inhibitors, in silico molecular docking studies are now becoming part of many pharmacological studies. This strategy produced a conceptual idea for a potential pharmacophore, followed by a targeted chemical synthesis. The application of a ligand-based method and focused chemical synthesis produced novel small synthetic molecules (MW < 500) [51]. After the virtual screening, the short-listed compounds were screened by the SwissADME modeling tool to discard any molecules with undesirable pharmacokinetics and therapeutic qualities. To identify the top binders of the target protein, iterative docking was also used for the drug-like compounds. For the evaluation of the dynamic behavior, stability of the protein–ligand complex, and binding affinity, binding free energy calculations were made, which led to the identification of possible inhibitors [52].

Previous studies indicated that *Shigella* infections were more common in Ethiopia and that they tended to be more resistant to conventional medications such as ampicillin, amoxicillin, and erythromycin [15,21]. A considerable prevalence combined with higher levels of resistance to the currently used drugs brings a demand for a natural-based antibiotic. WHO identified *Shigella* as a marked pathogen against which new drugs need to be formulated and proposed in silico approach to identify drug targets [17]. Considering these, we aimed to justify the anti-shigellosis activity of the aerial part of *A. integrifolia* and identify responsible components for this activity using in silico molecular docking.

2. Results

2.1. Antioxidant Activity (DPPH Assay)

The DPPH antioxidant activities of the methanol extracts of the aerial of *A. integrifolia* (AIA) and root part of *A. integrifolia* (AIR) were analyzed using eight concentration points. The antioxidant activity was confirmed first by its color change, which indicated the scavenging activities of the parts of the target medicinal plant. The IC_{50} values of the extracts of AIA and AIR were calculated after being transformed and normalized to the minima (zero) to maxima (plateaus) from log [inhibitor conc.], which are variable responses. The IC_{50} value for the aerial part is significantly lower than the root sample with similar test conditions (Table 1). The antioxidant activity of the aerial part is also much closer to the reference ascorbic acid. The difference in the % of radical scavenging activity of the aerial versus the root sample is observed at lower concentrations (Figure 1). The regression constants (R^2) for the sample extracts and the reference are similar and close to 1.

Table 1. Result summary of IC_{50} and R^2 values.

Sample	Log(inhibitor) vs. Normalized Response—Variable Slope		
	LogIC$_{50}$	**IC$_{50}$ (μg/mL)**	**R^2**
AIA	−0.053 ± 0.01	0.885 ± 0.02	0.991 ± 0.001
AIR	0.318 ± 0.004	2.080 ± 0.017	0.990 ± 0.002
Ascorbic acid	−0.908 ± 0.01	0.124 ± 0.01	0.991 ± 0.02

AIA: *A. integrifolia* aerial part, AIR: *A. integrifolia* root part.

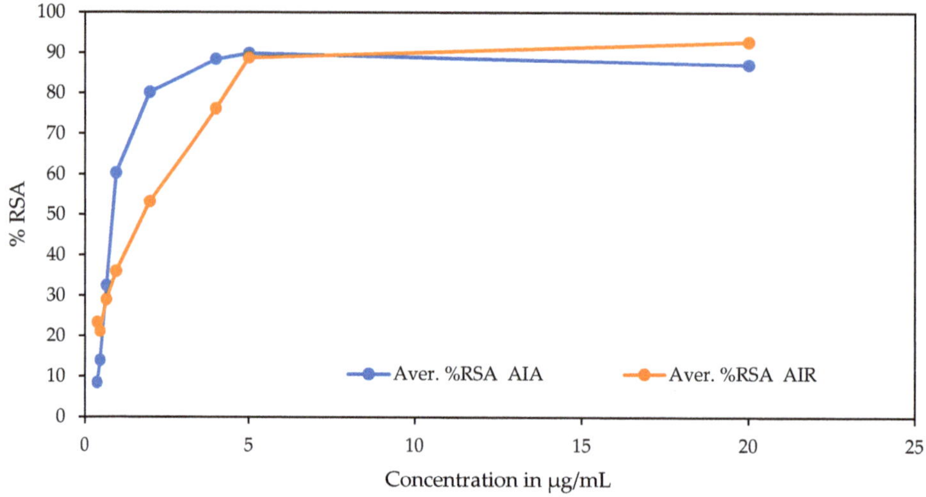

Figure 1. Graph of % free radical scavenging activity of DPPH by the extracts vs. concentration of sample extracts.

2.2. *Total Phenolic and Total Flavonoid Content*

Total phenolic content (TPC) and total flavonoid content (TFC) of the extracts of AIA and AIR samples were determined using Folin–Ciocalteu and aluminum chloride methods, respectively. The formation of yellow and blue colors was observed after the addition of the respective reagents for TPC and TFC determinations, respectively. The equation of the standard curve for the determination of TPC was $y = 0.0023x - 0.0693$, where $R^2 = 0.9981$, and for TFC, $y = 0.002x + 0.0399$, where $R^2 = 0.9971$. Both the TPC and TFC of AIA are almost double that of AIR (Table 2).

Table 2. Total phenol content and total flavonoid content for the sample extracts.

Sample	TPC—Total Phenolic Content (mg/100g)	TFC—Total Flavonoid Content (mg/100g)
AIA	868.98 ± 9.98	742.44 ± 24.47
AIR	475.79 ± 4.35	391.98 ± 6.44

AIA: *A. integrifolia* aerial part, AIR: *A. integrifolia* root part.

2.3. *HPLC Analysis*

HPLC analysis was carried out using the method mentioned above, and chromatograms were exported for qualitative identification of the presence of phenolic acids and flavonoids. The components from the sample extracts were identified using the reference standards' peak shape, retention time, and UV-Vis spectra. As shown in Figure 2, the reference peaks were identical and overlaid, leading to the confirmation of the presence of the components. After the identification of the components in the sample extracts, the components were quantified using external calibration standards. From the HPLC chromatogram (Figure 2), gallic acid (RT: 4.450 min), chlorogenic acid (RT: 4.160 min), and syringic acid (RT: 5.897 min) were found in 75% methanol extract; myricetin (RT: 8.117 min) and quercetin (RT: 11.673 min) were found in the methanol extract; and quercetin (RT: 12.600 min) and rutin (RT: 19.797 min) were found in the acetone dip instant extract of aerial part of *A. integrifolia*. As shown in Table 3, the concentration of the phenolic acids and flavonoids ranges from 1.56 to 11.26 mg/100 g for AIR, from 7.97 to 58.23 mg/100 g for methanol extract of AIA (AIA MeOH), from 110 to 220 mg/100 g for acetone extract of AIA (AIA acet), and from 2.07 to 60.13 mg/100 g for 75% methanol extract of AIA. The concentrations are higher for acetone extract as the extraction method is mainly for flavonoids. AIR showed lower concentrations.

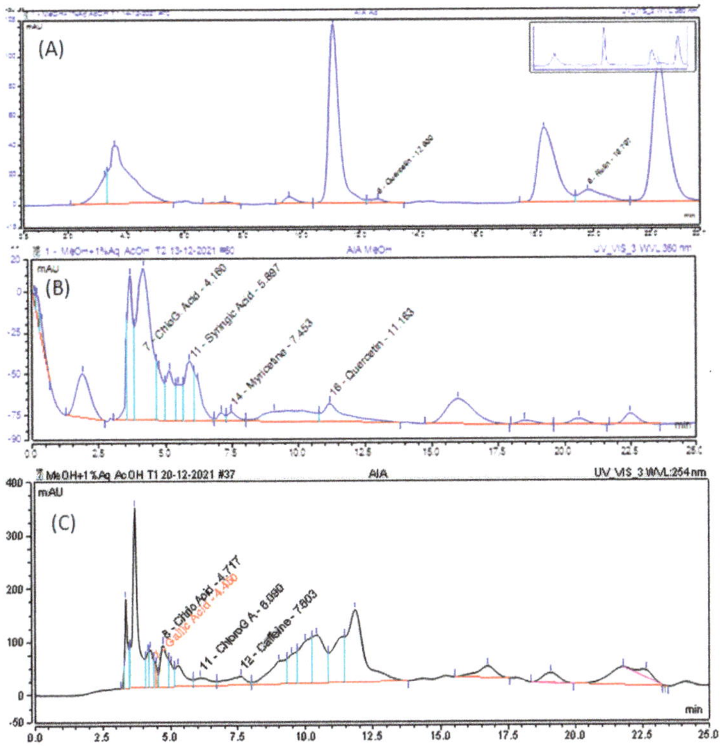

Figure 2. HPLC chromatograms for (**A**) acetone, (**B**) methanol, and (**C**) 75% methanol extract of the aerial part of *A. integrifolia*.

Table 3. HPLC quantitative result for *A. integrifolia* extracts.

Sample/Extracts	Mass	DF	Analytes	Ret. Time (min)	Amount (mg/L) AV ± SD	Concentration (mg/100 g) AV ± SD
AIA	5.0026	1	Chlorogenic Acid	4.237	23.66 ± 1.6	23.65 ± 1.6
AIA	5.0026	1	Gallic Acid **	4.060	60.16 ± 0	60.13 ± 0
AIA	5.0026	1	Myricetin	8.107	7.85 ± 1.07	7.84 ± 1.07
AIA	5.0026	1	Quercetin	11.967	2.07 ± 0.55	2.07 ± 0.55
AIA	5.0026	1	Syringic Acid	6.093	64.22 ± 5.48	64.18 ± 5.48
AIA	5.0026	1	Rutin	-	-	-
AIA Ac *	0.3	2	Chlorogenic Acid	4.287	6.61 ± 1.63	220.4 ± 54.25
AIA Ac *	0.3	2	Gallic Acid	-	-	-
AIA Ac *	0.3	2	Myricetin **	7.723	3.31 ± 0	110.27 ± 0
AIA Ac *	0.3	2	Quercetin	-	-	-
AIA Ac *	0.3	2	Syringic Acid	5.997	4.94 ± 2.2	164.59 ± 73.36
AIA Ac *	0.3	2	Rutin	-	-	-
AIA MeOH	5.0026	3	Chlorogenic Acid	4.183	19.42 ± 1.24	58.23 ± 3.72
AIA MeOH	5.0026	3	Gallic Acid **	4.030	8.12 ± 0	24.34 ± 0
AIA MeOH	5.0026	3	Myricetin	7.463	3.71 ± 0.72	11.12 ± 2.17
AIA MeOH	5.0026	3	Quercetin	11.873	3.8 ± 1.95	11.4 ± 5.86
AIA MeOH	5.0026	3	Syringic Acid	5.897	9.77 ± 1.97	29.3 ± 5.9
AIA MeOH	5.0026	3	Rutin	-	-	-
AIR	5.008	1	Chlorogenic Acid	4.31	7.98 ± 1.45	7.97 ± 1.45
AIR	5.008	1	Gallic Acid **	3.957	11.28 ± 0	11.26 ± 0
AIR	5.008	1	Myricetin	7.437	2.51 ± 0.91	2.51 ± 0.9
AIR	5.008	1	Quercetin	11.677	1.57 ± 0.56	1.56 ± 0.56
AIR	5.008	1	Syringic Acid	5.997	3.54 ± 1.81	3.53 ± 1.81
AIR	5.008	1	Rutin	-	-	-

* Concentration computed from extract mass. ** Quantitative determinations from a single measurement. DF, dilution factor; -: not applicable.

2.4. Antibacterial Study

The antimicrobial susceptibility test was managed with four strains, including *Shigella* spp. The strains are two Gram-negative and one Gram-positive, chosen based on availability. Methanol extract of AIA showed smaller activity for *E. coli* and a greater zone of inhibition (17.67 ± 1.47) for *Shigella* spp. (Table 4). The ciprofloxacin reference is used following the moderate resistance of *Shigella* spp. for chloramphenicol. AIR does not show any activity for all strains considered.

Table 4. Measurement of the zone of inhibition (mm) of test bacteria.

S. No.	*E. coli*	*P. aeruginosa*	*S. aureus*	*Shigella* spp.
AIA	3.33 ± 0.82	0	0	17.67 ± 1.47
AIR	0	0	0	0
Chloramphenicol	-	-	-	10.33 ± 0.82
Ciprofloxacin	-	-	-	20.33 ± 1.08

AIA: *A. integrifolia* aerial part, AIR: *A. integrifolia* root part.

2.5. In silico Molecular Docking Study

Molecular and physicochemical characteristics, including molecular weight (MW), molecular refractivity (MR), count of particular atom kinds, and topological polar surface area (TPSA), can be obtained via the SwissADME online system. Lipophilicity is often described by the partition coefficient between *n*-octanol and water. A soluble molecule facilitates drug development activities, as handling and formulation problems can be managed easily in oral administration, influencing absorption. Lipinski's Rule of Five and the Bioavailability Radar enable one to decide the drug-likeness of a molecule. Drug-likeness properties describe a molecule's potential to be an oral drug, or its bioavailability. The resolution of VcDHO (PDB ID: 5vgm) was to be 1.95 Å for the full validation report for the target protein. The overall quality factor retrieved from the PROHECK online server and full validation report was 97.4026%. From the Ramchandran plot, residues in the most favored regions were found to be 90.5% (Supplementary Material S1, Figure S1). No residues were in the disallowed regions. The Q-mean value after preparation was also 0.17 (z-score) (Supplementary Material S1, Figure S2). The resolution, overall quality, and Q-mean values are acceptable for the intended activity.

As shown in Table 5, Lipinski rule violations are tolerable for some molecules, and there were no violations for more than half of the ligands. Only rutin is beyond the tolerable limit of violation. The bioavailability score is above 0.5 for most and 0.1–0.2 for three molecules, including rutin. The synthetic accessibility is also >2.5 for most, including 6.52 for rutin. Medicinal chemistry friendliness is also reflected by lower pain alert and the absence of problematic fragments from Brenk filters [53].

The Bioavailability Radar shown in Figure 3 is used to predict a molecule's drug-likeness properties. The pink area on the radar represents the acceptable range for each property. The respective values are given as lipophilicity: $-0.7 < \text{XLOGP3} < +5.0$, size: $150 < \text{MW} < 500$ g/mol, polarity: $20 < \text{TPSA} < 130$ Å^2, solubility: $\log S < 6$, saturation: fraction of sp^3 hybridized carbons > 0.25, flexibility: number of rotatable bonds < 9. From the Bioavailability Radar (Figure 3), flexibility, lipophilicity, water insolubility, and size are in the acceptable range for most ligands. There exist some anomalies for saturation and polarity in some cases.

Table 5. Physicochemical and pharmacokinetic properties of ligands retrieved from SwissADME (Supplementary Material S2).

Molecule	MW	# Rotatable Bonds	# H-bond Acceptors	# H-bond Donors	MR	TPSA	iLOGP	GI Absorption	BBB Permeant	Pgp Substrate	CYP3A4 Inhibitor	Lipinski # Violations	Bioavailability Score	PAINS # Alerts	Synthetic Accessibility
Gallic Acid	170.12	1	5	4	39.47	97.99	0.21	High	No	No	Yes	0	0.56	1	1.22
Ciprofloxacin	331.34	3	5	2	95.25	74.57	2.24	High	No	Yes	No	0	0.55	0	2.51
Syringic Acid	198.17	3	5	2	48.41	75.99	1.54	High	No	No	No	0	0.56	0	1.7
Dihydromyricetin	320.25	1	8	6	76.78	147.68	0.67	Low	No	No	No	1	0.55	1	3.55
Taxifolin	304.25	1	7	5	74.76	127.45	0.71	High	No	No	No	0	0.55	1	3.51
Chlorogenic Acid	354.31	5	9	6	83.5	164.75	0.87	Low	No	No	No	1	0.11	1	4.16
Quercetin	302.24	1	7	5	78.04	131.36	1.63	High	No	No	Yes	0	0.55	1	3.23
Kaempferol	286.24	1	6	4	76.01	111.13	1.7	High	No	No	Yes	0	0.55	0	3.14
Myricetin	318.24	1	8	6	80.06	151.59	1.08	Low	No	No	Yes	1	0.55	1	3.27
Reptoside	390.38	5	10	5	87.44	155.14	1.92	Low	No	Yes	No	0	0.55	0	5.6
8-*O*-acetylharpagide	406.38	5	11	6	88.6	175.37	1.36	Low	No	No	No	2	0.17	0	5.75
Rutin	610.52	6	16	10	141.38	269.43	0.46	Low	No	Yes	No	3	0.17	1	6.52

#: number.

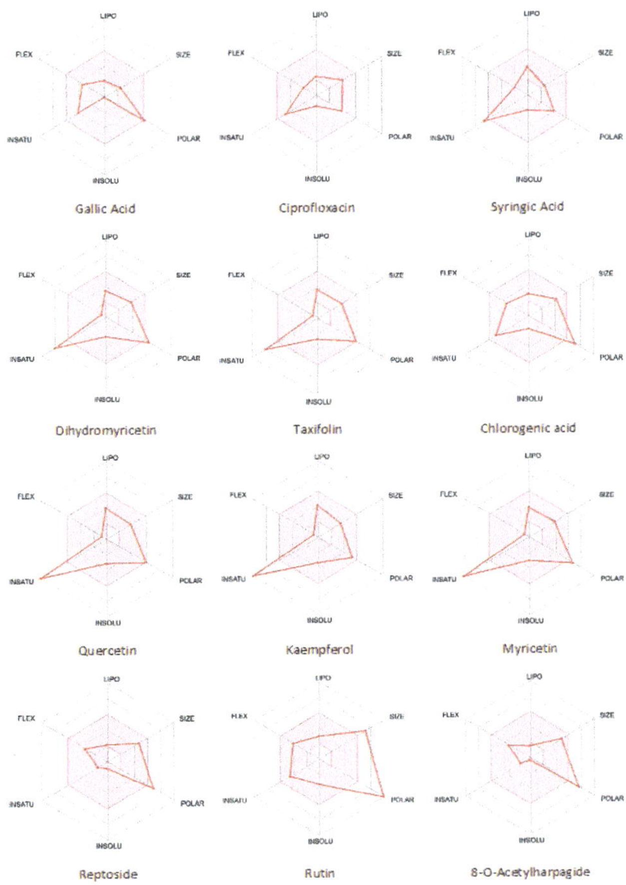

Figure 3. Bioavailability radar for the ligand molecules from SwissADME.

The binding energies and interaction study shows a better binding affinity for most as observed to have lower binding energies than the reference drug (Table 6). In addition to the binding energy, the RMSD and Ki of individual ligands are also convincing to consider them for the good inhibitory activity of the target protein. Active site residues identified from the pdbsum online server (http://www.ebi.ac.uk/thornton-srv/databases/cgi-bin/pdbsum/GetPage.pl; accessed 2 December 2022) considering the native ligands and the reference for the PDB data of the target protein [24] are His13, His15, His135, His173, and Asp246.

Table 6. Result summary of molecular docking analysis of ligands with VcDHO protein (PDB ID: 5vgm).

Ligand	Binding Energy (kcal/mol)	RMSD (Å)	KI (μM)	Interacting Residues	
				H-Bonding	Others
Chlorogenic acid	−8.1	3.416	1.15	**His15**, Arg17, **Asp246**, Pro101, Asn41, Ala262, Leu218, Glu137, His250	Ala248, **His135**, Thr139, Tyr100, Gly263,
Rutin	−8.1	2.264	1.15	Arg17, His250, Leu218, Tyr75, Pro101, Asn41, **Asp246**	**His135**, Tyr100, Ala262, Thr139, Ala248, Gly263
Taxifolin	−7.4	1.648	3.76	Arg17, **Asp246**, Ala262,Tyr75, Leu218	Asn41, **His15**, Ala248, Tyr100, His250
Dihydromyricetin	−7.3	0.546	4.45	Arg17, Ala262, **Asp246**, Tyr75, His250, Leu218	**His15**, Ala248, Asn41, Tyr100
Kaempferol	−7.2	1.491	5.27	Leu218, Pro101, Asn41, Glu137	**His135**, Tyr100, Thr139, Ala262
Reptoside	−7.1	1.852	6.24	Asn41, **His135**, Tyr75, Ala262, Leu218, **Asp246**	Tyr100, **His15, His173**
Ciprofloxacin	−7	2.267	7.39	Arg17, His250, Ala262	**His15**, Asn41, **His135**, Pro101, Glu137, Thr139, **Asp246**, Leu218, Tyr100, Ala248
Myricetin	−6.9	2.042	8.74	Pro101, His250, Ala262, Glu137, **Asp246**	Tyr100, **His15**, Thr139, Leu218, Ala248, **His135**
Quercetin	−6.6	1.977	14.51	Pro101, His250, Ala262, Glu137, **Asp246**	Tyr100, **His135, His15**, Thr139, Leu218, Ala248
8-*O*-Acetylharpagide	−6.3	1.121	24.07	Asn41, Leu218, Arg17	**His15**, Tyr100, **His135**, Ala262, Ala248, His250
Gallic acid	−6.2	0.509	28.5	**Asp246**, Arg17, Leu218, Asn41, **His135**, His250, **His173**, Ala262	**His15**, Tyr100, Ala248
Syringic acid	−5.4	0.146	109.99	Asn41, **Asp246,** His250, Arg17, Ala262	Tyr100, **His135**, Leu218, Ala248, Gly263, Cys217

The interaction diagrams from Discovery Studio and LigPlot software are similar. The interaction distance considered is less than or equal to 3 nm. Different interactions, including H-bonding, hydrophobic interactions or Van der Waals, Pi-alkyl, Pi-donor H-bond, and Pi-Pi stacked interactions shown in the interaction diagrams from Discovery Studio software in Figure 4.

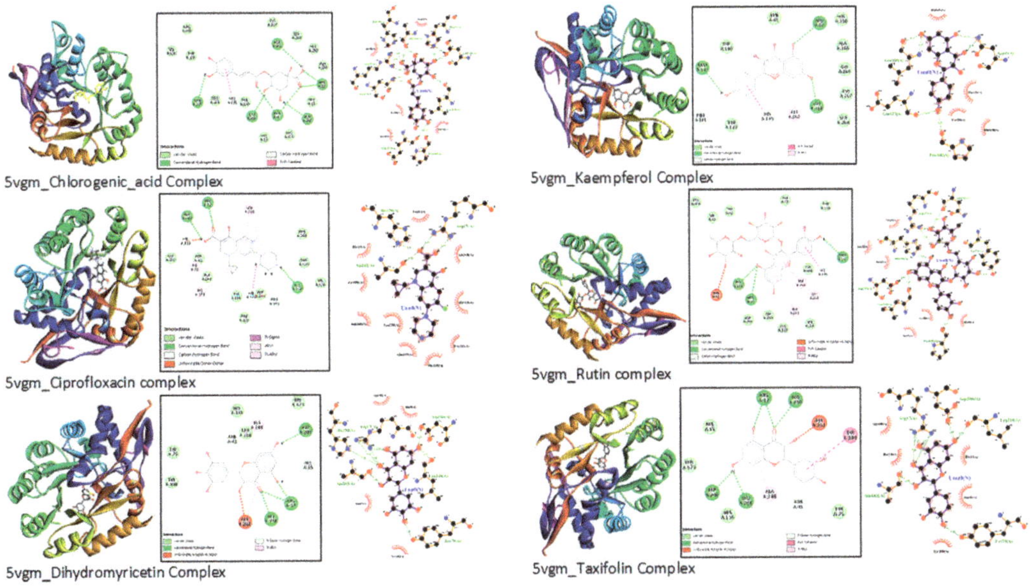

Figure 4. The 3D and 2D views of interactions from Discovery Studio and LigPlot+1.4.5.

3. Discussion

The antioxidant activities of AIA were observed to be nearly double those of AIR extracts. The IC_{50} value for AIA was found to be lower than the value reported by Lere Keshebo et al. [54]. The computed smaller IC_{50} values of the plant parts considered in our study reveal that this medicinal plant is effective in treating many illnesses [55] due to its high antioxidant activities. This is also shown in various ethnomedicinal studies [25–44] and dictates the presence of polyphenols [45,46] in the extracts more significantly in the aerial part, as demonstrated by Banothu et al. [56]. Similarly, measured TPC and TFC values for AIA were twice those of AIR. This indicates that there is a significant correlation between antioxidant activity and TPC/TFC values [55], and their health benefits are strongly associated. In this study, both TPC and TFC are estimated to be higher, as described in prior investigations [57]. Following higher values of TPC/TFC and antioxidant activity, the antimicrobial activity of AIA against *Shigella* spp. was found to be significant and comparable with other herbs [54] and with ciprofloxacin (Table 4).

The antioxidant activity study result (Table 1) and the total phenol and total flavonoid contents determined (Table 2) indicate the presence of flavonoids and phenolic acids in the aerial part rather than in the root sample. This is in support of the absence of antibacterial activity in the root sample. Phenolics and flavonoids are among the most common components of medicinal plants responsible for antioxidant activities and thus their therapeutic benefits [58]. As per the availability of standards, we checked the presence of flavonoids such as quercetin, myricetin, and rutin and phenolic acids such as chlorogenic acid, syringic acid, and gallic acid. The higher concentration of phenolics in the acetone extract confirmed the presence of flavonoids, as the method [59] is developed for the extraction of flavonoids. HPLC analysis showed that there are also phenolic acid components in the acetone extract in comparable amounts. The identified components from HPLC analysis (Table 3) and others from the literature [60] were considered in the molecular docking study towards screening responsible components for the anti-shigellosis activity observed in the antibacterial study (Table 4) and the therapeutic purpose indicated. The absence of antimicrobial activity for the root part of the plant agrees with the lower concentrations of the phenolics, as shown in Table 3.

Most ligands, except rutin, are known to be Lipinski-compliant and can be considered drug-like molecules. The oral bioavailability test using the Lipinski rule complements the information from the bioavailability radar. Bioavailability and ADMET testing suggest the acceptability of these ligands as a drug [61]. Molecular docking of 5vgm with identified flavonoids and phenolic acids shows that most ligands have better binding affinity than the reference drug (Ciprofloxacin). Binding energy ranges from −8.1 kcal/mole to −5.4 kcal/mol, which is −7.0 kcal/mol for the reference drug. RMSD for the molecular docking for each ligand ranges from 0.1 to 3.4 Å (Table 5). Lower Ki values indicate lower inhibition concentrations for the intended activity and are thus significant. In addition, the lower RMSD values imply consistent docking into the referred active site bounded in the Vina search space indicated above.

This indicates that the environment and parameters set for the molecular docking were reasonably reliable [62]. More robust and relatively stable interactions were observed for most ligands considered. In general, H-bonding interactions dominate over other interactions, including hydrophobic interactions. The role of H-bonding for the overall binding energy is significantly higher [63]. The dihydro-flavonoids, rutin, and chlorogenic acid take the lead in terms of binding affinity. The number of H-bonding interactions is also considerable. Kaempferol and reptoside are comparable to the reference drug in binding energies and RMSD. Myricetin and quercetin also have a suitable binding affinity, slightly lower than the reference drug. Most ligands interact with two or three of the active site residues. Binding interactions with the active site residues also confirm the interaction to be on the right binding site. Common residues such as His250, Arg17, Tyr100, Pro101, and Leu218 appear in most interactions (Table 6 and Figure 4).

Reptoside and 8-*O*-acetylharpagide were considered in the study only for comparison. The antibacterial study shows that the root does not show any significant activity for *Shigella* spp. However, regarding iridoid glycoside composition, both the root and aerial samples were compared when studied using TLC. Iridoid glycosides may not be responsible for the anti-shigellosis activity of the aerial part of *A. integrifolia*. From HPLC analysis, the root sample was found to have an insignificant concentration of flavonoids and phenolic acids. The top five responsible components for anti-shigellosis activity via inhibiting VcDHO are chlorogenic acid, rutin, taxifolin (dihydroquercetin), dihydromyricetin, and kaempferol. The 3-OH and 4-carbonyl functionalities are known to enhance antimicrobial activity. The number of hydroxyl groups is associated with hydrophobicity. Additive hydroxyl groups may lower hydrophobicity, but C3 charges may be increased, which is a clear sign of pharmacological activity [64,65]. The OH groups on the flavonoids are essential for bioactivity, and the change in the position or number of such groups affects biological potency. The plant-derived flavonoid quercetin is a broad-spectrum protein inhibitor [66]. The role of myricetin and quercetin is also not to be undermined following their lower binding energies and lower RMSD and Ki values. The synergic effect will also be considered as there is more than one binding active site for the target proteins and the varied composition of natural products. This demands further study and the activity study of the individual flavonoids and phenolic acids. The observed activity justifies the ethnomedical use of *Ajuga* spp. in traditional medicine. The minimum inhibition concentration (MIC) and minimum bactericidal concentration (MBC) could also be determined to complement the Ki values from the molecular docking study.

4. Materials and Methods

4.1. Chemicals and Reagents

All chemicals and reagents needed for the extraction, total phenolic content, and total flavonoid content determination and antibacterial activity were AR-grade. Standard references (phenolic acid standards: syringic acid, chlorogenic acid, and gallic acid; flavonoid standards: myricetin, quercetin, rutin, and kaempferol), DPPH, and ascorbic acid for the antioxidant test were all from Sigma (>99.9%, Sigma, Shanghai, China), While for HPLC analysis, all HPLC-grade reagents were used.

4.2. Plant Material Collection

The aerial part of *A. integrifolia* was collected from the Addis Ababa Science and Technology University campus and the nearby area of Koye Feche (VRP5 + 3W6, 8.8852 °N 38.8098 °E, elevation: 2840). The plant was identified and an herbarium sample was deposited (voucher number: FB-004/11) in the national herbarium at the College of Science, Addis Ababa University, Ethiopia.

4.3. Extraction

Methanol extract of the aerial parts was obtained via Soxhlet extraction using a method by Imoru et al. [67] with slight modifications. Using an ultrasonic probe (SJIA-950W, Sjiolab, Ningbo, China) 75% methanol (aqueous) extraction was performed. The method optimization of UAE followed a method reported by Zakaria et al. [68] with minor modifications. Acetone extract from dip instant extraction, specifically attempted for the extraction of flavonoids, was carried out using the method employed by Mawela [69].

4.4. Antioxidant Test—DPPH Assay Calorimetric Method

Modified protocol for the free radical scavenging effect using DPPH assay from Banothu et al. [56] was used, and Soxhlet methanol extracts were tested for antioxidant activity using DPPH as a reagent and ascorbic acid as a standard. We prepared 1000 ppm DPPH and 1000 ppm (mg/L) standard. Ascorbic acid was prepared by dissolving 0.10 g in 100 mL of methanol, and concentrations of 25, 50, 100, 150, 200, 250, 300, 350, 400, 450, and 500 mg/L were prepared for calibration. Absorbance was measured using a Jasco V-770 spectrophotometer (Jasco, Easton, MD, USA), using 1 mm path length in a rectangular cell holder. Sample extracts (0.1 g/mL) of the aerial part and root samples were diluted with CH_3OH using 5, 10, 25, 50, 100, 150, and 200 dilution factors. Absorbance was measured and recorded in triplicate after 30 min incubation time in a dark place at 517 nm. The proportion of sample to DPPH was 1:3, i.e., 750 mL of DPPH added to 250 mL of sample, which was modified for some cases. DPPH scavenging capacity was computed by using the following formula:

$$\text{Scavenging activity (\%)} = \left(\frac{\text{Absorbance}_{\text{control}} - \text{Absorbance}_{\text{sample}}}{\text{Absorbance}_{\text{control}}}\right) \times 100 \quad (1)$$

IC_{50} values were computed from the relation log [sample] vs. absorbance (normalized) using Graph pad prism 8 software, as suggested for better EC_{50} estimation [70].

4.5. Total Phenolic Content (TPC) Determination: FC Method

The total phenolic content of aerial and root samples of *A. integrifolia* was determined by a colorimetric method Folin–Ciocalteu (FC) assay, as described by McDonald et al. [71] with some modifications. First, 0.4 mL of sample extract and 0.4 mL of FC reagent (10× diluted) were mixed, and then 0.2 mL of 2% Na_2CO_3 was added after 5 min. Using a V-770 UV-Vis spectrophotometer, absorbance at 765 nm was measured after the mixture had been incubated at room temperature for 35 min. The control was methanol. A standard solution of gallic acid for the calibration curve was prepared in 1000 ppm (mg/L) and serially diluted to 6.25, 12.5, 25, 50, 100, 150, 200, 250, and 500 mg/L standard solutions. The calibration curve was created using the average absorbance values at the appropriate concentrations from each experiment performed in triplicate. TPC was calculated as the milligrams of gallic acid equivalent (GAE) for each extracted material gram (dry weight).

4.6. Total Flavonoid Content (TFC): Aluminum Chloride Method

With slight adjustments, the aluminum chloride colorimetric method reported by Chang et al. [72] was used to measure the extracts' total flavonoid concentration. Briefly, the mixture of 0.3 mL sample extract, 0.3 mL of 2% $AlCl_3$, 0.3 mL 1% sodium nitrite, and 0.3 mL 5% NaOH was mixed. The mixture was incubated at room temperature for a total of

30 min. Methanol was used as a control. Absorbance was measured at λ 314 nm using the spectrophotometer mentioned above. Quercetin prepared at 1000 ppm (mg/L) was used as a standard and calibration concentrations were prepared for TPC determination. TFC was computed as mg of quercetin equivalent (QE) per gram (dry weight) of sample extract.

4.7. HPLC (UHPLC-DAD) Analysis

Ultra-high-performance liquid chromatography coupled with a diode array detector (Ultimate-3000 UHPLC-DAD, Thermo Scientific Dionex, Sunnyvale, CA, USA) was used to determine the presence of flavonoids and phenolic acids. The column was in reverse phase with Fortis 5 mm C18 (4.6 × 250 mm column dimension). Methanol-acidified (1% acetic acid) ultra-pure water (60/40, *v/v*) at the flow rate of 0.8 mL/min was used as the mobile phase. Autosamplers and column temperature were set at 25 °C and 35 °C, respectively. Then, 10 µL of the purified sample extracts dissolved in the mobile phase mixture was injected into the column, and UV-Vis detection was attained at 254 nm, 272 nm, 360 nm, and 372 nm. The standard mixture of the phenolic acids syringic acid, chlorogenic acid, and gallic acid, and the flavonoids quercetin, myricetin, and rutin at the concentrations of 2.5, 10, 20, 40, 50 mg/mL were used as the external reference standard mixtures.

4.8. Antibacterial Tests

An antimicrobial efficacy study was conducted in the Ethiopian Biotechnology Institute, Ethiopia Microbiology Laboratory.

4.8.1. Test Organisms

Escherichia coli (ATCC 25922), *Pseudomonas aeruginosa* (ATCC 27853), and *Staphylococcus aureus* (ATCC 25923) were obtained from the Ethiopian Public Health Institute. *Shigella* spp. was obtained from clinical isolates selected by a researcher at Bio and Emerging Technology Institute (formerly EBTi), Ethiopia.

4.8.2. Antibacterial Activity

Selected test organisms were subjected to susceptibility tests for each extract using the agar well diffusion method as used by Lulekal et al. [73] with slight modifications. The susceptibility test was performed in triplicate for the methanol extract of the root and aerial part of *A. integrifolia*. Test organisms were swabbed onto sterile Mueller Hinton agar (HIMEDIA, Mumbai, India) plates using sterile cotton swabs. Mueller Hinton was swabbed, the agar was allowed to dry, and the blue points were used to drill 6 evenly spaced holes. Then, 100 µL of one sample extract (with a concentration of 0.1 g/mL) was placed in three holes, while 100 µL of the other extract was placed in the remaining three holes. For the negative and positive controls, equal volumes of sterile distilled water, methanol (diluent), chloramphenicol, and ciprofloxacin suspensions were used. The zones of inhibition (ZOI) (susceptibility or resistance) of the extracts and control for each test organism were then measured with a ruler and reported in millimeters after incubation at 37 °C for 24 h.

4.9. In Silico Study

4.9.1. Physicochemical and Pharmacokinetic Properties

The SwissADME web tool (http://www.swissadme.ch; accessed 2 December 2022), with free access to a pool of fast and robust predictive models including built-in methods such as the BOILEDEgg, iLOGP, and Bioavailability Radar, was used to retrieve properties such as physicochemical, pharmacokinetics, and drug-likeness. Medicinal chemistry friendliness was checked using two models: PAINS alert and the Brenk filter models [53]. Easy, efficient input formats and interpretation are made possible through a user-friendly interface.

4.9.2. Molecular Docking: Interaction Study

For the selection and preparation of target protein and ligand structures, DHO is preferred for its essential role in the proliferation of pathogens and as one of the key enzymes in the aforementioned pathway [24]. Lipowska and coworkers proposed two structures of DHOs, the plague-causing pathogen from Yersinia pestis (YpDHO) (PDB ID: 6CTY), and the causative agent of cholera from *Vibrio cholerae* (VcDHO) (PDB ID: 5VGM). We selected the latter one after checking the crystal structure resolution, total quality factor, and Q-mean values [74] and testing for docking with the reference drug on the respective active site. The ligands used in this study were the flavonoids and phenolic acids identified in the HPLC analysis and from the literature [75]. The structure of these ligands is shown in Figure 5. Ciprofloxacin was used as a reference after its first-line treatment for shigellosis [8,76]. Preparations of the receptor protein and ligands were managed after the structure retrieved from pdb (https://www.rcsb.org/structure/5VGM; accessed 2 December 2022) and PubChem (https://pubchem.ncbi.nlm.nih.gov/; accessed 2 December 2022) online servers using standard procedure [77]. The structures were minimized using USCF Chimera 1.15 software (Resource for Biocomputing, Visualization, and Informatics University of California, San Francisco, CA, USA) and ChemDraw 3D software (Version 12.0, PerkinElmer, Waltham, MA, USA). The prepared structures were saved as pdb files and made ready for molecular docking.

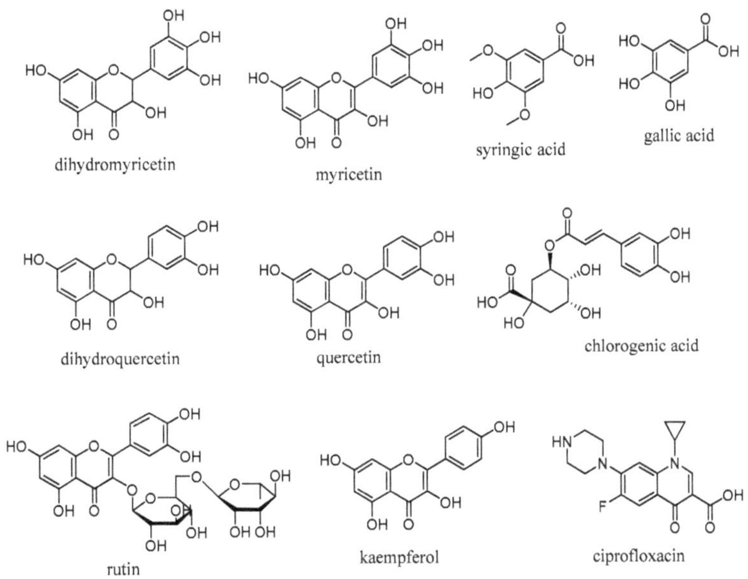

Figure 5. Structures of ligands selected for the in silico study.

4.9.3. Molecular Docking and Visualization

Prepared structures of the receptor macromolecule and ligand molecules were loaded in PyRx 8.0 software [63] and converted to the respective pdbqt files. Using the information about the residues of the active site, Vina search parameters were set as follows: exhaustiveness = 8; center x = 21.5380864417, y = 18.4743434105, z = 82.1235234094; dimensions x = 30.1832504022, y = 30.810349529, z = 30.3537531811. After completing the Vina, a CSV output file with pose binding energies and respective RMSD values was generated for each ligand on an Excel sheet and output for further interaction study using visualizing software such as PyMol 2.5 (Schrödinger, Mannheim, Germany), Discovery Studio (Dassault Systèmes, Vélizy-Villacoublay, France) [78], and LiPlot+1.4.5 (European Bioinformatics Institute, Cambridge, United Kingdom) [79]. From the visualizing software's interaction types, their distance and interacting residues were extracted and used for the discussion and conclusions.

5. Conclusions

The search for novel and effective therapeutic targets has become necessary due to the rise in organisms' antimicrobial resistance tendencies. The antioxidant activity of the aerial part of *A. integrifolia* was more significant than the root sample. This indicates the presence of phenolics and flavonoids in the aerial part rather than in the root sample, as was also confirmed for the antibacterial activity. Following the considerable activity of the aerial part against *Shigella* spp., potential anti-shigellosis activity was screened. VcDHO was selected as a drug target for its role in the proliferation of pathogenic bacteria. As responsible components, flavonoids and phenolic acids identified from HPLC analysis and others from the previous study in the aerial sample of *A. integrifolia* were considered. Most likely, chlorogenic acid, rutin, taxifolin (dihydroquercetin), dihydromyricetin, and kaempferol are potentially responsible for the anti-shigellosis activity and therapeutic potential. The synergic effect and specific activity of the compounds identified as potential inhibitors of VcDHO are essential to reaching a more robust conclusion.

Supplementary Materials: The following supporting information can be downloaded at: https://www.mdpi.com/article/10.3390/molecules28031111/s1, Supplementary Material S1: Figure S1: Main Ramachandran plot for 5VGM; Supplementary Material S1: Figure S2 Quality comparison for 5VGM; Supplementary Material S2: SwissADME data for ligands.

Author Contributions: Conceptualization, F.B.T.; Formal analysis, F.B.T. and T.B.A.; Funding acquisition, S.M.W. and K.A.D.; Investigation, T.G.T.; Methodology, F.B.T., T.B.A. and T.G.T.; Resources, M.G.T. and R.K.B.; Supervision, M.G.T. and R.K.B.; Validation, Y.H.G., T.B.A., A.B., D.P.P., S.M.W., K.A.D., I.Š., P.K., V.K. and S.A.F.; Writing—original draft, F.B.T.; Writing—review and editing, Y.H.G., T.B.A., A.B., D.P.P., S.M.W., K.A.D., I.Š., P.K., V.K., S.A.F. and R.K.B. All authors have read and agreed to the published version of the manuscript.

Funding: This work was funded by the Researchers Supporting Project Number (RSP2023R388) King Saud University, Riyadh, Saudi Arabia.

Institutional Review Board Statement: Not applicable.

Informed Consent Statement: Not applicable.

Data Availability Statement: This article and its supplemental information files contain all the data created or analyzed during this investigation.

Acknowledgments: This work was funded by the Researchers Supporting Project Number (RSP2023R388) King Saud University, Riyadh, Saudi Arabia. The authors acknowledge Belete Adefris for providing standard references and Addis Ababa Science and Technology University, Industrial Chemistry Department, and EBTi for the facilities to conduct the investigations. All individuals included in this section have consented to the acknowledgement.

Conflicts of Interest: The authors declare no conflict of interest.

Sample Availability: Not applicable.

References

1. Fichtl, R.; Adi, A. *Honeybee Flora of Ethiopia*; Edwards, S., Kelbessa, E., Eds.; Margraf Verlag: Weikersheim, Germany, 1994.
2. Hedberg, I. *Flora of Ethiopia and Eritrea. Vol. 5, Gentianaceae to Cyclocheilaceae*; Hedberg, I., Kelbessa, E., Edwards, S., Demissew, S., Persson, E., Eds.; The National Herbarium, Biology Department, Science Faculty, Addis Ababa University: Addis Ababa, Ethiopia, 2006.
3. Ullah, M.A.; Gul, F.Z.; Khan, T.; Bajwa, M.N.; Drouet, S.; Tungmunnithum, D.; Giglioli-Guivarc'h, N.; Liu, C.; Hano, C.; Abbasi, B.H. Differential Induction of Antioxidant and Anti-Inflammatory Phytochemicals in Agitated Micro-Shoot Cultures of *Ajuga Integrifolia* Buch. Ham. Ex D.Don with Biotic Elicitors. *AMB Express* **2021**, *11*, 1–13. [CrossRef] [PubMed]
4. Alene, M.; Abdelwuhab, M.; Belay, A.; Yazie, T.S. Evaluation of Antidiabetic Activity of *Ajuga Integrifolia* (Lamiaceae) Root Extract and Solvent Fractions in Mice. *Evid.-Based Complement. Altern. Med.* **2020**, *2020*, 1–11. [CrossRef] [PubMed]

5. Ahmad, G.H. Review of the Active Principles of Medicinal and Aromatic Plants and Their Disease Fighting Properties. In *Medicinal and Aromatic Plants: Expanding their Horizons through Omics*; Aftab, T., Hakeem, K.R., Eds.; Academic Press: Cambridge, MA, USA, 2021; pp. 1–36. ISBN 9780128195901.
6. Ferrari, B.; Castilho, P.; Tomi, F.; Rodrigues, A.I.; Do Ceu Costa, M.; Casanova, J. Direct Identification and Quantitative Determination of Costunolide and Dehydrocostuslactone in the Fixed Oil of *Laurus Novocanariensis* By13C-NMR Spectroscopy. *Phytochem. Anal.* **2005**, *16*, 104–107. [CrossRef] [PubMed]
7. Magozwi, D.K.; Dinala, M.; Mokwana, N.; Siwe-Noundou, X.; Krause, R.W.M.; Sonopo, M.; McGaw, L.J.; Augustyn, W.A.; Tembu, V.J. Flavonoids from the Genus *Euphorbia*: Isolation, Structure, Pharmacological Activities and Structure–Activity Relationships. *Pharmaceuticals* **2021**, *14*, 428. [CrossRef]
8. Mulatu, G.; Beyene, G.; Zeynudin, A. Prevalence of Shigella, Salmonella and Campylobacter Species and Their Susceptibility Patterns among under Five Children with Diarrhea in Hawassa Town, South Ethiopia. *Ethiop. J. Health Sci.* **2014**, *24*, 101–108. [CrossRef]
9. Tosisa, W.; Mihret, A.; Ararsa, A.; Eguale, T.; Abebe, T. Prevalence and Antimicrobial Susceptibility of *Salmonella* and *Shigella* Species Isolated from Diarrheic Children in Ambo Town. *BMC Pediatr.* **2020**, *20*, 1–8. [CrossRef]
10. Hunde, D.; Asfaw, Z.; Kelbessa, E. Use of Traditional Medicinal Plants by People of "Boosat" Sub District, Central Eastern Ethiopia. *Ethiop. J. Health Sci.* **2006**, *16*, 141–155.
11. Clarkson, K.A.; Talaat, K.R.; Alaimo, C.; Martin, P.; Bourgeois, A.L.; Dreyer, A.; Porter, C.K.; Chakraborty, S.; Brubaker, J.; Elwood, D.; et al. Immune Response Characterization in a Human Challenge Study with a *Shigella Flexneri* 2a Bioconjugate Vaccine. *EBioMedicine* **2021**, *66*, 103308. [CrossRef]
12. Ahmed, A.M.; Furuta, K.; Shimomura, K.; Kasama, Y.; Shimamoto, T. Genetic Characterization of Multidrug Resistance in *Shigella* Spp. from Japan. *J. Med. Microbiol.* **2006**, *55*, 1685–1691. [CrossRef]
13. Ud-Din, A.; Wahid, S. Relationship among *Shigella* Spp. and Enteroinvasive *Escherichia Coli* (EIEC) and Their Differentiation. *Braz. J. Microbiol.* **2014**, *45*, 1131–1138. [CrossRef]
14. Qasim, M.; Wrage, M.; Nüse, B.; Mattner, J. *Shigella* Outer Membrane Vesicles as Promising Targets for Vaccination. *Int. J. Mol. Sci.* **2022**, *23*, 994. [CrossRef] [PubMed]
15. Hussen, S.; Mulatu, G.; Yohannes Kassa, Z. Prevalence of *Shigella* Species and Its Drug Resistance Pattern in Ethiopia: A Systematic Review and Meta-Analysis. *Ann. Clin. Microbiol. Antimicrob.* **2019**, *18*, 1–11. [CrossRef] [PubMed]
16. Debas, G.; Kibret, M.; Biadglegne, F.; Abera, B. Prevalence and Antimicrobial Susceptibility Patterns of Shigella Species at Felege Hiwot Referral Hospital, Northwest Ethiopia. *Ethiop. Med. J* **2011**, *49*, 249–256. [PubMed]
17. Mukhopadhyay, S.; Ganguli, S.; Chakrabarti, S. *Shigella* Pathogenesis: Molecular and Computational Insights. *AIMS Mol. Sci.* **2020**, *7*, 99–121. [CrossRef]
18. Beyene, G.; Tasew, H. Prevalence of Intestinal Parasite, *Shigella* and *Salmonella* Species among Diarrheal Children in Jimma Health Center, Jimma Southwest Ethiopia: A Cross Sectional Study. *Ann. Clin. Microbiol. Antimicrob.* **2014**, *13*, 1–7. [CrossRef] [PubMed]
19. Mengistu, G.; Mulugeta, G.; Lema, T.; Aseffa, A. Prevalence and Antimicrobial Susceptibility Patterns of Salmonella Serovars and Shigella Species. *J. Microb. Biochem. Technol.* **2014**, *6*, S2-006. [CrossRef]
20. Lamboro, T.; Ketema, T.; Bacha, K. Prevalence and Antimicrobial Resistance in Salmonella and Shigella Species Isolated from Outpatients, Jimma University Specialized Hospital, Southwest Ethiopia. *Can. J. Infect. Dis. Med. Microbiol.* **2016**, *2016*, 4210760. [CrossRef] [PubMed]
21. Marami, D.; Hailu, K.; Tolera, M. Prevalence and Antimicrobial Susceptibility Pattern of *Salmonella* and *Shigella* Species among Asymptomatic Food Handlers Working in Haramaya University Cafeterias, Eastern Ethiopia. *BMC Res. Notes* **2018**, *11*, 7–12. [CrossRef]
22. Terfassa, A.; Jida, M. Prevalence and Antibiotics Susceptibility Pattern of *Salmonella* and *Shigella* Species among Diarrheal Patients Attending Nekemte Referral Hospital, Oromia, Ethiopia. *Int. J. Microbiol.* **2018**, *2018*, 9214689. [CrossRef]
23. Yang, G.; Wang, L.; Wang, Y.; Li, P.; Zhu, J.; Qiu, S.; Hao, R.; Wu, Z.; Li, W.; Song, H. Hfq Regulates Acid Tolerance and Virulence by Responding to Acid Stress in *Shigella Flexneri*. *Res. Microbiol.* **2015**, *166*, 476–485. [CrossRef]
24. Lipowska, J.; Miks, C.D.; Kwon, K.; Shuvalova, L.; Zheng, H.; Lewiński, K.; Cooper, D.R.; Shabalin, I.G.; Minor, W. Pyrimidine Biosynthesis in Pathogens— Structures and Analysis of Dihydroorotases from *Yersinia Pestis* and *Vibrio Cholerae*. *Int. J. Biol. Macromol.* **2019**, *136*, 1176–1187. [CrossRef] [PubMed]
25. Asres, K.; Bucar, F.; Kartnig, T.; Witvrouw, M.; Pannecouque, C.; Clercq, E. De Antiviral Activity against Human Immunodeficiency Virus Type 1 (HIV-1) and Type 2 (HIV-2) of Ethnobotanically Selected Ethiopian Medicinal Plants. *Phytochem. Res.* **2001**, *15*, 62–69.
26. Fullas, F. *Ethiopian Traditional Medicine: Common Medicinal Plants in Perspective*, 1st ed.; F. Fullas: Sioux City, IA, USA, 2001.
27. Suleman, S.; Alemu, T. A Survey on Utilization of Ethnomedicinal Plants in Nekemte Town, East Wellega (Oromia), Ethiopia. *J. Herbs Spices Med. Plants* **2012**, *18*, 37–41. [CrossRef]
28. El-Hilaly, J.; Amarouch, M.Y.; Morel, N.; Lyoussi, B.; Quetin-Leclercq, J. *Ajuga Iva* Water Extract Antihypertensive Effect on Stroke-Prone Spontaneously Hypertensive Rats, Vasorelaxant Effects Ex Vivo and in Vitro Activity of Fractions. *J. Ethnopharmacol.* **2021**, *270*, 113791. [CrossRef] [PubMed]
29. Bekele, D.; Asfaw, Z.; Petros, B.; Tekie, H. Ethnobotanical Study of Plants Used for Protection against Insect Bite and for the Treatment of Livestock Health Problems in Rural Areas of Akaki District, Eastern Shewa, Ethiopia. *Topclass. J. Herb. Med.* **2012**, *1*, 12–24.

30. Meragiaw, M.; Asfaw, Z. Review of Antimalarial, Pesticidal and Repellent Plants in the Ethiopian Traditional Herbal Medicine. *Res. Rev. J. Herb. Sci.* **2014**, *3*, 21–25.
31. Asnake, S.; Teklehaymanot, T.; Hymete, A.; Erko, B.; Giday, M. Survey of Medicinal Plants Used to Treat Malaria by Sidama People of Boricha District, Sidama Zone, South Region of Ethiopia. *Evid.-Based Complement. Altern. Med.* **2016**, *2016*, 9690164. [CrossRef]
32. Hailu, W.; Engidawork, E. Evaluation of the Diuretic Activity of the Aqueous and 80 % Methanolic Extracts of the Leaves of *Ajuga Remota* B. (Lamiaceae) in Mice. Ph.D. Thesis, Addis Ababa University, Addis Ababa, Ethiopia, 2011.
33. Chekole, G. Ethnobotanical Study of Medicinal Plants Used against Human Ailments in Gubalafto. *J. Ethnobiol. Ethnomed.* **2017**, *13*, 55. [CrossRef]
34. Abera, B. Medicinal Plants Used in Traditional Medicine by Oromo People, Ghimbi District, Southwest Ethiopia. *J. Ethnobiol. Ethnomed.* **2014**, *10*, 1–15. [CrossRef]
35. Tuasha, N.; Petros, B.; Asfaw, Z. Plants Used as Anticancer Agents in the Ethiopian Traditional Medical Practices: A Systematic Review. *Evid.-Based Complement. Altern. Med* **2018**, *2018*, 6274021. [CrossRef]
36. Gebrehiwot, M. An Ethnobotanical Study of Medicinal Plants in Seru Wereda, Arsi Zone of Oromia Region, Ethiopia. Ph.D. Thesis, Addis Ababa University, Addis Ababa, Ethiopia, 2010.
37. Regassa, R. Assessment of Indigenous Knowledge of Medicinal Plant Practice and Mode of Service Delivery in Hawassa City, Southern Ethiopia. *J. Med. Plants Res.* **2013**, *7*, 517–535. [CrossRef]
38. Teklay, A.; Abera, B.; Giday, M. An Ethnobotanical Study of Medicinal Plants Used in Kilte Awulaelo District, Tigray Region of Ethiopia. *J. Ethnobiol. Ethnomed.* **2013**, *9*, 1–23. [CrossRef] [PubMed]
39. Maryo, M.; Nemomissa, S.; Bekele, T. An Ethnobotanical Study of Medicinal Plants of the Kembatta Ethnic Group in Enset-Based Agricultural Landscape of Kembatta Tembaro (KT) Zone, Southern Ethiopia. *Pelagia Res. Libr. Asian J. Plant Sci. Res.* **2015**, *5*, 42–61. [CrossRef]
40. Gedif, T.; Hahn, H. The Use of Medicinal Plants in Self-Care in Rural Central Ethiopia. *J. Ethnopharmacol.* **2003**, *87*, 155–161. [CrossRef]
41. Parvez, N.; Yadav, S. Ethnopharmacology of Single Herbal Preparations of Medicinal Plants in Asendabo District, Jimma, Ethiopia. *Indian J. Tradit. Knowl.* **2010**, *9*, 724–729.
42. Gabriel, T.; Guji, T. Ethnopharmacological Survey of Medicinal Plants in Agaro District, Jimma Zone, South West Ethiopia. *Int J. Pharm. Sci. Res.* **2014**, *5*, 3551–3559. [CrossRef]
43. Tafesse, T.B.; Hymete, A.; Mekonnen, Y.; Tadesse, M. Antidiabetic Activity and Phytochemical Screening of Extracts of the Leaves of *Ajuga Remota* Benth on Alloxan-Induced Diabetic Mice. *BMC Complement. Altern. Med.* **2017**, *17*, 1–9. [CrossRef]
44. Rubnawaz, S.; Okla, M.K.; Akhtar, N.; Khan, I.U.; Bhatti, M.Z.; Duong, H.Q.; El-tayeb, M.A.; Elbadawi, Y.B.; Almaary, K.S.; Moussa, I.M.; et al. Antibacterial, Antihemolytic, Cytotoxic, Anticancer, and Antileishmanial Effects of *Ajuga Bracteosa* Transgenic Plants. *Plants* **2021**, *10*, 1894. [CrossRef]
45. Meite, S.; N'Guessan, J.D.; Bahi, C.; Yapi, H.F.; Djaman, A.J.; Guede Guina, F. Antidiarrheal Activity of the Ethyl Acetate Extract of *Morinda Morindoides* in Rats. *Trop. J. Pharm. Res.* **2009**, *8*, 201–207. [CrossRef]
46. Asrie, A.B.; Abdelwuhab, M.; Shewamene, Z.; Gelayee, D.A.; Adinew, G.M.; Birru, E.M. Antidiarrheal Activity of Methanolic Extract of the Root Bark of *Cordia Africana*. *J. Exp. Pharm.* **2016**, *8*, 53–59. [CrossRef]
47. Mishra, M.P.; Rath, S.; Swain, S.S.; Ghosh, G.; Das, D.; Padhy, R.N. In Vitro Antibacterial Activity of Crude Extracts of 9 Selected Medicinal Plants against UTI Causing MDR Bacteria. *J. King Saud. Univ. Sci.* **2017**, *29*, 84–95. [CrossRef]
48. Cushnie, T.P.T.; Lamb, A.J. Antimicrobial Activity of Flavonoids. *Int. J. Antimicrob. Agents.* **2005**, *26*, 343–356. [CrossRef] [PubMed]
49. Kumar, S.; Pandey, A. Chemistry and Biological Activities of Flavonoids: An Overview. *Sci. World J.* **2013**, *2013*, 162750. [CrossRef] [PubMed]
50. Obaid, R.J.; Mughal, E.U.; Naeem, N.; Sadiq, A.; Alsantali, R.I.; Jassas, R.S.; Moussa, Z.; Ahmed, S.A. Natural and Synthetic Flavonoid Derivatives as New Potential Tyrosinase Inhibitors: A Systematic Review. *RSC Adv.* **2021**, *11*, 22159–22198. [CrossRef] [PubMed]
51. Kranich, R.; Busemann, A.S.; Bock, D.; Schroeter-Maas, S.; Beyer, D.; Heinemann, B.; Meyer, M.; Schierhorn, K.; Zahlten, R.; Wolff, G.; et al. Rational Design of Novel, Potent Small Molecule Pan-Selectin Antagonists. *J. Med. Chem.* **2007**, *50*, 1101–1115. [CrossRef] [PubMed]
52. Batool, F.; Mughal, E.U.; Zia, K.; Sadiq, A.; Naeem, N.; Javid, A.; Ul-Haq, Z.; Saeed, M. Synthetic Flavonoids as Potential Antiviral Agents against SARS-CoV-2 Main Protease. *J. Biomol. Struct. Dyn.* **2022**, *40*, 3777–3788. [CrossRef] [PubMed]
53. Brenk, R.; Schipani, A.; James, D.; Krasowski, A.; Gilbert, I.H.; Frearson, J.; Wyatt, P.G. Lessons Learnt from Assembling Screening Libraries for Drug Discovery for Neglected Diseases. *ChemMedChem* **2008**, *3*, 435–444. [CrossRef]
54. Lere Keshebo, D.; Washe, A.P.; Alemu, F. Determination of Antimicrobial and Antioxidant Activities of Extracts from Selected Medicinal Plants. *Am Sci Res J Eng* **2016**, *16*, 212–222.
55. Škrovánková, S.; Mišurcová, L.; Machů, L. Antioxidant Activity and Protecting Health Effects of Common Medicinal Plants. In *Advances in Food and Nutrition Research*; Elsevier: Amsterdam, The Netherlands, 2012; Volume 67, pp. 75–135. ISBN 9780123945983.
56. Banothu, V.; Neelagiri, C.; Adepally, U.; Lingam, J.; Bommareddy, K. Phytochemical Screening and Evaluation of in Vitro Antioxidant and Antimicrobial Activities of the Indigenous Medicinal Plant *Albizia Odoratissima*. *Pharm. Biol.* **2017**, *55*, 1155–1161. [CrossRef]

57. Aryal, S.; Baniya, M.K.; Danekhu, K.; Kunwar, P.; Gurung, R.; Koirala, N. Total Phenolic Content, Flavonoid Content and Antioxidant Potential of Wild Vegetables from Western Nepal. *Plants* **2019**, *8*, 96. [CrossRef]
58. Bhatt, I.D.; Rawat, S.; Rawal, R.S. Antioxidants in Medicinal Plants. In *Biotechnology for Medicinal Plants*; Chandra, S., Lata, H., Varma, A., Eds.; Springer: Berlin/Heidelberg, Germany, 2013; ISBN 9783642299742.
59. Omosa, L.K.; Amugune, B.; Ndunda, B.; Milugo, T.K.; Heydenreich, M.; Yenesew, A.; Midiwo, J.O. Antimicrobial Flavonoids and Diterpenoids from *Dodonaea Angustifolia*. *S. Afr. J. Bot.* **2014**, *91*, 58–62. [CrossRef]
60. Israili, Z.H.; Lyoussi, B. Ethnopharmacology of the Plants of Genus Ajuga. *Pak. J. Pharm. Sci.* **2009**, *22*, 425–462. [PubMed]
61. Alam, J.; Roy, T. *In Silico* Screening for Novel Inhibitors of Invasion Protein Antigen IpaD of *Shigella Flexneri*—An Agent of Shigellosis. *Int. J. Comput. Bioinform. Silico Model* **2015**, *3*, 497–501.
62. Rabbi, M.F.; Akter, S.A.; Hasan, M.J.; Amin, A. In Silico Characterization of a Hypothetical Protein from *Shigella Dysenteriae* ATCC 12039 Reveals a Pathogenesis-Related Protein of the Type-VI Secretion System. *Bioinform. Biol. Insights* **2021**, *15*, 1–12. [CrossRef]
63. Trott, O.; Olson, A. AutoDock Vina: Improving the Speed and Accuracy of Docking with a New Scoring Function, Efficient Optimization, and Multithreading. *J. Comput. Chem.* **2012**, *31*, 455–461. [CrossRef]
64. Wang, T.Y.; Li, Q.; Bi, K. shun Bioactive Flavonoids in Medicinal Plants: Structure, Activity and Biological Fate. *Asian J. Pharm. Sci.* **2018**, *13*, 12–23. [CrossRef]
65. Shamsudin, N.F.; Ahmed, Q.U.; Mahmood, S.; Shah, S.A.A.; Khatib, A.; Mukhtar, S.; Alsharif, M.A.; Parveen, H.; Zakaria, Z.A. Antibacterial Effects of Flavonoids and Their Structure-Activity Relationship Study: A Comparative Interpretation. *Molecules* **2022**, *27*, 1149. [CrossRef]
66. Aderogba, M.A.; Ndhlala, A.R.; Rengasamy, K.R.R.; Staden, J. Van Antimicrobial and Selected In Vitro Enzyme Inhibitory Effects of Leaf Extracts, Flavonols and Indole Alkaloids Isolated from *Croton Menyharthii*. *Molecules* **2013**, *18*, 12633–12644. [CrossRef]
67. Imoru, A.; Onibi, G.E.; Osho, I.B. Nutritional and Biochemical Compositions of Turmeric (*Curcuma Longa* Linn) Rhizome Powder—A Promising Animal Feed Additive. *Int. J. Sci. Eng. Res.* **2018**, *9*, 424–429.
68. Zakaria, F.; Tan, J.-K.; Mohd Faudzi, S.M.; Rahman, M.; Ashari, E. Ultrasound-Assisted Extraction Condition Optimization Using Response Surface Methodology from *Mitragyna Speciosa* (Korth.) Havil Leaves. *Ultrason. Sonochem.* **2021**, *81*, 105851. [CrossRef]
69. Mawela, K.G. The toxicity and repellent properties of plant extracts used in ethnoveterinary medicine to control ticks. Ph.D. Thesis, University of Pretoria, Pretoria, South Africa, 2008; pp. 72–82. Available online: https://repository.up.ac.za/handle/2263/29224 (accessed on 11 December 2022).
70. Chen, Z.; Bertin, R.; Froldi, G. EC_{50} Estimation of Antioxidant Activity in DPPH Assay Using Several Statistical Programs. *Food Chem.* **2013**, *138*, 414–420. [CrossRef] [PubMed]
71. McDonald, S.; Prenzler, P.D.; Antolovich, M.; Robards, K. Phenolic Content and Antioxidant Activity of Olive Extracts. *Food Chem.* **2001**, *73*, 73–84. [CrossRef]
72. Chang, C.C.; Yang, M.H.; Wen, H.M.; Chern, J.C. Estimation of Total Flavonoid Content in Propolis by Two Complementary Colorimetric Methods. *J. Food Drug Anal.* **2002**, *10*, 178–182. [CrossRef]
73. Lulekal, E.; Rondevaldova, J.; Bernaskova, E.; Cepkova, J.; Asfaw, Z.; Kelbessa, E.; Kokoska, L.; Van Damme, P. Antimicrobial Activity of Traditional Medicinal Plants from Ankober District, North Shewa Zone, Amhara Region, Ethiopia. *Pharm. Biol.* **2014**, *52*, 614–620. [CrossRef]
74. Benkert, P.; Biasini, M.; Schwede, T. Toward the Estimation of the Absolute Quality of Individual Protein Structure Models. *Bioinformatics* **2011**, *27*, 343–350. [CrossRef] [PubMed]
75. Luan, F.; Han, K.; Li, M.; Zhang, T.; Liu, D.; Yu, L.; Lv, H. Ethnomedicinal Uses, Phytochemistry, Pharmacology, and Toxicology of Species from the Genus Ajuga L.: A Systematic Review. *Am. J. Chin. Med.* **2019**, *47*, 959–1003. [CrossRef] [PubMed]
76. Wambe, H.; Noubissi, P.A.; Fokam Tagne, M.A.; Foyet Fondjo, A.; Fankem, G.O.; Kamtchouing, I.; Ngakou Mukam, J.; Nguelefack, T.B.; Kamgang, R. Anti-Shigellosis Activity of Cola Anomala Water/Ethanol Pods Extract on *Shigella Flexneri*-Induced Diarrhea in Rats. *Biomed. Res. Int.* **2019**, *2019*, 6706230. [CrossRef]
77. Fitriah, A.; Holil, K.; Syarifah, U.; Fitriyah; Utomo, D.H. In Silico Approach for Revealing the Anti-Breast Cancer and Estrogen Receptor Alpha Inhibitory Activity of Artocarpus Altilis. In *AIP Conference Proceedings*; AIP Publishing: Melville, NY, USA, 2018; p. 070003.
78. Utami, W.; Aziz, H.A.; Fitriani, I.N.; Zikri, A.T.; Mayasri, A.; Nasrudin, D. In Silico Anti-Inflammatory Activity Evaluation of Some Bioactive Compound from *Ficus Religiosa* through Molecular Docking Approach. *J. Phys. Conf. Ser.* **2020**, *1563*, 1–9. [CrossRef]
79. Tanuja, J.; Priyanka, S.; Tushar, J.; Subhash, C. In Silico Screening of Anti-Inflammatory Compounds from Lichen by Targeting Cyclooxygenase-2. *J. Biomol. Struct. Dyn.* **2019**. [CrossRef]

molecules MDPI

Article

Molecular Docking of Bacterial Protein Modulators and Pharmacotherapeutics of *Carica papaya* Leaves as a Promising Therapy for Sepsis: Synchronising In Silico and In Vitro Studies

Juveria Usmani [1], Hina Kausar [2], Saleem Akbar [3], Ali Sartaj [4], Showkat R. Mir [2], Mohammed Jaseem Hassan [5], Manju Sharma [1], Razi Ahmad [6,*], Summaya Rashid [7] and Mohd Nazam Ansari [7,*]

1 Department of Pharmacology, School of Pharmaceutical Education and Research, Jamia Hamdard University, New Delhi 110062, India
2 Department of Pharmacognosy and Phytochemistry, School of Pharmaceutical Education and Research, Jamia Hamdard University, New Delhi 110062, India
3 Department of Pharmaceutical Chemistry, School of Pharmaceutical Education and Research, Jamia Hamdard University, New Delhi 110062, India
4 Department of Pharmaceutics, School of Pharmaceutical Education and Research, Jamia Hamdard University, New Delhi 110062, India
5 Department of Pathology, Jawaharlal Nehru Medical College, Aligarh Muslim University, Aligarh 202002, India
6 Department of Pharmacology, Hamdard Institute of Medical Sciences and Research, Jamia Hamdard University, New Delhi 110062, India
7 Department of Pharmacology & Toxicology, College of Pharmacy, Prince Sattam Bin Abdulaziz University, Al-Kharj 11942, Saudi Arabia
* Correspondence: rahmad50@gmail.com (R.A.); nazam.ansari@gmail.com (M.N.A.)

Citation: Usmani, J.; Kausar, H.; Akbar, S.; Sartaj, A.; Mir, S.R.; Hassan, M.J.; Sharma, M.; Ahmad, R.; Rashid, S.; Ansari, M.N. Molecular Docking of Bacterial Protein Modulators and Pharmacotherapeutics of *Carica papaya* Leaves as a Promising Therapy for Sepsis: Synchronising In Silico and In Vitro Studies. *Molecules* **2023**, *28*, 574. https://doi.org/10.3390/molecules28020574

Academic Editors: Raffaele Pezzani and Sara Vitalini

Received: 15 December 2022
Revised: 31 December 2022
Accepted: 2 January 2023
Published: 6 January 2023

Abstract: Sepsis is a serious health concern globally, which necessitates understanding the root cause of infection for the prevention of proliferation inside the host's body. Phytochemicals present in plants exhibit antibacterial and anti-proliferative properties stipulated for sepsis treatment. The aim of the study was to determine the potential role of *Carica papaya* leaf extract for sepsis treatment in silico and in vitro. We selected two phytochemical compounds, carpaine and quercetin, and docked them with bacterial proteins, heat shock protein (PDB ID: 4PO2), surfactant protein D (PDB ID: 1PW9), and lactobacillus bacterial protein (PDB ID: 4MKS) against imipenem and cyclophosphamide. Quercetin showed the strongest interaction with 1PW9 and 4MKS proteins. The leaves were extracted using ethanol, methanol, and water through Soxhlet extraction. Total flavonoid content, DPPH assay, HPTLC, and FTIR were performed. In vitro cytotoxicity of ethanol extract was screened via MTT assay on the J774 cell line. Ethanol extract (EE) possessed the maximum number of phytocomponents, the highest amount of flavonoid content, and the maximum antioxidant activity compared to other extracts. FTIR analysis confirmed the presence of N-H, O-H, C-H, C=O, C=C, and C-Cl functional groups in ethanol extract. Cell viability was highest (100%) at 25 μg/mL of EE. The present study demonstrated that the papaya leaves possessed antibacterial and cytotoxic activity against sepsis infection.

Keywords: *Carica papaya*; in silico; cytotoxicity; ethanol extract; sepsis

1. Introduction

Plants have been believed to be beneficial for humans since the days of yore. Traditional medicines and their active constituents are counted as potent sources of remedies for various ailments. Active ingredients in plant extracts like alkaloids, tannins, flavonoids, saponins, enzymes, volatile oils, etc., possess significant pharmacological activities [1]. *Carica papaya* is one such plant, popularly consumed around the world with the likelihood to provide numerous medicinal and nutraceutical benefits [2,3]. The papaya plant belongs to the family *Caricaceae*, is popularly cultivated in tropical and sub-tropical areas, and has been used historically in traditional medicine to cure and manage various disease conditions [4]. Different parts of the papaya plant such as its leaves, barks, roots, latex, fruit, flowers, and seeds are used and have considerable therapeutic benefits, including restoration of damaged skin, digestion of food, removal of dandruff, and pain relief [1,5]. Previous studies have reported anti-hypertensive, anthelminthic, antibacterial, diuretic, anti-fertility, hypolipidemic, antifungal, antitumor, antithrombocytopenic, and platelet-enhancing effects of *C. papaya* [6–11]. Studies on the extract of *C. papaya* have also shown immunomodulatory and anti-inflammatory potentials by inhibiting release of pro-inflammatory and anti-inflammatory cytokines [12]. Papaya leaves have also been reported to possess antioxidant properties and platelet-enhancing activity along with antithrombocytopenic properties [9,10]. In silico molecular docking is known to be one of the most efficient ways to reduce financial stress in research. Molecular docking is a useful tool for drug discovery campaigns, especially for virtual screening, and has many limitations. These molecular data provide insight into the formation of a ligand–receptor complex and its binding affinity [13]. According to a previous study, heat shock protein has exhibited a cytoprotective effect against stressful conditions, including inflammation, tissue injury, and oxidative stress [14]. Additionally, surfactant protein (PDB ID: 1PW9) is antimicrobial in nature and potentially inhibits cell proliferation of gram-negative bacteria by the action of improving cell membrane permeability [15]. Lactobacillus bacterial protein (PDB ID: 4MKS) was found to be the primary cause of septic urinary infection [16]. According to a study, 0.04 mg/g of quercetin was present in 0.25 mg/g of methanolic extract of papaya leaves [17]. Quercetin had the potential to inhibit the proliferation and activation of macrophages and cytokine stimulation induced by LPS, which allowed for the use of quercetin for the treatment of inflammatory bowel disease [18]. Quercetin exhibited antibacterial effects by significantly decreasing the presence of *Pseudomonas aeruginosa*, *Salmonella enterica*, *Staphylococcus aureus*, and *E. coli* [19]. On the other hand, carpaine contained in papaya leaves is responsible for antitumor, anticancer, and antimicrobial properties. The secondary metabolite contents of the papaya leaves, namely, carpaine alkaloids and pseudocarpaineam, are in the piperidine group of alkaloids. The piperidine-type alkaloid compounds have anticancer activity by virtue of inducing apoptosis [20].

Thus, the development of papaya leaf medications is needed through several in silico and in vitro studies.

Thin-layer chromatography is the simplest and most low-cost technique among other chromatographic methods [21]. High-performance thin-layer chromatography (HPTLC) has become a rational and important analytical option that is a simple, sophisticated, more powerful, and effective tool that works on the principle of the detection, separation, and authentication of metabolites and their products [22]. Therefore, the present study utilizes this technique for the detection, separation, and authentication of ethanol extract from *C. papaya* leaves.

In the past, studies have shown the cytotoxic effects of papaya leaves on several cell lines, demonstrating reduced cell viability in vitro [23]. In a study by Joseph et al., methanol extract from papaya leaves was found to be cytotoxic, whereas chloroform extract from the same plant was non-cytotoxic [24]. This demonstrates that the cytotoxic activity of a plant extract corresponds to the presence of different phytoconstituents. Despite the preceding cytotoxic studies on papaya leaves, there seems to be a paucity of information on the cytotoxic effect of ethanol extract from papaya leaves on the sepsis cell line. MTT assay is a widely used standard colorimetric assay to evaluate the cell viability of all types of cells in the culture media [25,26]. Considering the capabilities of the papaya plant, the objective of this study is to systematically evaluate the phytoconstituents present in various extracts of *C. papaya* leaves and explore their cytotoxic effect on a sepsis cell line, providing comprehensive insights that can be leveraged for therapeutic application in future.

2. Results

2.1. Molecular Docking In Silico Study

Molecular docking was performed in order to establish the binding ability of carpaine, quercetin, imipenem, and cyclophosphamide with heat shock protein, surfactant protein D, and lactobacillus bacterial protein. The docking scores of all compounds are presented in Table 1. All the compounds were found to exhibit several molecular interactions (hydrogen bond, pi–pi interaction, and hydrophobic interaction) with the target protein and were considered to be responsible for the antibacterial activity of the compounds. Among all the titled compounds tested for antibacterial activity, quercetin (test compound) was found to be most potent against two proteins, heat shock protein (PDB ID: 4PO2) and lactobacillus bacterial protein (PDB ID: 4MKS), and have the highest docking score (−6.04 and −5.86 Kcal/mol) as compared to imipenem and cyclophosphamide. Quercetin also exhibited a weak docking score against surfactant protein D (PDB ID:1PW9) with a comparable docking score (−4.48 Kcal/mol) (Table 1). Quercetin demonstrated a hydrogen bond with amino acid residues (GLU404, LEU439, GLN435, THR429, THR430; 1.985 Å) as well as pi–pi stacking with amino acid residue (TYR 228) against heat shock protein (PDB ID: 4PO2) (Figure 1). Carpaine also assumes a favourable orientation within the binding site by interacting with other residues (GLN473, ASN540) as shown in Figure 1. The standard compound (imipenem) revealed several hydrogen bonds with amino acid residues (GLU404, ALA406, TYR431, GLN435, and LEU439) and also had a good docking score (−6.64 Kcal/mol) against heat shock protein (PDB ID: 4PO2) (Table 1 and Figure 1). It also exhibits interaction with amino acid residues (THR405, GLY407, GLY408, PHE428, THR429, THR430, VAL438, and GLN441). The standard compound (cyclophosphamide) showed a hydrophobic interaction with the amino acid residue GLN435. It also resulted in a weak docking score (−4.97 Kcal/mol) and interaction with some other amino acid residues (VAL438, LEU439, ILE440, GLN441, LEU403, GLU404, THR405, ALA406, PHE428, THR429, THR430, and TYR431) against heat shock protein (PDB ID: 4PO2) (Table 1). Another test compound (carpaine) demonstrated weak interaction with the amino acid residues (HIS220, LEU221, ALA224, PHE225, TYR228, GLU232, ILE244, LYS246, ALA264, GLY265, and PHE355) and has the lowest docking score (−2.71 Kcal/mol) against surfactant protein D (PDB ID: 1PW9) (Table 1 and Figure 2). A molecular docking simulation was performed for the titled compounds and displayed good MMGBSA binding energies in the range of −51.31 to −11.03 Kcal/mol (Table 1). The test compounds carpaine and quercetin presented −24.65 and −40.54 Kcal/mol, respectively, binding free energies against surfactant protein D (PDB ID: 1PW9), −41.41 and −38.38 Kcal/mol, respectively, against heat shock protein (PDB ID: 4PO2), and −32.87 and −38.71 Kcal/mol, respectively, against lactobacillus bacterial protein (PDB ID: 4MKS). The redocking of a docked complex of standard (cyclophosphamide) with heat shock protein (PDB ID: 4PO2), surfactant protein D (PDB ID: 1PW9), and lactobacillus bacterial protein (PDB ID: 4MKS) exhibited a similar docking mode with RMSD values of 0.0598, 0.0172, and 0.000 Å, respectively. The proteins namely, surfactant protein D (PDB ID: 1PW9), heat shock protein (PDB ID: 4PO2), and lactobacillus bacterial protein (PDB ID: 4MKS) and ZINC00630526 was the only compound that passed the virtual screening, and the docking score of ZINC00630526 were −2.309 (1PW9), −2.127 (heat shock protein (PDB ID: 4PO2)), and −2.406 (lactobacillus bacterial protein (PDB ID 4MKS)), respectively, which are very low compared to known inhibitors (imipenem), i.e., −4.20 (surfactant protein D (PDB ID: 1PW9)), −6.64 (heat shock protein (PDB ID: 4PO2)), and −5.34 (lactobacillus bacterial protein (PDB ID: 4MKS)). After completing the experiment, the known inhibitors (imipenem) ranked one, one, one against surfactant protein D (PDB ID: 1PW9), heat shock protein (PDB ID: 4PO2), and lactobacillus bacterial protein (PDB ID: 4MKS), respectively, among the randomly selected non-inhibitor molecules.

Table 1. Detailed information on molecular docking of bacterial proteins with the test and standard compounds.

PDB ID	Compound	Docking Score	No. of Hydrogen Bond	Hydrogen Bond-Forming Residue	Another Interacting Residue	PRIME-MMGBSA Binding Free Energy (Kcal/mol)
1PW9	Carpaine	−2.71	n.f*	n.f*	TYR228, PHE225, ALA224, LEU221, HIS220, GLU232, ILE244, LYS246, ALA246, GLY265, PHE355	−24.65
	Quercetin	−4.48	02	TYR228	GLU232, LYS229, PHE225, TYR228, VAL231	−40.54
	Imipenem	−4.20	03	TYR228, LYS246	LEU233, GLU232, LYS229, PHE225, TYR228, VAL231, ILE244	−51.31
	Cyclophosphamide	−4.35	n.f*	n.f*	ILE244, GLU232, PHE225, TYR228, LYS229	−11.03
4PO2	Carpaine	−3.44	02	GLN473, ASN540	THR405, ALA406, VAL409, THR411, ARG533, VAL536, SER537, ASN540, ALA541, SER544, GLN426, ILE427, PHE428, THR429, ASN548, GLY470, ARG469	−41.41
	Quercetin	−6.04	05	GLU404, LEU439, GLN435, THR429, THR430	GLY408, GLY407, ALA406, THR405, GLU404, TYR431, PHE428, VAL438, GLN441	−38.38
	Imipenem	−6.64	05	GLU404, ALA406, TYR431, GLN435, LEU439	THR405, GLY407, GLY408, PHE428, THR429, THR430, VAL438, GLN441	−45.18
	Cyclophosphamide	−4.97	01	GLN435	VAL438, LEU439, ILE440, GLN441, LEU403, GLU404, THR405, ALA406, PHE428, THR429, THR430, TYR431	−39.28
4MKS	Carpaine	−4.36	01	SER246	ASN158, ASN159, VAL160, ASP161, GLY152, GLY153, LYS154, THR43, GLU292, GLU247, PHE248, TYR249, LYS251, THR258	−32.87
	Quercetin	−5.86	n.f*	n.f*	ILE4, VAL3, ALA122, TYR26, LEU29, ILE80, GLY81, LEU82, VAL84, ASP6, THR5, ACE2, GLU28, THR27, GLU125, THR85, ASP83	−38.71
	Imipenem	−5.34	03	GLU268, GLU269, ASN250	ARG263, THR266, TRP270, ASP291, LEU290, PRO289, ALA245, PHE248, TYR249, LYS251	−36.96
	Cyclophosphamide	−4.12	01	GLU269	GLU268, TRP270, LEU290, PRO289, ALA245, PHE248, TYR249, ASN250, LYS251, ASP252	−20.46

Abbreviation: n.f*; not found.

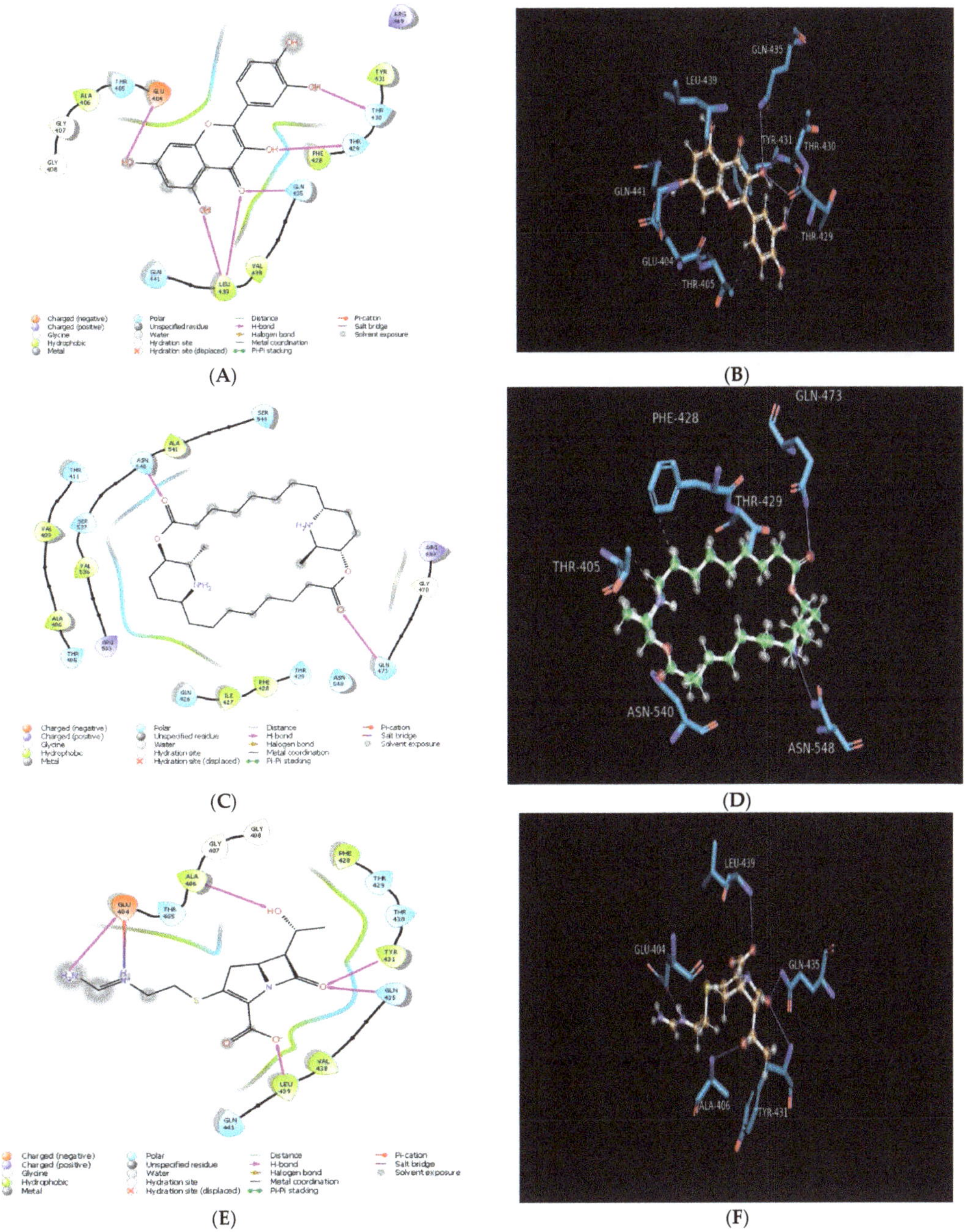

Figure 1. *Cont.*

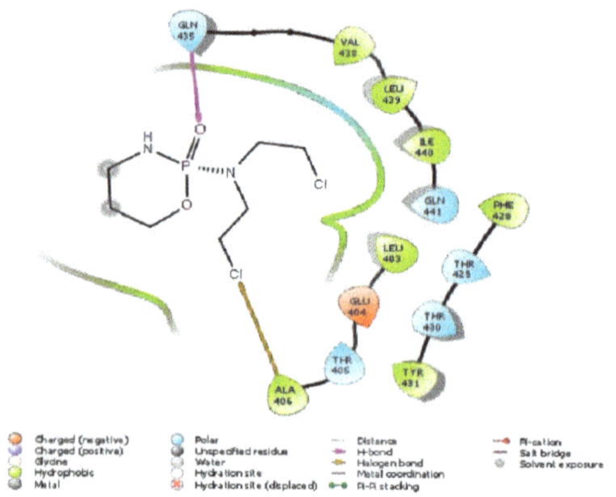

(G)

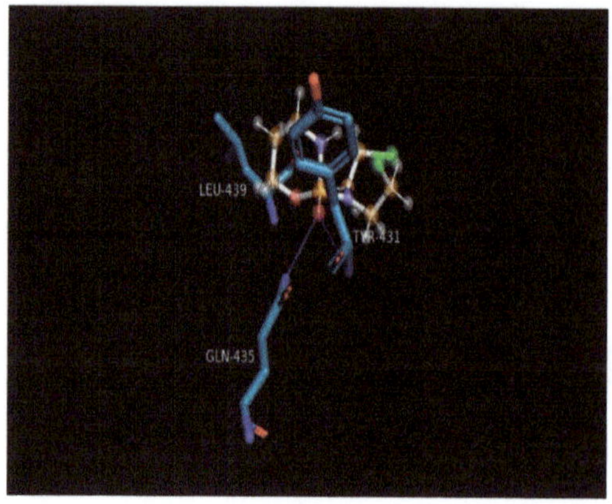

(H)

Figure 1. Molecular docking of compounds with heat shock protein (PDB ID: 4PO2): (**A**) 2D schematic diagram showing interactions of quercetin. (**B**) Cartoon view of heat shock protein with quercetin. (**C**) 2D schematic diagram showing interactions of carpaine. (**D**) Cartoon view of heat shock protein with carpaine. (**E**) 2D schematic diagram showing interactions of imipenem (standard). (**F**) Cartoon view of heat shock protein with imipenem (standard). (**G**) 2D schematic diagram showing interactions of cyclophosphamide (standard). (**H**) Cartoon view of heat shock protein with cyclophosphamide (standard). Residues involved in hydrogen bonding, van der Waals interactions, carbon–hydrogen, and pi–alkyl are represented in different colours indicated in the inset.

We screened test and reference compounds against lactobacillus bacterial protein (4MKS). After screening, we found that quercetin demonstrated very good results, with a docking score value of −5.86 Kcal/mol, and showed hydrophobic interaction with the amino acid residues (VAL3, ILE4, TYR26, LEU29, ILE80, GLY81, LEU82, VAL84, and ALA122) (Table 1). On the other hand, carpaine exhibited one hydrogen bond with the amino acid residue SER246 and had a lower docking score of −4.36 Kcal/mol compared to quercetin and imipenem, but a higher docking score compared to cyclophosphamide. The 2D and 3D interaction diagram of quercetin, carpaine, imipenem, and cyclophosphamide are represented in Figure 3.

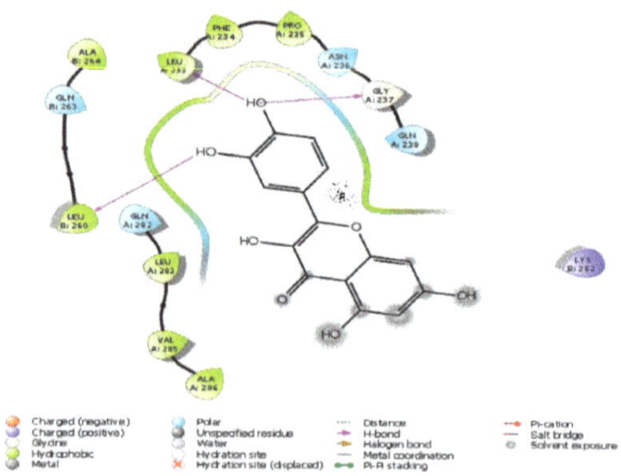

(A)

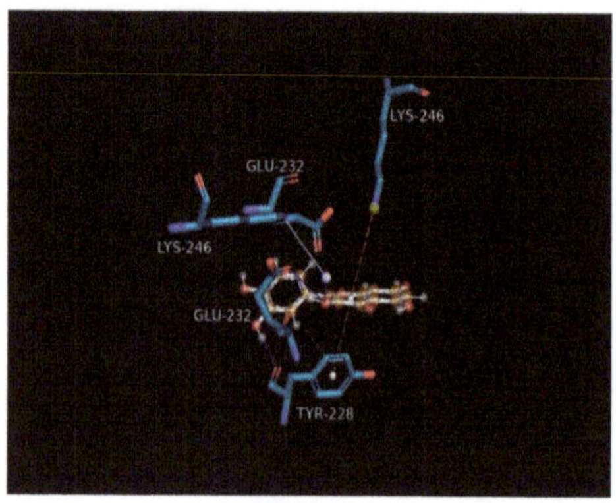

(B)

Figure 2. *Cont.*

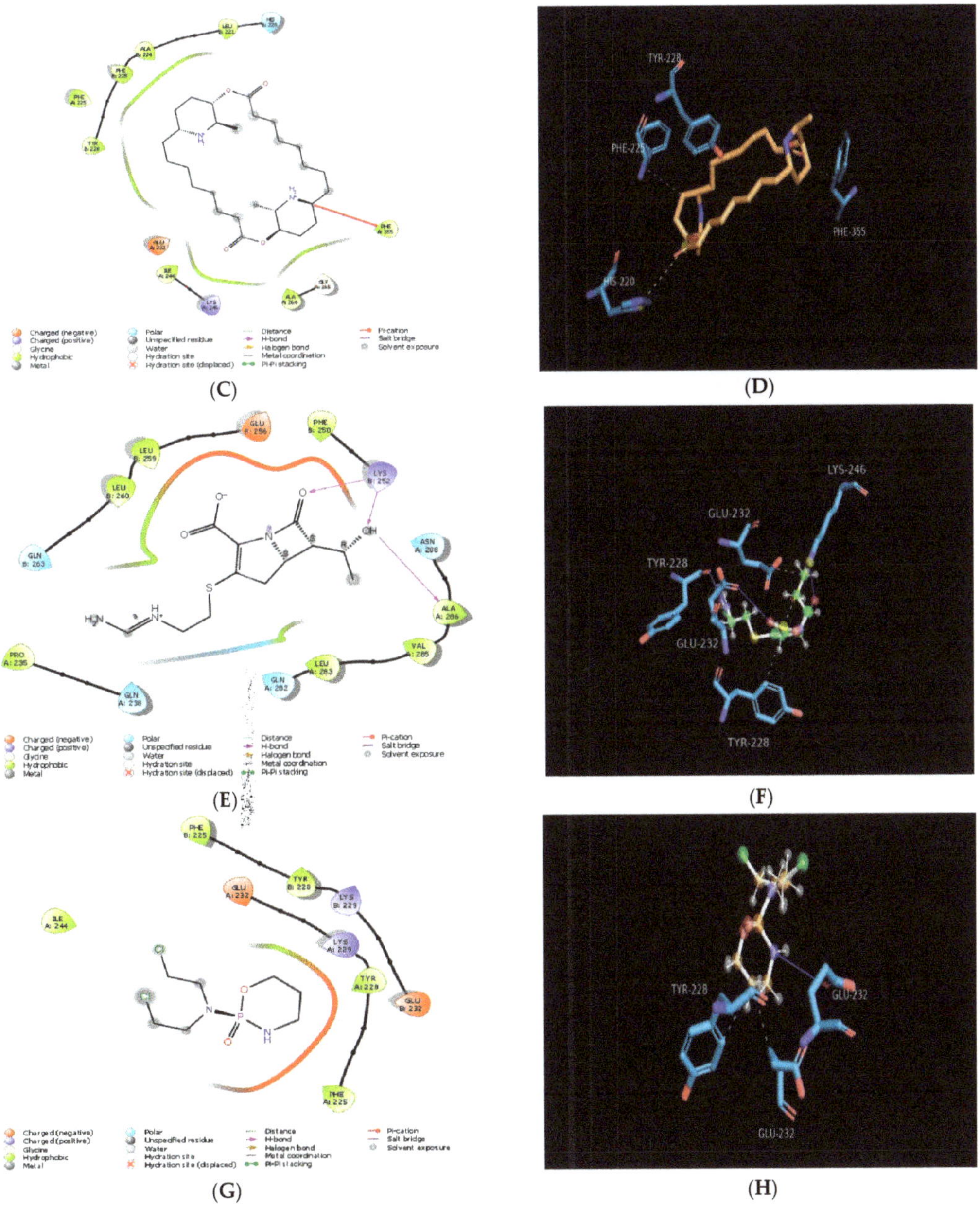

Figure 2. Molecular docking of compounds with surfactant protein D (PDB ID: 1PW9): (**A**) 2D schematic diagram showing interactions of quercetin. (**B**) Cartoon view of surfactant protein D with quercetin. (**C**) 2D schematic diagram showing interactions of carpaine. (**D**) Cartoon view of surfactant protein D with carpaine. (**E**) 2D schematic diagram showing interactions of imipenem (standard). (**F**) Cartoon view of surfactant protein D with imipenem (standard). (**G**) 2D schematic diagram showing interactions of cyclophosphamide (standard). (**H**) Cartoon view of surfactant protein D with cyclophosphamide (standard). Residues involved in hydrogen bonding, van der Waals interactions, carbon–hydrogen, and Pi–alkyl are represented in different colours indicated in the inset.

Among the tested compound, quercetin demonstrated the best interaction with heat shock protein (PDB ID: 4PO2) and lactobacillus bacterial protein (4MKS) as compared to surfactant protein D (PDB ID: 1PW9) and bound to the active binding domain of all three proteins as shown in the superimposed image of Figure 4.

2.2. Percentage Yield

The percentage yield of ethanol, methanol, and aqueous extracts of papaya leaves was calculated using a standard formula and found to be 13.1%, 12.2%, and 10.6%, respectively. The percentage yield of ethanol extract was the highest among all three extracts.

2.3. Phytochemical Screening of Extracts

The qualitative analysis of different extracts of papaya leaves indicates the presence of various phytoconstituents, as shown in Table 2. Ethanolic extract was found to contain a higher number of phytoconstituents as compared to aqueous and methanolic extract. Alkaloids, flavonoids, terpenoids, saponins, and glycosides were found in ethanol extract. Only alkaloids, flavonoids, and glycosides were present in the aqueous extract, whereas methanol extract was found to contain alkaloids, flavonoids, phenolic compounds, and saponins in trace amounts. Therefore, we proceeded with further analysis of the ethanol extract.

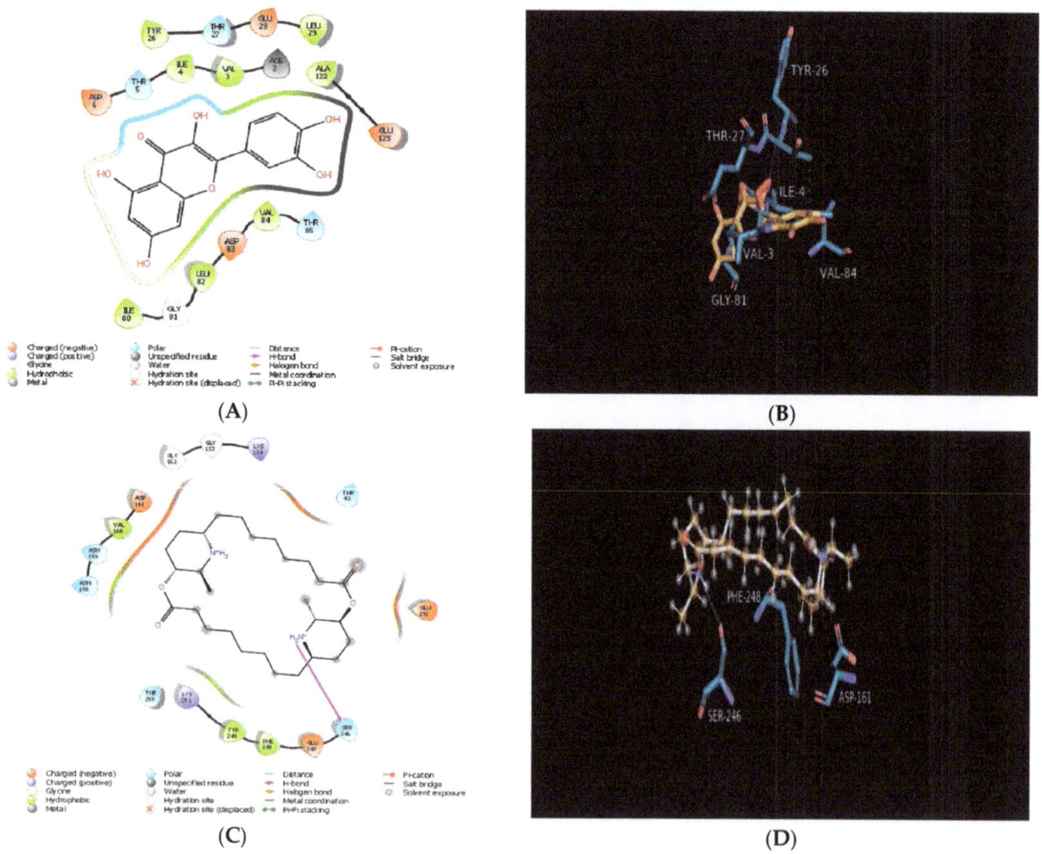

Figure 3. *Cont.*

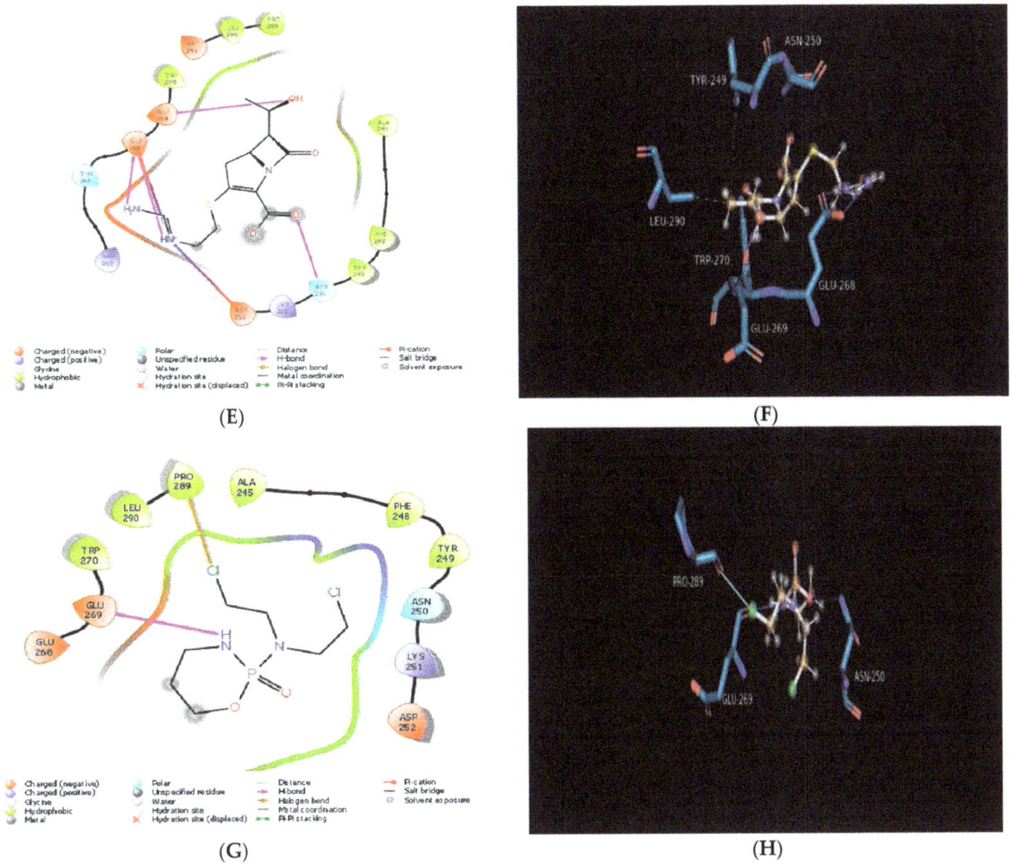

Figure 3. Molecular docking of compounds with lactobacillus bacterial protein (PDB ID: 4MKS): (**A**) 2D schematic diagram showing interactions of quercetin. (**B**) Cartoon view of lactobacillus bacterial protein with quercetin. (**C**) 2D schematic diagram showing interactions of carpaine. (**D**) Cartoon view of lactobacillus bacterial protein with carpaine. (**E**) 2D schematic diagram showing interactions of imipenem (standard). (**F**) Cartoon view of lactobacillus bacterial protein with imipenem (standard). (**G**) 2D schematic diagram showing interactions of cyclophosphamide (standard). (**H**) Cartoon view of lactobacillus bacterial protein with cyclophosphamide (standard). Residues involved in hydrogen bonding, van der Waals interactions, carbon–hydrogen, and pi–alkyl are represented in different colours indicated in the inset.

Table 2. Phytochemical screening of extracts of *Carica papaya* leaves.

Phytoconstituents	Extracts of *Carica papaya* Leaves		
	Aqueous	Methanol	Ethanol
Alkaloid	+	++	+++
Flavonoids	++	+	+++
Phenolic compound	−	+	−
Terpenoids	−	−	+++
Saponins	−	+	++
Glycosides	+	−	+

+, least positive; ++, more positive; +++, most positive; −, negative.

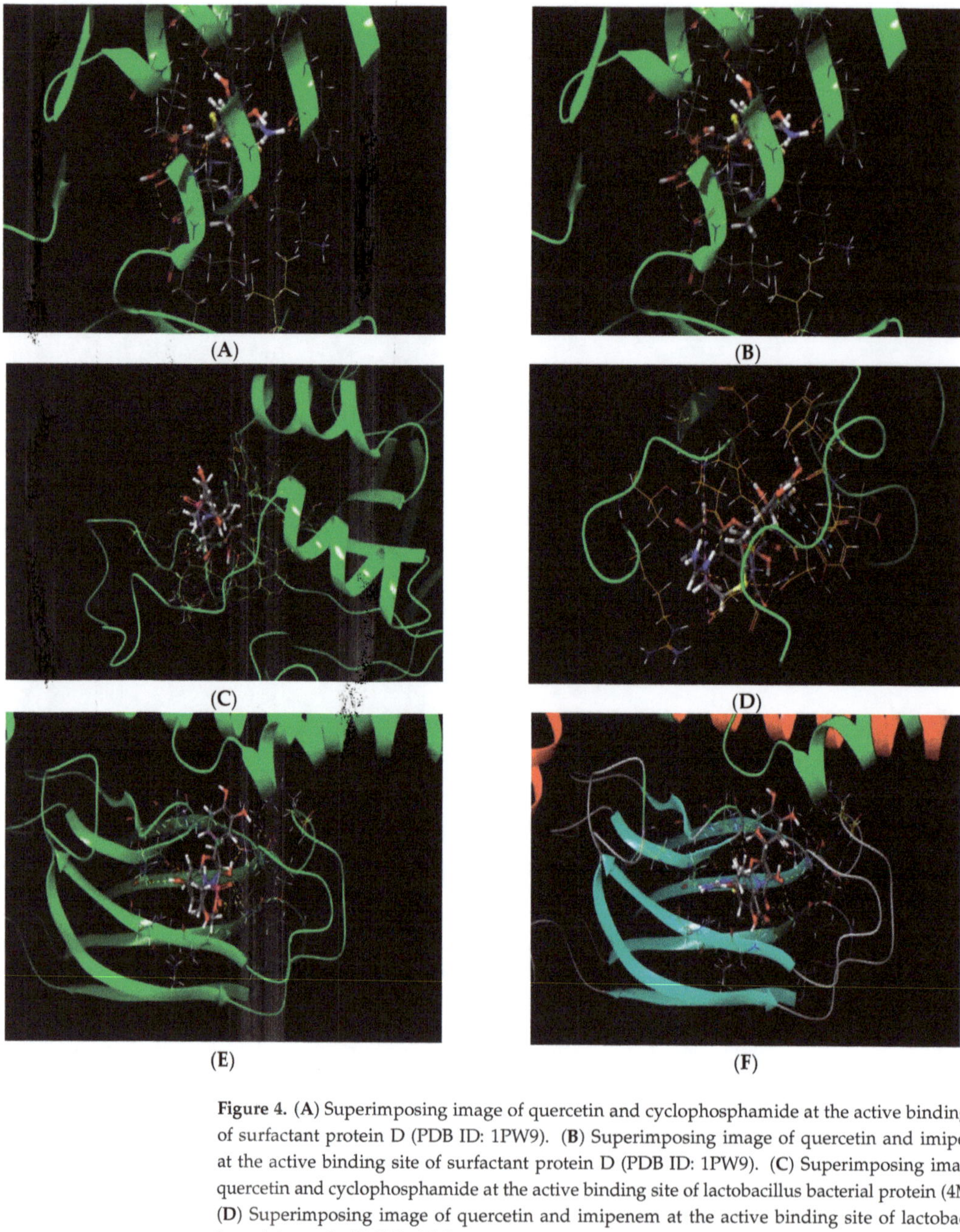

Figure 4. (**A**) Superimposing image of quercetin and cyclophosphamide at the active binding site of surfactant protein D (PDB ID: 1PW9). (**B**) Superimposing image of quercetin and imipenem at the active binding site of surfactant protein D (PDB ID: 1PW9). (**C**) Superimposing image of quercetin and cyclophosphamide at the active binding site of lactobacillus bacterial protein (4MKS). (**D**) Superimposing image of quercetin and imipenem at the active binding site of lactobacillus bacterial protein (4MKS). (**E**) Superimposing image of quercetin and cyclophosphamide at the active binding site of heat shock protein (PDB ID: 4PO2). (**F**) Superimposing image of quercetin and imipenem at the active binding site of heat shock protein (PDB ID: 4PO2).

2.4. Estimation of Total Flavonoids

The total flavonoid content was found to be 1.83 ± 0.16%, 1.16 ± 0.28%, and 2.23 ± 0.24% for aqueous, methanol, and ethanol extracts, respectively, of *Carica papaya* leaves. The given values are expressed as mean ± SD of three different determinations.

2.5. DPPH Radical Scavenging Activity

The examination of the antioxidant activity of different extracts of papaya leaves was carried out using ascorbic acid as standard in DPPH assays as presented in Table 3.

Table 3. DPPH analysis of different extracts of *C. papaya* leaves.

Concentration (µg/mL)	Types of Extracts			
	Aqueous (%)	Methanol (%)	Ethanol (%)	Ascorbic Acid (%)
20	51.26 ± 1.47	70.43 ± 3.47	77.86 ± 3.08	7.23 ± 1.67
40	65.90 ± 1.41	75.10 ± 0.53	83.73 ± 5.28	15.40 ± 2.24
60	69.30 ± 1.26	80.66 ± 2.46	85.23 ± 3.93	37.46 ± 7.18
80	77.33 ± 1.31	85.23 ± 4.60	88.76 ± 1.35	56.16 ± 7.98
100	80.66 ± 0.09	89.00 ± 1.20	89.63 ± 1.73	60.76 ± 2.28

Data from multiple groups of treatment were analysed by two-way ANOVA using the Bonferroni post-test. Values are expressed as mean ± SD (n = 3). A significant variation was observed in the three extracts of leaves as compared to ascorbic acid (Figure 5).

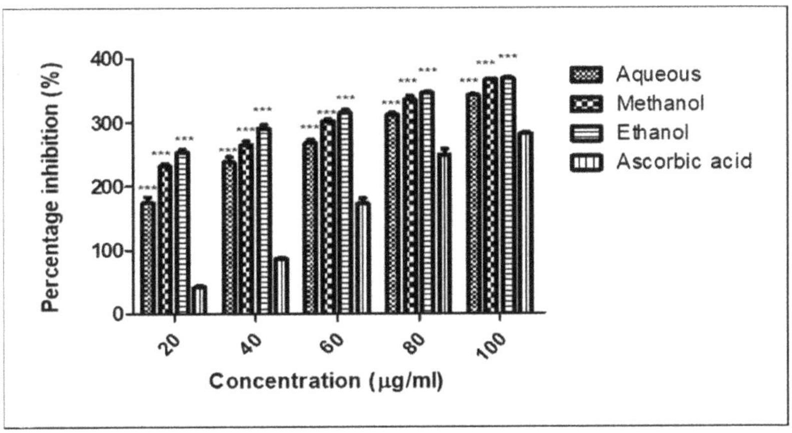

Figure 5. DPPH free radical scavenging activity of different extracts of *C. papaya* leaves. Different solvents like methanol, aqueous, and ethanol were used to obtain extracts of papaya leaves, which underwent DPPH assay at different concentrations to determine the free radical scavenging activity of these extracts and compare with ascorbic acid as standard antioxidant. A significant difference (*** $p < 0.001$) was observed between these extracts and the standard. The methanol and ethanol extracts showed higher significance, while the aqueous extract displayed comparatively less significant values indicative of antioxidant characteristics.

A significant variation was observed in the three extracts of leaves as compared to ascorbic acid (Figure 5). The results demonstrated dose-dependent free radical scavenging activity at a concentration of 20–100 mg/mL. The free radical scavenging activity of the aqueous, methanol, and ethanol extracts was found to be 51.26 ± 1.47–80.66 ± 0.09, 70.43 ± 3.47–89 ± 1.2, and 77.86 ± 3.08–89.63 ± 1.73 units, respectively, each of which was significantly higher than that of ascorbic acid extract, which was 7.23 ± 1.67–60.76 ± 2.28. Data from multiple groups of treatment were analysed by two-way ANOVA using the

Bonferroni post-test. A statistically significant difference was expressed as *** $p < 0.001$ between the groups.

2.6. TLC Analysis

TLC is one of the easiest and most common techniques based on the principle of identification and separation of various phytocomponents in an herbal drug. In our study, TLC was performed for ethanolic extract of papaya leaves where three spots were observed in the UV region as shown in Figure 6.

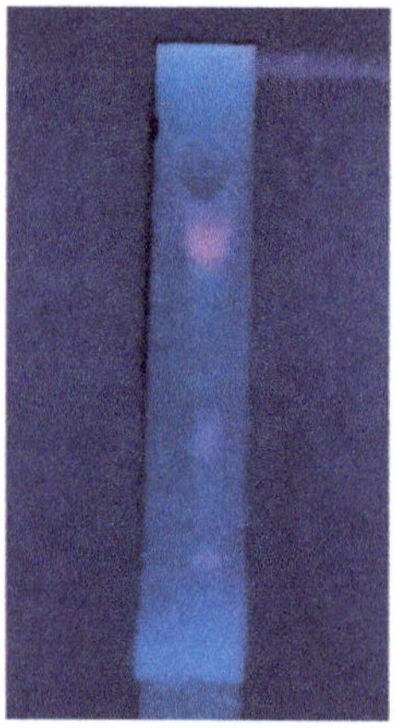

Figure 6. TLC plate observed in UV chamber.

2.7. HPTLC Fingerprinting

High-performance thin-layer chromatography (HPTLC) fingerprinting was carried out using toluene: ethylacetate: formic acid (8:1.5:0.5) as the mobile phase for the purpose of identification of phytochemical constituents in the papaya leaf extract. Peaks were observed at 254 nm with different Rf values in the HPTLC chromatogram (Figure 7). A total of 7 peaks with Rf values 0.05, 0.15, 0.27, 0.40, 0.58, 0.71, and 0.96 were obtained at 254 nm.

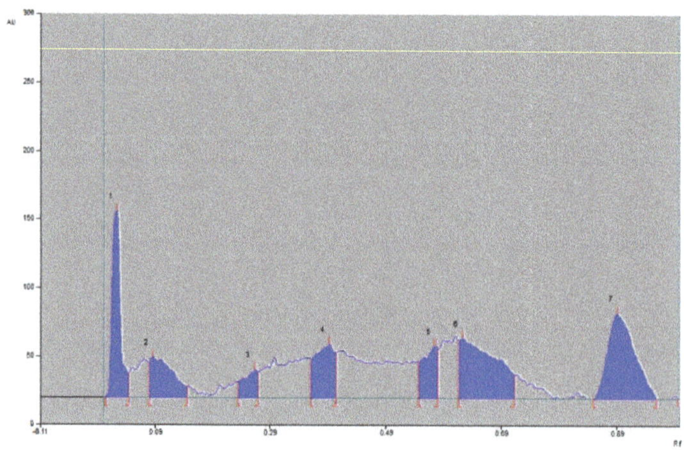

Figure 7. Chromatogram of ethanol extract of *C. papaya*.

2.8. FTIR Spectrum

As seen in Figure 8, the FTIR analysed the presence of different kinds of molecules in the papaya extract. Peaks observed at 3387.15/cm correspond to N-H stretch (amines) and O-H stretch (alcohols); peaks observed at 2919.39/cm relate to C-H stretch (alkanes) and O-H stretch (carboxylic acids); peaks observed at 2819.95/cm are C-H stretch (aldehydes) and O-H stretch (carboxylic acids); peaks observed at 1730.22/cm are associated with C=O stretch (aldehydes) and C=O stretch (ketones); peaks observed at 1651.03/cm are related to C=O stretch (amides) and C=C stretch (alkenes); peaks observed at 1616.12/cm are C=C stretch (alkenes) and C=C stretch (aromatic rings); while peaks observed at 825.57 and 719.18 are associated with =C-H bend (alkenes), C-H bend (aromatic compounds), and C-Cl stretch (alkyl and aryl halides).

2.9. *In Vitro* % Cell Viability

As represented in Figure 9, the percentage cell viability of J774 cells upon treatment with increasing concentrations (25, 50, 100, 200, 400, and 800 µg/mL) of ethanol extract of papaya leaves was found to be 100%, 99.8%, 98.2%, 94.3%, 84.2%, and 80.3%, respectively. Cell viability was found to be highest (100%) at 25 µg/mL and lowest (80.3%) at 800 µg/mL of ethanol extract (EE), indicating that the lowest concentration of extract is unable to inhibit the total population of J774 cells, whereas the highest concentration of extract can inhibit around 20% of the cell population. This indicates that 800 µg/mL of extract produces the highest cytotoxicity.

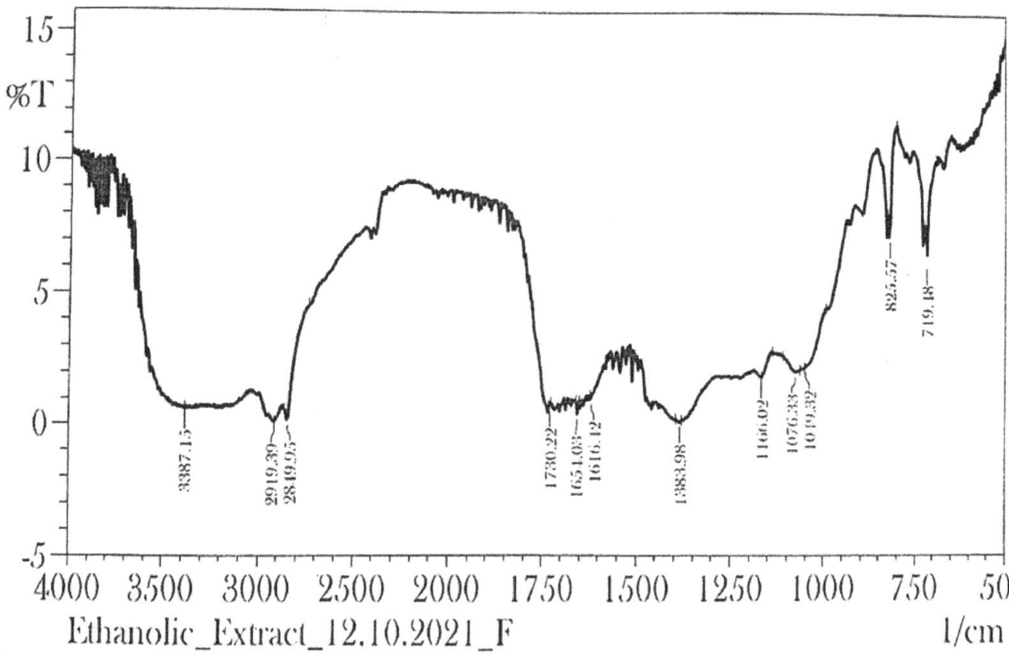

Figure 8. FTIR spectrum of ethanol extract of *Carica papaya* leaves.

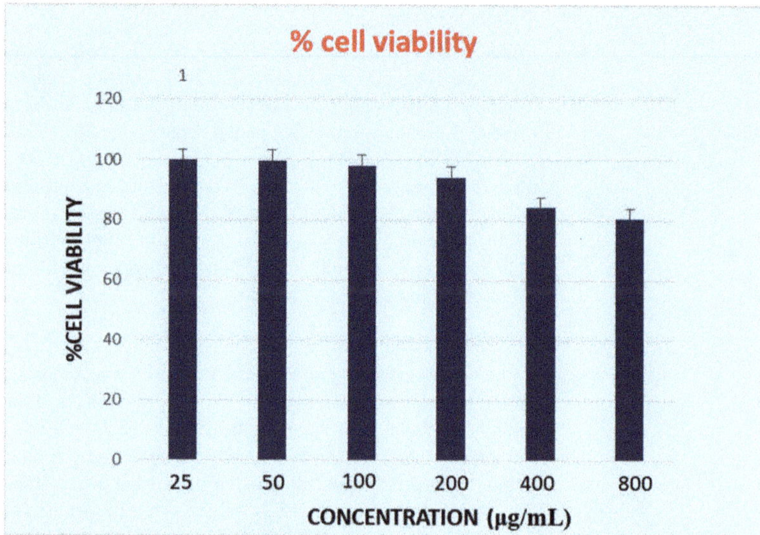

Figure 9. Effect of ethanol extract of *C. papaya* leaves on percentage viability of J774 cells.

3. Discussion

Infectious diseases have been a global burden and a major cause of death. Sepsis is an immunocompromised infection that occurs due to the host's response to injury. Treatment of sepsis commonly relies on antibiotic therapy; however, due to the increasing incidence of antibiotic resistance, the treatment approach remains limited [27]. In silico antibacterial activity of *C. papaya* leaves was performed by using bioinformatics tools. The flavonoid quercetin exhibited the highest interaction with the bacterial protein. The present study is suggestive of the fact that several van der Waals, covalent, carbon–hydrogen, pi–alkyl, and electrostatic interactions were observed to be the key forces for bonding of quercetin, carpaine, imipenem, and cyclophosphamide together with the heat shock protein (PDB ID: 4PO2), surfactant protein D (PDB ID: 1PW9), and lactobacillus bacterial protein (PDB ID: 4MKS). According to a previous study, heat shock protein (PDB ID: 4PO2) has exhibited a cytoprotective effect against stressful conditions, including inflammation, tissue injury, and oxidative stress [14]. Also, surfactant protein (PDB ID: 1PW9) is antimicrobial in nature and potentially inhibits cell proliferation of gram-negative bacteria by the action of improving cell membrane permeability [15]. Another study reported that the enteral surfactant protein D worsens mortality after CLP by enhancing bacterial colonization in the gut [28]. Lactobacillus bacterial protein (PDB ID: 4MKS) was found to be the primary cause of septic urinary infection [16]. In conjunction with this, the results from our study have shown an excellent docking score by producing the interaction of quercetin and carpaine with the heat shock protein (PDB ID: 4PO2) and lactobacillus bacterial protein (PDB ID: 4MKS), but comparatively less binding with surfactant protein D (PDB ID: 1PW9), thus confirming the beneficial cytoprotective and antibacterial activity of the titled compounds for sepsis treatment. Moreover, this study indicated that *C. papaya* might employ antibacterial activity, which could be a platform to investigate the role of test compounds against sepsis.

C. papaya is enriched with antioxidants including α-tocopherol, ascorbic acid, various flavonoids, phenolic compounds, glycosides, enzymes, etc., and is commonly used for the prevention and treatment of innumerable diseases [1,29]. An important therapeutic application of papaya leaves has been its use as an antithrombocytopenic drug to treat dengue fever [10,30]. Phytochemical screening of papaya leaves extracts has identified alkaloids, terpenoids, flavonoids, saponins, steroids, tannins, and phenols that have shown potent therapeutic significance against inflammation, oxidative stress, and hypoglycaemic conditions [12,29,31]. The present study also confirms the presence of these chemical compounds in *C. papaya* leaves and suggests that the amount of the phytochemicals such as alkaloids, flavonoids, terpenoids, saponins, and glycosides was highest in ethanol extracts as compared to aqueous and methanol extracts that remained the mainstream of the present study (Table 1). The highest percentage yield of papaya leaves was 13.1% in ethanol extract.

Flavonoids are important phytocomponents that possess antioxidant and anti-inflammatory properties. Alkaloids are also widely distributed phytoconstituents sought after for their anti-inflammatory, antimalarial, stimulant, narcotic, analgesic, antispasmodic, and antitumoral properties [31]. Therefore, the current study can be valuable in further assessing quantitative parameters of phytotherapeutically active molecules. The total flavonoid content was highest in ethanol extract as compared to methanol and aqueous extracts, which correlates with previous studies [32,33]. These findings confirm that *C. papaya* leaves contain a significant quantity of flavonoid compounds, which exert an anti-inflammatory effect.

Several types of assays are being included as potent tools to quantify the antioxidant potential of natural products. The DPPH free radical scavenging assay is usually preferred over other methods due to its stability, simplicity, reproducibility, feasibility, and efficiency [34]. The present study has displayed the potent oxidative stress-reducing potential of *C. papaya* leaf extracts. In vitro, a DPPH assay was carried out to determine the antioxidant potential of papaya leaves, demonstrating higher significance in ethanol and methanol extracts and lesser in the aqueous extract as compared to ascorbic acid, which is conducive to inhibiting oxidative stress levels corresponding to previous studies [32]. Taking into account the positive outcomes of ethanol extract, it was selected for further analysis and examination of cytotoxicity using a sepsis cell line.

HPTLC has become a rational and important analytical option that is a simple, sophisticated, more powerful, and effective tool for the detection, separation, and authentication of herbal drugs and their products [22]. HPTLC fingerprinting of ethanolic extract of *C. papaya* leaves demonstrated the presence of various phytoconstituents in the ethanol extract. The chromatogram observed 7 peaks at different Rf values detected at 254 nm in the toluene: ethylacetate: formic acid (8:1.5:0.5) solvent system, indicating the number of constituents in the extract that can be further utilized to evaluate its therapeutic efficacy. FTIR is the most significant tool for the identification of functional groups present in phytomedicines [34]. The FTIR analysis indicated the presence of numerous characteristic functional groups such as: N-H, indicated as amines; O-H, indicating the presence of amines or carboxylic acid; C=O, specified as aldehydes, ketones, or amides; C=C, indicating the presence of alkenes; and the aromatic ring in ethanol extract of papaya leaves, holding prolific medicinal properties. The presence of saponins in *C. papaya* leaves has shown cytotoxic effects through increased cell permeability [35]. This directly correlates with our results, indicating the presence of saponins in the leaf extract of the papaya plant. It is thus, necessary to analyse the cytotoxicity of the plant leaves. MTT assay is a standard colorimetric assay widely used to evaluate the cell viability of all types of cells in the culture media. The measurement of cell viability in this method corresponds to cellular respiration and the quantity of formazan produced, indicating the number of viable cells in the culture incorporated with the test/standard agent. The presence of a higher number of viable cells in the culture results in higher levels of formazan crystal formation, which helps in determining cell proliferation and thus the cytotoxicity of the treatment used [25,36]. No evidence is available to date on the cytotoxic effect of EE on the J774 sepsis cell lines. Hence, in vitro cytotoxicity was carried out using an MTT assay at different concentrations of ethanol extract of papaya leaves. The results revealed that the cell viability was at a maximum (100%) at the lowest concentration (25 μg/mL) and lowest (80.3%) at the highest concentration (800 μg/mL), which confirms that the cytotoxic activity of the plant was observed at the highest concentration (800 μg/mL) of EE tested. The results showed that the EE inhibits cell proliferation and produces significant sepsis managing potentials. This further necessitates in-depth investigation on exploring the potential of papaya leaves for managing sepsis and related complications.

4. Materials and Methods

4.1. Sample Collection and Authentication

Fresh leaves of *C. papaya* were collected from the campus of Jamia Hamdard, and a voucher specimen (BOT/DAC/2021/06) was deposited in the herbarium of the Department of Botany, School of Chemical and Life Sciences, Jamia Hamdard, New Delhi, where identification of leaves was carried out by a taxonomist.

4.2. In Silico Study

The molecular docking study was carried out to establish different interactions between the test compounds and the target protein. The 3D structure of heat shock protein (PDB ID: 4PO2), surfactant protein D (PDB ID: 1PW9), and lactobacillus bacterial protein (4MKS) was performed on a Maestro 12.5 program (Schrodinger Inc., New York, NY, USA) using a the 64-bit operating system [Intel (R) Core (TM) i3-7020U CPU @ 2.30 GHz, 8 GB RAM]. The X-ray crystal structure, heat shock protein (PDB ID: 4PO2) [37], surfactant protein D (PDB ID: 1PW9) [38], and lactobacillus bacterial protein (PDB ID: 4MKS) [39] with the known inhibitor (imipenem and cyclophosphamide), was retrieved from the RCSB protein Data Bank (http://www.pdb.org/pdb/home/home.do accessed on 26 July 2022). The protein obtained was first prepared using the protein preparation wizard module. Water molecules and all other undesirable residues were removed, and hydrogen atoms were added before subjecting to the docking process. Site mapping was performed by selecting minimized proteins using the sitemap module of Schrodinger. We have selected a particular site for docking based on a high site score among the generated binding site.

The grid was prepared using minimized protein, which indicates the drugs have binding sites related to the specific target. The prepared grid was used for further processing in the advanced docking process. The grid box was generated according to the active site of the protein (4PO2, 1PW9, and 4MKS) where the centre was X: 50.0, Y: 25.0, Z: 70.0 (coordinates), X: 20.0, Y: 25.0, Z: 10.0 (coordinates), and X: −30.0, Y: −50.0, Z: 10.0 (coordinates), respectively. With no restrictions, the van der Waals radius scaling factor was set to 1.0. Finally, the grid was created with a partial charge cut-off of around 0.25. The amino acid residues SER400, THR405, THR411, SER418, THR422, THR425, THR429, THR430, TYR443, THR450, THR491, THR502, THR504, SER537, and SER544 for the protein (PDB ID: 4PO2); SER226, TYR228, SER239, THR247, and THR262 for the protein (PDB ID: 1PW9); and SER15, THR144, SER174, THR185, THR188, TYR231, CYS243, SER285, SER335, TYR361, THR362, SER366, THR372, THR384, THR391, SER393, THR395, TYR403, TYR419, SER424, and TYR426 for the protein (PDB ID: 4MKS) were considered for the grid generation. The Receptor Grid Generation tool in Maestro was used to generate the grid. ChemDraw 12.0 software (PerkinElmer Inc., Cambridge, MA, USA) was used to draw the structure of ligand molecules as a mol file. Their energy was minimized using the LigPrep module of Maestro. All possible ionization states at pH 7.0 ± 2.0 were generated and minimized. Ligand molecules prepared were docked into the active site in extra precision mode (XP) using Glide. Docking of carpaine and quercetin as test compounds and imipenem and cyclophosphamide as reference compounds was performed into the active site of each heat shock protein, surfactant protein D, and lactobacillus bacterial protein. The visualization was done to analyse the interaction between ligands and residues of amino acid residues on each protein by PyMOL software. The binding energy calculation (MM-GBSA) was further conducted using Prime in Maestro to analyse the potential biological response of the free binding energy of the ligands that are binding to the active site of the protein in the docked complex, using XP docking mode [40–42].

For validation of the docking protocol, redocking of docked compounds of the standard (cyclophosphamide) with heat shock protein (PDB ID: 4PO2), surfactant protein D (PDB ID: 1PW9), and lactobacillus bacterial protein (PDB ID: 4MKS) was performed. The RMSD measurements were calculated to determine the stability of the docking poses, which demonstrates the structural variation and protein stability. The decoy database was obtained from the DUD (a Directory of Useful Decoys), released on 22 October 2006 [43].

Fifty molecules (non-inhibitors) from the dataset and one known inhibitor (imipenem) were randomly selected. Further, virtual screening of the randomly selected non-inhibitors was performed against surfactant protein D (PDB ID: 1PW9), heat shock protein (PDB ID: 4PO2), and lactobacillus bacterial protein (PDB ID: 4MKS) using the virtual screening workflow module of Schrodinger, and docking scores were compared with the known inhibitors.

4.3. Preparation of Extracts

Fresh leaves obtained were washed thoroughly to remove dirt and impurities and air-dried for 2–3 days. These were then crushed and reduced to powdered form using a mortar pestle. Dried powdered leaves weighing 58 g were extracted with different solvents like ethanol, water, and methanol (500 mL) in the Soxhlet apparatus for 24 h. The temperature of the solvent was kept above 78 °C. The extract obtained was evaporated to dryness by a rotary evaporator at 65 °C and then kept in an oven at 64 °C until crude extract was obtained. The final extracts were stored at 2–4 °C until further estimations were made. The percentage yield of extract was calculated using the formula [44]:

$$\%\text{ yield of extract} = \frac{\text{weight of extract obtained}}{\text{weight of powder material}} \times 100$$

4.4. Preliminary Estimation of Phytoconstituents

Analysis of phytoconstituents present in papaya leaves was made using water, methanol, and ethanol. Tests for chemical components like alkaloids, flavonoids, saponins, tannins, phenolic compounds, terpenoids, steroids, anthraquinones, cardiac glycosides, and volatile oils were performed following standard protocols from previous studies [45].

4.5. Total Flavonoid Content

The flavonoid content was determined by the method described in [46]. Ten milligrams of different extracts were weighed and mixed with respective solvents. The extracts were filtered with Whatman No. 42 filter paper. The filtrates were collected and evaporated to dryness on a water bath to a constant weight.

The flavonoid content was calculated using the following formula [32]:

$$\%\text{ flavonoids} = \frac{\text{weight of final filtrate}}{\text{weight of sample}} \times 100$$

4.6. DPPH Radical Scavenging Activity

Free radical scavenging activity of the sample extract was performed as described in a previous study [47]. DPPH is a commonly practiced assay for the evaluation of free radical scavenging potentials of the antioxidant content of pure compounds. This assay is considered a reliable, easy, and standard colorimetric method utilized for the characterization of antioxidant properties [48]. An amount of 0.004% w/v DPPH (2,2-diphenyl-l-picryl hydrazyl) solution was prepared in 95% methanol. 100 mL of stock solution of plant extracts in standard ascorbic acid was then prepared at a concentration of 100 μg/mL. From this stock solution, 2 mL, 4 mL, 6 mL, 8 mL, and 10 mL of this solution were mixed with methanol to make the final volume up to 10 mL, making the final concentration up to 20 μg/mL, 40 μg/mL, 60 μg/mL, 80 μg/mL, and 100 μg/mL, respectively. A freshly prepared DPPH solution (2 mL) was added to each test tube and left to be mixed in the dark for 15 min. Absorbance was measured at 523 nm against the blank using the UV spectrophotometer. For the control, the DPPH solution (2 mL) was mixed with methanol (10 mL). The assay was carried out in triplicate. DPPH free radical scavenging activity of the extracts was calculated as percentage inhibition (%) using the following formula:

$$\text{DPPH scavenging activity} = \frac{[1 - (\text{Abs sample} - \text{Abs blank sample})]}{\text{Abs control}} \times 100$$

4.7. Thin Layer Chromatography Analysis

The ethanol extract of papaya leaves was analysed using thin-layer chromatography (TLC) to separate fractions of different active constituents of the drug. TLC plates (Merck-silica gel 60 F254) were developed to confirm the presence of fractions of different phytopharmaceuticals in the drug. It was allowed to run in a glass chamber containing a mobile phase. Different combinations of mobile phases were allowed to run through the drug sample to obtain the best visuals of separated constituents. The most appropriate outcome was obtained by the combination of toluene: ethylacetate: formic acid (8:1.5:0.5) solvent system. The plate was then observed to identify the positions of spots in an ultraviolet chamber at 254 nm. The Rf value of the herbal extract was calculated using the standard formula [49]:

$$\text{Rf} = \frac{\text{Distance traveled by the solute}}{\text{Distance traveled by the solvent}}$$

4.8. HPTLC Fingerprinting Profile of Ethanolic Extract of Carica Papaya Leaves

High-performance thin-layer chromatography (HPTLC) fingerprinting was carried out using toluene: ethylacetate: formic acid (8:1.5:0.5) as a mobile phase for the purpose of identification of phytochemical constituents in the papaya leaf extract. Peaks were observed at 254 nm with different Rf values in the HPTLC chromatogram.

4.9. Fourier Transform Infrared Spectroscopy (FTIR)

FTIR is very economical, easy, fast, and requires a small amount of a sample. The functioning of the tool is based on the principle of identification of the presence of functional groups in the phytoconstituents [50,51]. A dried extract of ethanol extract of papaya leaves was taken for FTIR analysis. Pellets of leaf extract were prepared by mixing with 1–2 mg KBr powder to achieve a translucent powder, which was then compressed by the mechanical press to get the desired pellet. The spectrum was analysed by the means of IR solution software 3.50 build 214 (Shimadzu, Kyoto, Japan) to identify the presence of functional groups in the compound [52].

4.10. Cytotoxicity Study

The in vitro cell proliferation assay was performed using a 4,5-dimethylthiazole-2-yl)-2,5 diphenyltetrazolium bromide (MTT) assay. Briefly, J774 cells were seeded in a 96-well microplate. Cells were incubated at 37 °C until they attached to the bottom of wells. After 48 h of incubation, 40 µL RPMI (International PBI, Milan, Italy), taken as control, was removed, and cells were washed with 100 µL phosphate buffer. MTT (5 mg/mL) was subsequently added along with 1.6% DMSO as a positive control, and various concentrations of the EE (25, 50, 100, 200, 400, and 800 µg/mL) were added to 20 µL of J774 cell 106 suspensions (cells/mL). Imipenem as a positive control with different concentrations (0.25, 0.5, 1, 2, and 4 µg/mL) was added to other wells. The plates were incubated in a CO_2 incubator at 37 °C for 24 hrs. The media in the plate was removed by aspiration. The absorbance was measured at 595 nm using a microplate reader (Microplate reader 680, Bio-Rad, Hercules, CA, USA) following the addition of 100 µL of distilled water, and the % cell viability was calculated using the following equation [53,54]:

$$\text{Cell viability } (\%) = \frac{\text{OD of treated} - \text{OD of control}}{\text{OD of control}} \times 100$$

5. Conclusions

Several previous studies have demonstrated the antibacterial, antioxidant, anti-inflammatory, and cytotoxic activity of papaya leaves by inhibiting numerous pathways. Studies have reported the presence of the flavonoid quercetin to be responsible for the antioxidant, antibacterial, and anti-inflammatory properties. From our study, an excellent docking score of quercetin against surfactant protein D (PDB ID: 1PW9) and lactobacillus bacterial protein (PDB ID: 4MKS) was observed due to strong molecular interactions, such as hydrogen bonds, pi–pi interactions, and hydrophobic interactions, and favourable orientation within the binding site by interacting with other residues (GLU232, LYS229, TYR228, PHE225, VAL231, and CA404). This inhibits bacterial protein against infectious diseases, inflammation, and stressful conditions. Data also revealed that the papaya plant was a safe and effective therapeutic agent without possessing acute toxicity. Despite significant information available, including considerable in vitro cell line and in vivo studies, there is still a lack of evidence from clinical studies to determine the role of papaya leaves in the treatment of sepsis, as single-drug therapy, or as an adjuvant to immunomodulatory drugs. It is established by our study that *C. papaya* leaves have great antioxidant and cytotoxic potentials as identified by DPPH free radical scavenging activity and MTT assay. Additionally, the presence of flavonoids confers to the papaya leaves anti-inflammatory capabilities. Thus, ethanol extract of *Carica papaya* leaves can be a suitable candidate for future investigation as a potential herbal therapy for treating sepsis.

Author Contributions: Conceptualization, J.U. and R.A.; methodology, J.U. and R.A.; software, J.U., H.K., S.R., A.S. and S.A.; formal analysis, J.U., H.K. and S.A.; investigation, J.U., H.K. and S.A.; resources, S.R.M. and R.A.; writing—original draft preparation, J.U. and A.S.; writing—review and editing, J.U., R.A., H.K., S.A., M.J.H., M.S., S.R., A.S. and M.N.A.; supervision, R.A., S.R.M., M.J.H., M.S. and M.N.A.; funding acquisition, M.N.A. All authors have read and agreed to the published version of the manuscript.

Funding: This research received no external funding.

Institutional Review Board Statement: Not applicable.

Informed Consent Statement: Not applicable.

Data Availability Statement: Not applicable.

Conflicts of Interest: The authors declare no conflict of interest.

References

1. Rahmani, A.H.; Aldebasi, Y.H. Potential Role of *Carica Papaya* and Their Active Constituents in the Prevention and Treatment of Diseases Implication of PTEN, Akt and Bcl2 Expressions and Its Co-Relation with Apoptotic Pathways in Oral Squamous Cell Carcinoma View Project Natural Product. *Int. J. Pharm. Pharm. Sci.* **2016**, *8*, 11–15.
2. Priyadarshi, A.; Ram, B. A Review on Pharmacognosy, Phytochemistry and Pharmacological Activity of *Carica Papaya* (Linn.) Leaf. *Int. J. Pharm. Sci. Res.* **2018**, *9*, 4071–4078.
3. Santana, L.F.; Inada, A.C.; Santo, B.L.S.D.E.; Filiú, W.F.O.; Pott, A.; Alves, F.M.; Guimarães, R.D.C.A.; Freitas, K.D.C.; Hiane, P.A. Nutraceutical Potential of *Carica Papaya* in Metabolic Syndrome. *Nutrients* **2019**, *11*, 1608. [CrossRef] [PubMed]
4. Sagadevan, P.; Selvakumar, S.; Raghunath, M.; Megala, R.; Janarthanan, P.; Vinitha Ebziba, C.; Senthil Kumar, V. Medicinal Properties of *Carica Papaya* Linn: Review. *Madridge J. Nov. Drug Res.* **2019**, *3*, 120–125. [CrossRef]
5. Bhowmik, D. Traditional and Medicinal Uses of *Carica Papaya*. *J. Med. Plants Stud. Year* **2013**, *1*, 7–15.
6. Miean, K.H.; Mohamed, S. Flavonoid (Myricetin, Quercetin, Kaempferol, Luteolin, and Apigenin) Content of Edible Tropical Plants. *J. Agric. Food Chem.* **2001**, *49*, 3106–3112. [CrossRef] [PubMed]
7. Khuzhaev, V.U.; Aripova, S.F.; Shakirov, R.S. Dynamics of the Accumulation of the Alkaloids of Arundo Donax. *Chem. Nat. Compd.* **1995**, *30*, 637–638. [CrossRef]
8. Olafsdottir, E.S.; Bolt Jorgensen, L.; Jaroszewski, J.W. Cyanogenesis in Glucosinolate-Producing Plants: *Carica Papaya* and Carica Quercifolia. *Phytochemistry* **2002**, *60*, 269–273. [CrossRef]
9. Dharmarathna, S.L.C.A.; Wickramasinghe, S.; Waduge, R.N.; Rajapakse, R.P.V.J.; Kularatne, S.A.M. Does *Carica Papaya* Leaf-Extract Increase the Platelet Count? An Experimental Study in a Murine Model. *Asian Pac. J. Trop. Biomed.* **2013**, *3*, 720–724. [CrossRef]
10. Anjum, V.; Arora, P.; Ansari, S.H.; Najmi, A.K.; Ahmad, S. Antithrombocytopenic and Immunomodulatory Potential of Metabolically Characterized Aqueous Extract of *Carica Papaya* Leaves. *Pharm. Biol.* **2017**, *55*, 2043–2056. [CrossRef]
11. Kad, D.R.; Tambe, V.S. Phytochemical Screening and Evaluation of Platelet—Google Scholar. *Adv. Plants Agric. Res.* **2018**, *8*, 531–535.
12. Pandey, S.; Cabot, P.J.; Shaw, P.N.; Hewavitharana, A.K. Anti-Inflammatory and Immunomodulatory Properties of *Carica Papaya*. *J. Immunotoxicol.* **2016**, *13*, 590–602. [CrossRef] [PubMed]
13. Renganathan, S.; Aroulmoji, V.; Shanmugam, G.; Devarajan, G.; Rao, K.V.; Rajendar, V.; Park, S.H. Silver Nanoparticle Synthesis from *Carica Papaya* and Virtual Screening for Anti-Dengue Activity Using Molecular Docking. *MRE* **2019**, *6*, 035028. [CrossRef]

14. Jiang, B.; Liang, P.; Deng, G.; Tu, Z.; Liu, M.; Xiao, X. Increased Stability of Bcl-2 in HSP70-Mediated Protection against Apoptosis Induced by Oxidative Stress. *Cell Stress Chaperones* **2010**, *16*, 143–152. [CrossRef] [PubMed]
15. Wu, H.; Kuzmenko, A.; Wan, S.; Schaffer, L.; Weiss, A.; Fisher, J.H.; Kim, K.S.; McCormack, F.X. Surfactant Proteins A and D Inhibit the Growth of Gram-Negative Bacteria by Increasing Membrane Permeability. *J. Clin. Invest.* **2003**, *111*, 1589–1602. [CrossRef]
16. Dickgießer, U.; Weiss, N.; Fritsche, D. Lactobacillus Gasseri as the Cause of Septic Urinary Infection. *Infection* **1984**, *12*, 14–16. [CrossRef]
17. Canini, A.; Alesiani, D.; D'Arcangelo, G.; Tagliatesta, P. Gas Chromatography-Mass Spectrometry Analysis of Phenolic Compounds from *Carica Papaya*, L. Leaf. *J. Food Compos. Anal.* **2007**, *20*, 584–590. [CrossRef]
18. Comalada, M.; Camuesco, D.; Sierra, S.; Ballester, I.; Xaus, J.; Gálvez, J.; Zarzuelo, A. In Vivo Quercitrin Anti-Inflammatory Effect Involves Release of Quercetin, Which Inhibits Inflammation through down-Regulation of the NF-KB Pathway. *Eur. J. Immunol.* **2005**, *35*, 584–592. [CrossRef]
19. Wang, S.; Yao, J.; Zhou, B.; Yang, J.; Chaudry, M.T.; Wang, M.; Xiao, F.; Li, Y.; Yin, W. Bacteriostatic Effect of Quercetin as an Antibiotic Alternative In Vivo and Its Antibacterial Mechanism In Vitro. *J. Food Prot.* **2018**, *81*, 68–78. [CrossRef] [PubMed]
20. Avinash Shinde, A.; Hase, D. Isolation of Carpaine from *Carica Papaya* Leaves by Using LCMS. *J. Med. Plants Stud.* **2020**, *8*, 1–5.
21. Sherma, J. *Basic TLC Techniques, Materials, and Apparatus*; CRC Press: Boca Raton, FL, USA, 2003; pp. 25–85. [CrossRef]
22. Ram, M.; Abdin, M.Z.; Khan, M.A.; Jha, P. HPTLC Fingerprint Analysis: A Quality Control for Authentication of Herbal Phytochemicals. In *High-Performance Thin-Layer Chromatogr*; Springer: Berlin/Heidelberg, Germany, 2011; pp. 105–116. [CrossRef]
23. Yuliani, R.; Syahdeni, F. Cytotoxicity of Ethanolic Extract of Papaya Leaves (*Carica Papaya*) and Its Fractions on T47D Cells. *Pharma. J. Farm. Indones.* **2020**, *17*, 17–23. [CrossRef]
24. Joseph, B.; Sankarganesh, P.; Ichiyama, K.; Yamamoto, N. In Vitro Study on Cytotoxic Effect and Anti-DENV2 Activity of *Carica Papaya*, L. Leaf. *Front. Life Sci.* **2015**, *2*, 18–22. [CrossRef]
25. Mahajan, S.D.; Law, W.C.; Aalinkeel, R.; Reynolds, J.; Nair, B.B.; Yong, K.T.; Roy, I.; Prasad, P.N.; Schwartz, S.A. Nanoparticle-Mediated Targeted Delivery of Antiretrovirals to the Brain. *Methods Enzymol.* **2012**, *509*, 41–60. [CrossRef] [PubMed]
26. Grela, E.; Kozłowska, J.; Grabowiecka, A. Current Methodology of MTT Assay in Bacteria—A Review. *Acta Histochem.* **2018**, *120*, 303–311. [CrossRef]
27. Cecconi, M.; Evans, L.; Levy, M.; Rhodes, A. Sepsis and Septic Shock. *Lancet* **2018**, *392*, 75–87. [CrossRef] [PubMed]
28. Varon, J.; Arciniegas Rubio, A.; Amador-Munoz, D.; Corcoran, A.; DeCorte, J.A.; Isabelle, C.; Pinilla Vera, M.; Walker, K.; Brown, L.; Cernadas, M.; et al. Surfactant Protein D Influences Mortality During Abdominal Sepsis by Facilitating Escherichia Coli Colonization in the Gut. *Crit. Care Explor.* **2022**, *4*, e0699. [CrossRef]
29. Usmani, J.; Khan, T.; Ahmad, R.; Sharma, M. Potential Role of Herbal Medicines as a Novel Approach in Sepsis Treatment. *Biomed. Pharmacother.* **2021**, *144*, 112337. [CrossRef]
30. Zunjar, V.; Dash, R.P.; Jivrajani, M.; Trivedi, B.; Nivsarkar, M. Antithrombocytopenic Activity of Carpaine and Alkaloidal Extract of *Carica Papaya* Linn. Leaves in Busulfan Induced Thrombocytopenic Wistar Rats. *J. Ethnopharmacol.* **2016**, *181*, 20–25. [CrossRef]
31. Baskaran, C.; Bai, V.R.; Velu, S.; Kumaran, K. The Efficacy of *Carica Papaya* Leaf Extract on Some Bacterial and a Fungal Strain by Well Diffusion Method. *Asian Pacific J. Trop. Dis.* **2012**, *2*, S658–S662. [CrossRef]
32. Agada, R.; Usman, W.A.; Shehu, S.; Thagariki, D. In Vitro and in Vivo Inhibitory Effects of *Carica Papaya* Seed on α-Amylase and α-Glucosidase Enzymes. *Heliyon* **2020**, *6*, e03618. [CrossRef] [PubMed]
33. Abdel-Halim, S.; Ibrahim, M.; Abdel Mohsen, M.; Abou-Setta, L.; Sleem, A.; El-Missiry, M. The Influence of the Extraction Method on Polyphenols, Flavonoids Composition and Anti-Hyperlipidemic Properties of Papaya Leaves (*Carica Papaya* Linn.). *Bull. Natl. Res. Cent.* **2021**, *45*, 85. [CrossRef]
34. Gulcin, İ. Antioxidants and Antioxidant Methods: An Updated Overview. *Arch. Toxicol.* **2020**, *94*, 651–715. [CrossRef] [PubMed]
35. Alorkpa, E.J.; Boadi, N.O.; Badu, M.; Saah, S.A. Phytochemical Screening, Antimicrobial and Antioxidant Properties of Assorted *Carica Papaya* Leaves in Ghana. *J. Med. Plants Stud.* **2016**, *4*, 193–198.
36. Winikoff, S.E.; Zeh, H.J.; DeMarco, R.; Lotze, M.T. Cytolytic Assays. In *Measuring Immunity: Basic Science and Clinical Practice*; Elsevier: Amsterdam, The Netherlands, 2011; pp. 341–343.
37. Zhang, P.; Leu, J.I.J.; Murphy, M.E.; George, D.L.; Marmorstein, R. Crystal Structure of the Stress-Inducible Human Heat Shock Protein 70 Substrate-Binding Domain in Complex with Peptide Substrate. *PLoS ONE* **2014**, *9*, e103518. [CrossRef] [PubMed]
38. Shrive, A.K.; Tharia, H.A.; Strong, P.; Kishore, U.; Burns, I.; Rizkallah, P.J.; Reid, K.B.M.; Greenhough, T.J. High-Resolution Structural Insights into Ligand Binding and Immune Cell Recognition by Human Lung Surfactant Protein D. *J. Mol. Biol.* **2003**, *331*, 509–523. [CrossRef] [PubMed]
39. Raghunathan, K.; Harris, P.T.; Spurbeck, R.R.; Arvidson, C.G.; Arvidson, D.N. Crystal Structure of an Efficacious Gonococcal Adherence Inhibitor: An Enolase from Lactobacillus Gasseri. *FEBS Lett.* **2014**, *588*, 2212–2216. [CrossRef] [PubMed]
40. Singh, V.; Dhankhar, P.; Dalal, V.; Tomar, S.; Kumar, P. In-Silico Functional and Structural Annotation of Hypothetical Protein from Klebsiella Pneumonia: A Potential Drug Target. *J. Mol. Graph. Model.* **2022**, *116*, 108262. [CrossRef]
41. Singh, V.; Dhankhar, P.; Dalal, V.; Tomar, S.; Golemi-Kotra, D.; Kumar, P. Drug-Repurposing Approach To Combat Staphylococcus Aureus: Biomolecular and Binding Interaction Study. *ACS Omega* **2022**, *7*, 38448–38458. [CrossRef]
42. Akbar, S.; Das, S.; Iqubal, A.; Ahmed, B. Synthesis, Biological Evaluation and Molecular Dynamics Studies of Oxadiazine Derivatives as Potential Anti-Hepatotoxic Agents. *J. Biomol. Struct. Dyn.* **2021**, *40*, 9974–9991. [CrossRef]

43. DUD—A Directory of Useful Decoys. Available online: http://dud.docking.org/ (accessed on 1 December 2022).
44. Shuhel, M.A.; Easwari, T.S.; Sen, S. Stability Study and Haematological Profile of Aqueous Leaves Extract of *Carica Papaya*. *Der Pharm. Lett.* **2016**, *8*, 182–187.
45. Evans, W.C. *Trease and Evans' Pharmacognosy*, 16th ed.; Elsevier: Amsterdam, The Netherlands, 2009; Volume 16, pp. 1–603.
46. Boham, B.A.; Kocipai-Abyazan, A.C. Flavonoids and Condensed Tannins from Seed of *Carica Papaya*. *Pac. Sci.* **1994**, *8*, 458–463.
47. Lee, S.E.; Hwang, H.J.; Ha, J.S.; Jeong, H.S.; Kim, J.H. Screening of Medicinal Plant Extracts for Antioxidant Activity. *Life Sci.* **2003**, *73*, 167–179. [CrossRef]
48. Mishra, K.; Ojha, H.; Chaudhury, N.K. Estimation of Antiradical Properties of Antioxidants Using DPPH Å Assay: A Critical Review and Results. *Food Chem.* **2011**, *130*, 1036–1046. [CrossRef]
49. De Britto, J.A.; Roshan Sebastian, S.; Sujin, M.R. Phytochemical Analysis of Eight Medicinal Plants of Lamiaceae. *J. Res. Plant Sci.* **2011**, *1*, 001–006.
50. Hussain, S.Z.; Razvi, N.; Ali, S.I.; Hasan, S.M.F. Development of Quality Standard and Phytochemical Analysis of *Carica Papaya* Linn Leaves. *Pak J. Pharm Sci.* **2018**, *31*, 2169–2177.
51. Hernández-Martínez, M.; Gallardo-Velázquez, T.; Osorio-Revilla, G.; Almaraz-Abarca, N.; Castañeda-Pérez, E. Application of MIR-FTIR Spectroscopy and Chemometrics to the Rapid Prediction of Fish Fillet Quality. *CyTA J. Food* **2014**, *12*, 369–377. [CrossRef]
52. Fanelli, S.; Zimmermann, A.; Totóli, E.G.; Salgado, H.R.N. FTIR Spectrophotometry as a Green Tool for Quantitative Analysis of Drugs: Practical Application to Amoxicillin. *J. Chem.* **2018**, *2018*, 3920810. [CrossRef]
53. Jayasinghe, C.D.; Gunasekera, D.S.; De Silva, N.; Jayawardena, K.K.M.; Udagama, P.V. Mature Leaf Concentrate of Sri Lankan Wild Type *Carica Papaya* Linn. Modulates Nonfunctional and Functional Immune Responses of Rats. *BMC Complement. Altern. Med.* **2017**, *17*, 230. [CrossRef] [PubMed]
54. Verma, A.K.; Singh, S. Phytochemical Analysis and in Vitro Cytostatic Potential of Ethnopharmacological Important Medicinal Plants. *Toxicol. Reports* **2020**, *7*, 443–452. [CrossRef] [PubMed]

molecules

MDPI

Article

Almond [*Prunus dulcis* (Mill.) DA Webb] Processing Residual Hull as a New Source of Bioactive Compounds: Phytochemical Composition, Radical Scavenging and Antimicrobial Activities of Extracts from Italian Cultivars ('Tuono', 'Pizzuta', 'Romana')

Simona Fabroni *, Angela Trovato, Gabriele Ballistreri, Susanna Aurora Tortorelli, Paola Foti, Flora Valeria Romeo and Paolo Rapisarda

Council for Agricultural Research and Economics (CREA)—Research Center for Olive, Fruit and Citrus Crops, Corso Savoia 190, 95024 Acireale, Italy

* Correspondence: simona.fabroni@crea.gov.it

Abstract: In this study we developed a new extract, by the use of conventional solid-solvent extraction and a food-grade hydroalcoholic solvent, rich in phenolic and triterpenoid components from almon hull to be employed as functional ingredient in food, pharma and cosmetic sectors. Two autochthonous Sicilian cultivars ('Pizzuta' and 'Romana') and an Apulian modern cultivar ('Tuono') have been tested for the production of the extract. Results showed that the two Sicilian varieties, and in particular the 'Romana' one, present the best characteristics to obtain extracts rich in triterpenoids and hydroxycinnamic acids, useful for the production of nutraceutical supplements. About triterpenoids, the performance of the hydroalcoholic extraction process allowed to never go below 46% of recovery for 'Pizzuta' samples, with significantly higher percentages of recovery for 'Tuono' and 'Romana' extracts (62.61% and 73.13%, respectively) while hydroxycinnamic acids were recovered at higher recovery rate (84%, 89% and 88% for 'Pizzuta', 'Romana' and 'Tuono' extracts, respectively). Invitro antioxidant and antimicrobial activities exerted by the extracts showed promising results with *P. aeruginosa* being the most affected strain, inhibited up to the 1/8 dilution with 'Romana' extract. All the three tested extracts exerted an antimicrobial action up to 1/4 dilutions but 'Romana' and 'Pizzuta' extracts always showed the greatest efficacy.

Keywords: almond [*Prunus dulcis* (Mill.) DA Webb] hull; extract; triterpenoid acids; hydroxycinnamic acids; antioxidant activity; antimicrobial activity

Citation: Fabroni, S.; Trovato, A.; Ballistreri, G.; Tortorelli, S.A.; Foti, P.; Romeo, F.V.; Rapisarda, P. Almond [*Prunus dulcis* (Mill.) DA Webb] Processing Residual Hull as a New Source of Bioactive Compounds: Phytochemical Composition, Radical Scavenging and Antimicrobial Activities of Extracts from Italian Cultivars ('Tuono', 'Pizzuta', 'Romana'). *Molecules* **2023**, *28*, 605. https://doi.org/10.3390/molecules28020605

Academic Editors: Raffaele Pezzani and Sara Vitalini

Received: 30 November 2022
Revised: 31 December 2022
Accepted: 2 January 2023
Published: 6 January 2023

1. Introduction

Almond [*Prunus dulcis* (Mill.) DA Webb] is one of the most popular nut crop all over the world. Worldwide production of almonds (in shell) is estimated at 4,140,043 tonnes in 2020 with America (including Argentina, Brazil, Chile, Mexico, Peru and USA) being the leading producer accounting to more than 50% of the global production per year. Asia (648,111 tonnes) and Europe (561,453 tonnes) follow with an estimated production which covers, for each continent, almost the 15% of the worldwide production [1] per year. Almond production in Italy is estimated at almost 85,000 tonnes per year on a cultivated area accounting to more than 50,000 ha [2]. The production is concentrated in Sicily (60%) and Apulia (30%) with negligible productions in other regions of southern Italy (Calabria, Basilicata and Sardinia). In Sicily the cultivation of the almond tree is developed in the southern area and in the south-eastern part of the region; among the native varieties, 'Pizzuta' and 'Romana' prevail. In recent years, the market demand for the consumption of almonds as dried fruit and as an ingredient in food products has significantly increased [3]. The almonds processing produces a significant quantity of waste and byproducts, which entail considerable difficulties in disposal, with the main residue being represented by the green hull which constitutes about 50% of the weight of the fresh fruit. Currently the almond by-products are mostly exploited for energetic purpose or in the livestock sector [4]. The green hull has been recently employed mostly in animal husbandry for testing its use in lamb diet [5], in broiler diet [6], on lactating cows [7] and on laying hens [8]. Beyond its use in animal husbandry, the use of almond hull for bioenergy and biofuels purposes has been also proposed with encouraging results [9,10]. Furthermore, very recently, an innovative synergistic hydrothermal co-valorization approach for biofuels production through coprocessing of almond hulls and FFP2 (Filtering Face Piece 2) disposable face masks in seawater has been proposed [11]. All these applications, even if very promising, might represent only a partial solution to the huge environmental impact of the residues of the almond dehulling process, by the way generating only low value further incomes for growers and producers. Based on its high content in bioactive compounds, including triterpenoids, lactones, phenolic compounds and fibres [12–16], a profitable alternative to the disposal of the raw almond hull is represented by its use in the food and pharma industries, which are high added value supply chains [17,18]. Most of the research has been focused on the chemical characterization as well as bioactivity of extracts from almond hulls. Takeoka and Dao [13] have characterized a methanolic extract of almond hull by HPLC-PDA. Furthermore, the extract was tested for its ability to inhibit the oxidation of methyl linoleate at 40 °C, and the result showed that, at the same concentration, it had a greater antioxidant activity than α-tocopherol. Studies carried out on almond hull have led to the identification of triterpenoid acids, such as betulinic, oleanolic and ursolic acids [12]. There is a growing interest in natural triterpenoids, mostly related to their broad spectrum of biological activities; they are, in fact, bactericidal, fungicidal, antiviral, cytotoxic, analgesic, antitumor, spermicidal, cardiovascular, antiallergic, anti-inflammatory and antidiabetic agents [19–34]. Moreover, ursolic acid exerts hepatoprotective activity and induction of cell apoptosis [35,36] as well as other pentacyclic triterpenoids such as betulinic acid, imberbic acid, oleanolic acid and zeylasteral have been reported to possess antimicrobial activity [37–39], but there are only few reports regarding the antibacterial mechanism of triterpenoids. Ursolic acid showed a synergistic effect with ampicillin and tetracycline against several bacterial pathogens [37], this makes almonds a promising source of natural antibacterial compounds, despite the mechanisms of triterpenoids action has not been clarified.

The present study was aimed at developing a new extract, rich in phenolic and triterpenoid components to be employed as functional ingredient in food, pharma and cosmetic sectors. The obtained extracts were characterized with respect to their triterpenoids and hydroxycinnamic acids quali-quantitative profile and further antiradical and antimicrobial activities were tested and validated with promising results.

2. Results and Discussion

2.1. HPLC-PDA Analysis of Triterpenoids

In Table 1 the results of the HPLC-PDA analysis of triterpenoids in 'Pizzuta', 'Romana' and 'Tuono' hull raw samples (H) and extracts (E) are reported.

Total triterpenoids in H samples ranged between 412.67 ± 0.27 mg/100 g DW in 'Tuono' samples to 457.59 ± 2.78 mg/100 g DW in 'Pizzuta' samples with average and not statistically different values for 'Romana' variety (446.95 ± 10.88 mg/100 g DW). The quantitative analysis of the three varieties showed that 'Romana' and 'Pizzuta', the two autochthonous Sicilian varieties, present higher concentrations of total triterpenoids with a predominance of oleanolic acid for 'Romana' variety and a predominance of ursolic acid for 'Pizzuta'. Looking at the qualitative composition of the individual compounds, the triterpenoid profile was different in each of the three variety. Indeed, 'Pizzuta' hull samples showed a higher presence of ursolic and betulinic acids (162.83 ± 1.61 mg/100 g DW and 160.94 ± 1.04 mg/100 g DW, respectively) with regard to oleanolic acid (133.82 ± 0.13 mg/100 g DW). Conversely, 'Romana' and 'Tuono' hull samples showed higher levels of oleanolic acid respect to betulinic and ursolic acids. In general, statistical analysis of the acquired data showed that betulinic and ursolic acids were significantly higher in 'Pizzuta' H samples while lower values were recorded for 'Tuono' samples and average levels in 'Romana' samples. On the other hand, oleanolic acid was most represented in 'Tuono' and 'Romana' hull samples respect to the other variety.

The quali-quantitative triterpenoid profile in E samples of the three varieties showed higher values of oleanolic acid in 'Romana' samples while betulinic acid was mostly represented in 'Pizzuta' samples and ursolic acid was equally represented in 'Pizzuta' and 'Romana' samples respect to 'Tuono' samples where lower values were recorded for this compound (41.69 ± 0.34 mg/100 mL). In general, total triterpenoids acids resulted higher in 'Romana' samples, showing the highest % of recovery of the bioactive triterpenoids respect to the hull raw triterpenoid content (73.13%). Results showed that, on average, the performance of the hydroalcoholic extraction process allowed to never go below 46% of recovery for 'Pizzuta' samples, with significantly higher percentages of recovery for 'Tuono' and 'Romana' extracts (62.61% and 73.13%, respectively). Moreover, quantitatively, 'Romana' extracts resulted the richest in triterpenoid acids with an average value equal to 326.86 ± 6.34 mg/100 mL. It can be concluded that the results of this study show that the two Sicilian varieties, and in particular the 'Romana' one, present the best characteristics to obtain extracts rich in triterpenoids, useful for the production of nutraceutical supplements.

As far as the recovery of terpenoids from almond residual hull is concerned, the first study which led to the identification of triterpenoid acids, such as betulinic, oleanolic and ursolic acids, from almond hull was that of Takeoka et al. [12]. In this study, the authors report an estimated overall amount equal to ~1% of the hulls for the three triterpenoids comprised together, but no reference to the cultivar or variety of origin of the investigated hull samples is reported. With their results, the authors claimed that almond hulls could be considered as a rich source of these promising anti-inflammatory, anti-HIV and anticancer agents. Further research was published on antiproliferative terpenoids from almond hulls [40]. Our results show that hydroalcoholic extraction could be proposed as a valuable and efficient process to quantitatively recover triterpenoids from raw almond hull, especially for the 'Romana' cultivar which showed the most promising % of recovery associated with a valuable quali-quantitative triterpenoid profile.

Table 1. HPLC-PDA analysis of triterpenoids in 'Pizzuta', 'Romana' and 'Tuono' hull raw samples (H) and extracts (E).

	Hull Raw Samples *				Extracts **				
	Oleanolic Acid (mg/100 g DW)	Betulinic Acid (mg/100 g DW)	Ursolic Acid (mg/100 g DW)	Total Triterpenoids (mg/100 g DW)	Oleanolic Acid (mg/100 mL)	Betulinic Acid (mg/100 mL)	Ursolic Acid (mg/100 mL)	Total Triterpenoids (mg/100 mL)	Recovery (%)
'Pizzuta'	133.82 ± 0.13 [B]	160.94 ± 1.04 [A]	162.83 ± 1.61 [A]	457.59 ± 2.78 [a]	60.26 ± 2.09 [C]	75.29 ± 1.99 [A]	75.41 ± 1.11 [A]	210.96 ± 5.02 [C]	46.11
'Romana'	276.03 ± 5.67 [A]	73.97 ± 1.78 [B]	96.95 ± 3.44 [B]	446.95 ± 10.88 [ab]	204.88 ± 3.29 [A]	50.27 ± 0.96 [B]	71.72 ± 2.21 [A]	326.86 ± 6.34 [A]	73.13
'Tuono'	285.71 ± 0.19 [A]	57.98 ± 0.05 [C]	68.97 ± 0.12 [C]	412.67 ± 0.27 [b]	178.28 ± 3.41 [B]	38.39 ± 1.04 [C]	41.69 ± 0.34 [B]	258.37 ± 4.31 [B]	62.61

Data are expressed as means of three analytical replicates ± standard deviation; Means in the same column followed by different letters are significantly different: $p \leq 0.01$—capital letters; $p \leq 0.05$—small letters. *: abbreviated in the text as H samples; **: abbreviated in the text as E samples.

2.2. HPLC-PDA Analysis of Hydroxycinnamic Acids

In Table 2 the results of the HPLC-PDA analysis of hydroxycinnamic acids in 'Pizzuta', 'Romana' and 'Tuono' hull raw samples (H) and extracts (E) are reported.

Total hydroxycinnamic acids in almond H samples showed relevant values ranging from 138.39 ± 9.42 mg/100 g DW to 303.22 ± 21.87 mg/100 g DW, respectively, for 'Tuono' and 'Pizzuta' varieties, showing that 'Pizzuta' variety resulted the richest in total hydroxycinnamic acids. 'Romana' hull samples showed average values (180.96 ± 9.63 mg/100 g DW) which did not produce any statistical difference with regard to the other two varieties. As observed for total triterpenoid acids, the two autochthonous Sicilian cultivars, with particular reference to 'Pizzuta', showed quantitatively higher values of total hydroxycinnamic acids respect to the 'Tuono' variety, confirming the high potential of the almond hull deriving from the dehulling process of these varieties, as a functional ingredient. The predominant compounds were represented by chlorogenic, neochlorogenic and cryptochlorogenic acids whose levels showed a similar trend in all the samples from the three varieties. Indeed, chlorogenic acid resulted to be the predominant hydroxycinnamic compound in all of the three varieties with values which ranged from 12.51 ± 0.18 mg/100 g DW to 122.03 ± 17.69 mg/100 g DW for 'Tuono' and 'Pizzuta' cultivars, respectively. Cryptochlorogenic showed to be the second most present compound in 'Pizzuta' (6.56 ± 0.12 mg/100 g DW) and 'Tuono' (5.85 ± 0.32 mg/100 g DW) samples with the exception of the 'Romana' variety in which neochlorogenic acid quantitatively followed the most abundant chlorogenic acid and preceded cryptochlorogenic acid.

The quali-quantitative profile of hydroxycinnamic acids in E samples reflected the same trend recorded for the H samples of the three investigated varieties. 'Pizzuta' extract was the richest in total hydroxycinnamic acids with 254.70 ± 18.37 mg/100 mL, recording a high recovery rate equal to 84% with respect to the content of the almond H samples. The 'Romana' extract, the other autochthonous Sicilian cultivar, quantitatively followed with an average content equal to 161.05 ± 8.57 mg/100 mL of total hydroxycinnamic acids, showing an 89% recovery rate, considering the original content of the H samples deriving from this cultivar. The 'Tuono' variety showed not significantly lower values (121.78 ± 8.29 mg/100 mL) with respect to the 'Romana' variety, with an average recovery rate equal to 88%. As observed in almond H samples, chlorogenic, neochlorogenic and cryptochlorogenic acids resulted to be the most represented compounds with chlorogenic acid showing higher values in all of the three varieties. Chlorogenic acid showed values ranging from 111.05 ± 16.10 mg/100 mL to 11.64 ± 0.17 mg/100 mL, for 'Pizzuta' and 'Tuono', respectively, with 'Romana' showing an intermediate concentration equal to 63.12 ± 0.59 mg/100 mL. Cryptochlorogenic acid showed higher values in 'Romana' hull extract with a concentration equal to 9.36 ± 0.26 mg/100 mL and lower and comparable values in 'Pizzuta' (5.71 ± 0.10 mg/100 mL) and 'Tuono' (5.15 ± 0.28 mg/100 mL) hull extracts. 'Romana' hull extract showed higher concentration of neochlorogenic acid (11.68 ± 0.13 mg/100 mL), if compared to the other two varieties (1.57 ± 0.02 mg/100 mL for 'Pizzuta' and 1.61 ± 0.05 mg/100 mL for 'Tuono' hull extracts, respectively). As noted in H samples, 'Pizzuta' and 'Tuono' hull extracts showed the following trend: chlorogenic acid > cryptochlorogenic acid > neochlorogenic acid while in 'Romana' hull extract neochlorogenic acid quantitatively followed the most abundant chlorogenic acid and preceded cryptochlorogenic acid.

The previous literature has shown the great potential of almond hull extracts as valuable bioactive ingredients in food, feed or nutra-pharma formulations. Indeed, almond hull is a rich source of terpenoids and phenolics. Takeoka and Dao, in 2003 [13], analysed by HPLC-PDA a methanolic extract of almond hull belonging to 'Nonpareil' variety finding that chlorogenic acid was the main phenolic compound in almond hull while cryptochlogenic was the second most abundant, followed by neochlorogenic. Our results confirm this finding showing that chlorogenic acid is the most represented phenolic compound in all of the three Italian investigated cultivars. The quali-quantitative profile of phenolic compounds in almond hull may be also affected by the cultivar, by way of the genotype that may influence the antioxidants profile, as previously reported by Sfahaln et al. [41]. In 2006, Wijeratne et al. [42] demonstrated that a green almond hull ethanolic extract (variety not reported) performed better than almond skin and seed extracts in inhibiting the formation of both primary and secondary oxidation products. In the same study, the authors analysed almond hull extract performing a HPLC-PDA analysis on free phenolic acids and those liberated from soluble esters and glycosides. Our results are also in line with this finding confirming that the main phenolic compounds in almond hull are cinnamic acids derivatives. Esfahlan et al. [43] successively reported a comprehensive review on antioxidants from almond and its byproducts, showing that hydroxybenzoic acids and aldehydes (with the only exception of protocatechuic acid reported in traces by Wijeratne et al. [44]), anthocyanidins, procyanidins and flavanone glycosides have not ever been reported in almond hull. Conversely, the authors reported that the main phenolic compounds in almond hull are represented by hydroxycinnamic acids, specifically the most represented being caffeoylquinic derivatives. The previous available literature reports traces of kaempferol-3-O-rutinoside and isorhamnetin-3-O-glucoside, as flavonol glycosides and quercetin, quercitrin and isorhamnetin as flavonol aglycons and morin as flavanone aglycone in almond hull [44]. More recently, Kahlaoui et al. [45] characterized by HPLC-PDA the polyphenolic compounds extracted by a hydroalcoholic solvent from different varieties of almond hulls, including the Sicilian 'Pizzuta' and 'Romana' ones, by comparing ultrasound assisted extraction (UAE) with conventional solid-solvent extraction (CSE). Results of this study have confirmed that chlorogenic acid is the most abundant phenolic acid in almond hull with the 'Pizzuta' variety being the richest. This study also highlighted that other minor phenolic compounds (catechin, protocatechuic acid, quercetin-3-glucoside, *p*-coumaric acid and epicatechin) have been identified in trace amounts only in some varieties and only in UAE, being not detectable in CSE. The output of our research, performed on hydroalcoholic extracts obtained by conventional solid-solvent extraction, confirm this finding.

Table 2. HPLC-PDA analysis of hydroxycinnamic acids in 'Pizzuta', 'Romana' and 'Tuono' hull raw samples (H) and extracts (E).

	Hull Raw Samples *				Extracts **				
	Neochlorogenic Acid (mg/100 g DW)	Chlorogenic Acid (mg/100 g DW)	Cryptochlorogenic Acid (mg/100 g DW)	Total Hydroxycinnamic Acids (mg/100 g DW)	Neochlorogenic Acid (mg/100 mL)	Chlorogenic Acid (mg/100 mL)	Cryptochlorogenic Acid (mg/100 mL)	Total Hydroxycinnamic Acids (mg/100 mL)	Recovery (%)
'Pizzuta'	1.85 ± 0.03 B	122.03 ± 17.69 A	6.56 ± 0.12 B	303.22 ± 21.87 A	1.57 ± 0.02 B	111.05 ± 16.10 A	5.71 ± 0.10 B	254.70 ± 18.37 A	84
'Romana'	13.28 ± 0.15 A	70.92 ± 0.66 AB	10.34 ± 0.28 A	180.96 ± 9.63 AB	11.68 ± 0.13 A	63.12 ± 0.59 AB	9.36 ± 0.26 A	161.05 ± 8.57 B	89
'Tuono'	1.75 ± 0.06 B	12.51 ± 0.19 B	5.85 ± 0.32 B	138.39 ± 9.42 B	1.61 ± 0.05 B	11.64 ± 0.17 B	5.15 ± 0.28 B	121.78 ± 8.29 B	88

Data are expressed as means of three analytical replicates ± standard deviation; Means in the same column followed by different letters are significantly different: $p \leq 0.01$—capital letters. *: abbreviated in the text as H samples; **: abbreviated in the text as E samples.

2.3. *Antioxidant Activity Assays*

2.3.1. Folin–Ciocalteu Reagent Assay

The results of the Folin–Ciocalteu Reagent (FCR) colorimetric assay of the hull extract of the 'Pizzuta', 'Romana' and 'Tuono' varities are shown in Table 3. Results showed that the total phenolic content of the Sicilian varieties ('Pizzuta' and 'Romana') is significantly higher compared to the Apulian one ('Tuono'). Indeed, 'Pizzuta' hull extract accounted for 22,593.33 ± 187.53 mg GAE/100 mL while 'Romana' variety recorded a total phenolic content equal to 21,284.76 ± 195.24 mg GAE/100 mL, highlighting the high antioxidant value of the hull as the main almond processing byproduct. 'Tuono' variety showed a significantly lower total phenolic content (18,307.26 ± 94.15 mg GAE/100 mL) if compared with the two Sicilian autochthonous varieties; however, recording a relevant total phenolic content. Correlation analysis has shown a significant direct correlation between total phenolic content and chlorogenic acid (r = 0.98; $p \leq 0.05$), betulinic acid (r = 0.91; $p \leq 0.05$) and ursolic acid (r = 0.98; $p \leq 0.05$) contents. One of the first paper reporting total phenolic content of an ethanolic extract of almond hull (variety not shown) is the work by Siriwardhana et al. [46], where the authors, in addition to determining the total phenolic content of almond seed, skin and hull extracts, examined the free radical-scavenging activity of almond extract by using different in vitro assays. Later on, Wijeratne et al. [39] evaluated the antioxidant activity of almond hull extract while in a contemporary work by the same authors [44] demonstrated that the total phenolic content of almond hull ethanolic extract was nine times higher than that of the whole seed. In 2009, Sfahlan et al. [41] investigated on the total phenolic content and related antiradical activity of almond (*Amygdalus communis* L.) hulls and shells methanolic extracts belonging to 18 different genotypes. The authors found significant differences in phenolic contents and radical scavenging capacities among the different genotypes. More recently, Kahlaoui et al. [45]

showed the total phenolic content values of almond hull ethanolic extracts from different varieties, including the Sicilian 'Pizzuta' and 'Romana' ones, obtained by UAE and CSE. The authors showed that hull extracts from 'Pizzuta' displayed the highest content of polyphenolic compounds with concentrations which are comparable with those reported in our study. Qureshi et al. [47] recently reported a concentration of 1% *w*/*w* of polyphenolic compounds of a dried 70% ethanolic hull extract from almonds collected in China. Taking into consideration data from the previous literature, our results highlight the high added value of hulls from Italian almond varieties, particularly those autochthonous from Sicily, and put in evidence that a conventional solid extraction process, by using hydroalcoholic food grade solvents, can be conveniently used to recover relevant amount of polyphenolic bioactive compounds.

Table 3. Total Phenolic Content (TPC) and Oxygen Radical Absorbance Capacity (ORAC) units in 'Pizzuta', 'Romana' and 'Tuono' hull extracts.

	TPC (mg GAE/100 mL)	**ORAC Units (µmoles Trolox Equiv./100 mL)**
'Pizzuta'	22,593.33 ± 187.53 [A]	44,424 ± 1376.37 [A]
'Romana'	21,284.76 ± 195.24 [A]	41,966 ± 1596.59 [A]
'Tuono'	18,307.26 ± 94.15 [B]	29,250 ± 591.76 [B]

Data are expressed as means of three analytical replicates ± standard deviation; Means in the same column followed by different letters are significantly different: $p \leq 0.01$—capital letters.

2.3.2. ORAC Assay

Differently from FCR assay which is based on an electron-transfer (ET) reaction, the ORAC assay is based on a hydrogen atom transfer (HAT) reaction mechanism and measures the antioxidant scavenging activity against peroxyl radicals induced by AAPH radical. Therefore, it is manly correlated with those compounds of the extract which exert their antioxidant activity thanks to the transfer of a hydrogen. So, it can be said that ORAC units measure antioxidant capacity of a sample. The higher the ORAC units, the higher the antioxidant potential of the sample. Our results (Table 3) showed that the Sicilian varieties present a definitely higher antioxidant activity compared to the Apulian one. The two Sicilian autochthonous varieties showed significantly greater ORAC units (44.424 ± 1376.37 µmoles Trolox equiv./100 mL and 41.966 ± 1596.59 µmoles Trolox equiv./100 mL for 'Pizzuta' and 'Romana' almond hull extracts, respectively) respect to the 'Tuono' almond hull extract which accounted to 29.250 ± 591.76 µmoles Trolox equiv./100 mL. Correlation analysis has shown a significant direct correlation between ORAC units and chlorogenic acid (r = 0.94; $p \leq 0.05$), betulinic acid (r = 0.84; $p \leq 0.05$) and ursolic acid (r = 1.00; $p \leq 0.05$) contents. Previous studies aimed at pointing out the antioxidant activity of almond hull extracts have revealed that it is active in (i) inhibiting the formation of both primary and secondary oxidation products [42], (ii) inhibiting human low-density lipoprotein oxidation, DNA scission and inducing metal ion chelation [44], (iii) scavenging nitrite, hydrogen peroxide and superoxide radicals [41] and (iv) inhibiting tyrosine phosphatase-1B (PTP1B) enzyme [47]. Our results reveal that almond hull extracts provide a great and relevant intake of antioxidants, also assuring an outstanding increase in ORAC units. The high antioxidant value of the hull may be ascribed to the fact that almond hull constitutes a natural protection of the oil-rich internal almond seed from oxidation induced by sunlight in the presence of atmospheric oxygen, anyway the favourable in vitro potential of any compound or extract, the safety level and the beneficial outcomes must be supported by in vivo toxicological studies [48].

2.4. Antimicrobial Activity

In general, triterpenoid compounds such as betulinic, ursolic and oleanolic acids, in addition to their antiviral, antitumor or anti-inflammatory properties, have been shown to exhibit antimicrobial properties against human pathogens [49]. Although the mechanism of their antimicrobial activity has not yet been elucidated, recent studies indicate that the antibacterial action of triterpene acids is due to pleiotropic effects, destroying the integrity of bacterial membranes and inhibiting metabolic pathways and protein synthesis [50,51]. Furthermore, Oloyede et al. [52] demonstrated that antimicrobial action, particularly against *Escherichia coli*, *Pseudomonas areuginosa* and *Staphylococcus aureus*, could be linked to bacterial cell death through the generation of reactive oxygen species. In Table 4, the antimicrobial activity results of the tested extracts are shown. The inhibition halos obtained with the extracts showed a dose-dependent antimicrobial action against some pathogenic strains such as *P. aeruginosa*, *S. aureus* and *E. coli*. Regarding the *S. aures* strain, these results agree with Mandalari et al. [53] and Smeriglio et al. [54] that reported that phenolic fraction of almond skins possess significant antimicrobial activity against Gram-positive bacterial strains [55]. On the contrary, the extracts showed no inhibition against *Candida albicans* as confirmed by Foti et al. [56] with different natural extracts. These results could be due to a different hydrophilic/lipophilic ratio balance that could increase cellular uptake, enhancing antioxidant or antimicrobial activities, as suggested by Diallinas et al. [57]. In detail, *P. aeruginosa* was the most affected strain showing the largest halos and the only one inhibited up to the 1/8 dilution with 'Romana' extract. All the three tested extracts exerted an antimicrobial action up to 1/4 dilutions but 'Romana' and 'Pizzuta' extracts always showed the greatest efficacy.

Table 4. Results of antimicrobial activity assays of the extracts of the Romana (R), Pizzuta (P), Tuono (T) almond hulls.

Microorganism	Cultivar	RE	D1	D2	D3	D4	Control
P. aeruginosa	R	11	10	8	7	0	0
	P	10	10	10	0	0	0
	T	9	9	8	0	0	0
S. aureus	R	9	8	8	0	0	0
	P	10	7	0	0	0	0
	T	8	0	0	0	0	0
E. coli	R	10	8	7	0	0	0
	P	8	8	7	0	0	0
	T	7	7	0	0	0	0
L. innocua	R	9	9	7	0	0	0
	P	8	8	7	0	0	0
	T	6	0	0	0	0	0
C. albicans	R	0	0	0	0	0	0
	P	0	0	0	0	0	0
	T	0	0	0	0	0	0

Data are expressed as diameter of the inhibition halos (mm). Legend: RE = raw extract; Control = water; D1 = extract 1/2 diluted; D2 = extract 1/4 diluted; D3 = extract 1/8 diluted; D4 = extract 1/16 diluted.

Since, the raw extracts showed a mean value of 16% dry matter (DM) content, their MIC value expressed as DM activity was 40 mg/mL for *S. aureus*, *E. coli* and *L. innocua* and 20 mg/mL of extract (for 'Romana' extract) for *P. aeruginosa*. However, those MIC values expressed as total triterpenoids (Table 1) correspond to a range between 0.527 mg/mL and 0.646 mg/mL for 'Pizzuta' and 'Tuono' extracts (D2 dilution), respectively, and 0.409 mg/mL for 'Romana' extract (D3 dilution). In addition, other authors have shown antimicrobial activity of compounds derived from hydroxycinnamic acids. In detail, chlorogenic, neochlorogenic and cryptochlorogenic acids showed an inhibitory action against *S. aureus* with a MIC value of 5 mg/mL, exhibiting bactericidal activity and inhibiting *S. aureus* biofilm formation [58,59]. Whereas, analysing the results of the present study, the MIC expressed as total hydroxycinnamic acids (Table 2) showed values between 0.304 and 0.637 mg/mL for 'Tuono' and 'Pizzuta' extracts (D2 dilution), respectively, and 0.201 mg/mL for 'Romana' extract (D3 dilution). The interactions among the different phytochemicals found in the 'Romana', 'Pizzuta' and 'Tuono' almond extracts could synergistically affect the microorganisms' growth by increasing the inhibitory effect.

Further studies should be performed with purified isolated fractions of the raw extracts with the aim of determining the main compounds responsible of the detected antimicrobial activity. Furthermore, in order to investigate if other extracting solvents/techniques could potentially increase their antimicrobial effect, by increasing total triterpenoids and/or total hydroxycinnamic acids contents, further investigations will be planned in the near future.

3. Materials and Methods

3.1. Chemical Reagents and Standards

Triterpenoid acids (betulinic, oleanolic and ursolic acids) and hydroxycinnamic acids (chlorogenic, cryptochlorogenic and neochlorogenic acids) were purchased from Sigma Chemical Co., (St. Louis, MO, USA). Food-grade ethanol was used for the hydroalcoholic extraction. All other chemicals were of analytical grade, and the solvents used for chromatography were HPLC grade (Merck KGaA, Darmstadt, Germany).

3.2. Plant Material

Almond hull samples from 'Pizzuta' and 'Romana' autochthonous Sicilian cultivars were sampled at the organic farm AgribioLeone located at Noto (SR) while samples from 'Tuono' cultivar were sampled at the organic farm Sorelle Di Pino located at Belpasso (CT). The hull samples were collected immediately after the hulling process which almonds were subjected to after harvesting from the trees. The hull samples were taken to the laboratory and subjected to drying in a ventilated drying oven at 45 ± 1 °C in order to avoid the onset of mould proliferation or other deterioration processes. Once dried (residual humidity equal to 2 ± 0.1%), the hull samples were ground with a domestic grinder to a fine powder to further proceed with the chemical characterization of the triterpenoid and hydroxycinnamic compounds and the further hydroalcoholic extraction of the bioactive molecules.

3.3. Extraction by Hydroalcoholic Solvent

Taking into consideration that the final aim of the study was to obtain extracts from almond residual hull rich in bioactive molecules to be characterized and potentially suggested to be employed in food or cosmetic industries, only food-grade solvents have been used. Following preliminary lab trials performed using hydroalcoholic solutions at different v/v ratios, the best extraction efficiency and yield was obtained with the ethanol:water solution 80:20 (v/v). The extraction of the bioactive molecules was carried out by placing 10 g of hull samples with 100 mL of an ethanol:water solution 80:20 (v/v) in flasks fitted with a cap and kept stirring at 120 rpm for 24 h at a temperature of 25 °C. At the end of 24 h, the hydroalcoholic extract was filtered and set aside; the residual hull was stirred again with another 100 mL aliquot of solvent, under the same experimental conditions; after 24 h this extract was also filtered and added to the first. The combined extracts were then subjected to distillation under vacuum in a rotavapor lab system in order to remove the residual ethanol. The extracts rich in bioactive molecules were then employed for further determinations of their relative content in triterpenoids and hydroxycinnamic acids, antioxidant and antimicrobial activities. The concentrated aqueous extracts had on average a dry matter content equal to 16% (16.48% for 'Romana', 15.98% for 'Pizzuta' and 16.03% for 'Tuono' extracts).

3.4. HPLC-PDA Analysis of Triterpenoids

Triterpenoid acids have been determined in hull raw samples (H) and in aqueous extracts after the hydroalcoholic extraction phase of the hull samples and the subsequent removal of the residual ethanol (E).

The H samples were characterized by HPLC-PDA analysis respect to their native triterpenoids quali-quantitative composition, prior to the hydroalcoholic extraction phase, in order to subsequently determine the extraction yield of these compounds. In brief, 10 g of the powdered hull samples were extracted with 100 mL of methanol at ambient temperature under stirring at 120 rpm for 24 h in flasks fitted with a cap. After filtration, the residual hull powder was stirred again with another 100 mL aliquot of methanol solvent, under the same experimental conditions. Then the two methanolic extracts were combined, filtered with 0.45 μm PTFE filters and injected into the chromatographic system for the determination of triterpenoid acids.

After the hydroalcoholic extraction phase of the hull samples and the subsequent removal of the residual ethanol, 500 μL of the aqueous extracts were diluted to 25 mL with methanol, then filtered with 0.45 μm PTFE filters and injected into the chromatographic system for the determination of triterpenoid acids.

All the samples were analysed by HPLC-PDA using the method of Taralkar and Chattopadhyay [60], with some modifications. The HPLC analyses were performed using the Alliance Waters 2695 instrumentation, equipped with a PDA detector. The separation was achieved on a Luna (Phenomenex, Torrance, CA, USA) C18 analytical column (250 mm × 4.6 mm, 5 μm). The mobile phase was an acetonitrile:methanol solution (80:20, *v:v*) and the flow rate was set at 0.5 mL/min in isocratic mode. All analyses were performed with the column temperature maintained at 35 °C and the injection volume of each sample was 20 μL. The data were analysed at a wavelength of 210 nm. All analyses were repeated three times. Quantification of triterpenoid compounds was performed using an external standard calibration curve and expressed as mg/100 g of dried almond hull or mg/100 mL of concentrated aqueous extract.

3.5. HPLC-PDA Analysis of Hydroxycinnamic Acids

Hydroxycinnamic acids have been determined in hull raw samples (H) and in aqueous extracts after the hydroalcoholic extraction phase of the hull samples and the subsequent removal of the residual ethanol (E).

The H samples were characterized by HPLC-PDA analysis respect to their native hydroxycinnamic quali-quantitative composition, prior to the hydroalcoholic extraction phase, in order to subsequently determine the extraction yield of these compounds. In brief, 10 g of the powdered hull samples were extracted with 100 mL of methanol (containing 0.5% of hydrochloric acid) at ambient temperature under stirring at 120 rpm for 24 h in flasks fitted with a cap. After filtration, the residual hull powder was stirred again with another 100 mL aliquot of the same methanol solvent, under the same experimental conditions. Then the two methanolic extracts were combined, filtered with 0.45 μm PTFE filters and injected into the chromatographic system for the determination of hydroxycinnamic acids.

After the hydroalcoholic extraction phase of the powdered hull samples and the subsequent removal of the residual ethanol, 500 μL of the aqueous extracts were diluted to 25 mL with methanol (containing 0.5% of hydrochloric acid), then filtered with 0.45 μm PTFE filters and injected into the chromatographic system for the determination of hydroxycinnamic acids.

All the samples were analysed by HPLC-PDA using the same instrument described above. The separation was achieved on a Luna (Phenomenex, Torrance, CA, USA) C18 analytical column (250 mm × 4.6 mm, 5 μm). The eluents used were solvent A: water:formic acid (99.7:0.3 *v*/*v*) and solvent B: acetonitrile:formic acid (99.7:0.3 *v*/*v*) with a gradient transition as follows: (0 min) 95% solvent A; (50 min) 72% solvent A; (57 min) 57% solvent A; (75 min) 57% solvent A; (80 min) 95% solvent A; (90 min) 95% solvent A. Flow rate was set at 1 mL/min and all analyses were performed with the column temperature maintained at 35 °C and the injection volume of each sample was 20 μL. The data were analysed at a wavelength of 320 nm. All analyses were repeated three times. Quantification of the hydroxycinnamic compounds was performed using an external standard calibration curve and expressed as mg/100 g of dried almond hull or mg/100 mL of concentrated aqueous extract.

3.6. Antioxidant Activity Assays

There are several in vitro assays to determine the antioxidant activity including ORAC, TEAC (or ABTS), DPPH, FRAP and FCR. In order to evaluate the antioxidant activity of the aqueous extracts to be potentially employed in food or cosmetic industries two methods were used: the ORAC and FCR assays. The ORAC assay represents a hydrogen atom transfer (HAT) reaction mechanism and measures the antioxidant scavenging activity against peroxyl radicals induced by AAPH, while FCR is based on an electron-transfer (ET) reaction with the oxidant as an indicator of the reaction endpoint [61].

3.6.1. Folin–Ciocalteu Reagent Assay

The FCR colorimetric method was applied as described by Amenta et al. [62] with some modifications. The colorimetric method [63] usually used to determine total phenolics, was applied in this work to evaluate antioxidant activity of hull extracts. After the hydroalcoholic extraction phase of the powdered hull samples and the subsequent removal of the residual ethanol, appropriately diluted concentrated aqueous extracts (1:200, *v*/*v*) samples (1 mL) were mixed with 5 mL of FCR commercial reagent (previously diluted with water 1:10 *v*/*v*) and 4 mL of a 7.5% sodium carbonate solution. The mixture was stirred for 2 h at room temperature while avoiding strong light exposure. The absorbance of the resulting blue solution was measured spectrophotometrically at 740 nm, and the concentration of total phenolics was expressed as (±) gallic acid equivalents (mg/100 mL).

3.6.2. ORAC Assay

The ORAC assay was used as described by Cao et al. [64], and improved by Ou, Hampsch-Woodill and Prior [65], adapted and modified. Briefly, the measurements were performed on a Wallac 1420 Victor III 96-well plate reader (EG & Wallac, Turku, Finland) equipped with fluorescence filters (excitation 485 nm, emission 535 nm). Fluorescein (116 nM) was the target molecule for free radical attack from AAPH (153 mM) as the peroxyl radical generator. The reaction was conducted at 37 °C, at pH 7.0 with Trolox (1 μM) as the control standard and 75 mM phosphate buffer (pH 7.0) as the blank. All solutions were freshly prepared prior to analysis. After the hydroalcoholic extraction phase of the powdered hull samples and the subsequent removal of the residual ethanol, concentrated aqueous extracts were diluted with phosphate buffer (1:200, *v*/*v*) prior to analysis, and results were reported as micromoles of Trolox equivalents per 100 mL of extract.

3.7. Antimicrobial Activity

The inhibitory activity of the extracts was tested on five relevant potentially pathogenic strains such as: *Pseudomonas aeuroginosa* DSM 1117, *Staphylococcus aureus* DSM 30862, *Escherichia coli* DSM 1103, *Listeria innocua* DSM 20649 and *Candida albicans* DSM 1386 (Leibniz-Institute DSMZ, German collection). The media used for the test were PDA (Potato Dextrose Agar, Oxoid, UK) for Candida albicans, while MHA (Muller Hinton Agar Base, Merck, Germany) medium was used for all other strains. Before plating, microorganism inocula were standardized to a turbidity of 0.5 McFarland standard (10^6 yeasts or 10^8 bacteria cells/mL). In each plate containing the selective medium, 1 mL of inoculum was spatulated, allowed to dry, and then 6 mm cellulose discs soaked with four different dilutions of extracts were placed to determine their minimal inhibitory concentration (MIC). For each extract, serial twofold dilutions indexed to the base 2 starting from each raw extract [66,67] were prepared for the test. Distilled water was used as a negative control. Plates were incubated at specific temperatures for 48 h and results expressed as diameter of the inhibition halo (mm). Each plate was set up in triplicate.

3.8. Statistical Analysis

Data were statistically analysed by using STATISTICA 6.0 (StatSoft Italia srl, Vigonza, Padova, Italy). The statistical differences were assessed by variance analysis (ANOVA) and means partitioning was carried out by the Tukey's HSD test. To determine the relationships between the evaluated parameters, Pearson correlation coefficients (r; $p \leq 0.05$) were used.

4. Conclusions

In this study, we have proposed almond processing residual hull as a new source of bioactive compounds by the application of a conventional solid-solvent extraction associated with the use of a food-grade hydroalcoholic solvent solution in order to obtain a phytoextract rich in bioactive components. Indeed, our results have showed that a concentrated phytoextract rich in phenolic and triterpenoid components may be obtained starting from two autochthonous Sicilian cultivars ('Pizzuta' and 'Romana') and an Apulian modern cultivar ('Tuono'). More specifically, our results have showed that hydroalcoholic extraction could be proposed as an easy and efficient process to quantitatively recover triterpenoids from raw almond hull, especially for the 'Romana' cultivar which showed the most promising % of recovery associated with a valuable quali-quantitative triterpenoid profile. Furthermore, the quali-quantitative profile of hydroxycinnamic acids in concentrated extracts have showed that the 'Pizzuta' extract was the richest in total hydroxycinnamic acids while 'Romana' have showed the highest recovery rate. Taking into consideration previous works aimed at recovering polyphenolic compounds or, alternatively, triterpenoid compounds from almond hull by the use of conventional or assisted extraction methods, our results allow to demonstrate that high recovery rates for both triterpenoids and polyphenolic compounds could be achieved in a single extract by the use of a simple and not expensive hydroalcoholic solvent extraction process. Due to the presence of the hydrophobic triterpenoid backbone, triterpenoids compounds, even if generally water soluble, could exert multiple hydrophobic interactions and this is the reason why previous works have been targeted on their selective recovery without taking into consideration the possibility to simultaneously recover polyphenolic compounds. To the best of our knowledge, for the first time our work have showed that the proposed extraction method allow to obtain a combined extract rich in both triterpenoids and hydroxycinnamic acids which are recovered at high recovery rates. Further in vitro antioxidant assays have confirmed that almond hull extracts derived from the three investigated cultivars may be used as food supplement providing a great and relevant intake of antioxidants, also assuring an outstanding increase in ORAC units. Finally, the in vitro antimicrobial activity of the extracts was evaluated, and results highlighted the great potential of the new phytoextract to be employed as antimicrobic agent for different applications. *P. aeruginosa* was the most affected strain and 'Romana' and 'Pizzuta' extracts showed the greatest antimicrobial efficacy.

In conclusion, our results showed the great potential of almond green hull to be used for the recovery of valuable bioactive compounds with a wide range of applications in the food, pharma and cosmetic sectors. Further studies should be addressed aiming at validating the use of the new phytoextract as ingredient in innovative food or nutraceutical formulations.

Author Contributions: Conceptualization, S.F.; methodology, S.F., A.T., G.B., S.A.T., P.F., F.V.R. and P.R.; software, S.F.; validation, S.F., G.B. and F.V.R.; formal analysis, S.F.; resources, S.F. and A.T.; data curation, S.F., P.F. and F.V.R.; writing—original draft preparation, S.F., P.F. and F.V.R.; writing—review and editing, S.F., P.F., G.B. and F.V.R.; supervision, S.F., G.B. and F.V.R.; project administration, S.F. and P.R.; funding acquisition, P.R. All authors have read and agreed to the published version of the manuscript.

Funding: This research was funded by Regione Sicilia in the framework of the INNOVIA project "Prodotti e processi innovativi nei settori del vivaismo ortofloricolo e agroalimentare", Programma Operativo del Fondo Sociale Europeo Regione Siciliana 2014–2020. DDG n. 6067, 3 August 2017.

Data Availability Statement: Data are available from the authors.

Acknowledgments: Authors want to acknowledge the technical support given by Patrizia Rita Pepe.

Conflicts of Interest: The authors declare no conflict of interest.

Sample Availability: Samples of the compounds are available from the authors.

References

1. FAOSTAT Database. Available online: https://www.fao.org/faostat/en/#home (accessed on 10 October 2022).
2. ISTAT Database. Available online: http://dati.istat.it/Index.aspx?DataSetCode=DCSP_COLTIVAZIONI (accessed on 10 October 2022).
3. Özcan, M.M.; Ünver, A.; Erkan, E.; Arslan, D. Characteristics of some almond kernel and oils. *Sci. Hortic.* **2011**, *127*, 330–333. [CrossRef]
4. Chen, P.; Cheng, Y.; Deng, S.; Lin, X.; Huang, G.; Ruan, R. Utilization of almond residues. *Int. J. Agric. Biol. Eng.* **2010**, *3*, 1.
5. Scerra, M.; Bognanno, M.; Foti, F.; Caparra, P.; Cilione, C.; Mangano, F.; Natalello, A.; Chies, L. Influence of almond hulls in lamb diets on animal performance and meat quality. *Meat Sci.* **2022**, *192*, 108903. [CrossRef] [PubMed]
6. Wang, J.; Wu, C.; Kong, F.; Kim, W.K. Effect of almond hulls on the growth performance, body composition, digestive tract weight, and liver antioxidant capacity of broilers. *J. Appl. Poult. Res.* **2021**, *30*, 100149. [CrossRef]
7. Swanson, K.L.; Bill, H.M.; Asmus, J.; Heguy, J.M.; DePeters, E.J. Feeding high amounts of almond hulls to lactating cows. *J. Dairy Sci.* **2021**, *104*, 8846–8856. [CrossRef]
8. Wang, J.; Kong, F.; Kim, W.K. Effect of almond hulls on the performance, egg quality, nutrient digestibility, and body composition of laying hens. *Poult. Sci.* **2021**, *100*, 101286. [CrossRef]
9. Salgado-Ramos, M.; Martí-Quijal, F.J.; Huertas-Alonso, A.J.; Sánchez-Verdú, M.P.; Barba, F.J.; Moreno, A. Almond hull biomass: Preliminary characterization and development of two alternative valorization routes by applying innovative and sustainable technologies. *Ind. Crops Prod.* **2022**, *179*, 114697. [CrossRef]
10. Remón, J.; Sevilla-Gasca, R.; Frecha, E.; Pinilla, J.L.; Suelves, I. Direct conversion of almond waste into value-added liquids using carbon-neutral catalysts: Hydrothermal hydrogenation of almond hulls over a Ru/CNF catalyst. *Sci. Total Environ.* **2022**, *825*, 154044. [CrossRef]
11. Remón, J.; Zapata, G.; Oriol, L.; Pinilla, J.L.; Suelves, I. A novel 'sea-thermal', synergistic co-valorisation approach for biofuels production from unavoidable food waste (almond hulls) and plastic residues (disposable face masks). *J. Chem. Eng.* **2022**, *449*, 137810. [CrossRef]
12. Takeoka, G.; Dao, L.; Teranishi, R.; Wong, R.; Flessa, S.; Harden, L. Identification of three triterpenoids in almond hulls. *J. Agricul Food Chem.* **2000**, *48*, 3437–3439. [CrossRef]
13. Takeoka, G.R.; Dao, L.T. Antioxidant constituents of almond [*Prunus dulcis* (Mill.) D.A. Webb] hulls. *J. Agricul Food Chem.* **2003**, *51*, 496–501. [CrossRef]
14. Sang, S.; Cheng, X.; Fu, H.Y.; Shieh, D.E.; Bai, N.; Lapsley, K. New type sesquiterpene lactone from almond hulls (*Prunus amygdalus* Batsch). *Tetrahedron Lett.* **2002**, *43*, 2547–2549. [CrossRef]
15. Sang, S.; Lapsley, K.; Jeong, W.S.; Lachence, P.A.; Ho, C.T.; Rosen, R.T. Antioxidative phenolic compounds isolated from almond skins (*Prunus amygdalus* Batsch). *J. Agricul Food Chem.* **2002**, *50*, 2459–2463. [CrossRef] [PubMed]
16. Sang, S.; Lapsley, K.; Rosen, R.T.; Ho, C.T. New prenylated benzoic acid and other constituents from almond hulls (*Prunus amygdalus* Batsch). *J. Agricul Food Chem.* **2002**, *50*, 607–609. [CrossRef] [PubMed]
17. Najari, Z.; Khodaiyan, F.; Yarmand, M.S.; Hosseini, S.S. Almond hulls waste valorization towards sustainable agricultural development: Production of pectin, phenolics, pullulan, and single cell protein. *J. Waste Manag.* **2022**, *141*, 208–219. [CrossRef] [PubMed]
18. Kahlaoui, M.; Bertolino, M.; Barbosa-Pereira, L.; Ben Haj Kbaier, H.; Bouzouita, N.; Zeppa, G. Almond hull as a functional ingredient of bread: Effects on physico-chemical, nutritional, and consumer acceptability properties. *Foods* **2022**, *11*, 777. [CrossRef] [PubMed]
19. Patočka, J. Biologically active pentacyclic triterpenes and their current medicine signification. *J. Appl. Biomed.* **2003**, *1*, 7–12. [CrossRef]
20. Ghante, M.H.; Jamkhande, P.G. Role of pentacyclic triterpenoids in chemoprevention and anticancer treatment: An overview on targets and underling mechanisms. *J. Pharmacopunct.* **2019**, *2*, 55–67. [CrossRef]
21. Furtado, N.A.J.C.; Pirson, L.; Edelberg, H.; Miranda, L.M.; Loira-Pastoriza, C.; Preat, V.; Larondelle, Y.; André, C.M. Pentacyclic triterpene bioavailability: An overview of in vitro and in vivo studies. *Molecules* **2017**, *22*, 400. [CrossRef]
22. Kessler, J.H.; Mallauer, F.B.; de Roo, G.M.; Medema, J.P. Broad in vitro efficacy of plant-derived betulinic acid against cell lines derived from the most prevalent human cancer types. *Cancer Lett.* **2007**, *251*, 132–145. [CrossRef]
23. Rzeski, W.; Stepulak, A.; Szymański, M.; Sifringer, M.; Kaczor, J.; Wejksza, K.; Zdzisińska, B.; Kandefer-Szerszen, M. Betulinic acid decreases expression of bcl-2 and cyclin D1, inhibits proliferation, migration and induces apoptosis in cancer cells. *Naunyn-Schmeiedeberg's Acrh. Pharmacol.* **2006**, *374*, 11–20. [CrossRef] [PubMed]
24. Basu, S.; Ma, R.; Boyle, P.J.; Mikulla, B.; Bradley, M.; Smith, B.; Basu, M.; Banerjee, S. Apoptosis of human carcinoma cells in the presence of potential anti-cancer drugs: III. Treatment of Colo-205 and SKB3 cells with: Cis-platin, tamoxifen, melphalan, betulic acid, L-PDMP, L-PPMP, and GD3 ganglioside. *Glycoconj. J.* **2004**, *20*, 563–577. [CrossRef] [PubMed]

25. Janicsák, G.; Veres, K.; Kakasy, A.Z.; Máthé, I. Study of the oleanolic and ursolic acid contents of some species of the *Lamiaceae*. *Biochem. Syst. Ecol.* **2006**, *34*, 392–396. [CrossRef]
26. Fai, Y.M.; Tao, C.C. A review of presence of oleanolic acid in natural products. *Nat. Prod. Med.* **2009**, *2*, 77–290.
27. Heinzen, H.; de Vries, J.X.; Moyna, P.; Remberg, G.; Martinez, R.; Tietze, L.F. Mass spectrometry of labelled triterpenoids: Thermospray and electron impact ionization analysis. *Phytochem. Anal.* **1996**, *7*, 237–244. [CrossRef]
28. Wang, X.; Ye, X.L.; Liu, R.; Chen, H.L.; Bai, H.; Liang, X.; Zhang, X.D.; Wang, Z.; Li, W.L.; Hai, C.X. Antioxidant activities of oleanolic acid in vitro: Possible role of Nrf2 and MAP kinases. *Chem. Biol. Interact.* **2010**, *184*, 328–337. [CrossRef]
29. Yoo, S.R.; Jeong, S.J.; Lee, N.R.; Shin, H.K.; Seo, C.S. Quantification analysis and In vitro anti-inflammatory effects of 20-hydroxyecdysone, momordin ic, and oleanolic acid from the fructus of *Kochia scoparia*. *Pharmacogn. Mag.* **2017**, *13*, 339–344.
30. Liu, J. Pharmacology of oleanolic acid and ursolic acid. *J. Ethnopharmacol.* **1995**, *49*, 57–68. [CrossRef]
31. Novotny, L.; Vachalkova, A.; Biggs, D. Ursolic acid: An antitumorigenic and chemopreventive activity. Minireview. *Neoplasma* **2001**, *48*, 241–246.
32. Manez, S.; Recio, M.C.; Giner, R.M.; Rios, J.L. Effect of selected triterpenoids on chronic dermal inflammation. *Eur. J. Pharmacol.* **1997**, *334*, 103–105. [CrossRef]
33. Verma, A.K.; Slaga, T.J.; Wertz, P.W.; Mueller, G.C.; Boutwell, R.K. Inhibition of skin tumor promotion by retinoic acid and its metabolite 5,6-epoxyretinoic acid. *Cancer Res.* **1980**, *40*, 2367–2371. [PubMed]
34. Suh, N.; Honda, T.; Finlay, H.J.; Barchowsky, A. Novel triterpenoids suppress inducible nitric oxide synthase (iNOS) and inducible cyclooxygenase (COX-2) in mouse macrophages. *Cancer Res.* **1998**, *58*, 717–723. [PubMed]
35. Waladkhani, A.R.; Clemens, M.R. Effect of dietary phytochemicals on cancer development (review). *Int. J. Mol. Med.* **1998**, *1*, 747–753. [CrossRef]
36. Kelloff, G.J. Perspectives on cancer chemoprevention research and drug development. *Adv. Cancer Res.* **2000**, *78*, 199–334. [PubMed]
37. Wang, C.M.; Chen, H.T.; Wu, Z.Y.; Jhan, Y.L.; Shyu, C.L.; Chou, C.H. Antibacterial and synergistic activity of pentacyclic triterpenoids isolated from Alstonia scholaris. *Molecules* **2016**, *21*, 139. [CrossRef] [PubMed]
38. de Leon, L.; Beltran, B.; Moujir, L. Antimicrobial activity of 6-oxophenolic triterpenoids. Mode of action against *Bacillus subtilis*. *Planta Med.* **2005**, *71*, 313–319. [CrossRef] [PubMed]
39. Katerere, D.R.; Gray, A.I.; Nash, R.J.; Waigh, R.D. Antimicrobial activity of pentacyclic triterpenes isolated from African Combretaceae. *Phytochemistry* **2003**, *63*, 81–88. [CrossRef]
40. Amico, V.; Barresi, V.; Condorelli, D.; Spatafora, C.; Tringali, C. Antiproliferative terpenoids from almond hulls (*Prunus dulcis*): Identification and structure−activity relationships. *J. Agric. Food Chem.* **2006**, *54*, 810–814. [CrossRef]
41. Sfahlan, A.J.; Mahmoodzadeh, A.; Hasanzadeh, A.; Heidari, R.; Jamei, R. Antioxidants and antiradicals in almond hull and shell (*Amygdalus communis* L.) as a function of genotype. *Food Chem.* **2009**, *115*, 529–533. [CrossRef]
42. Wijeratne, S.S.K.; Amarowicz, R.; Shahidi, F. Antioxidant activity of almonds and their by-products in food model systems. *J. Amer Oil Chem. Soc.* **2006**, *83*, 223–230. [CrossRef]
43. Esfahlan, A.J.; Jamei, R.; Esfahlan, R.J. The importance of almond (*Prunus amygdalus* L.) and its by-products. *Food Chem.* **2010**, *120*, 349–360. [CrossRef]
44. Wijeratne, S.S.K.; Abou-zaid, M.M.; Shahidi, F. Antioxidant polyphenols in almond and its coproducts. *J. Agricul Food Chem.* **2006**, *54*, 312–318. [CrossRef] [PubMed]
45. Kahlaoui, M.; Borotto Dalla Vecchia, S.; Giovine, F.; Ben Haj Kbaier, H.; Bouzouita, N.; Barbosa Pereira, L.; Zeppa, G. Characterization of polyphenolic compounds extracted from different varieties of almond hulls (*Prunus dulcis* L.). *Antioxidants* **2019**, *8*, 647. [CrossRef] [PubMed]
46. Siriwardhana, S.S.K.W.; Shahidi, F. Antiradical activity of extracts of almond and its by-products. *J. Amer Oil Chem. Soc.* **2002**, *79*, 903–908. [CrossRef]
47. Qureshi, M.N.; Numonov, S.; Aisa, H.A. Chemical and pharmacological evaluation of hulls of *Prunus dulcis* nuts. *Int. J. Anal. Chem.* **2019**, *2019*, 5861692. [CrossRef]
48. Vedeanu, N.; Voica, C.; Magdas, D.A.; Kiss, B.; Stefan, M.G.; Simedrea, R.; Georgiu, C.; Berce, C.; Vostinaru, O.; Boros, R.; et al. Subacute co-exposure to low doses of ruthenium (III) changes the distribution, excretion and biological effects of silver ions in rats. *Environ. Chem.* **2020**, *17*, 163–172. [CrossRef]
49. Spivak, A.Y.; Khalitova, R.R.; Nedopekina, D.A.; Gubaidullin, R.R. Antimicrobial properties of amine- and guanidine-functionalized derivatives of betulinic, ursolic and oleanolic acids: Synthesis and structure/activity evaluation. *Steroids* **2020**, *154*, 108530. [CrossRef]
50. Wang, C.M.; Jhan, Y.L.; Tsai, S.J.; Chou, C.H. The pleiotropic antibacterial mechanisms of ursolic acid against methicillin-resistant *Staphylococcus aureus* (MRSA). *Molecules* **2016**, *21*, 884. [CrossRef]
51. Chung, P.Y. Novel targets of pentacyclic triterpenoids in *Staphylococcus aureus*: A systematic review. *Phytomedicine* **2020**, *73*, 152933. [CrossRef]
52. Oloyede, H.O.B.; Ajiboye, H.O.; Salawu, M.O.; Ajiboye, T.O. Influence of oxidative stress on the antibacterial activity of betulin, betulinic acid and ursolic acid. *Microb. Pathog.* **2017**, *111*, 338–344. [CrossRef]

53. Mandalari, G.; Tomaino, A.; Arcoraci, T.; Martorana, M.; LoTurco, V.; Cacciola, F.; Rich, G.T.; Bisignano, C.; Saija, A.; Dugo, P. Characterization of polyphenols, lipids and dietary fibre from almond skins (*Amygdalus communis* L.). *J. Food Comp. Anal.* **2010**, *23*, 166–174. [CrossRef]
54. Smeriglio, A.; Bisignano, C.; Filocamo, A.; Barreca, D.; Bellocco, E.; Trombetta, D. Polyphenolic content and biological properties of Avola almond (*Prunus dulcis* Mill. D.A. Webb) skin and its industrial byproducts. *Ind. Crops Prod.* **2016**, *83*, 283–293. [CrossRef]
55. Prgomet, I.; Gonçalves, B.; Dominguez-Perles, R.; Santos, R.; Saavedra, M.J.; Aires, A.; Pascual-Seva, N.; Barros, A. Irrigation deficit turns almond by-products into a valuable source of antimicrobial (poly)phenols. *Ind. Crops Prod.* **2019**, *132*, 186–196. [CrossRef]
56. Foti, P.; Occhipinti, P.S.; Romeo, F.V.; Timpanaro, N.; Musumeci, T.; Randazzo, C.L.; Randazzo, C.; Caggia, C. Phenols recovered from olive mill wastewater as natural booster to fortify blood orange juice. *Food Chem.* **2022**, *393*, 133428. [CrossRef] [PubMed]
57. Diallinas, G.; Rafailidou, N.; Kalpaktsi, I.; Komianou, A.C.; Tsouvali, V.; Zantza, I.; Mikros, E.; Skaltsounis, A.L.; Kostakis, I.K. Hydroxytyrosol (HT) analogs act as potent antifungals by direct disruption of the fungal cell membrane. *Front. Microbiol.* **2018**, *9*, 2624. [CrossRef] [PubMed]
58. Chang, S.-Y.; Xiao, K.; Zhang, J.-Q. Antibacterial and Antibiofilm Effects of Zanthoxylum bungeanum Leaves against Staphylococcus aureus. *Nat. Prod. Commun.* **2018**, *13*, 1001–1006. [CrossRef]
59. Khan, F.; Bamunuarachchi, N.I.; Tabassum, N.; Kim, Y.M. Caffeic acid and its derivatives: Antimicrobial drugs toward microbial pathogens. *J. Agric. Food Chem.* **2021**, *69*, 2979–3004. [CrossRef] [PubMed]
60. Taralkar, S.V.; Chattopadhyay, S. A HPLC method for determination of ursolic acid and betulinic acids from their methanolic extracts of *Vitex negundo* Linn. *J. Anal. Bioanal. Tech.* **2012**, 3, 134. [CrossRef]
61. Huang, D.; Boxin, O.; Prior, R.L. The chemistry behind antioxidant capacity assay. *J. Agric. Food Chem.* **2005**, *53*, 1841–1856. [CrossRef]
62. Amenta, M.; Ballistreri, G.; Fabroni, S.; Romeo, F.V.; Spina, A.; Rapisarda, P. Qualitative and nutraceutical aspects of lemon fruits grown on the mountainsides of the Mount Etna: A first step for a protected designation of origin or protected geographical indication application of the brand name 'Limone dell'Etna'. *Food Res. Int.* **2015**, *74*, 250–259. [CrossRef]
63. Singleton, V.L.; Orthofer, R.; Lamuela-Raventos, R.M. Analysis of total phenols and other oxidation substrates and antioxidant by means of Folin–Ciocalteu reagent. *Methods Enzymol.* **1999**, *299*, 152–178.
64. Cao, G.; Alessio, H.M.; Cutler, R.G. Oxygen-radical absorbance capacity assay for antioxidants. *Free Radic. Biol. Med.* **1993**, *14*, 303–311. [CrossRef] [PubMed]
65. Ou, B.; Hampsch-Woodill, M.; Prior, R. Developing and validation of an improved oxygen radical absorbance capacity assay using fluorescein as the fluorescent probe. *J. Agricul Food Chem.* **2001**, *49*, 4619–4626. [CrossRef] [PubMed]
66. CLSI. Methods for dilution antimicrobial susceptibility tests for bacteria that grow aerobically: Approved standard. In *Clinical and Laboratory Standards Institute Document M7-A7*, 7th ed.; CLSI: Wayne, PA, USA, 2006; ISBN 1-56238-587-9.
67. CLSI. Reference method for broth dilution antifungal susceptibility testing of yeasts: Approved standard. In *Clinical and Laboratory Standards Institute Document M27-A3*, 3rd ed.; CLSI: Wayne, PA, USA, 2008; ISBN 1-56238-666-2.

molecules

MDPI

Article

Synchronous Extraction, Antioxidant Activity Evaluation, and Composition Analysis of Carbohydrates and Polyphenols Present in Artichoke Bud

Xiao Lin [1,†], Xian-Kun Lu [2,†], Kai-Hao Zhu [1], Xin-Yang Jiang [1], Jiong-Chao Chen [1], Pei-Zheng Yan [1] and Dong-Sheng Zhao [1,*]

1 College of Pharmacy, Shandong University of Traditional Chinese Medicine, Jinan 250355, China
2 School of Pharmaceutical Engineering, Shenyang Pharmaceutical University, Shenyang 110016, China
* Correspondence: 60030102@sdutcm.edu.cn; Tel./Fax: +86-531-8962-8200
† These authors contributed equally to this work.

Abstract: This study investigated the optimization of ultrasonic-assisted aqueous two-phase synchronous extraction of carbohydrates and polyphenols present in artichoke bud, evaluated their antioxidant activities in vitro, and analyzed the composition of carbohydrates and polyphenols by high-performance liquid chromatography (HPLC). The powder mass, ultrasonic time, ammonium sulfate concentration, and alcohol–water ratio were considered the influencing factors based on the single-factor experiment results, and a dual-response surface model was designed to optimize the synchronous extraction process to extract carbohydrates and polyphenols. The antioxidant activity was evaluated by measuring the scavenging capacity of $ABTS^{+}\cdot$ and $DPPH\cdot$ and the reducing capacity of Fe^{3+}. The optimal process conditions in this study were as follows: the powder mass of 1.4 g, ammonium sulfate concentration of 0.34 g/mL, alcohol–water ratio of 0.4, and ultrasonic time of 43 min. The polyphenol content in artichoke bud was 5.32 ± 0.13 mg/g, and the polysaccharide content was 74.78 ± 0.11 mg/g. An experiment on in vitro antioxidant activity showed that both carbohydrates and polyphenols had strong antioxidant activities, and the antioxidant activity of polyphenols was stronger than that of carbohydrates. The HPLC analysis revealed that the carbohydrates in artichoke bud were mannose, rhamnose, glucuronic acid, galacturonic acid, glucose, galactose, and arabinose, and the molar ratio was 10.77:25.22:2.37:15.74:125.39:48.62:34.70. The polyphenols comprised chlorogenic acid, 4-dicaffeoylquinic acid, caffeic acid, 1,3-dicaffeoylqunic acid, isochlorogenic acid B, isochlorogenic acid A, cynarin, and isochlorogenic acid C, and the contents were 0.503, 0.029, 0.022, 0.017, 0.008, 0.162, 1.621, 0.030 mg/g, respectively. This study also showed that the carbohydrates and polyphenols in artichoke bud could be important natural antioxidants, and the composition analysis of HPLC provided directions for their future research. Carbohydrates and polyphenols in artichoke buds can be separated and enriched using the optimized process technology, and it is an effective means of extracting ingredients from plants.

Keywords: artichoke bud; synchronous extraction; aqueous two-phase system; dual-response surface model; antioxidant activity; carbohydrates; polyphenols

Citation: Lin, X.; Lu, X.-K.; Zhu, K.-H.; Jiang, X.-Y.; Chen, J.-C.; Yan, P.-Z.; Zhao, D.-S. Synchronous Extraction, Antioxidant Activity Evaluation, and Composition Analysis of Carbohydrates and Polyphenols Present in Artichoke Bud. *Molecules* **2022**, *27*, 8962. https://doi.org/10.3390/molecules27248962

Academic Editor: Claudio Ferrante

Received: 20 November 2022
Accepted: 14 December 2022
Published: 16 December 2022

1. Introduction

Artichoke (*Cynara scolymus* L.), also known as chrysanthemum thistle, vegetable thistle, and so forth, is a perennial herb of the genus *Cynara* of the family Asteraceae [1]. This herb is native to the western and central Mediterranean regions [2] and is also cultivated in the Shanghai, Yunnan, and Zhejiang provinces of China. Nowadays, artichoke is widely used in pharmaceutical [3], food [4], and other industries. Modern pharmacological studies have shown that artichoke and its chemical components have various biological activities, such as antioxidant, antibacterial, antitumor, hepatoprotective, and cholesterol-lowering activities [1,5–8]. Artichoke not only has significant health benefits, it is also rich in a variety of functional components, such as flavonoids, carbohydrates, polyphenols, and terpenoids [3,9,10]. Besides its applications as a functional food and its bioactive ingredient sources, artichoke is recognized as a sustainable feedstock for biofuel production.

Reactive oxygen radicals are produced by the oxidative respiration metabolism of cells and have high reactivity because their outermost orbitals have unpaired electrons [11,12]. The activity of antioxidant enzymes in a body gradually decreases with increasing age, resulting in an accumulation of reactive oxygen radicals produced by cellular metabolism in the body. The accumulation of reactive oxygen radicals in the body leads to lipid peroxidation reactions [13,14], which can cause cell damage or even death. The body is prone to Alzheimer's disease, Parkinson's disease, diabetes, arthritis, cardiovascular diseases, tumors, and other diseases [15–19]. The research and development of antioxidants have become significant nowadays with experts trying to find ways to reduce damage caused by the presence of excessive reactive oxygen radicals in the human body, and natural drugs are gaining attention for their excellent antioxidant properties. Carbohydrates and polyphenolic compounds are the main natural components in artichokes with significant antioxidant activity [20,21], which is why they have become a significant topic of research for their use in pharmaceuticals, functional foods, and bioactive ingredient sources.

Carbohydrates or polyphenols are mainly extracted using the traditional solvent extraction method [22,23], enzymatic method [24], microwave method [25,26], ultrasonic method [25], and so forth. Although the traditional solvent extraction method is easy to operate, it is time-consuming and has low extraction efficiency [23]; the enzymatic method affects the extraction efficiency because of the easy inactivation of enzymes [27]; and the ultrasonic method uses cavitation and mechanical and thermal effects to significantly increase the extraction efficiency of active ingredients [28]. The dual-phase aqueous system can use environmentally friendly phase-forming components, such as ethanol and inorganic salts and can be rapidly phase-formed at room temperature, depending on the different partition coefficients of different substances in the two phases, thus enabling the rapid partitioning of the extracted substances in the two phases of this system [29–31]. The dual-phase aqueous extraction technique compared with the conventional extraction techniques used for the simultaneous extraction of active plant ingredients can reduce the associated energy consumption and has the advantages of high extraction efficiency, mild conditions, a nontoxic and environmentally-friendly nature, and simple operation [32,33]. Considering these advantages, the ultrasonic-assisted aqueous two-phase synchronous extraction method was used to extract carbohydrates and polyphenols out of artichoke bud.

The research on the extraction of the active components of artichoke currently focuses only on optimizing the extraction process targeting a single component (polyphenols) as an index [34,35]. In the absence of a multiple indicator extraction optimization process, the extraction of other active substances in artichoke cannot be performed, limiting the efficient development and utilization of artichoke resources. In this study, an ultrasonic-assisted ethanol–ammonium sulfate aqueous two-phase system was used to synchronously extract carbohydrates and polyphenols from artichoke bud. The extraction process was optimized using a dual-response surface model, and the composition and antioxidant activity in vitro were analyzed. It provided a theoretical basis for the in-depth study of the bioactive components of artichoke and also provided scientific guidance for the comprehensive development and utilization of artichoke.

2. Results and Discussion

2.1. Effects of the Investigated Factors on the Carbohydrate and Polyphenol Contents

2.1.1. Effects of Ultrasonic Powers on the Carbohydrate and Polyphenol Contents in Artichoke Bud

When the ultrasonic power was in the range of 300–350 W, the extraction rate of carbohydrates from artichoke bud increased significantly ($p < 0.05$), but when the ultrasonic power continued to increase, the carbohydrates content decreased significantly ($p < 0.05$) (Figure S1A). Within this range, the variation in extraction trends of polyphenols and carbohydrates was roughly the same, but not as significant as that of carbohydrates. It was mainly because the cavitation effect of ultrasound could effectively dissolve carbohydrates and polyphenols present in artichoke bud, but the powerful ultrasonic power destroys the cell wall, causing the dissolved polyphenols to interact with carbohydrates to form insoluble aggregates, resulting in the reduction of content [36,37]. Taking into consideration the polyphenol and carbohydrate contents and the energy they used, 350 W was considered the fixed power.

2.1.2. Effects of Different Powder Mass on the Carbohydrate and Polyphenol Contents in Artichoke Bud

When the powder mass was in the range of 0.5–1.5 g, the carbohydrate content increased significantly ($p < 0.05$), but when the powder mass continued to increase, the polyphenol and carbohydrate contents always showed an overall decreasing trend ($p < 0.05$) (Figure S1B). It was mainly because, in the aqueous two-phase system, the volume of the upper alcohol phase was small and the volume of the lower water phase was large. When 0.5 g of powder was added, the polyphenols in the upper phase reached a saturated state of dissolution while the carbohydrates did not reach the saturated state. If the mass of the material was increased continuously, other alcohol-soluble components were dissolved and distributed in the upper phase, which affected the extraction of polyphenols, but this had no significant effect on the extraction of carbohydrates in the lower phase [38,39]. A powder mass of 1.0–2.0 g was considered the next factor for response surface optimization taking into consideration the polyphenol and carbohydrate contents.

2.1.3. Effects of Different Extraction Temperatures on Carbohydrate and Polyphenol Contents in Artichoke Bud

The polyphenol and carbohydrate contents showed an increase within the ultrasonic temperature range, and this increasing trend was not altered ($p < 0.05$) (Figure S1C). The study for the extraction of polyphenols and carbohydrates was conducted at 80 °C and not beyond because of the ultrasonic instrument's maximum allowed temperature limit at 80 °C.

2.1.4. Effects of Different Ultrasonic Times on the Carbohydrate and Polyphenol Contents in Artichoke Bud

The carbohydrate and polyphenol contents showed first an increasing trend and then a decreasing trend with the increase in ultrasonic time and reached their highest levels after 40 min (Figure S1D). This was because, in the early stage of extraction, the cavitation effect of ultrasonic waves promoted the transfer of solutes, accelerated the dissolution of polyphenols and carbohydrates, and increased the content [40]. A continuous increase in the ultrasonic time after the diffusion equilibrium did not cause the dissolution of carbohydrates and polyphenols. Instead, the content was decreased because, with increased ultrasonic power, polyphenols are more easily dissolved, interacting with dissolved carbs to produce insoluble aggregates [36,37]. Therefore, 30–50 min was considered a factor for further response surface optimization.

2.1.5. Effects of Different Concentrations of Ammonium Sulfate on the Carbohydrate and Polyphenol Contents in Artichoke Bud

When the concentration of ammonium sulfate was in the range of 0.27–0.33 g/mL, the carbohydrate content increased significantly ($p < 0.05$), but when the mass concentration of ammonium sulfate continued to increase, the extracted carbohydrate content decreased significantly ($p < 0.05$) (Figure S1E). The variation in the extraction trend of both polyphenols and carbohydrates in this range was roughly the same. The main reason behind it was that the aqueous two-phase system gradually became stable with the continuous increase in the concentration of ammonium sulfate, resulting in a significant increase in the polysaccharide and polyphenol contents. A continuous increase in the concentration of ammonium sulfate increased its solubilization effect, making it easier for polyphenols to distribute in the water phase, which in turn affected the extraction of upper-phase relative polyphenols and lower-phase relative carbohydrates [41]. Hence, the concentration of ammonium sulfate between 0.30 and 0.36 g/mL was considered a factor for response surface optimization.

2.1.6. Effects of Different Alcohol–Water Ratios on the Polysaccharide and Polyphenol Contents in Artichoke Bud

When the alcohol–water ratio was between 0.2 and 0.4, the polysaccharide and polyphenol contents increased significantly ($p < 0.05$), but when the alcohol–water ratio continued to increase, the carbohydrates and polyphenols decreased significantly ($p < 0.05$) (Figure S1F). It was mainly because, with the continuous increase in the volume of ethanol, the aqueous two-phase system was gradually stabilized and the polyphenol and polysaccharide contents increased. However, when the volume of ethanol continued to increase, some other substances (such as proteins) were precipitated, which adsorbed carbohydrates and polyphenols and reduced the polysaccharide and polyphenol contents [42]. Hence, the alcohol–water ratio of 0.3–0.5 was considered a factor for response surface optimization.

2.2. *Box–Behnken Response Surface Experimental Results*

Based on the results of the single-factor experiment, the powder mass (A), ultrasonic time (B), ammonium sulfate concentration (C), and alcohol–water ratio (D) were considered the influencing factors for further response surface optimization, and the polyphenol content (Y_1) and polysaccharide content (Y_2) were examined to design a response surface experiment. The response surface factor levels are presented in Table 1, and the experimental scheme and results are provided in Table 2.

Table 1. Response surface factor levels.

Levels	Factors			
	A Powder Mass (g)	B Ultrasound Time (min)	C Ammonium Sulphate Mass Concentration (g/mL)	D Alcohol–Water Ratio
−1	1.0	30	0.30	0.3
0	1.5	40	0.33	0.4
1	2.0	50	0.36	0.5

The software Design-Expert 11 was used to perform multiple regression analysis and binomial analysis using the data presented in Table 3. The multiple regression equations were: $Y_1 = 5.24 - 0.2067\ A + 0.0933\ B - 0.0608\ C + 0.8308\ D + 0.0325\ AB - 0.3125\ AC + 0.1925\ AD - 0.1613\ BC - 0.1137\ BD + 0.3688\ CD - 0.5608\ A^2 - 0.5071\ B^2 - 0.7533\ C^2 - 0.6558\ D^2$; $Y_2 = 74.72 - 2.16\ A + 1.13\ B + 3.33\ C - 2.82\ D + 0.5938\ AB + 2.12\ AC + 3.00\ AD - 0.6425\ BC - 0.3100\ BD - 2.37\ CD - 7.27\ A^2 - 3.74\ B^2 - 3.22\ C^2 - 5.62\ D^2$; the correlation coefficients r of Y_1 and Y_2 were 0.9772 and 0.9389, respectively, indicating that the independent variable could reflect the change of 97.72% polyphenol content and the change of 93.89% polysaccharide content. The absolute values of the previous coefficients showed how much each factor affected the examined value: the quadratic coefficients of the two models were negative, the parabolic opening of the three-dimensional response surface for each model was downward, and because this could reach the maximum value, we knew that the optimal process could be optimized [43,44].

Table 2. Response Surface Experiment Scheme and Results.

No.	A	B	C	D	The Content of Polyphenols (mg/g)	The Content of Carbohydrates (mg/g)
1	1	30	0.3	0.3	2.17	56.61
2	2	30	0.3	0.3	1.56	26.60
3	1	50	0.3	0.3	3.14	52.37
4	2	50	0.3	0.3	2.68	50.80
5	1	30	0.36	0.3	1.64	57.57
6	2	30	0.36	0.3	1.45	63.11
7	1	50	0.36	0.3	1.82	67.23
8	2	50	0.36	0.3	1.18	56.64
9	1	30	0.3	0.5	2.56	46.12
10	2	30	0.3	0.5	3.91	46.85
11	1	50	0.3	0.5	2.45	48.81
12	2	50	0.3	0.5	4.31	50.01
13	1	30	0.36	0.5	4.53	46.76
14	2	30	0.36	0.5	3.36	53.08
15	1	50	0.36	0.5	4.32	54.57
16	2	50	0.36	0.5	3.46	57.61
17	0.5	40	0.33	0.4	4.00	57.69
18	2.5	40	0.33	0.4	1.88	44.41
19	1.5	20	0.33	0.4	3.14	68.72
20	1.5	60	0.33	0.4	3.17	61.63
21	1.5	40	0.27	0.4	2.28	66.85
22	1.5	40	0.39	0.4	2.06	67.61
23	1.5	40	0.33	0.2	0.89	67.77
24	1.5	40	0.33	0.6	4.23	47.53
25	1.5	40	0.33	0.4	5.21	74.77
26	1.5	40	0.33	0.4	5.22	74.69
27	1.5	40	0.33	0.4	5.23	74.72
28	1.5	40	0.33	0.4	5.28	74.76
29	1.5	40	0.33	0.4	5.23	74.68
30	1.5	40	0.33	0.4	5.24	74.72

Table 3. Analysis of variance for polyphenols content.

Source	Sum of Squares of Deviation from Mean	Degrees of Freedom	Mean Square	*F* Value	*p* Value
Model	53.44	14	3.82	15.99	<0.0001
A	1.03	1	1.03	4.29	0.0559
B	0.2091	1	0.2091	0.8755	0.3643
C	0.0888	1	0.0888	0.3719	0.5511
D	16.57	1	16.57	69.38	0.0001
AB	0.0169	1	0.0169	0.0708	0.7938
AC	1.56	1	1.56	6.54	0.0218
AD	0.5929	1	0.5929	2.48	0.1359
BC	0.4160	1	0.4160	1.74	0.2066
BD	0.2070	1	0.2070	0.8670	0.3665
CD	2.18	1	2.18	9.11	0.0086
A^2	8.63	1	8.63	36.13	<0.0001
B^2	7.05	1	7.05	29.54	<0.0001
C^2	15.57	1	15.57	65.19	<0.0001
D^2	11.80	1	11.80	49.41	<0.0001
Residual	3.58	15	0.2388	—	—
Lack of fit	1.29	10	0.1290	4.23	0.1969
Pure Error	0.0030	5	0.0006	—	—
Sum error	57.03	29	—	—	—

The results of the analysis of variance are presented in Tables 3 and 4. The quadratic regression model of polyphenols had strong significance ($p < 0.0001$), the quadratic regression model of carbohydrates also had strong significance ($p < 0.05$), and the lack-of-fit term was not significant ($p > 0.05$), indicating that the two models were well fitted [45]. AC and CD had significant effects on polyphenol content ($p < 0.05$), and D, A^2, B^2, C^2, and D^2 had extremely significant effects on polyphenol content ($p < 0.0001$) (Table 3). Hence, the effect of polyphenols content was not just an ordinary linear relationship. Table 4 shows that A^2, B^2, and D^2 all significantly affected the carbohydrates content ($p < 0.05$). According to the p value, the degree of influence of each factor on Y_1 was D (alcohol–water ratio) > A (powder mass) > B (ultrasonic time) > C (ammonium sulfate concentration), and the degree of influence on Y_2 was C (ammonium sulfate concentration) > D (alcohol–water ratio) > A (powder mass) > B (ultrasonic time).

Table 4. Analysis of variance for carbohydrates content.

Source	Sum of Squares of Deviation from Mean	Degrees of Freedom	Mean Square	*F* Value	*p* Value
Model	3138.99	14	224.21	3.48	0.0112
A	112.23	1	112.23	1.74	0.2069
B	30.74	1	30.74	0.4766	0.5005
C	266.13	1	266.13	4.13	0.0603
D	190.41	1	190.41	2.95	0.1063
AB	5.64	1	5.64	0.0875	0.7715
AC	72.08	1	72.08	1.12	0.3071
AD	143.52	1	143.52	2.23	0.1565
BC	6.60	1	6.60	0.1024	0.7534
BD	1.54	1	1.54	0.0238	0.8793
CD	89.97	1	89.97	1.40	0.2559
A^2	1449.35	1	1449.35	22.47	0.0003
B^2	383.23	1	383.23	5.94	0.0277
C^2	285.13	1	285.13	4.42	0.0528
D^2	866.06	1	866.06	13.43	0.0023
Residual	967.32	15	64.49	—	—
Lack of fit	1.56	10	0.1560	1.62	0.3191
Pure Error	0.0065	5	0.0013	—	—
Sum error	4106.32	29	—	—	—

The plane contour map and the three-dimensional response surface map are presented in Figure 1, revealing that the interaction of factors had a significant impact on the examined value. The change in the response surface of powder mass was steeper than that of the ammonium sulfate concentration, indicating that the effect of powder mass on polyphenol content was more significant than that of ammonium sulfate concentration. The change in the response surface of the alcohol–water ratio was steeper than that of the ammonium sulfate concentration, indicating that the effect of the alcohol–water ratio on the polyphenol content was more significant than that of the ammonium sulfate concentration. The contour lines on the graph were dense and elliptical in shape, and the corresponding response surface was steep, indicating that the interaction between the powder mass, ammonium sulfate concentration, and alcohol–water ratio had a significant effect on the polyphenol content [46]. The other contour plots and the response surface plots showed that although the overall trend was higher and then lower, the contour lines were sparse, and the response surface was moderate. Hence, the corresponding two factors did not have a significant effect on them. The effect of the two factors on the response values was not significant. The results of this analysis were consistent with the results of the binomial analysis.

The polyphenol content under the optimal process using the ethanol–ammonium sulfate aqueous two-phase system as the solvent in this study compared with the optimization of the extraction process carried out in other studies was significantly higher than those of the other traditional solvents ($p < 0.05$) [35,47]. Furthermore, the polysaccharide content in artichoke bud was significantly higher than that of the polyphenols ($p < 0.05$), which was consistent with the results of Bao et al. [48].

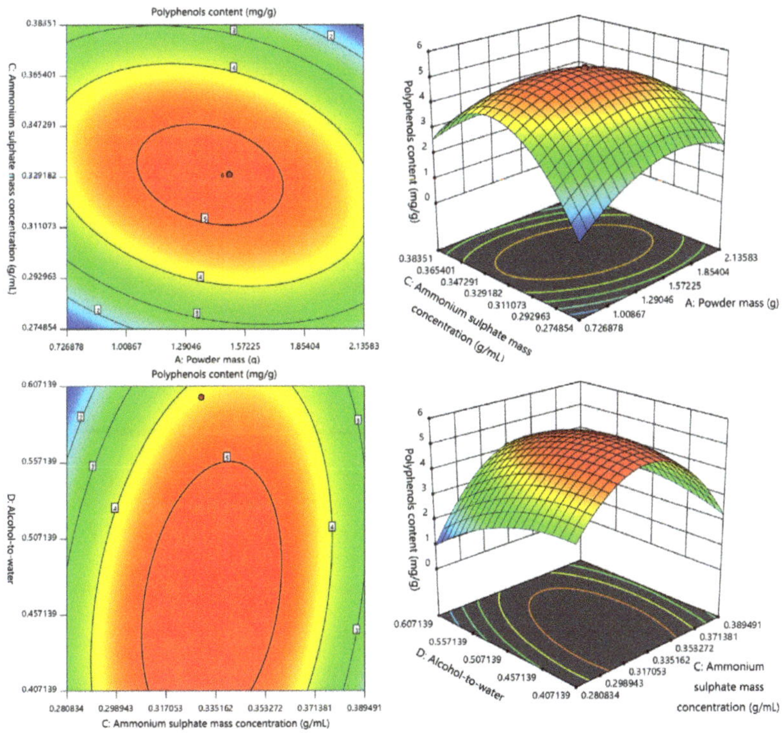

Figure 1. Effect of interaction between the powder mass (A) and the ammonium sulfate concentration (C) or the alcohol–water ratio (D) on the polyphenol content.

2.3. Optimal Process Validation

The optimal process obtained by the response surface software had the powder mass of 1.397 g, ammonium sulfate concentration of 0.335 g/mL, alcohol–water ratio of 0.409, ultrasonic time of 42.60 min, polyphenol content of 5.282 mg/g, and polysaccharide content of 74.845 mg/g. According to the actual situation, the powder mass was 1.4 g, ammonium sulfate concentration was 0.34 g/mL, alcohol–water ratio was 0.4, and ultrasonic time was 43 min. The polyphenol content in artichoke bud was 5.32 ± 0.13 mg/g and the polysaccharide content was 74.78 ± 0.11 mg/g, which were close to the predicted values of the software. Hence, carbohydrates and polyphenols could be effectively taken out of artichoke bud using this improved process.

2.4. Evaluation of Antioxidant Activity In Vitro

The higher the free radical scavenging rate, the stronger the antioxidant activity of the sample against the free radicals. Figure S2A,B show that polyphenols reached a scavenging rate of 97.8% at 0.06 mg/mL, while VC at the same concentration reached a scavenging rate of 97.4%. The IC_{50} value of polyphenols substances was 0.01108 mg/mL, and that of standard VC was 0.01294 mg/mL. The IC_{50} values of polyphenols were slightly lower than VC. Therefore, the ability of polyphenols to scavenge $ABTS^{+}\cdot$ was extremely strong, which was slightly stronger than that of VC. Carbohydrates reached a scavenging rate of 99.16% at 1.5 mg/mL while VC at the same concentration reached a scavenging rate of 99.86%, which meant that it had a remarkable ability to scavenge $ABTS^+$, but it was weaker than that of VC.

The magnitude of the free radical scavenging rate can indicate the strength of the antioxidant activity of the sample against free radicals. Figure S2C,D shows that polyphenols reached a scavenging rate of 70.10% at 0.1 mg/mL while VC at the same concentration reached a scavenging rate of 39.57%. The IC_{50} value of polyphenols was 0.04131 mg/mL, while the IC_{50} value of standard VC was 0.174 mg/mL. The IC_{50} value of polyphenols was significantly lower than that of VC. Therefore, the ability of polyphenols to scavenge DPPH· was strong, which was stronger than that of VC. Carbohydrates reached a scavenging rate of 45.35% at 1.5 mg/mL, while VC at the same concentration reached a scavenging rate of 60.38%, which meant that it had a good ability to scavenge DPPH but it was weaker than that of VC. However, different concentrations of DPPH solutions will lead to different IC_{50} values and low reference-ability. Therefore, we standardized the IC_{50} values of substances inhibiting the DPPH clearance rate according to the methods reported in the references [49], and the standardized results showed that the standardized IC_{50} value of polyphenols was 24.43 ng/nmol, the standardized IC_{50} value of VC was 102.9 ng/nmol, and that of carbohydrates was 1544.35 ng/nmol. These results indicated that the ability of polyphenols to inhibit the DPPH radical was much greater than that of VC, while that of carbohydrates was much less than that of VC.

The level of absorbance can reflect the strength of the reducing power of the sample. The greater the absorbance, the stronger the reducing power of the sample. Figure S2E,F shows that the Fe^{3+}-reducing power of polyphenols and carbohydrates and the dose–effect relationship of the concentration of the experimental solution were good. The absorbance of polyphenols at 0.4 mg/mL was 1.3439, while VC at the same concentration reached an absorbance value of 1.2781. The IC_{50} value of polyphenols was 0.06345 mg/mL, and that of standard VC was 0.06976 mg/mL. The IC_{50} value of polyphenols is slightly lower than that of VC. Thus, the ability to reduce Fe^{3+} was extremely strong, which was stronger than that of VC. The absorbance of carbohydrates at 3.5 mg/mL was 1.2672, while VC at the same concentration reached an absorbance value of 3.132. The ability of polyphenols to reduce Fe^{3+} was strong but weaker than that of VC.

The smaller the IC_{50} value of the in vitro antioxidant activity, the stronger the antioxidant activity of the substance and the weaker the antioxidant activity of the substance. Table 5 shows that the antioxidant activity of polyphenols was extremely strong, the ability to scavenge $ABTS^{+}\cdot$ was the strongest, and the ability to reduce Fe^{3+} was the weakest. Furthermore, the antioxidant activity of carbohydrates was stronger, and the ability to scavenge $ABTS^{+}\cdot$ was the strongest, with the weakest ability to scavenge DPPH. The IC_{50} values of various in vitro antioxidant experiments of polyphenols were lower than those of carbohydrates, indicating that the antioxidant activity of polyphenols was stronger than that of carbohydrates, which was consistent with the findings of Bao et al. [48].

Table 5. In vitro antioxidant experimental IC_{50} values.

Experiment Name	Carbohydrates	Polyphenols	VC
Clear $ABTS^+$	0.4141 mg/mL	0.01108 mg/mL	0.01294 mg/mL
Clear DPPH	2.6110 mg/mL	0.04131 mg/mL	0.17400 mg/mL
Fe^{3+} reducing power	0.9363 mg/mL	0.06345 mg/mL	0.06976 mg/mL

2.5. Results of HPLC Analysis of Polysaccharide and Polyphenol Compositions

Each monosaccharide, alduronic acid, and phenolic acid standard had a good resolution in the mixed standard, and the peaks of the standard and the sample might be in one-to-one correspondence (Figures 2 and 3). Figure 4A shows the different carbohydrates soluble in ammonium sulfate aqueous solution present in artichoke bud: mannose, rhamnose, glucuronic acid, galacturonic acid, glucose, galactose, and arabinose; the molar ratio was 10.77:25.22:2.37:15.74:125.39:48.62:34.70. Figure 4B shows the constituents of polyphenols present in artichoke bud: chlorogenic acid, 4-dicaffeoylquinic acid, caffeic acid, 1,3-dicaffeoylqunic acid, isochlorogenic acid B, isochlorogenic acid A, cynarin, and isochlorogenic acid C; the contents are 0.503, 0.029, 0.022, 0.017, 0.008, 0.162, 1.621, 0.030 mg/g, respectively. Compared with mass spectrometry, ultraviolet detector has poor sensitivity, so some components with low content may not be detected. With in-depth study, mass spectrometer will be applied to improve the limits of quantification.

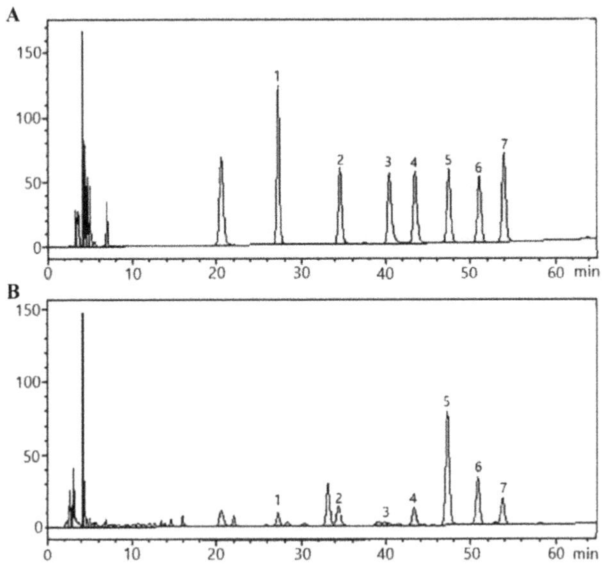

Figure 2. HPLC profile of monosaccharide, alduronic acid mixed standard (**A**) and polysaccharide samples (**B**). 1. Mannose, 2. rhamnose, 3. glucuronic acid, 4. galacturonic acid, 5. glucose, 6. galactose, 7. arabinose.

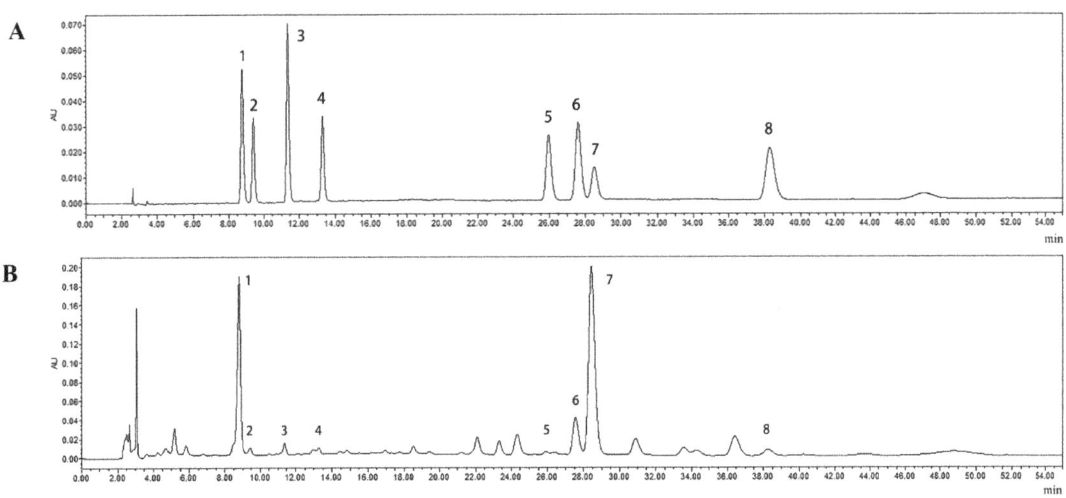

Figure 3. HPLC profile of phenolic acid mixed standard (**A**) and polyphenol samples (**B**). 1. Chlorogenic acid, 2. 4-dicaffeoylquinic acid, 3. caffeic acid, 4. 1,3-dicaffeoylqunic acid, 5. isochlorogenic acid B, 6. isochlorogenic acid A, 7. cynarin, 8. isochlorogenic acid C.

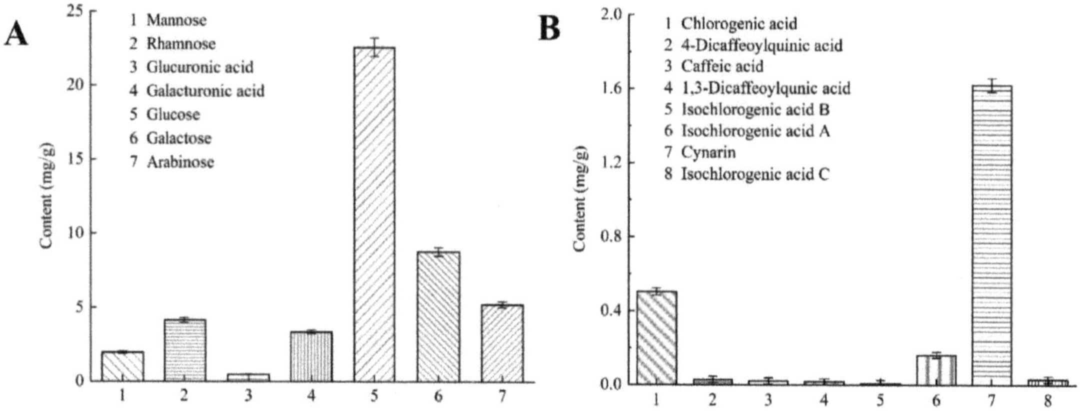

Figure 4. Concentrations of monosaccharides and alduronic acid in carbohydrates (**A**) and concentrations of eight phenolic acids in polyphenols (**B**).

As described in Section 2.4, polyphenols and carbohydrates in artichoke bud had strong antioxidant activity. Hence, the combinatory use of carbohydrates and polyphenols as a potential therapeutic alternative for treating relevant disorders with hyperoxidative conditions, such as aging, senescence, etc. Additionally, chlorogenic acid and cynarin were related to the antioxidant activity of artichoke polyphenols [50], but whether other components affected the antioxidant activity of polyphenols is still unknown. Although artichoke carbohydrates had strong antioxidant activity, related studies were few and the components that affected its antioxidant activity needed further exploration. In this study, the HPLC analysis was used to figure out how polyphenols and carbohydrates were made, and it gave us a reference point to learn more about their antioxidant activities.

3. Materials and Methods

3.1. Materials

Artichoke bud (originating in Yunnan) was obtained from Kunming Mingjian'an Biotechnology Co., Ltd. (Yunnan, China) and identified as dried bud of artichoke (*Cynara scolymus* L.) by Professor Li Jia of Shandong University of Traditional Chinese Medicine. Gallic acid (purity $\geq$ 98%) was purchased from Shanghai yuanye BioTechnoligy Co., Ltd. (Shanghai, China). Anhydrous Glucose (purity $\geq$ 99%) was obtained from Shanghai Macklin Biochemical Co., Ltd. (Shanghai, China). Folin phenol reagent, sodium carbonate, phenol, sulfuric acid, ethanol absolute, ammonium sulfate, DPPH, ABTS, potassium persulfate, potassium ferricyanide, trichloroacetic acid, iron trichloride hexahydrate, sodium dihydrogen phosphate, dibasic sodium phosphate, vitamin C (VC), trichloromethane, potassium dihydrogen phosphate, sodium hydroxide, acetic acid, 3-methyl-1-phenyl-2-pyrazolin-5-one, and trifluoroacetic acid are AR grade. Methanol and acetonitrile are HPLC grade. Mannose, Rhamnose, Glucuronic acid, Galacturonic acid, Glucose, Galactose, Arabinose, Chlorogenic acid, 4-Dicaffeoylquinic acid, Caffeic acid, Cynarin, Isochlorogenic acid B, Isochlorogenic acid A, 1,5-Dicaffeoylqunic acid, and Isochlorogenic acid C were purchased from Chengdu Must Biotech Co., LTD (Chengdu, China).

3.2. Synchronous Extraction Process

The synchronous extraction process for carbohydrates and polyphenols from artichoke bud is depicted in Figure 5. The sample solution was prepared as follows: a 50-mL centrifuge tube was filled with 1 g of artichoke bud powder. Then, a 35 mL ethanol-ammonium sulfate two-phase system was prepared by adding ammonium sulfate with a mass concentration of 0.33 g/mL and an alcohol–water ratio of 0.4. The extraction was carried out under fixed conditions of ultrasonic power of 500 W, an ultrasonic time of 30 min and an ultrasonic temperature of 60 °C in the ultrasonic cleaner. Suction filtration and centrifugation were at 7000 rpm for 10 min.

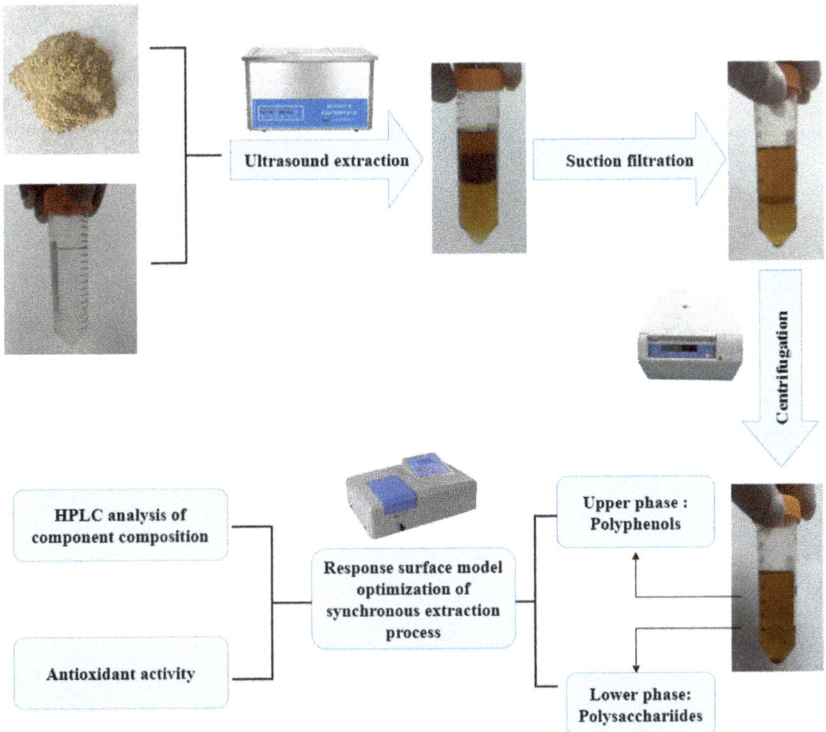

Figure 5. The synchronous extraction process for carbohydrates and polyphenols from artichoke bud.

3.3. Determination of Carbohydrate Content

The phenol–sulfuric acid method was used with minor modifications to determine the carbohydrate content [51]. The precisely pipetted 1.0 mL of the lower-phase solution diluted 50 times, 1.0 mL of distilled water, and 1.0 mL of a 5% phenol solution were mixed well, making up to 8.0 mL with concentrated sulfuric acid. The mixture was then left for 10 min at room temperature, placed in a water bath at 40 °C for 15 min, and cooled at room temperature. The absorbance was measured at 490 nm, and the average value was taken in triplicate. Anhydrous glucose was used as a standard and the results were expressed as milligrams of anhydrous glucose equivalents per gram artichoke bud sample.

3.4. Determination of Polyphenol Content

The polyphenol content was determined using the Folin–Ciocalteu colorimetric method [34]. The precisely pipetted 0.5 mL of the upper-phase solution and 2.0 mL of the Folin–Ciocalteu reagent with a volume fraction of 50% were mixed well. The mixture was left at room temperature for 5 min, and 4.0 mL of 0.2 g/L Na_2CO_3 solution was added to the mixture, making up to 10 mL with distilled water. The mixture was kept away from light at room temperature for 2 h. The absorbance was measured at 760 nm, and the average value was taken in triplicate. Gallic acid was used as a standard and the results were expressed as milligrams of gallic acid equivalents per gram artichoke bud sample.

3.5. Single-Factor Experiment

The effects of various factors on carbohydrates and polyphenols were examined using on ultrasonic power (300, 350, 400, 450, and 500 W), powder mass (0.5, 1.0, 1.5, 2.0, and 2.5 g), extraction temperature (40, 50, 60, 70, and 80 °C), ultrasonic time (20, 30, 40, 50, and 60 min), ammonium sulfate mass concentration (0.27, 0.30, 0.33, 0.36, and 0.39 g/mL), and alcohol–water ratio (0.2, 0.3, 0.4, 0.5, and 0.6).

3.6. Response Surface Experiment

A response surface experiment was performed for optimization based on the high and low levels in the single-factor experiment using carbohydrates and polyphenol contents as variables.

3.7. Determination of Antioxidant Activity In Vitro

3.7.1. Determination of $ABTS^{+}\cdot$ Scavenging Capacity

Referring to the method in the literature [52], we used 0.3 mL of specific concentration of lower-phase solution and upper-phase solution of 0.3 mL was aspirated to measure the $ABTS^{+}$ scavenging capacity. The curves depicting the scavenging rates of carbohydrates and polyphenols to $ABTS^{+}\cdot$ were drawn, and the IC_{50} value was calculated. The same concentration of VC was used as a positive control.

3.7.2. Determination of DPPH· Scavenging Capacity

We followed the method available in the literature with minor modifications to determine the scavenging capacity of DPPH [53]. For this, 0.3 mL of a specific concentration of the lower-phase solution and upper-phase solution was aspirated and mixed with 2.0 mL of DPPH· solution (0.1 mg/mL), making up to 10 mL with distilled water. The reaction was performed at room temperature for 30 min in the dark. The absorbance A_1 was measured at 517 nm. Next, 0.3 mL of the lower-phase solution and upper-phase solution was mixed with 2.0 mL of absolute ethanol, making up to 10 mL with distilled water. The reaction was performed at room temperature for 30 min in the dark and then the absorbance A_2 was measured at 517 nm. The curves depicting the scavenging rates of carbohydrates and polyphenols to DPPH· were drawn and IC_{50} values were calculated. The same concentration of VC was used as a positive control.

$$\text{DPPH}\cdot \text{ scavenging rate } (\%) = \left(1 - \frac{A_1 - A_2}{A_0}\right) \times 100\%$$

3.7.3. Determination of Fe^{3+}-Reducing Power

The method available in the literature was followed with minor modifications to determine the reducing capacity of Fe^{3+} [54]. Further, 0.3 mL of the lower-phase solution or upper-phase solution, 2.5 mL of phosphate buffer solution (pH = 7.4), and 2.5 mL of 1% K_3 $[Fe(CN)_6]$ solution were mixed. The mixture was then placed in a water bath for 30 min at 50 °C. It was taken out and cooled to room temperature. Further, 1.0 mL of TCA was added, and the mixture was allowed to stand for 20 min before adding 2.5 mL of the supernatant, 0.5 mL of 0.1% $FeCl_3$, and 2.5 mL of distilled water to it. The mixture was mixed well and left for 10 min. The absorbance was measured at 700 nm. The curves depicting Abs of carbohydrates and polyphenols were drawn, and IC_{50} values were calculated. The same concentration of VC was used as a positive control.

3.8. High-Performance Liquid Chromatography Analysis

3.8.1. Standard Solution Preparation

Appropriate masses of monosaccharides and uronic acid standards were accurately weighed and dissolved to prepare a 1 mg/mL mixed standard solution.

An appropriate mass of each phenolic acid standard was accurately weighed and dissolved to prepare a 0.2 mg/mL mixed standard solution.

3.8.2. Sample Solution Preparation

The lower-phase carbohydrate stock solution was freeze-dried as described in Section 3.2, and an appropriate mass of freeze-dried powder was weighed and dissolved to prepare the carbohydrate solution. To obtain the mixed standard solution and sample solution, 2 mL of 4 mol/L TFA solution was added to the carbohydrate solution and mixed well. The mixture was then hydrolyzed in an oven at 110 °C for 2 h, taken out, cooled to room temperature, and adjusted to neutral with 4 mol/L NaOH, making up to 5 mL with distilled water. We then took 200 μL of this hydrolyzed solution and added 200 μL of 0.3 mol/L NaOH solution and 200 μL of 0.5 mol/L PMP methanol solution and mixed the solution well. The solution was then placed in a water bath for 70 min at 70 °C. The solution was allowed to cool to room temperature. The pH of the solution was then adjusted by adding 200 μL of a 0.3 mol/L acetic acid solution and 200 μL of a 0.1 mol/L KH_2PO_4 (pH = 6.0) solution. We then added 2 mL of $CHCl_4$ and vortexed the solution well to remove the chloroform phase. It was done thrice. The chromatographic analysis was performed on the solution after passing it through a 0.45-μm aqueous film.

The polyphenol solution was obtained using an appropriate volume of the upper-phase stock solution prepared as described in Section 3.2.

3.8.3. Chromatographic Conditions

The high-performance liquid chromatography (HPLC) conditions of carbohydrates and polyphenols are listed in Supplementary Table S1.

3.9. Statistical Analyses

Origin Pro 8 (OriginLab Corporation, Northampton, MA, USA) was used to analyze and map the experimental data, and Design-Expert 11 (Stat-Ease, Inc., Minneapolis, MN, USA) was used to design and analyze response surface experiments. IC_{50} calculations were performed using Origin Pro 8 software.

4. Conclusions

In this study, the powder mass, ammonium sulfate concentration, alcohol–water ratio, and ultrasonic time were considered the optimizing factors, and the optimum process conditions were obtained by constructing a double-response surface model: powder mass of 1.4 g, ammonium sulfate concentration of 0.34 g/mL, alcohol–water ratio of 0.4, and ultrasonic time of 43 min. Under these conditions, the polyphenol content in artichoke bud was 5.32 ± 0.13 mg/g and the polysaccharide content was 74.78 ± 0.11 mg/g. The HPLC spectrum analysis revealed that the carbohydrates in artichoke bud comprised mannose, rhamnose, glucuronic acid, galacturonic acid, glucose, galactose, and arabinose, with a molar ratio of 10.77:25.22:2.37:15.74:125.39:48.62:34.70. The polyphenols comprised chlorogenic acid, 4-dicaffeoylquinic acid, caffeic acid, 1,3-dicaffeoylqunic acid, isochlorogenic acid B, isochlorogenic acid A, cynarin, and isochlorogenic acid C, and the contents were 0.503, 0.029, 0.022, 0.017, 0.008, 0.162, 1.621, 0.030 mg/g, respectively. The antioxidant activity of $ABTS^{+}\cdot$ and DPPH· scavenging and Fe^{3+}-reducing abilities was determined in vitro. The results showed that the antioxidant activity of carbohydrates and polyphenols in artichoke bud was strong, and the antioxidant activity of polyphenols was stronger than those of carbohydrates.

In an aqueous two-phase system, the combination of ethanol and saline solution can be considered an environmentally friendly method for extracting biologically active substances. These mixtures are not only low-cost and simple to prepare but also provide a mild and nontoxic environment. In addition, replacing traditional solvents with an aqueous two-phase system has shown great benefits in extracting natural plant components, such as shorter extraction times and higher extraction efficiencies. Its use in combination with ultrasonic power provides a rapid and effective method to extract polyphenols and carbohydrates from artichoke bud. The optimal extraction process obtained by optimizing the dual-response surface model was stable and feasible, the carbohydrates and polyphenols obtained by crude extraction had good antioxidative effects in vitro, and the material composition was obtained by HPLC analysis. These advantages provided a theoretical basis for exploring the biological activities of carbohydrates and polyphenols, and for resource development and utilization in more detail.

Supplementary Materials: The following supporting information can be downloaded at: https://www.mdpi.com/article/10.3390/molecules27248962/s1, Table S1 HPLC conditions of carbohydrates and polyphenols; Figure S1 Influence of ultrasonic power (A), powder mass (B), extraction temperature (C), ultrasonic time (D), mass concentration of ammonium sulphate (E), and alcohol-to-water ratio (F) on the extraction of carbohydrates and polyphenols from artichoke bud; Figure S2 In vitro antioxidant experimental evaluation of carbohydrates and polyphenols.

Author Contributions: Conceptualization, D.-S.Z.; methodology, X.L. and X.-K.L.; software, K.-H.Z., X.-Y.J.; investigation, X.L., K.-H.Z. and X.-K.L.; writing—original draft preparation, X.L. and X.-K.L.; writing—review and editing, D.-S.Z.; visualization, J.-C.C.; funding acquisition, P.-Z.Y. and D.-S.Z. All authors have read and agreed to the published version of the manuscript.

Funding: The authors thank the following funders: the Focus on Research and Development Plan in Shandong Province (2022TZXD0036), the National Natural Science Foundation of China (82004233), and the TCM Science and Technology Development Plan of Shandong Province (2019-0030).

Institutional Review Board Statement: Not applicable.

Informed Consent Statement: Not applicable.

Data Availability Statement: The data are contained within the article.

Conflicts of Interest: The authors declare no conflict of interest.

Sample Availability: Samples of the compounds chlorogenic acid, caffeic acid, isochlorogenic acid B, isochlorogenic acid A, cynarin, and isochlorogenic acid C are available from the authors.

References

1. Wang, Z.S.; Chen, S.M.; Xin, S.Y.; Sheng, H.Y.; Jiang, H.T. Total Phenols and Flavonoids and Antioxidant Activity of Artichoke (*Cynara scolymus* L.) Bud Juices before and after Gastrointestinal Digestion in Vitro. *Food Sci.* **2019**, *40*, 136–142.
2. Sgroi, F.; Fodera, M.; Di Trapani, A.M.; Tudisca, S.; Testa, R. Profitability of Artichoke Growing in the Mediterranean Area. *Hortscience* **2015**, *50*, 1349–1352. [CrossRef]
3. Zhao, Y.M.; Wang, C.D.; Zhang, R.; Hou, X.J.; Zhao, F.; Zhang, J.J.; Wang, C. Study on literature of artichoke and properties of traditional Chinese medicine. *China J. Chin. Mater. Med.* **2020**, *45*, 3481–3488.
4. Boubaker, M.; El Omri, A.; Blecker, C.; Bouzouita, N. Fibre concentrate from artichoke (*Cynara scolymus* L.) stem by-products: Characterization and application as a bakery product ingredient. *Food Sci. Technol. Int.* **2016**, *22*, 759–768. [CrossRef] [PubMed]
5. Mileo, A.M.; Di Venere, D.; Linsalata, V.; Fraioli, R.; Miccadei, S. Artichoke polyphenols induce apoptosis and decrease the invasive potential of the human breast cancer cell line MDA-MB231. *J. Cell. Physiol.* **2012**, *227*, 3301–3309. [CrossRef]
6. Zhu, X.F.; Zhang, H.X.; Lo, R. Antifungal activity of *Cynara scolymus* L. extracts. *Fitoterapia* **2005**, *76*, 108–111. [CrossRef]
7. Wang, M.F.; Simon, J.E.; Aviles, I.F.; He, K.; Zheng, Q.Y.; Tadmor, Y. Analysis of antioxidative phenolic compounds in artichoke (*Cynara scolymus* L.). *J. Agric. Food Chem.* **2003**, *51*, 601–608. [CrossRef]
8. Speroni, E.; Cervellati, R.; Govoni, P.; Guizzardi, S.; Renzulli, C.; Guerra, M.C. Efficacy of different *Cynara scolymus* preparations on liver complaints. *J. Ethnopharmacol.* **2003**, *86*, 203–211. [CrossRef]
9. Sisto, A.; Luongo, D.; Treppiccione, L.; De Bellis, P.; Di Venere, D.; Lavermicocca, P.; Rossi, M. Effect of *Lactobacillus paracasei* Culture Filtrates and Artichoke Polyphenols on Cytokine Production by Dendritic Cells. *Nutrients* **2016**, *8*, 635. [CrossRef]
10. Ruiz-Cano, D.; Perez-Llamas, F.; Frutos, M.J.; Arnao, M.B.; Espinosa, C.; Lopez-Jimenez, J.A.; Castillo, J.; Zamora, S. Chemical and functional properties of the different by-products of artichoke (*Cynara scolymus* L.) from industrial canning processing. *Food Chem.* **2014**, *160*, 134–140. [CrossRef]
11. Hourieh, A. A Review on Free Radicals and Antioxidants. *Infect. Disord. Drug Targets* **2020**, *20*, 16–26.
12. Zhao, D.; Liu, X.L.; Zheng, X.Q.; Fan, L. Antioxidant activity of pea protein hydrolysates in vitro and on HepG2 cells. *Food Ferment. Ind.* **2022**, *48*, 1–11.
13. Li, M.R.; Zhou, Y.Z.; Du, G.H.; Qin, X.M. Research progress about the anti-aging effect and mechanism of flavonoids from traditional Chinese medicine. *Acta Pharm. Sin.* **2019**, *54*, 1382–1391.
14. Ademowo, O.S.; Dias, H.K.I.; Burton, D.G.; Griffiths, H.R. Lipid (per) oxidation in mitochondria: An emerging target in the ageing process? *Biogerontology* **2017**, *18*, 859–879. [CrossRef]

15. Li, J.M.; Han, Y.Q.; Wu, T.; Zhang, T.C.; Dong, X.H. Lipid peroxidation in ferroptosis and its relationship with Alzheimer disease. *Chin. J. Pathophysiol.* **2022**, *38*, 1142–1147.
16. Simunkova, M.; Alwasel, S.H.; Alhazza, I.M.; Jomova, K.; Kollar, V.; Rusko, M.; Valko, M. Management of oxidative stress and other pathologies in Alzheimer's disease. *Arch. Toxicol.* **2019**, *93*, 2491–2513. [CrossRef] [PubMed]
17. Moloney, J.N.; Cotter, T.G. ROS signalling in the biology of cancer. *Semin. Cell Dev. Biol.* **2018**, *80*, 50–64. [CrossRef]
18. Incalza, M.A.; D'Oria, R.; Natalicchio, A.; Perrini, S.; Laviola, L.; Giorgino, F. Oxidative stress and reactive oxygen species in endothelial dysfunction associated with cardiovascular and metabolic diseases. *Vasc. Pharmacol.* **2018**, *100*, 1–19. [CrossRef]
19. Chandrasekaran, A.; Idelchik, M.D.S.; Melendez, J.A. Redox control of senescence and age-related disease. *Redox Biol.* **2017**, *11*, 91–102. [CrossRef]
20. Xie, J.H.; Liu, H.M.; Liu, M.H.; Zheng, M.Z.; Xu, Q.; Liu, J.S. Extraction Process Optimization and Antioxidant Activity Analysis of Polyphenols from Azuki Bean Coats (Vigna angularis). *J. Chin. Inst. Food Sci. Technol.* **2020**, *20*, 147–157.
21. Zheng, Q.R.; Li, W.F.; Zhang, H.; Gao, X.X.; Tan, S. Optimizing synchronous extraction and antioxidant activity evaluation of polyphenols and polysaccharides from Ya'an Tibetan tea (Camellia sinensis). *Food Sci. Nutr.* **2020**, *8*, 489–499. [CrossRef] [PubMed]
22. Yang, S.M.; Guo, W.H.; Wang, Z.J.; Deng, L. Optimization of Polysaccharides Extraction from Lycopus Lucidus Turcz. by Response Surface Methodology and Study on Its Antioxidant Activity. *Chin. J. Mod. Appl. Pharm.* **2021**, *38*, 2554–2559.
23. Zhang, Z.Y.; Zhao, Z.H.; Chen, X.; Liu, F.C.; Wang, J.G. Comparison of Polyphenols of Ziziphus Jujuba Extracted by Different Methods. *Food Res. Dev.* **2020**, *41*, 153–159.
24. Wang, Y.H.; Wu, Q.; Chi, Y.L.; Yao, K.; Jia, D.Y. Properties of crude polysaccharides extracted from *Tremella fuciformis* by acid, alkali and enzyme-assisted methods. *Food Sci. Technol.* **2019**, *44*, 200–204.
25. Jin, W.Q.; Peng, J.Y.; Wang, Y.S.; Tang, Y.J.; Tang, Z.Z.; Chen, H.; Sun, R.; Zheng, T.R.; Liu, M.Y.; Sun, W.J. Optimization of Response Surface Methodology for Microwave-assisted and Ultrasonic-assisted Extraction of Polysaccharides from *Amaranthus hybridus* L. *Genom. Appl. Biol.* **2019**, *38*, 757–765.
26. Sahin, S.; Samli, R.; Tan, A.S.B.; Barba, F.J.; Chemat, F.; Cravotto, G.; Lorenzo, J.M. Solvent-Free Microwave-Assisted Extraction of Polyphenols from Olive Tree Leaves: Antioxidant and Antimicrobial Properties. *Molecules* **2017**, *22*, 1056. [CrossRef] [PubMed]
27. Chen, L.; He, W.Q.; Cao, Q.Q. Optimization of Enzymatic Extraction Technology of Polysaccharide from Portulaca oleracea by Response Surface Methodology. *Food Res. Dev.* **2020**, *41*, 79–84.
28. Anticona, M.; Blesa, J.; Frigola, A.; Esteve, M.J. High Biological Value Compounds Extraction from Citrus Waste with Non-Conventional Methods. *Foods* **2020**, *9*, 811. [CrossRef]
29. Wang, J.H. Extraction of Flavonoid in Osmanthus Leaves by Aqueous Two-phase System Assisted Inner Ebullition Method. *Food Res. Dev.* **2022**, *43*, 22–28.
30. Zhang, S.M.; Chen, M.Z. Extraction of Polysaccharides from *Porphyra haitanensis* Using Ethanol-Ammonium Sulfate Aqueous Two-Phase System. *Food Sci.* **2014**, *35*, 46–49.
31. Sadeghi, R.; Jamehbozorg, B. The salting-out effect and phase separation in aqueous solutions of sodium phosphate salts and poly (propylene glycol). *Fluid Phase Equilib.* **2009**, *280*, 68–75. [CrossRef]
32. Rang, F.J.; Liu, W.; Ouyang, Y. Aqueous Two-phase Extraction Process and Antioxidant Activity of the Polyphenols and Polysaccharides of Physalis alkekengi Fruit. *Food Res. Dev.* **2021**, *42*, 114–120.
33. Hou, Q.Z.; Xin, H.J.; Liu, Z.J. Study on Aqueous Two-phase System Extraction of Flavones and Polysaccharides from Pineapple Peel. *Food Res. Dev.* **2018**, *39*, 72–76.
34. Singleton, V.L.; Rossi, J.A. Colorimetry of Total Phenolics with Phosphomolybdic-Phosphotungstic Acid Reagents. *Am. J. Enol. Vitic.* **1965**, *16*, 144.
35. Zhao, Y.Y.; Wang, Q.Z.; Zhang, J.H.; Feng, X.; Wang, M. Optimization of Extraction Conditions for Phenolic Acid of *Cynara scolymus* L. by Orthogonal Design. *Food Res. Dev.* **2013**, *34*, 15–17.
36. Fernandes, P.A.R.; Le Bourvellec, C.; Renard, C.M.G.C.; Wessel, D.F.; Cardoso, S.M.; Coimbra, M.A. Interactions of arabinan-rich pectic polysaccharides with polyphenols. *Carbohydr. Polym.* **2020**, *230*, 115644. [CrossRef]
37. Renard, C.M.G.C.; Watrelot, A.A.; Le Bourvellec, C. Interactions between polyphenols and polysaccharides: Mechanisms and consequences in food processing and digestion. *Trends Food Sci. Technol.* **2017**, *60*, 43–51. [CrossRef]
38. Ye, Z.W.; Ye, R.; Hao, D.X.; Ren, S.W.; Li, J.P.; Chen, Q. Optimization of polysaccharide extraction from Xi *Pinellia ternate* by response surface methodology and its antioxidant activity. *China Food Addit.* **2022**, *33*, 90–98.
39. Su, Y.L.; Zhang, J.H. Optimization of Ultrasonic-assisted Ethanol Extraction of Polyphenols from Kelp and Their Antioxidant Activities. *Chin. Condiment* **2020**, *45*, 174–180.
40. Wu, Y.H.; Lu, W.J.; Liu, M.H.; Zhang, J.P.; Chen, A.H.; Shao, Y.; Wang, C.K.; Liu, E.Q. Optimization of ultrasonic-assisted aqueous two-phase extraction of burdock polysaccharide by response surface design and its antioxidant activities. *Food Ferment. Ind.* **2020**, *46*, 215–223.
41. Liu, L.L.; Li, L.; Zhai, L.M.; Xiao, B.Z. Ultrasonic—assisted Aqueous Two—phase Extraction of Naringin from Pomelo Peel by Orthogonal Experimental Design. *Sci. Technol. Food Ind.* **2018**, *39*, 159–163, 171.
42. Gong, X.H.; Li, M.C.; Xin, M.H.; Zhao, X.J.; Lu, G.; Xu, J.; Zhao, S.Y. Optimization of the Extraction Technology of Tea Polyphenols from Tea Waste by Aqueous Two—phase Systems with Ionic Liquids as Additives. *Sci. Technol. Food Ind.* **2020**, *41*, 158–166.

43. Majd, M.H.; Rajaei, A.; Bashi, D.S.; Mortazavi, S.A.; Bolourian, S. Optimization of ultrasonic-assisted extraction of phenolic compounds from bovine pennyroyal (Phlomidoschema parviflorum) leaves using response surface methodology. *Ind. Crops Prod.* **2014**, *57*, 195–202. [CrossRef]
44. Sahin, S.; Aybastier, O.; Isik, E. Optimisation of ultrasonic-assisted extraction of antioxidant compounds from Artemisia absinthium using response surface methodology. *Food Chem.* **2013**, *141*, 1361–1368. [CrossRef]
45. Guo, C.; Li, C.; Hou, M.M.; Bai, M.Y.; Zhou, C.C.; Zhang, H.Y. Extraction optimization and its inoxidizability and hypoglycemic properties in vitro of polysaccharide from *Salvia plebeia* R. Br. *Sci. Technol. Food Ind.* **2022**, *43*, 211–219.
46. Li, B.; Lei, Y.; Meng, X.J.; Jiao, X.Y.; Gao, N.X.; Zhao, Y.; Zhang, J.C. Optimization of Ultrasonic-Assisted Extraction of Polyphenols from Haskap Berries (*Lonicera caerulea* L.) Using Response Surface Methodology and Their Antioxidant Capacity. *Food Sci.* **2015**, *36*, 33–39.
47. Song, S.; He, H.; Tang, X.; Wang, W. Determination of Polyphenols and Chlorogenic Acid in Artichoke. *Life Sci. Instrum.* **2007**, *5*, 47–49.
48. Bao, F.; Yuan, H.Z.; Lin, X.; Liu, L.L.; Zhuang, H.; Zhao, D.S. Comparative study on the different parts of artichoke based on the correlation of "components-antioxidant". *Cereals Oils* **2022**, *35*, 142–145.
49. Martinez-Morales, F.; Alonso-Castro, A.J.; Zapata-Morales, J.R.; Carranza-Álvarez, C.; Aragon-Martinez, O.H. Use of standardized units for a correct interpretation of IC50 values obtained from the inhibition of the DPPH radical by natural antioxidants. *Chem. Pap.* **2020**, *74*, 3325–3334. [CrossRef]
50. Garbetta, A.; Capotorto, I.; Cardinali, A.; D'Antuono, I.; Linsalata, V.; Pizzi, F.; Mineruini, F. Antioxidant activity induced by main polyphenols present in edible artichoke heads: Influence of in vitro gastro-intestinal digestion. *Funct. Foods* **2014**, *10*, 456–464. [CrossRef]
51. DuBois, M.; Gilles, K.A.; Hamilton, J.K.; Rebers, P.A.; Smith, F. Colorimetric Method for Determination of Sugars and Related Substances. *Anal. Chem.* **1956**, *28*, 350–356. [CrossRef]
52. Re, R.; Pellegrini, N.; Proteggente, A.; Pannala, A.; Yang, M.; Rice-Evans, C. Antioxidant activity applying an improved ABTS radical cation decolorization assay. *Free. Radic. Biol. Med.* **1999**, *26*, 1231–1237. [CrossRef] [PubMed]
53. Brand-Williams, W.; Cuvelier, M.E.; Berset, C. Use of a free radical method to evaluate antioxidant activity. *LWT Food Sci. Technol.* **1995**, *28*, 25–30. [CrossRef]
54. Benzie, I.F.F.; Strain, J.J. Ferric reducing/antioxidant power assay: Direct measure of total antioxidant activity of biological fluids and modified version for simultaneous measurement of total antioxidant power and ascorbic acid concentration. *Methods Enzymol.* **1999**, *28*, 15–27.

molecules

MDPI

Article

The Detailed Pharmacodynamics of the Gut Relaxant Effect and GC-MS Analysis of the *Grewia tenax* Fruit Extract: In Vivo and Ex Vivo Approach

Najeeb Ur Rehman [1,*], Mohd Nazam Ansari [1], Wasim Ahmad [2] and Mohd Amir [3]

[1] Department of Pharmacology & Toxicology, College of Pharmacy, Prince Sattam Bin Abdulaziz University, Al-Kharj 11942, Saudi Arabia
[2] Department of Pharmacy, Mohammed Al-Mana College for Medical Sciences, Dammam 34222, Saudi Arabia
[3] Department of Natural Products and Alternative Medicine, College of Clinical Pharmacy, Imam Abdulrahman Bin Faisal University, Dammam 31441, Saudi Arabia
* Correspondence: n_rehman5@hotmail.com; Tel.: +966-11-5886035

Abstract: The study was performed to assess and rationalize the traditional utilization of the fruit part of *Grewia tenax* (*G. tenax*). The phytoconstituents present in the methanolic extract were analyzed using Gas-Chromatography-Mass Spectroscopy (GC-MS), while the anti-diarrheal activity was investigated in the Swiss albino mice against castor oil-provoked diarrhea in vivo. The antispasmodic effect and the possible pharmacodynamics of the observed antispasmodic effect were determined in an isolated rat ileum using the organ bath setup as an ex vivo model. GC-MS findings indicate that *G. tenax* is rich in alcohol (6,6-dideutero-nonen-1-ol-3) as the main constituent (20.98%), while 3-Deoxy-d-mannoic lactone (15.36%) was detected as the second major constituents whereas methyl furfural, pyranone, carboxylic acid, vitamin E, fatty acid ester, hydrocarbon, steroids, sesquiterpenes, phytosterols, and ketones were verified as added constituents in the methanolic extract. In mice, the orally administered *G. tenax* inhibited the diarrheal episodes significantly ($p < 0.05$) at 200 mg/kg (40% protection), and this protection was escalated to 80% with the next higher dose of 400 mg/kg. Loperamide (10 mg/kg), a positive control drug, imparted 100% protection, whereas no protection was shown by saline. In isolated rat ileum, *G. tenax* completely inhibited the carbamylcholine (CCh; 1 μM) and KCl (high K^+; 80 mM)-evoked spasms in a concentrations-mediated manner (0.03 to 3 mg/mL) by expressing equal potencies ($p > 0.05$) against both types of evoked spasms, similar to papaverine, having dual inhibitory actions at phosphodiesterase enzyme (PDE) and Ca^{2+} channels (CCB). Similar to papaverine, the inhibitory effect of *G. tenax* on PDE was further confirmed indirectly when *G. tenax* (0.1 and 0.3 mg/mL) preincubated ileal tissues shifted the isoprenaline-relaxation curve towards the left. Whereas, pre-incubating the tissue with 0.3 and 1 mg/mL of *G. tenax* established the CCB-like effect by non-specific inhibition of $CaCl_2$–mediated concentration-response curves towards the right with suppression of the maximum peaks, similar to verapamil, a standard CCB. Thus, the present investigation revealed the phytochemical constituents and explored the detailed pharmacodynamic basis for the curative use of *G. tenax* in diarrhea and hyperactive gut motility disorders.

Keywords: *G. tenax*; GC-MS; anti-spasmodic; Ca^{2++} channel blockers; phosphodiesterase

Citation: Rehman, N.U.; Ansari, M.N.; Ahmad, W.; Amir, M. The Detailed Pharmacodynamics of the Gut Relaxant Effect and GC-MS Analysis of the *Grewia tenax* Fruit Extract: In Vivo and Ex Vivo Approach. *Molecules* **2022**, *27*, 8880. https://doi.org/10.3390/molecules27248880

Academic Editors: Raffaele Pezzani and Sara Vitalini

Received: 16 November 2022
Accepted: 11 December 2022
Published: 14 December 2022

Publisher's Note: MDPI stays neutral with regard to jurisdictional claims in published maps and institutional affiliations.

1. Introduction

Both diarrhea and hyperactive gut are gastrointestinal motility disorders responsible for morbidity and mortality, especially in developing countries [1]. Diarrhea is characterized by three or more episodes of defecation with unusually loose and watery stools. The severity of the diarrhea is scored based on its frequentness, constancy, and mass of the stool [2]. It is usually associated with gastric pain, fecal urgency, perianal discomfort, and bowel incontinence [3]. Diarrhea is not a disease in and of itself but rather a sign and symptom caused by a combination of factors [4,5].

Diarrhea is both preventable and curable. Its treatment is non-specific and is more focused on decreasing the discomfort and inconvenience of frequent bowel motions. Even with a wide range of interventions for the management of diarrhea, three-quarters of the population in developing countries still depend on traditional medicine for essential health care, including the management of diarrhea [6]. The currently used drugs on the market to manage diarrhea have limitations associated with their adverse effects and contraindications. For instance, the occurrence of vomiting, bronchospasm, and fever is associated with the administration of racecadotril and loperamide (used to cure secretory diarrhea). Additionally, loperamide is reported to cause intestinal obstruction in children below 6 years of age and hence is contraindicated in children [7].

A major proportion of the population in underdeveloped and emergent nations relies on the utilization of traditional medicines for the treatment of several disease conditions, including diarrhea [8,9]. Hence, the search for a plant with promising antidiarrheal potential draws lots of attention from researchers. In the traditional system of medicine, multiple medicinal plants have been reported for antidiarrheal potential. However, many of these plants are not backed by scientific evaluation. As a result, scientific research exploring the curative prospects of these herbs is critical to search for new anti-diarrheal agents with distinctive mode(s) of action and with synergistic and/or side-effect neutralizing components.

The genus Grewia (family: Tiliaceae) is distributed in the tropics and sub-tropics globally and contains approximately 150 species. It is also the only member of the family that produces eatable fruits [10]. *Grewia tenax (Forsk.) Fiori* "Guddaim" is a largely valued wild shrub in Sudan, particularly in the Northern and Middle Sudan, majorly distributed in arid and semi-arid zones [11]. The fruit (per 100 g) has a substantial amount of potassium (817 mg) and iron (20.8–22.3 mg), and hence it is given as an iron supplement to treat anemia in children. Compared to other *Grewia* spp., *G. tenax* contains a significant amount of reducing sugar (13.8%) and starch (44.4%) [12]. The composition of *G. tenax* fruit in % is D-fructose (24.3), D-glucose (21.0), starch (15.1), fiber (8.1), protein (6.3), ash (4.5), sucrose (1.6), and fat (0.4) [13]. *G. tenax* fruits are predominant in sugars, starch, tannins, phenolic compounds, flavonoids, steroids, vitamin C, and protein [14,15].

The fruits are sweet and can be consumed whole, in parts, or as juice. Traditionally, the Guddaim juice is usually indicated for breastfeeding women as it promotes their health and lactation. It is considered a good iron source for children less than 8 years of age [16]. It is extensively used for curing several common illnesses, such as diarrhea, dysentery, cough, fever, jaundice, and rheumatism [17]. The seeds and the green leaves are utilized during the animal delivery season [18]. The powder of the fruit is mixed with the milk and is used to cure fractures and bone swelling [19].

Despite multiple reported activities such as antimicrobial, antibacterial, antifungal [20], hepatoprotective and antimalarial [21], to date, to the best of our knowledge, the antidiarrheal and antispasmodic potential of *G. tenax* fruit with detailed mechanistic has not been evaluated. Consequently, the study is directed to evaluate the anti-diarrheal and antispasmodic potential of the *G. tenax* extract using ex vivo and in vivo experimental models, whereas the phytochemical composition was analyzed using the GC-MS technique.

2. Results

2.1. Methanolic Extract Yield (%)

The finally concentrated methanolic extract yield of *G. tenax* fruit was found to be 25.97%.

2.2. GC-MS Phytochemical Profiling

Table 1 presents the detailed phytoconstituents of *G. tenax* fruit, identified by GC-MS. A total of -28 constituents were identified. A typical GC-MS chromatogram is shown in Figure 1. Alcohol (6,6-dideutero-nonen-1-ol-3) was recorded as the main constituent of the methanolic extract (20.98%), while 3-Deoxy-d-mannoic lactone was recorded as the second most abundant occurring constituent (15.36%). Additionally, methyl furfural, pyranone, carboxylic acid, vitamin E, fatty acid ester, hydrocarbon, steroids, sesquiterpenes, phytosterols, and ketones were verified as other additional constituents in the extract. Further details are available in Table 1 and Figure 1.

Table 1. List of phytoconstituents present in methanolic extract of *G. tenax* fruits.

S. No.	Compound Name	RT (Min)	% Area	Nature of Compound
1	1,2,3-Propanetriol	5.420	11.40	Phenol
2	6-Amino-5-nitroso-1H-pyrimidine-2,4-dione	5.974	4.31	Pyrimidinediones
3	4H-Pyran-4-one, 2,3-dihydro-3,5-dihydroxy-6-methyl	6.821	8.27	Pyranone
4	2-Pentanone	7.976	1.85	Ketone
5	5-Hydroxymethylfurfural	8.429	11.01	Hydroxymethylfurfural
6	1-Tetradecene	10.230	0.11	Hydrocarbon
7	1,2-Cyclobutanedicarboxylic acid	11.427	5.72	Carboxylic acid
8	6,6-Dideutero-nonen-1-ol-3	12.334	20.98	Alcohol
9	3-Deoxy-d-mannoic lactone	14.484	15.36	Lactone
10	Hexadecanoic acid, methyl ester	16.473	0.62	Fatty Acid Ester
11	Hexadecanoic acid, ethyl ester	17.138	0.23	Fatty Acid Ester
12	9,12-Octadecadienoic acid (Z,Z)-, methyl ester	18.114	1.50	Fatty Acid Ester
13	9-Octadecenoic acid, methyl ester, (E)-	18.172	1.84	Fatty Acid Ester
14	Octadecenoic acid, methyl ester	18.408	0.49	Fatty Acid Ester
15	Ethyl (9Z,12Z)-9,12-octadecadienoate	18.721	0.17	Fatty Acid
16	Glycidyl palmitate	19.950	0.44	Fatty Acid Ester
17	1,8,11-Heptadecatriene, (Z,Z)-	21.395	0.87	Hydrocarbon
18	Glycidyl oleate	21.430	1.20	Ester
19	15-Hydroxypentadecanoic acid	21.629	0.43	Fatty Acid
20	9,12-Octadecadienoic acid (Z,Z)-, 2-hydroxyl-1-(hydroxymethyl)ethyl ester	23.19	2.05	Fatty Acid Ester
21	9-Octadecenamide, (Z)-	23.880	0.56	Fatty Acid Primary amide
22	Squalene	24.021	1.02	Hydrocarbon
23	2H-1-benzopyran-6-OL, 4-dihydro-2,5,7,8-tetramethyl-2-(4,8,12-trimethyltridecyl)	27.846	3.79	Vitamin E acetate
24	Beta-sitosterol	30.127	2.12	Phytosterol
25	4,4-Dimethyl-5.alpha.-D1-androstane-3.beta	31.105	1.02	Acetate
26	Ergost-5-en-3-ol,(3.beta.)	31.406	1.54	Steroid
27	Stigmasta-4,22-dien-3-one	33.005	0.32	Steroid
28	1,1,6-Trimethyl-3-methylene-2-(3,6,10,13,14–pentamethyl-3-ethenyl-pendadec-4-enye)cyclohexane	33.450	0.35	Sesquiterpenoid
			99.66	

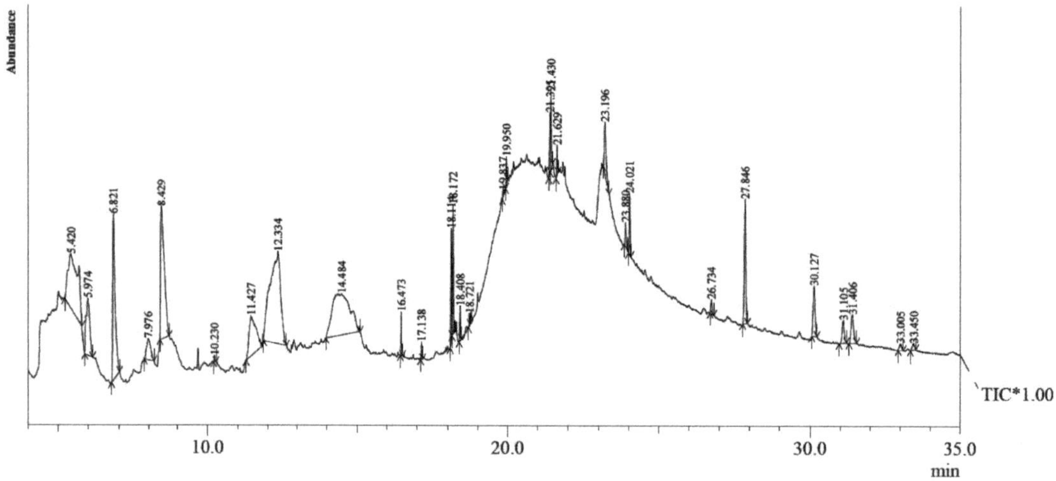

Figure 1. GC-MS total ion chromatogram of *G. tenax* fruits methanolic extract.

2.3. In Vivo Antidiarrheal Effect

Oral administration of *G. tenax* extract at both lower and higher doses in mice imparted marked protection against castor oil-evoked copious diarrhea as compared to the animals of the saline group (Table 2). The mice treated with a lower dose of 200 mg/kg showed 40% protection as two out of five mice did not drop loose and watery stools, while the next higher dose of 400 mg/kg produced 80% protection, whereas the mice treated with loperamide (10 mg/kg) did not show any spots in their cages, suggesting 100% protection against diarrhea (Table 2).

Table 2. Antidiarrheal activity of the methanolic extract of *G. tenax* fruits against castor oil (10 mL/kg)-evoked diarrhea in mice.

Treatment (p.o.), Dose (mg/kg)	No. of Mice with Diarrhea	% Protection
Saline (10 mL/kg) + Castor oil	5/5	0
G. tenax (200 mg/kg) + Castor oil	3 */5	40
G. tenax (400 mg/kg) + Castor oil	1 */5	80
Loperamide (10 mg/kg) + Castor oil	0 **/5	100

* $p < 0.05$ and ** $p < 0.01$ vs. Saline + Castor oil treated group (χ^2-test).

2.4. Ex Vivo Antispasmodic Effects

G. tenax resulted in complete relaxation of both types of CCh and high K^+ mediated spasms in isolated rat ileum, with resultant EC_{50} values of 1.22 mg/mL (0.98–1.58, 95% CI, n = 4) and 1.34 mg/mL (1.04–1.36, 95% CI, n = 4) respectively (Figure 2A). Similarly, comparable relaxant effects were shown by the standard drug, papaverine, with resultant EC_{50} values 8.28 µM (7.86–8.94, 95% CI, n = 4) and 8.64 µM (8.12–8.72, 95% CI, n = 4), (Figure 2B). On the other hand, verapamil showed significantly higher potency against high K^+ compared to CCh-mediated spasm, with EC_{50} values of 0.18 µM (0.14–0.21, 95% CI, n = 4) and 2.42 µM (2.14–2.85, 95% CI, n = 4), respectively (Figure 2C).

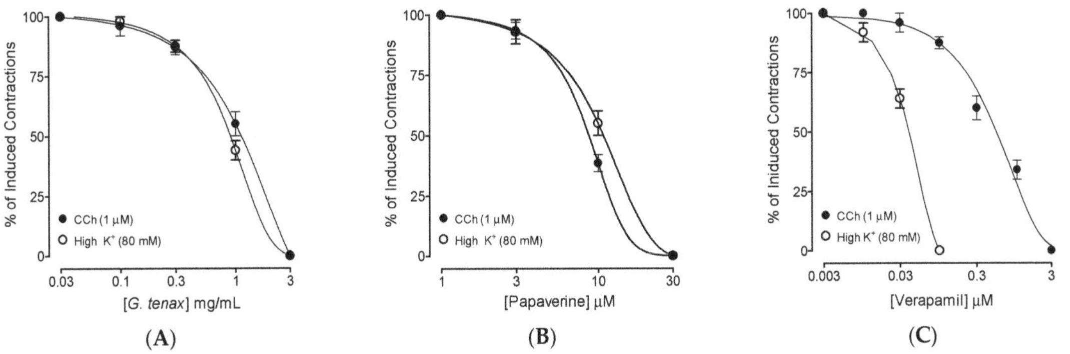

Figure 2. Concentration-response curves presenting a comparison of the inhibitory effect of (**A**) methanolic extract of *Grewia tenax* fruits (*G. tenax*), (**B**) papaverine, and (**C**) verapamil against carbachol (CCh; 1 μM) and high K^+ (80 mM)-evoked contractions in isolated rat ileum preparations. Values are presented as mean ± SEM, $n = 4$.

2.5. Phosphodiesterase Enzyme (PDE) Inhibitory-like Effect

The possible PDE-inhibitory effect of the extract was established by pre-treating the tissue with *G. tenax* (0.1 and 0.3 mg/mL), which resulted in the shift of isoprenaline-induced inhibitory CRCs towards the left (Figure 3A), thus depicting its potentiating effect. With the positive control drug, papaverine (1 and 3 μM) similar leftward isoprenaline curve shift was observed (Figure 3B), whereas verapamil did not show a potentiating effect (Figure 3C).

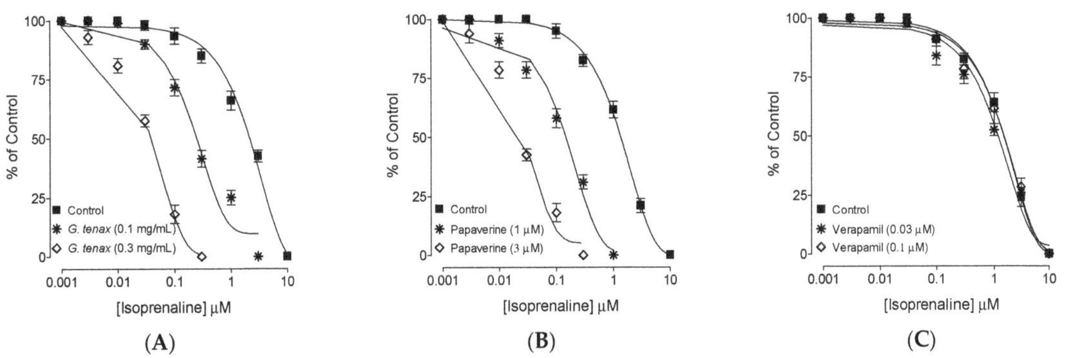

Figure 3. Inhibitory concentration-response curve of isoprenaline against carbachol (CCh)-induced contraction in the absence and presence of various concentrations of (**A**) methanolic extract of *Grewia tenax* fruit (*G. tenax*), (**B**) papaverine, and (**C**) verapamil in isolated rat ileum preparations. Values are expressed as mean ± SEM, $n = 4$.

2.6. Calcium Channel Blocking (CCB)-like Effect

The Ca^{2+} inhibitory effect of *G. tenax* methanolic extract was confirmed by preincubating ileum tissue with extract (0.3 and 1 mg/mL), which resulted in Ca^{2+} CRCs curve shifts towards the right with suppression of the maximum response (Figure 4A), similar to verapamil (0.01 and 0.03 μM; Figure 4B). Papaverine (1 and 3 μM) also displaced the Ca^{2+} curve towards the right, as shown in Figure 4C.

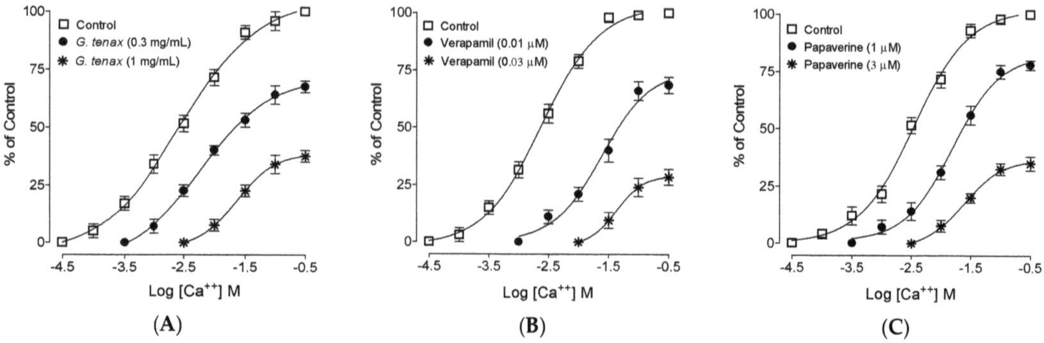

Figure 4. Concentration-response curves of Ca^{2+} in the absence and presence of the increasing concentrations of the (**A**) methanolic extract of *Grewia tenax* fruits (*G. tenax*), (**B**) verapamil, and (**C**) papaverine in isolated rat ileum preparations. Values shown are mean ± SEM, $n = 5$.

3. Discussion

The present study aimed to test the traditional claim of the plant extract of *G. tenax* in diarrhea and hyperactive gut disorders and to further explore the detailed pharmacodynamics for the observed antispasmodic effect in rodents, whereas the phytochemical constituents were determined using the GC-MS technique. The extract of *G. tenax* was tested against castor-oil-evoked diarrhea, where it showed a dose-mediated antidiarrheal effect by inhibiting diarrheal drops, compared to the animals of the saline-treated control group in which copious diarrhea was observed in all the animals. Castor oil-induced diarrhea model is a well-established assay to test the unknown samples for antidiarrheal potential where it evokes diarrhea in animals as it gets hydrolyzed in the gut into ricinoleic acid, causing severe irritation in the bowel and thus leading to spasms in the gut [22]. Pretreatment of the *G. tenax* extract in mice protected them against diarrhea in a dose-mediated manner similar to loperamide, a positive control drug. For evaluation of possible pharmacodynamics involved in the observed diarrhea protection, the methanolic extract of *G. tenax* in cumulative concentration was assessed in isolated rat ileum, considered a standard assay for screening of antispasmodic effect [23]. Based on our previously reported studies, we observed that antispasmodic drugs usually act by blocking Ca^{2+} channels [24,25] and/or PDE inhibition [26]. Based on these reports, we tested *G. tenax* extract in rat ileum against CCh and high K^+-induced contractions [27]. Interestingly, *G. tenax* inhibited both types of contraction with comparable inhibitory CRCs patterns against CCh and high K^+.

Likewise, papaverine, a dual inhibitor of Ca^{2+} channels and PDE-enzyme [28], also inhibited CCh and high K^+ mediated spasms at comparable concentrations, whereas verapamil, a standard CCB [29,30], selectively showed increased potency against high K^+ compared to CCh, a typical characteristic of Ca^{2+} channel blocker. Therefore, like papaverine, *G. tenax* exhibits dual inhibitory activity against PDE and Ca^{2+} channels. The PDE-inhibitory-like effect of *G. tenax* was further authenticated indirectly when the pre-incubated ileal tissues with increasing concentrations of *G. tenax* deflected and potentiated the isoprenaline-mediated CRCs towards the left, and this potentiation is most probably considered the resulted increase in cAMP levels in the tissue [31] Phosphodiesterase enzymes converts cAMP (active form) to AMP (inactive form); thus any substance that has an inhibitory effect on PDE will cause an increase in cAMP and will thus produce relaxation of smooth muscle [32]. A similar type of potentiation of inhibitory CRCs of isoprenaline towards lower doses (leftward) was seen in pre-incubated tissues with papaverine, a standard PDE inhibitor [31]. Our current findings are further supported by the previous studies that reported CCh-mediated smooth muscle spasms inhibited by PDE inhibitors [33,34].

High K^+ (>30 mM) produces contraction by depolarizing the tissue via activation of voltage-gated L-type Ca^{2+} channels, and a substance that reverses this contraction is known to possess CCBs-like activity [35]. Hence, the CCB-like action of *G. tenax* was assumed when it inhibited K^+-80 mM-induced contractions, and this CCB-like effect was further confirmed when preincubation of ileum tissues with increasing concentrations of the extract in Ca^{2+} free tissues caused rightward deflection of Ca^{2+}-CRC with suppression of the maximum peak similar to verapamil, a standard CCB, thus confirming the CCB-like effect. The phytoconstituents of *G. tenax* fruit methanolic extract identified by GC-MS show 28 constituents, and their % yield showed alcohol (6,6-dideutero-nonen-1-ol-3) as the main constituent of the methanolic extract (20.98%), whereas, 3-Deoxy-d-mannoic lactone (15.36%) was recognized as the next major constituents. Methyl furfural, pyranone, carboxylic acid, vitamin E, fatty acid ester, hydrocarbons, steroids, sesquiterpenes, phytosterols, and ketones were verified as other constituents in the extract.

4. Materials and Methods

4.1. Extraction of Plant Material

The fruits of *G. tenax* were collected from the local market of Dammam (Saudi Arabia) and authenticated with a voucher specimen of PL/045/2020-21/P-011 by the Department of Pharmacognosy, College of Clinical Pharmacy, Taif University, Saudi Arabia. A total of 40 gm of dried fruit powder was extracted using a soxhlet unit using 200 mL of methanol, and the concentrated extra was kept at 5–10 °C in an airtight glass container until further usage.

The % yield of the extract was calculated using the given formula. The reported methods were used for phytochemical investigations of the extract [36,37].

$$\%\ \text{Extraction yield} = \frac{\text{Weight of dried extract}}{\text{Weight of drug sample}} \times 100$$

4.2. Chemicals

Carbamylcholine (carbachol; CCh), loperamide, acetylcholine perchlorate (Ach), isoprenaline, verapamil, and papaverine were obtained from Sigma Company, St. Louis, MO, USA. Salts for a physiological salt solution such as potassium chloride (Sigma Co.), calcium chloride, glucose, magnesium sulfate, potassium dihydrogen phosphate, sodium bicarbonate, and sodium chloride were obtained from Merck, Germany. All the chemicals were of analytical grade. Castor oil was procured from a nearby pharmacy.

4.3. Animals

Rats (200–250 g) for ex-vivo and Swiss albino mice (30–35 g) for in vivo experiments were procured from the Animal Care Unit, College of Pharmacy, Prince Sattam bin Abdulaziz University, Saudi Arabia. They were kept in optimum conditions of temperature (22 ± 1 °C), relative humidity (55 ± 5 °C), and balanced light/dark cycle exposure. All the animals were fed, including a standard pellet diet and water *ad libitum*. Experimental animals, i.e., rats, were kept on fasting for 24 h before performing ex vivo experiments, and cervical dislocation was used after light anesthesia. The NRC instructions were followed while conducting the experiments on animals [38]. The approval of the study protocol was obtained by the Bio-Ethical Research Committee (BERC) at Prince Sattam Bin Abdulaziz University with the reference number BERC-004-12-19.

4.4. GC-MS Analysis

Gas-chromatography mass spectroscopy technique was employed for the phytochemical investigation of the methanolic extract of *G. tenax*. Phytoconstituents separation was attained on Agilent GCMS (Agilent Technologies, Santa Clara, CA, USA) the capillary column 60 M TRX 5-MS (30 m × 250 μm film) by utilizing a 2 μL aliquot of sample. The oven temperature program was as follows: 80 °C initially for 3 min and then ramped at a rate of 10 °C/min to 280 °C for 19 min. Helium was used as a carrier gas with a flow rate of 1.21 mL/min. The set temperature for the injector and source was 260 °C and 220 °C, respectively. The electron ionization energy system was employed with 70 eV at a multiplier voltage of 380 within the m/z range.

Individual phytoconstituents were identified by referring to the Institute of Standards and Technologies (NIST) libraries [37,39].

4.5. In Vivo Antidiarrheal Study

The assessment of the possible potential of the extract to protect mice from diarrhea was done according to the previously described method with a slight modification [40]. A random grouping of twenty mice was followed with an equal number of mice into four groups, and the mice were kept on fasting for 24 h. The Group 1 mice were treated with normal saline (10 mL/kg, oral) and categorized as a negative control group. The mice of the tested group 2 and 3 were administered with the 2 increasing doses of *G. tenax* 200 and 400 mg/kg, respectively. On the basis of an acute toxicity test conducted in mice, *G. tenax* extract was found safe at 4.0 g/kg, and thus 1/20th and 1/10th (200 and 400 mg/kg, respectively) were chosen for the possible antidiarrheal effect. Group 4 was categorized as a positive control group and administered 10 mg/mL of loperamide. Each mouse was placed in a separate cage with a blotting paper placed at the base of the cage to identify and mark the absence and presence of diarrhea by a blind observer. All the experimental animals were administered 10 mL/kg of castor oil orally, using a 1 mL syringe after 1 h for the respective groups of saline, extract, and loperamide treatment. All the blotting papers in each cage were examined after 4 h for typical diarrheal dropping. The protection was observed if the blotting paper lacked the diarrheal drops, as reported by Rehman et al. [41].

4.6. Ex Vivo Experiments on Isolated Rat Ileum

The rats were sacrificed, and their ileum was isolated as per the method of Rehman et al. [42]. Further, 2–3 cm ileum tissue segments were freed from adjoining tissues and fecal material. The mounting of the ileum was then done with emkaBath (France), fitted with a transducer and IOX software. The tissue bath (20 mL) was filled with fresh tyrode solution and gassed with carbogen while the temperature was maintained at 37 °C. The composition of the concentration (mM) of Tyrode's solution was as follows: NaCl 136.9, KCl 2.68, $MgCl_2$ 1.05, $NaHCO_3$ 11.90, $CaCl_2$ 1.8, NaH_2PO_4 0.42, and Glucose 5.55 (pH 7.4). To apply a tension of 1 g to the tissue, the transducer knob was rotated clockwise, and for stabilization, it was left for 30 min with several exposures of acetylcholine (0.3 μM). After stabilization, the tissue was exposed to spasmogens; CCh and high K^+ to evoke sustained contraction. Afterward, the *G. tenax* and standard drugs, from low to higher concentrations, were added to the bath solution till complete inhibition was observed. Once the tested samples' relaxant effects against CCh and high K^+-mediated contractions were found, further experiments were conducted to explore the detailed mechanism(s) of *G. tenax* on inhibitory effects on voltage-gated Ca^{2+} channels and/or PDE inhibition by following the earlier assays reported by Godfrained et al. [43]. When multiple smooth muscles are exposed to K^+ (>30 mM), it will cause activation of the L-type C^{++} channel, leading to depolarization, and hence sustained contractions are observed in smooth muscles. Moreover, any test material that will inhibit sustained contraction induced by CCh and high K^+ with similar potencies is labeled as a PDE inhibitor [44].

4.7. Ca^{2+} Inhibitory Confirmation

Once in the preliminary experiments, the plant extract relaxant effects against high K^+ were observed, further confirmation of Ca^{2+} channel blockade (CCB) was determined by making tissues Ca^{2+} free by 45 min incubation in Ca^{2+}-free Tyrode's solution with a chelating agent (EDTA, 0.1 mM). To further deplete the intracellular stores of Ca^{2+}, Tyrode solution without Ca^{2+} was replaced with K^+-rich and Ca^{2+}-free Tyrode's solution with the following concentrations (mM): potassium chloride: 50, sodium chloride: 91.03, sodium dihydrogen phosphate dehydrates: 0.32, sodium bicarbonate: 11.9, magnesium chloride hexahydrate: 0.50, glucose: 5.05. At the end of 45 min incubation period of ileal tissue in this solution, cumulative Ca^{2+} CRCs in the absence and presence of increasing concentrations of *G. tenax* were constructed by adding exogenous $CaCl_2$ in the bath, and the obtained CRCs of the plant and/or papaverine was compared with the standard CCB agent, verapamil [30].

4.8. PDE Inhibitory Confirmation

The relaxant effect of *G. tenax* against CCh and high K^+ at similar potencies is an indication of PDE inhibition [45]; therefore, indirect confirmation for PDE inhibition by *G. tenax* was evaluated by constructing dose-mediated inhibitory CRCs of isoprenaline against CCh in the absence (control) and presence of *G. tenax*. A leftward shift in the inhibitory curves indicates potentiation and is considered the plant/test material inhibitory potential on PDE [46].

4.9. Statistical Analysis

The obtained results are mentioned as the mean ± standard error of the mean (SEM), and "*n*" represents the number of the experiment. The examination of median effective concentration (EC50) was done with a 95% (CI) Confidence interval. The multiple comparisons of CRCs with respective controls were done by applying test mainly Student's test or two-way ANOVA followed by Bonferroni's post-test. Additionally, the Chi-square (χ^2) test was used to evaluate protection from diarrhea, and a comparison was made between all groups and the saline control group. $p < 0.05$ was considered statistically significant. For regression analysis, GraphPad Prism (Version 4) was employed.

5. Conclusions

The methanolic extract of *G. tenax* protected mice in a dose-mediated way from castor oil-induced diarrhea and also showed concentration-dependent anti-spasmodic effects tested in isolated rat ileum. The spasmolytic effect observed against both CCh and high K^+ at similar concentrations indicates dual inhibitory effects of *G. tenax* against PDE enzymes and Ca^{2+} channels, similar to papaverine. The PDE inhibition was further confirmed by the leftward shift of isoprenaline inhibitory CRCs, whereas the rightward deflection of Ca^{2+} CRCs towards the right authenticated the CCB-like effect of *G. tenax*. GC-MS analysis shows the identification of 28 phytoconstituents in the extract of *G. tenax* fruits where 6,6-dideutero-nonen-1-ol-3 (20.98%) was recorded as the main constituent and 3-Deoxy-d-mannoic lactone (15.36%) as a second major constituent in addition to the methyl furfural, pyranone, carboxylic acid, vitamin E, fatty acid ester, hydrocarbon, steroids, sesquiterpenes, phytosterols, and ketones. Thus, this study provides a rationale for the medicinal use of *G. tenax* fruits in addressing gastrointestinal hyperactivity-related disorders with detailed pharmacodynamics explored. We recommend further in-depth molecular assays to develop *G. tenax* for clinical use in the future.

Author Contributions: Conceptualization, N.U.R., M.N.A. and W.A.; methodology, N.U.R., M.A., W.A. and M.N.A.; software, N.U.R., W.A. and M.N.A.; formal analysis and investigation, N.U.R., M.A. and M.N.A.; resources, N.U.R., M.A., W.A. and M.N.A.; writing—original draft preparation, N.U.R., W.A. and M.N.A.; writing—review and editing, N.U.R., M.A., W.A. and M.N.A.; supervision, N.U.R. and M.N.A.; project administration, N.U.R. and M.N.A.; funding acquisition, N.U.R. and M.N.A. All authors have read and agreed to the published version of the manuscript.

Funding: The authors extend their appreciation to the Deputyship for Research & Innovation, Ministry of Education in Saudi Arabia, for funding this research work through project number (IF2/PSAU/2022/03/20431).

Institutional Review Board Statement: This statement is not applicable to our study. Moreover, for animal ethical approval, we already mentioned it in Section 4.3. Animals.

Informed Consent Statement: Not Applicable.

Data Availability Statement: Not Applicable.

Conflicts of Interest: The authors declare no conflict of interest.

Sample Availability: Samples of the compounds are available from the authors.

References

1. Haque, M.A.; Abdullah, C.S.; Romana, B.; Rafique, M.B.; Zia-ulHuda, G.M.; Hossain, S.F.; Begum, B. Evaluation of anti-diarrheal and anti-diabetic activities of the stem, barks and leaves of the plant *Vernonia cinerea* (family: Asteraceae). *J. Appl. Pharm. Sci.* **2013**, *3*, 69–72.
2. Sinan, K.I.; Chiavaroli, A.; Orlando, G.; Bene, K.; Zengin, G.; Cziáky, Z.; Jekő, J.; Mahomoodally, M.F.; Picot-Allain, M.; Menghini, L.; et al. Biopotential of *Bersama abyssinica* fresen stem bark extracts: UHPLC profiles, antioxidant, enzyme inhibitory, and antiproliferative propensities. *Antioxidants* **2020**, *9*, 163. [CrossRef] [PubMed]
3. Dinesh, M.R. An investigation in to the anti-diarrheal property of monteleukast. *Int. J. Pharm. Sci.* **2014**, *6*, 147–148.
4. Sweetser, S. Evaluating the patient with diarrhea: A case-based approach. *Mayo Clin. Proc.* **2012**, *87*, 596–602. [CrossRef]
5. Wansi, S.L.; Nguelefack-Mbuyo, E.P.; Nchouwet, M.L.; Miaffo, D.; Nyadjeu, P.; Wabo, J.P.; Mbiantcha, M.; NKengEfouet, P.A.; Nguelefack, T.B.; Kamanyi, A. Antidiarrheal activity of aqueous extract of the stem bark of *Sapium ellipticum (Euphorbiaceae). Trop. J. Pharm. Res.* **2014**, *13*, 929–935. [CrossRef]
6. Alam, A.; Rehman, N.U.; Ansari, M.N.; Palla, A.H. Effects of essential oils of *Elettaria cardamomum* grown in India and Guatemala on Gram-negative bacteria and gastrointestinal disorders. *Molecules* **2021**, *26*, 2546. [CrossRef]
7. Tadesse, W.T.; Hailu, A.E.; Gurmu, A.E.; Mechesso, A.F. Experimental assessment of antidiarrheal and antisecretory activity of 80% methanolic leaf extract of *Zehneria scabra* in mice. *BMC Complement. Altern. Med.* **2014**, *14*, 460. [CrossRef]
8. Ansari, M.N.; Bhandari, U. Effect of an Ethanol Extract of *Embelia ribes* Fruits on Isoproterenol-Induced Myocardial Infarction in Albino Rats. *Pharm. Biol.* **2008**, *46*, 928–932. [CrossRef]
9. Umer, S.; Tekewe, A.; Kebede, N. Antidiarrhoeal and antimicrobial activity of *Calpurnia aurea* leaf extract. *BMC Complement. Altern. Med.* **2013**, *13*, 21. [CrossRef]
10. Ullah, W.; Uddin, G.; Siddiqui, B.S. Ethnic uses, pharmacological and phytochemical profile of genus *Grewia*. *J. Asian Nat. Prod. Res.* **2012**, *14*, 186–195. [CrossRef]
11. Gebauer, J.; El-Siddig, K.; El-Tahir, B.A.; Salih, A.A.; Ebert, G.; Hammer, K. Exploiting the potential of indigenous fruit trees: *Grewia tenax* in Sudan. *Genet. Resour. Crop Evol.* **2007**, *54*, 1701–1708. [CrossRef]
12. Elhassan, G.M.; Yagi, S.M. Nutritional composition of Grewia species (*Grewia tenax* (Forsk.) Fiori, *G. flavescens* Juss and *G. villosa* Willd) fruits. *Adv. J. Food Sci. Technol.* **2010**, *2*, 159–162.
13. Abdelmutti, O.M.S. Biochemical of Nutritional Evaluation of Famine Food of Sudan. Doctor Dissertation, University of Khartoum, Khartoum, Sudan, 1991.
14. Gupta, M.K.; Sharma, P.K.; Ansari, S.H.; Lagarkha, R. Pharmacognostical evaluation of *Grewia asiatica* fruits. *Int. J. Plant. Sci.* **2006**, *1*, 249.
15. Sharma, N.; Patni, V. *Grewia tenax* (frosk.) Fiori.—A traditional medicinal plant with enormous economic prospective. *Asian J. Pharm. Clin. Res.* **2012**, *5*, 28–32.
16. Kehlenbeck, K.; Asaah, E.; Jammadass, R. Diversity of indigenous fruit trees and their contribution to nutrition and livelihoods in sub Saharan Africa: Examples from Kenya and Cameroon. In *Diversifying Food and Diets: Using Agricultural Biodiversity to Improve Nutrition and Health Issues in Agricultural Biodiversity*; Fanzo, J., Hunter, D., Borelli, T., Mattei, F., Eds.; Earthscan: London, UK, 2013; pp. 257–269.
17. Goyal, P.K. Phytochemical and pharmacological properties of the genus Grewia: A review. *Int. J. Pharm. Pharm. Sci.* **2012**, *4*, 72–78.
18. Aboagarib, E.A.A.; Yang, R.; Hua, X.; Siddeeg, A. Chemical compositions, nutritional properties and volatile compounds of Guddaim (*Grewia tenax.* Forssk) Fiori fruits. *J. Food Nutr. Res.* **2014**, *2*, 187–192. [CrossRef]
19. Shekhawat, D.; Batra, A. House hold remedies of Keshavraipatan tehsil in Bundi district, Rajasthan. *Indian J. Tradit. Knowl.* **2006**, *5* (Suppl. 3), 362–367.
20. Karim, A.M.; Azim, A.I.; Sufian, A. GC-MS analysis and antimicrobial activity of sudanese *Grewia tenax* forssk (tiliaceae) fixed oil. *World J. Pharm. Sci.* **2019**, *5*, 37–41.
21. Saleh, I.A.; Shams, K.A.; Tawfik, W.A.; Habib, A.A.; Hassan, R.A.; Shahat, A.A.; Aboutab, E.A.; Hammouda, F.M.; Abdel-azim, N.S.; Investigation of the lipid and carbohydrate contents of *Grewia tenax* forssk. Fruits & evaluation of hepatoprotection activity. *Int. J. Pharm. Sci.* **2015**, *7*, 179–182.
22. Croci, T.; Landi, M.; Elmonds-Alt, X.; Le-Fur, G.; Maffrand, J.P.; Manara, L. Role of tachykinins in castor oil-induced diarrhoea in rats. *Br. J. Pharmacol.* **1997**, *121*, 375–380. [CrossRef]
23. Iwao, I.; Terada, Y. On the mechanism of diarrhea due to castor oil. *Jpn. J. Pharmacol.* **1962**, *12*, 137–145. [CrossRef]
24. Palla, A.; Gilani, A.H.; Bashir, S.; Rehman, N.U. Multiple mechanisms of Flaxseed-effectiveness in Inflammatory Bowel disease. *Evid.-Based Complement. Altern. Med.* **2020**, *2020*, 7974835. [CrossRef] [PubMed]
25. Rehman, N.U.; Gilani, A.H.; Khan, A.; Nazneen, M.; El Gamal, A.A.; Fawzy, G.A.; Al-Ati, H.Y.; Abdel-kader, M.S. Antidiarrheal and Antispasmodic Activities of *Buddleja polystachya* are Mediated Through Dual Inhibition of Ca^{++} Influx and Phosphodiesterase Enzyme. *Phytother. Res.* **2015**, *29*, 1211–1218. [CrossRef] [PubMed]

26. Khan, M.; Khan, A.U.; Rehman, N.U.; Gilani, A.H. Pharmacological basis for medicinal use of *Lens culinaris* in gastrointestinal and respiratory disorders. *Phytother. Res.* **2014**, *28*, 1349–1358. [CrossRef] [PubMed]
27. Imam, F.; Rehman, N.U.; Ansari, M.N.; Qamar, W.; Afzal, M.; Al-Harbi, K.S. Effect of Roflumilast in airways disorders via dual inhibition of Phosphodiesterase and Ca^{2+}-channel. *Saudi Pharm. J.* **2020**, *28*, 698–702. [CrossRef]
28. Ansari, M.N.; Rehman, N.U.; Karim, A.; Bahta, T.; Abujheisha, K.Y.; Ahamad, S.R.; Imam, F. Evaluation of Bronchodilator and antimicrobial activities of *Otostegia fruticose*: A multi-mechanistic approach. *Saudi Pharm. J.* **2020**, *28*, 281–289. [CrossRef]
29. Fleckenstein, A. Specific pharmacology of Ca^{++} in myocardium, cardiac pacemakers and vascular smooth muscle. *Rev. Pharmacol. Toxicol.* **1977**, *17*, 149–166. [CrossRef]
30. Downie, J.W.; Twiddy, D.A.; Awad, S.A. Antimuscarinic and non-competitive antagonist properties of dicyclomine hydrochloride in isolated human and rabbit bladder muscle. *J. Pharmacol. Exp. Ther.* **1977**, *201*, 662–668.
31. Choo, L.K.; Mitchelson, F. Antagonism of cholinomimetics by troxypyrrolidinium in guinea-pig atria and longitudinal ileal muscle: Comparison with hemicholinium-3. *Eur. J. Pharmacol.* **1978**, *52*, 313–322. [CrossRef]
32. Boswell-Smith, V.; Spina, D.; Page, C.P. Phosphodiesterase inhibitors. *Br. J. Pharmacol.* **2006**, *147* (Suppl. 1), S252–S257. [CrossRef]
33. Uddin, S.B.; Mahbub-Uz-Zaman, M.; Akhtar, R.; Ahmed, N.U. Antidiarrheal activity of ethanolic bark extract of *Mitragyna diversifolia*. *Bangladesh J. Pharmacol.* **2009**, *4*, 144–146.
34. Kaneda, T.; Takeuchi, Y.; Matsui, H.; Shimizu, K.; Urakawa, N.; Nakajyo, S. Inhibitory mechanism of papaverine on carbachol-induced contraction in bovine trachea. *J. Pharmacol. Sci.* **2005**, *98*, 275–282. [CrossRef] [PubMed]
35. Hamilton, T.C.; Weir, S.W.; Weston, A.H. Comparison of the effects of BRL 34915 and verapamil on electrical and mechanical activity in rat portal vein. *Br. J. Pharmacol.* **1986**, *88*, 103–111. [CrossRef]
36. Ahmad, W.; Tamboli, E.T.; Ali, A.; Amir, M.; Zaidi, S.M.A.; Ahmad, S. *Didymocarpous pedicellatus* R. Br.: Qualitative and Quantitative GCMS Approach for Quality Control in Traditional Poly-herbal Formulation with In vitro Antioxidant and Antimicrobial Activity. *Orient. J. Chem.* **2019**, *35*, 648–657. [CrossRef]
37. Ahmad, W.; Parveen, R.; Mujeeb, M.; Zaidi, S.M.A. Comparative Fingerprint Profiling of Unani Polyherbomineral (Safoof-e-Pathar Phori) Formulation by HPTLC, HPLC, and GC-MS. *J. AOAC Int.* **2020**, *103*, 659–668. [CrossRef]
38. National Research Council. *Guide for the Care and Use of Laboratory Animals*; National Academy Press: Washington, DC, USA, 1996; pp. 1–7.
39. Khan, W.; Chester, K.; Anjum, V.; Ahmad, W.; Ahmad, S.; Narwaria, A.; Katiyar, C.K. Chromatographic profiling of Pancharishta at different stages of its development using HPTLC, HPLC, GC–MS and UPLC–MS. *Phytochem. Lett.* **2017**, *20*, 391–400. [CrossRef]
40. Rehman, N.U.; Ansari, M.N.; Ahmad, W.; Amir, M. GC–MS Analysis and *In Vivo* and *Ex Vivo* Antidiarrheal and Antispasmodic Effects of the Methanolic Extract of *Acacia nilotica*. *Molecules* **2022**, *27*, 2107. [CrossRef]
41. Rehman, N.U.; Ansari, M.N.; Ahmad, W.; Ahamad, S.R. Dual Inhibition of Phosphodiesterase and Ca^{++} Channels Explains the Medicinal Use of *Balanites aegyptiaca* (L.) in Hyperactive Gut Disorders. *Plants* **2022**, *11*, 1183. [CrossRef]
42. Rehman, N.U.; Ansari, M.N.; Samad, A. In Silico, Ex Vivo and In Vivo Studies of Roflumilast as a Potential Antidiarrheal and Antispasmodic agent: Inhibition of the PDE-4 Enzyme and Voltage-gated Ca^{++} ion Channels. *Molecules* **2020**, *25*, 1008. [CrossRef]
43. Godfraind, T.; Miller, R.; Wibo, M. Calcium antagonism and calcium entry blockade. *Pharmacol. Rev.* **1986**, *38*, 321–416.
44. Gilani, A.H.; Khan, A.; Subhan, F.; Khan, M. Antispasmodic and bronchodilator activities of St. John's wort are putatively mediated through dual inhibition of calcium influx and phosphodiesterase. *Fundam. Clin. Pharmacol.* **2005**, *19*, 695–705. [CrossRef] [PubMed]
45. Hsu, Y.T.; Liao, G.; Bi, X.; Oka, T.; Tamura, S.; Baudry, M. The PDE10A inhibitor, papaverine, differentially activates ERK in male and female rat striatal slices. *Neuropharmacology* **2011**, *61*, 1275–1281. [CrossRef] [PubMed]
46. Rang, H.P.; Dale, M.M.; Ritter, J.M. *Pharmacology*, 4th ed.; Churchill Livingstone: New York, NY, USA, 1999; pp. 289–290.

molecules

Review

Plant Spices as a Source of Antimicrobial Synergic Molecules to Treat Bacterial and Viral Co-Infections

Nathália Barroso Almeida Duarte [1,2] and Jacqueline Aparecida Takahashi [2,*]

1 Department of Food Science, Faculty of Pharmacy, Universidade Federal de Minas Gerais, Av. Antônio Carlos, 6627, Belo Horizonte CEP 31270-901, Brazil
2 Chemistry Department, Universidade Federal de Minas Gerais, Av. Antônio Carlos, 6627, Belo Horizonte CEP 31270-901, Brazil
* Correspondence: jacqueline@ufmg.br

Abstract: The COVID-19 pandemic exposed the lack of antiviral agents available for human use, while the complexity of the physiological changes caused by coronavirus (SARS-CoV-2) imposed the prescription of multidrug pharmacotherapy to treat infected patients. In a significant number of cases, it was necessary to add antibiotics to the prescription to decrease the risk of co-infections, preventing the worsening of the patient's condition. However, the precautionary use of antibiotics corroborated to increase bacterial resistance. Since the development of vaccines for COVID-19, the pandemic scenario has changed, but the development of new antiviral drugs is still a major challenge. Research for new drugs with synergistic activity against virus and resistant bacteria can produce drug leads to be used in the treatment of mild cases of COVID-19 and to fight other viruses and new viral diseases. Following the repurposing approach, plant spices have been searched for antiviral lead compounds, since the toxic effects of plants that are traditionally consumed are already known, speeding up the drug discovery process. The need for effective drugs in the context of viral diseases is discussed in this review, with special focus on plant-based spices with antiviral and antibiotic activity. The activity of plants against resistant bacteria, the diversity of the components present in plant extracts and the synergistic interaction of these metabolites and industrialized antibiotics are discussed, with the aim of contributing to the development of antiviral and antibiotic drugs. A literature search was performed in electronic databases such as Science Direct; SciELO (Scientific Electronic Library Online); LILACS (Latin American and Caribbean Literature on Health Sciences); Elsevier, SpringerLink; and Google Scholar, using the descriptors: antiviral plants, antibacterial plants, coronavirus treatment, morbidities and COVID-19, bacterial resistance, resistant antibiotics, hospital-acquired infections, spices of plant origin, coronaviruses and foods, spices with antiviral effect, drug prescriptions and COVID-19, and plant synergism. Articles published in English in the period from 2020 to 2022 and relevant to the topic were used as the main inclusion criteria.

Keywords: bacterial resistance; combination therapy; COVID-19; medicinal plants; new drugs; synergism

Citation: Duarte, N.B.A.; Takahashi, J.A. Plant Spices as a Source of Antimicrobial Synergic Molecules to Treat Bacterial and Viral Co-Infections. *Molecules* **2022**, 27, 8210. https://doi.org/10.3390/molecules27238210

Academic Editors: Raffaele Pezzani and Sara Vitalini

Received: 28 October 2022
Accepted: 22 November 2022
Published: 25 November 2022

1. Introduction

The approval of the first antiviral drug for human use (idoxuridine) occurred in 1963 and, despite all scientific and technological development, the list of antiviral drugs available as medicines is still insufficient [1]. Research in this area is necessary, especially because viruses use the host cells to multiply and antiviral substances usually cause multiple side effects [2]. Many viruses are endemic in various parts of the world, such as Ebola, HIV, hepatitis B, dengue, zika, flu, and, more recently, coronavirus (SARS-CoV-2). Vaccination reduced the spread of pandemic and endemic viral diseases, but a relevant number of individuals are still infected by viruses daily. Although antiviral drugs can be useful in the treatment of some viruses, the development of less toxic effective drugs is desirable, especially considering the 2019 coronavirus pandemic scenario, which exposed the lack of effective and safe antiviral drugs and the need for advances in research to develop new drugs with faster results and less side effects. Remdesivir prescribed to severely ill patients resulted in faster recovery; however, for higher effectiveness, anti-inflammatory or immunomodulators such as baricitinib, tocilizumab, and bamlanivimab must be co-administrated to target both viral proliferation and the hyperimmune response [3]. Remdesivir's safety and scope of action is still the target of several clinical studies, although some side effects such as cardiotoxicity have already been reported in an in vitro human cardiac model. This side effect must be further evaluated to avoid fatalities and long-term sequelae, as cardiac-compromised individuals have a worse prognostic of COVID-19 [3]. Besides remdesivir, molnupiravir, fluvoxamine and Paxlovid are also oral antiviral drugs with a positive reduction in the mortality and hospitalization rates of COVID-19 patients, and they also feature good overall safety [4]. Molnupiravir has the advantage of being effective in five days and avoiding major events, as observed in clinical trials, although one study pointed out that this drug caused mutations. Molnupiravir is effective in the first days after the initial symptoms and is recommended by the OMS, but it is recommended to treat mild cases. Fluvoxamine's mechanism of action has drawn attention, while Paxlovid, a SARS-CoV-2 protease inhibitor, was designed to work directly on the novel coronavirus-specific protease [4]. However, side effects and the need for broader coverage led to clinical studies aimed at repurposing several drugs as antivirals to fight COVID-19, but most of them failed in the clinic studies or showed inconsistent results.

In addition, the complexity of the physiological changes caused by coronavirus has created several challenges, and adjustments were added to the antiviral pharmacotherapy scenario at the beginning of the COVID-19 pandemic. The treatment of individuals hospitalized with COVID-19 and severe acute respiratory syndrome (SRAG) was unsuccessful with the single prescription of antiviral agents and, in most cases, the administration of a pool of medicines was required. Among these medicines, antibiotics were frequently prescribed to prevent coinfections. Due to the pandemic dimension, this huge increase in the use of antibiotics has contributed to the development of acquired bacterial resistance [5]. This resistance appears when bacteria are able to make changes in the active site targeted by the antibiotic, reducing the infiltration of the active compound in the bacterial cell or even releasing enzymes capable of causing structural degradation in the antibiotic structure [6]. By 2030, the economic reflection of antimicrobial resistance problems could lead to extreme poverty for up to 24 million people [7]. According to the UN ad hoc Group on Interagency Coordination on Antimicrobial Resistance, drug resistance could cause 10 million deaths per year [7]. In this scenario, the development of new antibacterial drugs, together with antiviral drugs, is a world priority, in order to address the still ongoing demand caused by COVID-19, but also to decrease the number of deaths associated with bacterial infections worldwide. New antibacterial and antiviral drugs are expected to cause less side effects and are intended for several groups of individuals that still cannot rely on safe medicines, such as patients with comorbidities or immunocompromised and pregnant women. In addition, new antimicrobials are expected to have mechanisms of action, therefore, making it possible to fight resistant bacteria and viruses. Moreover, low cost is extremely desirable, in order to increase access to the treatment.

Plants remain an interesting source of novel molecules due to their proven efficacy, adequate bioavailability and easy applicability in human consumption, as well as the positive history of plant products that have become a source of modern prescription medicines [1,8]. Artemisinin is one of the most recent and successful examples of natural products that have reached the international market as a finished pharmaceutical product. Artemisinin and its derivatives have been adopted by dozens of countries as a first-line anti-malarial drug, and it has a significative bulk export production. Among the advantages of artemisinin over other anti-malarial drugs, this natural product provides a rapid decrease in parasitemia and delays drug resistance in combination therapy [9]. The good outcomes of new medicines from plants in several fields has led to an intense search for new drug leads to combat the coronavirus and to treat illnesses associated to COVID-19 [10].

However, drug development from plant metabolites is a slow process, contrasting with the urgent demand for new antiviral and antibacterial agents. In addition, once a promising molecule is identified, toxicity screenings, formulation studies, bioavailability and, later, the scale up for pilot industrial production can be limiting factors for reaching pre-clinical and clinical phases [11]. A faster strategy, likely to be successful, is based on the research of plant species already consumed by humans (i.e., with known toxicity) and industrially produced (i.e., available on a large-scale basis and with industrial processing already established) [12]. Spices of vegetable origin are among the plant sources that fit into this profile. A huge number of spices produced on an industrial scale and which are safe for human consumption and used in different countries are described in the scientific literature as promising antibacterial and antiviral agents [13,14].

In this review, we initially present data demonstrating the need for new antiviral and antibiotic drugs, especially corroborated within the scenario of the recent COVID-19 pandemic. The second part discusses the potential of plants widely used as spices, with antibacterial, antiviral and related properties described in the literature, for the development of new antiviral and antibacterial drugs. The pharmacological potential of the secondary metabolites produced by these species, the synergic effects and the current research on antivirals are also discussed.

2. COVID 19: Context, Treatment and New Drugs Demand

2.1. COVID-19 Scenario

COVID-19 originated from the SARS-CoV-2 virus of the Coronaviridae family. The organization of SARS-CoV-2 has been associated with the presence of 16 non-structural and four structural proteins: peak (S), envelope (E), membrane (M) and nucleocapsid (N). The S protein binds to the ACE2 receptor, which is located on the host cell surface, and allows the insertion of RNA into the host cell cytoplasm. This interaction also requires the priming of the S protein that takes place when SARS-CoV-2 binds to the protease. As a result, the binding of SARS-CoV-2 to the ace2 receptor allows the virus to enter the host cell, mainly by endocytosis, where it initiates a deregulated immune response, resulting in acute injuries [15].

The most common clinical symptoms in individuals affected by COVID-19 disease are fever (98% of cases), followed by cough (76%), dyspnea (55%) and myalgia or fatigue (44%). The recurrent laboratory characteristic among those infected with the virus is leukopenia (25% of cases), a worrying condition because the reduction in white blood cells in the bloodstream decreases the body's ability to fight infections, resulting in immunity output [16]. The clinical symptoms of COVID-19 were classified as mild, moderate or severe, with the need to prescribe antibiotics predominantly for severe and moderate conditions [17]. Dyspnea is the main motivation for intubation, a recurrent scenario related to the clinical complications caused by COVID-19. In this process, a cannula is positioned through the gloat to the region of the arytenoid cartilages vocal process. These arytenoids are coated with a thin layer of the mucosa perichondrium, and intubation makes these cartilages susceptible to trauma, excessive cough, upper respiratory tract-related infections, and pneumonia, clinical conditions that also require the prescription of antibiotics [18].

Comorbidities and advanced age were earlier associated with hospitalization and severe symptoms among patients with COVID-19. The immune debit caused by some comorbidities decreases the individual's resistance to SARS-CoV-2, making these individuals prone to more severe infections than those without comorbidities. These conditions predispose to the development of bacterial infections, increasing the possibility of coinfection and hindering the convalescence of patients [19]. A study with 315 patients who tested positive for COVID-19 revealed that 95 (30%) had comorbidities such as chronic obstructive pulmonary disease, hypertension, cardiovascular alterations, diabetes mellitus, chronic kidney disease and cancer. Among these 95 individuals, 69 presented superinfections caused by bacteria, impacting the length of hospitalization (30 days), while the hospitalization period of those that did not develop this condition was 11 days [20].

COVID-19-related morbidities fall under the class of chronic non-communicable diseases (NCCD). NCCDs are responsible for 70% of deaths worldwide, of which 80% occur in low- and middle-income countries (72.6% in Brazil), and the need to prescribe antibiotics for patients in this group greatly increased with the COVID-19 pandemic [21]. In a study conducted in China, it was reported that among a group of 856 patients diagnosed with COVID-19, approximately 30–50% had one or more comorbidities, mainly hypertension (30–50%), diabetes (8–20%), cardiovascular disease (5–20%), chronic liver disease (1–5%) and chronic kidney disease (1–4%). The study proved that the higher the number of comorbidities, the greater the risk of developing severe conditions and bacterial infections [22]. Among the factors that aggravated the number of people affected by COVID-19, physical inactivity, alcohol abuse, inadequate diet and smoking are frequently cited [21].

Clinical trials have been conducted on dozens of vaccines worldwide, and some of them have received emergency authorization to be used for active and preventive immunization against COVID-19. Some vaccines act by stimulating the production of neutralizing antibodies responsible for protecting the immune system [23]. Immunization through the vaccine has had a positive effect, but the disease is far from being controlled, especially by the emergence of more lethal or more contagious variants [24].

2.2. *Combination Pharmacotherapy for Treatment of Patients with COVID-19*

Effective drug therapies for the treatment of people infected with the SARS-CoV-2 virus have been discussed since the beginning of the pandemic. Combinations of drugs with active ingredients already registered for the treatment of other diseases with symptoms correlated with COVID-19 have been widely explored [25]. This approach expedites the release of drugs when early-stage clinical trials have already been conducted and the drugs are already supplied on the market [26]. However, there is still concern about patient safety, due to the adverse effects associated with the combined use of these drugs, since most of them have already caused side effects when used individually [27].

A main target of drugs within the scope of COVID-19 is the SARS-CoV-2 protein, and the proposed therapies act on viral enzymes or functional proteins, RNA synthesis and replication, blocking the binding of the virus to human cell receptors. They target structural proteins, restore host innate immunity, inhibit virulence factor, or act on host-specific receptors, thus, preventing viral entry [4,28,29].

In numerous completed or ongoing clinical trials, various antiviral and immunomodulatory molecules have been administered to patients with severe COVID-19. In an extensive review, Salasc and colleagues [27] described the combination of treatments used worldwide to combat COVID-19 through randomized clinical trials, such as remdesivir plus baricitinib, lopinavir-ritonavir plus ribavirin and interferon beta-1a (IFNb-1b), as well as dexamethasone together with azithromycin or remdesivir. Remdesivir, lopinavir/ritonavir, and favipiravir are capable of inhibiting viral enzymes or functional proteins, RNA synthesis and replication, thus, exhibiting an effect against SARS-CoV-2 [30].

Some medicines of the corticosteroid class—including cortisone, prednisone and methylprednisolone—were also implemented in combination therapies as steroidal anti-inflammatory drugs, as well as prescribed antibiotics, for example, azithromycin, ceftriaxone, moxifloxacin and imipenem [31]. The chemical structures of the main antiviral and antibiotic drugs can be found in Figure 1. The synergistic effect of combination therapies delays the onset of drug resistance, due to the additive effect, increasing the safety and efficacy of the treatment [32].

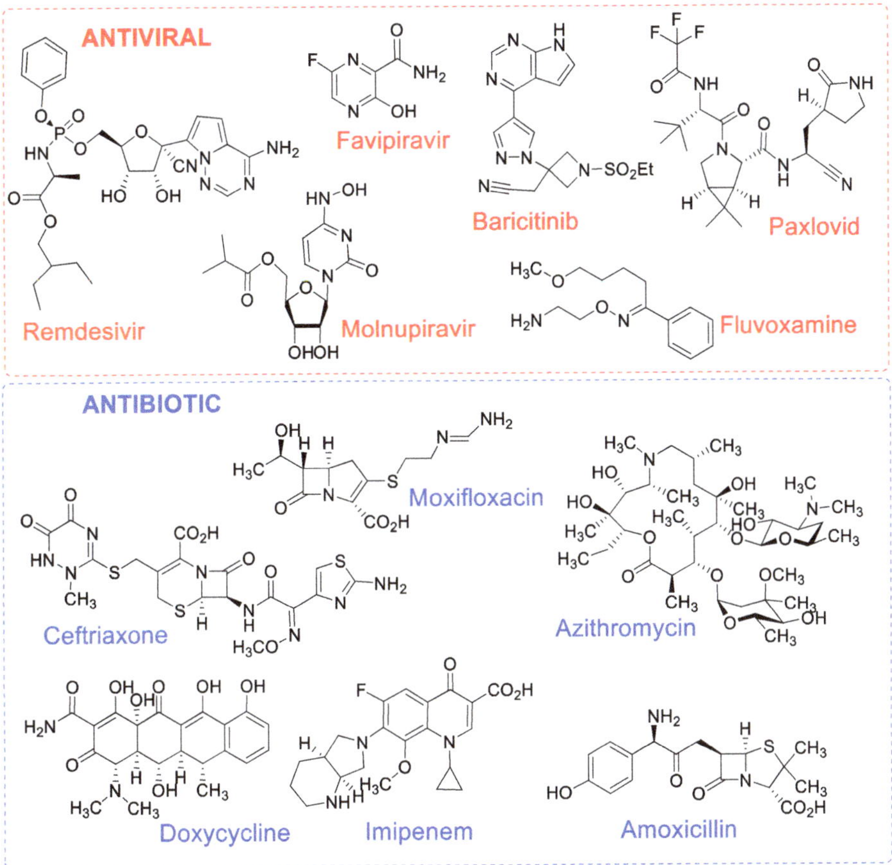

Figure 1. Chemical structures of antiviral and antibiotic drugs with different mechanisms of action used in mono or combination therapy in the treatment of patients with COVID-19.

2.3. *Prescription of Antibiotics for Patients with COVID-19 and Bacterial Resistance*

Currently, there are multiple demands for new antibiotics, some of which are due to the increasing number of serious post-surgery infections caused by resistant bacteria [33]. WHO guidelines did not indicate the prescription of antibiotics for patients with suspected or confirmed COVID-19 with mild symptomatology, nor for those with low suspicion of bacterial infection. However, in severe cases where diagnosis, treatment and prognosis are difficult, the administration of antibiotics is necessary to increase the chance of patients' survival. This precautionary measure contributes to the emergence of multidrug-resistant strains and to the reduction in the efficacy of the antibiotics currently available [34]. As the patient's condition becomes severe, the antibiotic prescription increases, as shown in Figure 2, and is 64.4% higher for patients requiring mechanical ventilation, as bacterial and fungal coinfections are common in critically ill patients [33]. According to Su et al. [35], 50% of deaths in severe patients with COVID-19 occur in patients with secondary infections. Specifically, in patients with COVID-19, the bacterial co-infection rate reaches 7.7% of cases, and it is mainly related to *Acinetobacter baumannii*, *Escherichia coli*, *Pseudomonas aeruginosa* and *Enterococcus* sp. Patients usually receive broad-spectrum antibiotics such as azithromycin, ceftriaxone, imipenem and moxifloxacin. However, resistant bacteria are a huge problem. Imipenem, a broad-spectrum β-lactam antibiotic, significantly reduces inflammatory cytokines, with better results for the treatment of COVID-19 patients with nosocomial bacterial infections. Resistance to carbapenem, an antibiotic of the class of imipenem, has been reported in patients co-infected with *A. baumannii* (55.6%), imposing treatment difficulties and increasing the likelihood of septic shock [36].

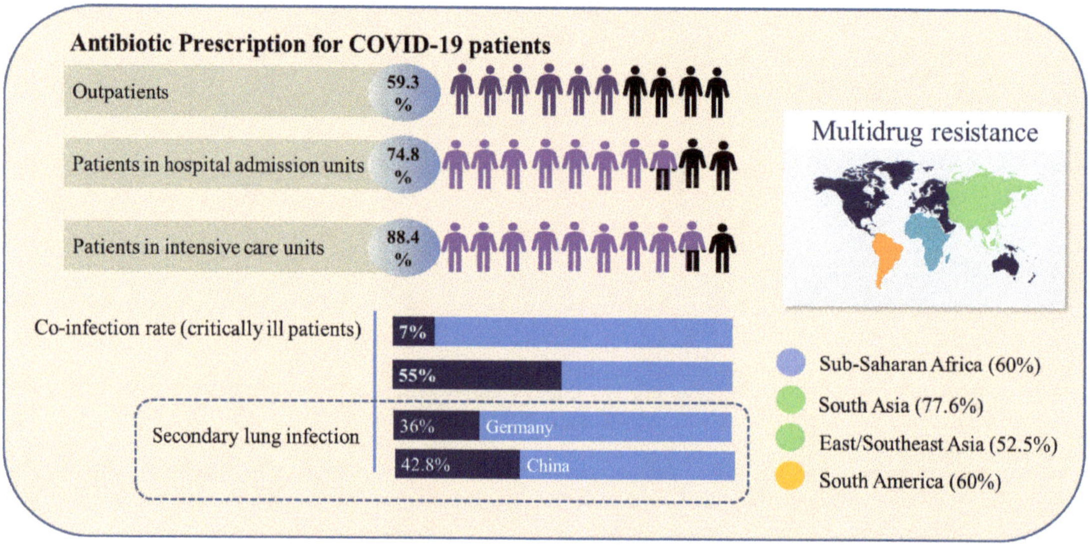

Figure 2. Comparative overview of antibiotic prescription, co-infection and secondary infection profile and geographic incidence of multidrug resistance related to COVID-19 patients.

The incidence of secondary lung infections caused by bacteria in COVID-19 patients reached 36% in Germany and 42.8% in China [37]. The geographic distribution of multidrug-resistant bacteria which are not affected by the action of conventional antibiotics is presented in Figure 2 [38]. Infections by multidrug-resistant bacteria in hospital environments are serious, due to the concentration of individuals with fragile health that are more susceptible to infections caused by opportunistic pathogens. The main pathogens related to these infections are identified with the acronym ESKAPE (*Enterococcus faecium*, *Staphylococcus aureus*, *Klebsiella pneumoniae*, *A. baumannii*, *P. aeruginosa* and *Enterobacter species*), which is characterized by bacteria capable of "escaping" the actions of antibiotics [36]. Beyond the issues associated with the mortality derived from ESKAPE bacteria, bacterial triggers associated with these organisms are supposed to participate in the manifestation of some autoimmune diseases, such as rheumatoid arthritis and multiple sclerosis [39–41].

2.4. Medicinal and Spice Plants with Antibiotic Activity and Their Synergistic Effects with Industrialized Antibiotics

Industrialized antibiotic accessibility is a huge problem in many parts of the world, due to their cost and the need for a strict uptake routine. Other issues such as stability also contribute to this scenario. To overcome stability issues, some antibiotics must be commercialized as dry powders to ensure activity and extend the shelf life, imposing a step of rehydration for their reconstitution that may become a further problem under several circumstances [42,43]. These issues can be minimized with the use of medicinal plants. Plants with antibiotic power can be used in natura or after simple processing steps, such as extraction, pulverization and drying, or boiling, without a loss of the pharmacological effects [44]. In addition, using the whole plants or their parts instead of commercial antibiotics has the advantage of reduced cost, making this alternative treatment more accessible for the population unable to use prescription antibiotics. A review on the world trends in medicinal plants research, published in 2020 [45], showed that over 110,000 studies on this theme were published between 1960 and 2019, most of which were classified within the scope of pharmacology, toxicology and pharmaceutics. It was estimated that 10% of all vascular plants are used due to their medicinal properties aiming at preventive, control and curative activities [46].

The mechanism of plant extracts' antibacterial action depends on several factors, such as the characteristics of the target bacterium, extract composition, and the chemical features of the phytoconstituents. In general, this mechanism usually involves disruption and lysis of the microorganism's cell wall, the release of cellular content, protein binding domain disruption, inhibition of microbial DNA replication and nucleic acid transcription, inhibition of the biosynthesis of compounds toxic to the host and other effects that lead to cell death or a decrease in deleterious effects [47]. The treatment of infections with plant extracts has not been suggested as a major factor in the development of bacterial resistance [48,49], and a number of plant extracts are active against resistant bacteria, as shown in Table 1. Among the plants cited in Table 1, *C. englerianum*, *E. depauperate*, *M. chamomilla*, *T. zygis*, and *T. willdenowii* showed high IZ values, therefore, pointing to the inhibition of resistant bacterial strains. In some studies, minimum inhibitory concentration (MIC) determination provided additional details, such as selectivity [50–55], and may be preferable to determine the full antimicrobial activity [56]. For example, IZ values observed for *A. aspera* in the presence of different resistant strains do not vary significantly (6.0–6.3 mm), but MIC data demonstrated that this plant is more active against MRSA and MRKP in relation to MrPA [50]. In another study, *L. inermis* presented higher IZ values for MRSA and MRPA, suggesting that these bacteria are more susceptible to the extract than MRKP, which was confirmed by MIC data [50].

Table 1. Scope of the activity of some plants against resistant bacteria.

Plant Species [Botanical Family]	Active Against	Scope of Activity	Reference
Achyranthes aspera [Amaranthaceae]	MRSA ATCC 43300 MRPA ATCC 27853 MRKP ATCC 00603	IZ 6.3 ± 0.6 mm; MIC 42.0 ± 14.4 mg/mL IZ 6.2 ± 0.3 mm; MIC 200.0 ± 0.0 mg/mL IZ 6.0 ± 0.0 mm; 50.0 ± 0.0 mg/mL	[50]
Acokathera oppositifolia [Apocynaceae]	MRKP ATCC 33495	MIC 6.25 ± 0.0 mg/mL	[51]
Ageratina adenophora [Compositae]	MRSA ATCC 25923	IZ 10 ± 0.0 mm; MIC 12.5 mg/mL	[52]
Areca catechu [Arecaceae]	MRPA CCARM 2092 MRAB CCARM 12005	IZ 6.4 ± 0.5–16.3 ± 1.5 mm; MIC 5.6 μg/mL IZ 6.0 ± 0.0–17.7 ± 1.2 mm; MIC 5.6 μg/mL	[52]
Artemesia vulgaris [Compositae]	MRSA ATCC 25923	IZ 10 ± 0.1 mm; MIC 12.5 mg/mL	[53]
Azadirachta indica [Meliaceae]	MRSA ATCC 43300 MRPA ATCC 27853 MRKP ATCC 00603	IZ 6.2 ± 0.3 mm; MIC 33.3 ± 14.4 mg/mL IZ 6.4 ± 0.4 mm; MIC 50.0 ± 0.0 mg/mL IZ 6.1 ± 0.2 mm; MIC 41.7 ± 144 mg/mL	[50]
Cirsium englerianum [Asteraceae]	MRSA ATCC 25923	IZ 28 ± 0.04 mm; MIC 16 μg/mL	[53]
Euphorbia depauperata [Euphorbiaceae]	MRSA ATCC 25923	IZ 26 ± 0.02 mm; MIC 4 μg/mL	[53]
Hydrastis canadensis. [Ranunculaceae]	MRSA AH1677	MIC 75 μg/mL	[54]
Kalanchoe fedtschenkoi [Crassulaceae]	MRAB CDC0033 MREC CDC08	MIC 256 μg/mL MIC > 256 μg/mL	[55]
Lawsonia inermis [Lythracea]	MRSA ATCC 43300 MRPA ATCC 27853 MRKP ATCC 00603	IZ 15.5 ± 0.5 mm; MIC 4.2 ± 2.0 mg/mL IZ 12.5 ± 0.5 mm; MIC 4.2 ± 1.8 mg/mL IZ 7.6 ± 0.5 mm; MIC 12.5 ± 0.0 mg/mL	[50]
Lippia adoensis [Verbenaceae]	MRSA ATCC 25923	IZ 27 ± 0.56 mm; MIC 64 μg/mL	[49]
Lippia javanica [Verbenaceae]	MRPA ATCC 9721	MIC 6.25 ± 3.2 mg/mL	[49]
Matricaria chamomilla [Asteraceae]	MRSA ATCC 43300 MRPA ATCC 27853	IZ 30 ± 2 mm; MIC 0.781 mg/mL IZ 13.66 ± 1.52 mm; MIC 0.590 mg/mL	[57] [57]
Morella kandtiana [Myricaceae]	MRAB CDC 0033 MBKP CDC 0076	MIC > 256 μg/mL MIC 256 μg/mL	[58]

Table 1. *Cont.*

Plant Species [Botanical Family]	Active Against	Scope of Activity	Reference
Mentha sp [Lamiaceae]	MRAB CI MRKP CI MRPA CI	MIC > 2 mg/mL MIC >2 mg/mL MIC 2 mg/mL	[59]
Ocimun basilicum [Lamiaceae]	MRAB CI MRKP CI MRPA CI	MIC > 2 mg/mL MIC > 2 mg/mL MIC > 2 mg/mL	[59]
Oxalis corniculata [Oxalidaceae]	MRKP CDC 0076	IZ 11 ± 0.0 mm; MIC 25 mg/mL	[53]
Plectranthus barbatus [Lamiaceae]	MRAB CI MRKP CI MRPA CI	MIC > 2 mg/mL MIC 1 mg/mL MIC 2 mg/mL	[59]
Punica granatum [Punicaceae]	MRKP CDC 0076	IZ 19–45 ± 0.7 mm	[60]
Salvia triloba [Lamiaceae]	MRSA ATCC 6538 P	IZ 9.5 mm	[61]
Scutellaria barbata [Lamiaceae]	MRAB CDC 0033	IZ 14–18 ± 0.0 mm; MIC 6.4 mg/mL	[62]
Thymus zygis L. [Lamiaceae]	MRSA ATCC 43300 MRAB CDC 0033	IZ 75 ± 00 mm; MIC 02 ± 0.0009 μL/mL IZ 71.5 ± 0.1 mm; MIC 02 ± 0.001 μL/mL	[63]
Thymus willdenowii [Lamiaceae]	MRSA ATCC 43300 MRAB CDC 0033	IZ 33 ± 0.2 mm; MIC 04 ± 00 μL/mL IZ 30 ± 00 mm; MIC 04 ± 0.001 μL/mL	[63]
Zanthoxylum chalybeum [Rutaceae]	MRSA ATCC 1677 MREF ATCC 0044	MIC 16 μg/mL MIC 32 μg/mL	[58]

MRPA: multidrug-resistant *P. aeruginosa;* MRAB: multidrug-resistant *A. baumannii;* MRE: multidrug-resistant *Enterobacter* ssp.; MRSA: multidrug-resistant *S. aureus;* MREF: multidrug-resistant *E. faecium;* MRKP: multidrug-resistant *K. pneumoniae;* IZ: inhibition zone; MIC: minimal inhibitory concentration; CCARM: Collection of Cultures of Antimicrobial Resistant Microbes; CDC: Centers for Disease Control and Prevention; ATCC: American Type Culture Collection; CI: clinical isolates.

Activity against resistant bacteria through new strategies has been described for plant extracts, such as the quorum-sense mechanism, which is considered a promising alternative to antibiotics. This mechanism, mediated by signal molecules, can be applied to multiple pathogenic bacteria and is based on the action of auto-inducers' self-excreted molecules capable of regulating the expression of virulence genes [58]. Overall, quorum-sensing acts in the host defense, controlling biofilm formation and the production of compounds that are toxic to the bacteria. The early inhibition of biofilm formation was reported over *S. aureus* and *S. pyogenes*. The restriction of van der Walls and electrotactic interactions, avoiding the attachment of pathogenic bacterial cells, was a mechanism of action that explains the decrease in infection [49].

Technological development has been successfully used to improve antibacterial activity, as in the development of clusters of natural products with silver ions. The clusters attach to the surface of the bacterial cell wall, damaging the cell surface. Moreover, silver ions disable the microbial growth process, as shown for the *A. catechu*-Ag cluster [58].

Complementing the new mechanisms of action, other major benefits of using the whole plant concern the well-known synergic and/or complementary effects of the different metabolites present in plants/plant extracts. In cumin, the antimicrobial activity is related to the synergy among terpenes and some minor constituents such as limonene, eugenol and pinene (Bazaka et al., 2015). The isolation of the bioactive metabolites may be unnecessary, since the phytoconstituents can be less active once administered alone, due to the lack of synergic compounds [41,64]. In this way, the concomitant use of industrialized antibiotics and bioactive plant extracts—the so-called antibiotic-adjunct combination strategy—has been proposed to overcome problems related to resistant bacteria [65]. The efficacy of combination therapy can be exemplified by the use of penicillinase and β-lactamase inhibitors as co-drugs to overcome bacterial resistance to some antibiotics. Combinatorial plant-drug therapies have already been indicated to treat viral infectious diseases, such as AIDS [66]. Even plant-drug combinations lacking synergic effects do not interfere in the individual activity of the antibiotic, which is highly encouraging for research and the application of combinational therapies [30,41].

The synergism between antibiotics (ampicillin, penicillin, tetracycline, methicillin, etc.) and several plant extracts (green tea, khat, pomegranate, basil, lemon balm, grape pomace and oregano), plant metabolites (curcumin, epigallocatechin-gallate, flavonoids, alkaloids and terpenes), as well as essential oils is widely described, and some examples are shown in Table 2. In general, synergy can be achieved by the blockage of multidrug resistance pumps, consequently facilitating the entrance and traffic of antibacterial agents into the bacterial cell [8]. Therefore, combinational synergistic approaches may be a good tool to combat resistant bacterial strains [41]. Simple and general methods exist that are capable of evaluating the synergistic effect of drug combination, enabling measurement of the qualitative and quantitative physiological effects of two or more drugs [67].

Table 2. Synergy effect of plant species with antibiotics towards bacterial pathogens.

Target Pathogen	Plant Species	Synergy Effect	Reference
Aggregatibacter actinomycetemcomitans	*Salvadora persica*	More than doubled the activity combined with metronidazole	[68]
B. cereus, S. aureus, E. coli, and *P. aeruginosa*	*Ficus nitida*	Antibacterial activity was enhanced in the presence of tetracycline	[69]
E. coli and *K. pneumoniae*	*Centaurea damascena*	Synergetic effect combined with gentamicin (ineffective for *E. coli*), vancomycin, ampicillin and chloramphenicol (ineffective for *K. pneumoniae*)	[70]
MDRAB and MDRPsA	*Pithecellobium clypearia*	Synergistic effect with imipenem and tetracycline [a]	[71]
MDRPsA	*Coriandrum sativum*	Synergism in the presence of antibiotics including mezlocillin, cefoperazone, cefotaxime and levofloxacin	[72]
MRSA 1485279	*Vernonia condensata*	High MIC reduction combined with ampicillin [a]	[73]
Multidrug-resistant enteric bacteria	*Carum copticum*	Reduced up to 64-fold MIC against *E. coli* with ciprofloxacin	[74]
P. mirabilis	*Petalostigma* spp.	Synergistic activity with penicillin-G, chloramphenicol and erythromycin	[41]
S. aureus ATCC 12600	*Origanum vulgare* and *Hypericum perforatum*	Combined extracts (1:1) increased inhibition over 3 times more than the individual extracts	[75]
S. aureus ATCC 25923 and *E. coli* ATTC 25922	*Vatica diospyroides*	Increased ampicillin efficacy; reduced the required antibiotic concentration by eight times	[76]
S. aureus strains 3993 and 4125	*Salvia officinalis, Senna macranthera,* and *Plectranthus ornatus*	Up to 8-fold reductions in the MIC, especially associated to ampicillin, kanamycin and gentamicin	[77]
Treponema denticola	*Cinnamomum zeylanicum*	More than doubled the activity combined with amoxicillin	[68]

MDRAB: multidrug-resistant *A.baumannii;* MDRPsA: multidrug-resistant *P. aeruginosa;* MRSA: methicillin-resistant *S. aureus;* MIC: minimum inhibitory concentration; [a] alteration in membrane permeability

As observed in Table 2, the association of plants with antibiotics lead to synergistic effects; this is important, as it allows the administration of lower doses of the industrialized antibiotic, with the same or even higher antibiotic effects, contributing to postponing the phenomenon of resistance. The inhibition of efflux pumps by some phytochemicals consequently avoids bacterial exposure to sub-therapeutic concentrations of antibiotics, which is one of the causes of acquired resistance [71]. Plant metabolites that favor morphological disruptions on the bacteria cell wall help to decrease the number of functional bacterial cells to be attacked by the antibiotic, as postulated for the synergistic interaction of thymol–ciprofloxacin [74]. The antimicrobial activity mechanism of thymol has been linked to the presence of free hydroxyl groups that are involved in membrane depolarization [74]. The action of this phytochemical plays an important role in potentializing the antibiotic effect of commercial drugs such as ciprofloxacin. In some examples, plant metabolites increase their outer membrane permeability to allow antibiotic activity. Flow cytometry was used to show that *V. diospyroides'* synergistic action in combination with ampicillin is the result of cell membrane granularity disruption [76]. Plants can act directly in specific intracellular enzymes or deactivate the production of enzymes produced by bacteria to degrade the antibiotic chemical structure, therefore, helping the antibiotic to enter the bacterium cell to exercise its activity [71]. The synergic multi-target action of phytochemicals can also influence the production of enzymes necessary for endogenous energy production and protein synthesis.

2.5. Potential of Plant Spices with Antibiotic Activity as Antiviral Agents

Plants used as spices have some interesting particularities that can be explored in the search for natural remedies to treat patients requiring the simultaneous use of antibacterial and antiviral agents. First, the widespread consumption of plant-derived spices over time has demonstrated their safety for human consumption [78]. Second, spices are likely composed of stable compounds, as they are frequently used in hot dishes, and concomitantly with acid compounds in food preparation, such as vinegar and lemon. Third, the use of such spices as adjuncts, including in COVID-19 prevention or treatment, would rely on the patient's consent, as these plants are already traditionally consumed [79]. Moreover, these plants are usually commercialized as finely sprayed powders with a long shelf-life, facilitating the development of multiple formulations. The safety, authentication and traceability of condiments are well studied, and effective non-destructive vibrational and atomic spectral analytical techniques are available for quality control and meet the increasing industrial demand [80,81]. The global condiment market, estimated to be worth USD 76.6B by 2020, is projected to reach USD 103.7B by 2026 [82]. China, Turkey, Vietnam and Indonesia are the markets leaders in the growth of this sector and, in the coming years, they should be influenced by trends related to health, gourmet flavors and ethical values.

Several plant spices with antibiotic activity have gained attention as antiviral agents, especially against SARS-CoV-2. The consumption of plant spices with antibiotic and antiviral properties by patients with mild and moderate cases of viral infections deserve deeper studies, as they could assist in avoiding and combating concomitant infections. Some secondary metabolites present in these plants have been proven as potent antiviral compounds. The chemical structures of some relevant antiviral natural products in the context of COVID-19 are presented in Figure 3.

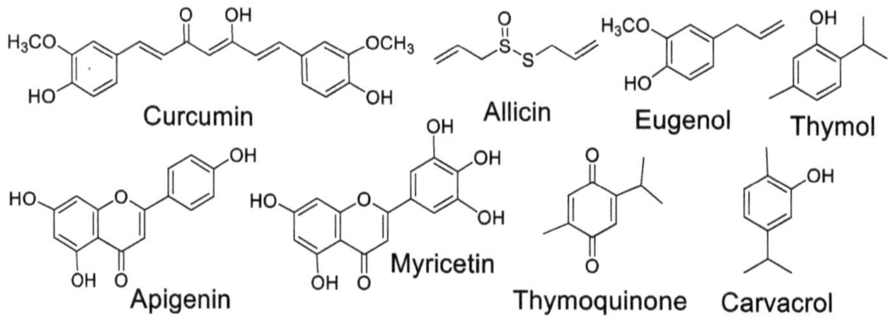

Figure 3. Chemical structures of relevant antiviral natural products in the context of COVID-19.

Curcumin, a natural component of *Curcuma longa*, is one of the most prominent examples of a worldwide consumed condiment which has both antibacterial [83] and antiviral properties [84] (Figure 4). Curcumin diminishes fatigue and bronchoconstriction, helping to alleviate the symptoms of COVID-19 [85]. The antiviral action mechanism of curcumin was recently reviewed [84]. The administration of curcumin in patients with COVID-19 on a formulation containing piperine (a compound that promotes the absorption of curcumin) improved some symptoms, compared to the control group, on a randomized clinical trial [86]. Curcumin is also suggested as a promising prophylactic candidate for the treatment of COVID-19 and is a target of other studies towards the development of nano formulations and supplements [87–94].

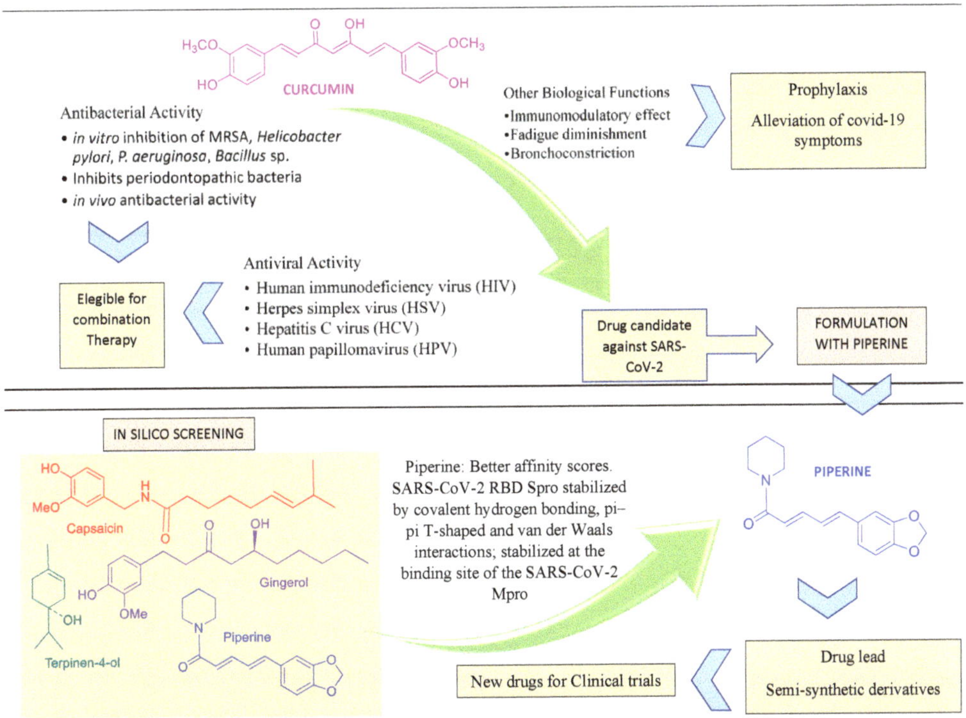

Figure 4. Curcumin and piperine features compatible with their use in combined pharmacotheraphy.

Allicin [95,96], eugenol [97,98], thymoquinone [99], apigenin [100,101], carvacrol [102,103] and thymol [104] are some of the most studied metabolites of spice plants with antiviral effects and consequent applications in the efforts to combat COVID-19 (Table 3).

Garlic is a condiment native to Asia and has been used as a medicine since ancient times, as in Egypt, where it was consumed during the construction of the pyramids. These and other historical features, as well as their antibacterial properties—especially those related to the organosulfur metabolites of garlic—were recently reviewed [105]. Garlic also has immunomodulatory, anti-inflammatory, anticancer, antitumor, antidiabetic and cardioprotective effects. Allicin, the major component of raw garlic, is a broad-spectrum antimicrobial agent. The inhibition of specific bacteria by garlic has been studied, as in the case of *Streptococcus agalactiae* (Group B *Streptococcus*), an invasive bacterium suggested to be the major cause of neonatal morbidity and mortality during the first weeks of life [106]. Garlic has been reported as active against over twenty viruses, including adenovirus, herpes simplex virus, influenza A virus and human immunodeficiency virus [107]. In silico studies targeting host receptor angiotensin-converting enzyme 2 (ACE2) protein related to coronavirus resistance showed strong interactions of garlic essential oils with the amino acids of the ACE2 protein and the main protease of SARS-CoV-2 (PDB6LU7). In addition, synergistic interactions of components were observed, resulting in good inhibition of the ACE2 and PDB6LU7 proteins [108]. On a broad docking screening with 75 metabolites present in traditional Indian spices, one of the highlights was the flavonoid myricetin, a metabolite produced by *A. sativum*. The activity of myricetin was more pronounced against main proteases (M^{pro}) than against SARS-CoV-2 spike proteins (SP) [109]. Pre-clinical and clinical studies of the effect of garlic on viral infections, including those caused by SARS-CoV, have shown the effectiveness of garlic's antiviral capacity by means of several mechanisms, such as downregulation of the extracellular-signal-regulated kinase (ERK) and mitogen-activated protein kinase (MAPK) signaling pathway, the blockage of viral entry into host cells, and inhibition of viral RNA polymerase [107].

Another spice, *N. sativa* (*black cumin*)—a healing herb of the Ranunculaceae family and much appreciated in North Africa—is traditionally used to treat various diseases including hypertension, asthma, inflammation, diabetes, cough, headache, bronchitis, eczema, dizziness and fever. Black cumin's pharmacological properties, such as immunomodulatory, anti-inflammatory, antimicrobial, antioxidant and anticancer activity are described [110], in addition to the inhibition of SARS-CoV-2 [111,112]. In a broader context, the immunomodulatory components present in traditional plants, such as basil (*O. sanctum*) and cinnamon (*C. verum*), can potentiate the effect of antibiotics when administered together (Table 3). Basil has an anti-inflammatory and antimicrobial agent named eugenol, while cinnamon is used in traditional medicine for various lung-related disorders, including pneumonia, infectious diseases and pleural effusion [113]. Other essential oils from spices have also been reported as antimicrobial agents [114].

Table 3. Plant spices and condiments and their antiviral profile.

Scientific Name [Popular Name]	Main Component	Antiviral Effects/COVID-19 Applications	Reference
C. longa [Turmeric]	Curcumin	Attenuation of poly(I:C)-induced immune and inflammatory responses by inhibiting the TLR3/TBK1/IFN-β cascade	[90]
		Enhancement of oral drug delivery system (Labrasol® and tween 80 bicelles)	[91]
		Molecular docking studies showed reliable ADME profile	[92]
		Analogues as dual inhibitor of SARS-CoV-2	[93]
		Development of nanoformulations	[87,94]
Allium sativum [Garlic]	Allicin	Suppresses production and secretion of pro-inflammatory cytokines and stimulates of immune system cells (NK, lymphocytes, eosinophils and macrophages)	[95]
		Suppression of pro-inflammatory cytokines TNF-α and CRP	[96]
Cinnamomum verum [Dalchini]	Eugenol	Inhibition of specific immune responses to allergens, reduces side effects of some anti-inflammatory drugs, antioxidant properties	[97]
		Increases the bioavailability of antiviral drug saquinavir	[98]
Nigella sativa [Black cumin]	Thymoquinone	Inhibitory effects on viral spike protein with cellular angiotensin-converting enzyme 2 (ACE2)	[99]
		Inhibition of RdRp of SARS-CoV-2, especially α-hederin; ongoing drug development strategy against SARS-CoV-2	[99]
O. basilicum [Basil]	Apigenin	The phytoconstituents vicenin, sorientin and ursolic acid inhibit SARS-CoV-2 M^{pro}	[100]
		Development of gellan gum hydrogel with basil oil nanoemulsion	[101]
O. vulgare [Oregano]	Carvacrol	Inhibition of viral replication and activity of SARS-CoV-2 $3CL^{PRO}$	[102]
		Potent inhibition of SARS-CoV-2 replication (modeling studies)	[103]
Thymus vulgaris [Thyme]	Thymol	Inhibits the viral spike protein, preventing SARS-CoV-2 entry	[103]
		Essential oils induce cytopathogenic effect against SARS-CoV in Vero-E6 cells	[104]

The therapeutic properties of *O. vulgare*, popularly known as oregano, are associated with terpenes and flavonoids. Antiviral studies demonstrated strong inhibition of viral replication and SARS-CoV-2 $3CL^{pro}$ activity, while in molecular modeling assays, terpe-noids were highlighted as potent inhibitors of SARS-CoV-2 replication [102,103]. Thyme (*T. vulgaris*) is also reported to be active against fungi, viruses and bacteria. It is used as a flavoring agent for cheese and beverages, and in traditional medicine to cure melancholic conditions, skin and respiratory lesions [115]. Thyme essential oil has been shown to be effective against several RNA viruses, including coronaviruses, and it induces a cytopathogenic effect against SARS-CoV in Vero-E6 cells [100,104].

Figure 5 shows a scheme on some key steps in SARS-CoV-2 penetration, replication and exit, as well as some medicines and spices that act in several points.

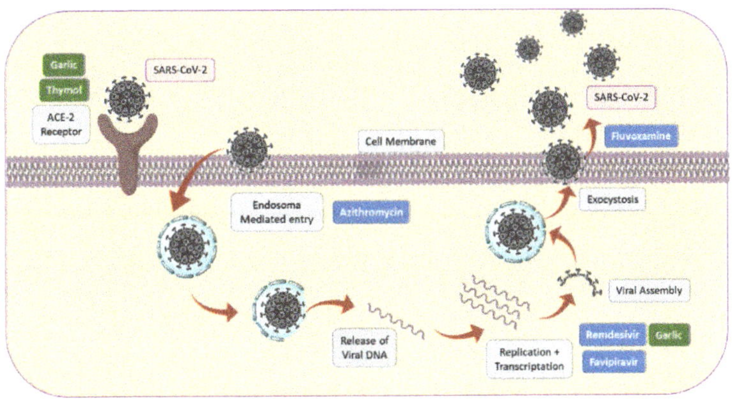

Figure 5. Some key steps of SARS-CoV-2 replication, together with action sites of some drugs and spices.

2.6. Recent Research in Spice-Derived Metabolites in COVID-19 Context

The recent literature shows a huge number of studies on traditional plant spices to enable the search for drugs that protect against COVID-19 and other possible viral pandemics. It is unnecessary to highlight the need to support these studies to avoid, as much as possible, other viral pandemics. Some of these studies and related review papers are listed in Table 4.

Table 4. Traditional plant species that are promising in the search for new antiviral medicines.

Scope	Reference
Indian Spices and Ayurvedic Herbs	
Spices with anti-inflammatory properties with suggested beneficial action in the prevention and treatment of COVID-19 associated cytokine storm.	[116]
Spices useful for future design of new protease inhibitors effective against SARS-CoV-2.	[117]
Antiviral activities of spices, herbs, and derivatives, mechanisms of action, and prospects for future studies.	[118]
Mechanism of action of spices regularly used for cooking purpose to enhance the taste of food in India.	[98]
In silico evaluation of Indian traditional spices with medicinal properties for their inhibitory activity against SARS-CoV-2 spike proteins (SP) and main proteases (Mpro).	[119]
Immune impact of various Indian spices, potential to tackle the novel coronavirus, safety and toxicity aspects.	[120]

Table 4. *Cont.*

Scope	Reference
Traditional herbs used for protection against COVID-19 in North India.	[121]
Modulation of host immune responses by spice-derived bioactive components with protective immunity in COVID-19.	[122]
Preventive effect of Trikadu (mixture of *Zingiber officinale*, *Piper nigrum* and *Piper longum*) by action in the immune system.	[123]
Docking of gingerol, thymol, thymohydroquinone, cyclocurcumin, hydrazinocurcumin, components of Indian medicinal plants (ginger, black cumin, turmeric) against initially deposited spike structural proteins (PDB ID 6WPT) and mutant variant D-614G (PDB ID 6XS6).	[124]
Quick screening of traditional herbs/spices phytoconstituents by in silico study in polyherbal/Ayurvedic formulations.	[125]
Indonesian herbal medicines	
Several healthy drinks related to the COVID-19 pandemic.	[126]
Tanzanian Traditional Medicine	
Phytochemical screening of medicinal plants used to combat COVID-19 in Tanzania.	[127]
Persian Traditional Medicine	
New traditional Persian medicine-based drug, efficacy and safety assessment in COVID-19 patients with major symptoms.	[128]
Other	
Available and affordable complementary treatments for COVID-19.	[129]
Scientific evidence on potential role of spices in offering innate and adaptive immunity to human body.	[130]
Role of functional foods through modulating the host immune system and promoting the synthesis of agents effective against the coronavirus.	[131]

These review papers demonstrate that endophyte fungi as a source of new drug leads to treat individuals with COVID-19 is a major research theme all over the world [116,117]. The research is mainly associated with host plants that have key biological activities, including species already used to fight COVID-19 symptoms [121,127]. Some works searched metabolites to be used in the prevention of viral infections, by means of strengthening the immune system [122,123]. Pure natural components of plant spices have also been reviewed [124], and several other approaches have been developed. The number of studies listed in Table 4, and the information gathered in this review clearly shows the potential of plants, and specifically, plant spices, as antibacterial and antiviral agents, alone or associated with industrialized medicines. Although it is important to study new plants to discover novel metabolites, there is much that can be done, and more quickly, using the scientific knowledge already available in the literature about specific plants. Although plant metabolites have a natural origin, problems such as the development of allergies, the use of ineffective dosages, toxicity caused by over-dosage, the form of administration, and standardization must be considered at some point. In addition, it must be considered that some spices have high added value and may undergo adulteration during marketing by adding other plants, flour, etc., diluting the active ingredient content.

However, the volume of scientific information proving the pharmacological attributes of hundreds of plants is highly disproportionate to their use as medicines, especially in Western countries. The pharmaceutical industry is expected to invest further in this area to produce more affordable and sustainable antiviral and antibacterial medicines.

3. Conclusions

The development of combination therapies that include antivirals, antibiotics and other drugs was very useful for the treatment of patients infected with coronavirus during the pandemic. However, the development of new antiviral drugs with fewer side effects is still necessary, in order to be prepared against other viruses that may emerge, as well as those circulating endemically in several countries. Among the various problems inherited by the current pandemic, we pointed to the increase in resistant bacterial strains, as the result of the prescription of antibiotics for patients with COVID-19 as a prophylactic measure. Medicinal plants are generally safe and highly accessible for consumption and have been used in many countries to prevent or cure the symptoms of COVID-19, although there is no consensus on the effectiveness of bioactive compounds in low concentrations. Among medicinal plants, plant spices frequently contain components with antiviral and antimicrobial activities, and the synergic interaction of multiple metabolites can improve the biological activity. The industrial availability of many plant spices increases the perspective of the large-scale production of extracts and compounds for clinical trials. In addition, plant extracts have demonstrated promising synergistic effects when used in conjunction with existing antibiotic drugs to combat several types of bacteria, including some resistant strains. Clinical trials using plant extracts combined with other oral medications, in the various existing formulations, are needed to better understand their pharmacokinetics, pharmacodynamics and therapeutic potential. In a broader scenario, the development of new antiviral and antibacterial drugs can benefit a significant number of people affected by several viral and infectious diseases, many of which are endemic in various regions of the world.

Author Contributions: Conceptualization, J.A.T.; investigation, writing—original draft preparation, N.B.A.D.; writing—review and editing, J.A.T. and N.B.A.D.; funding acquisition, J.A.T. All authors have read and agreed to the published version of the manuscript.

Funding: This research was funded by Fundação de Amparo à Pesquisa do Estado de Minas Gerais (FAPEMIG PPM-00255-18), Conselho Nacional de Desenvolvimento Científico e Tecnológico (CNPq Grant 304922/2018-8) and Pró-Reitoria de Pesquisa (PRPq)/UFMG.

Conflicts of Interest: The authors declare no conflict of interest.

References

1. Tompa, D.R.; Immanuel, A.; Srikanth, S.; Kadhirvel, S. Trends and strategies to combat viral infections: A review on FDA approved antiviral drugs. *Int. J. Biol. Macromol.* **2021**, *172*, 524–541. [CrossRef] [PubMed]
2. Kausar, S.; Said, K.F.; Ishaq, M.; Akram, M.; Riaz, M.; Rasool, G.; Malik, A. A review: Mechanism of action of antiviral drugs. *Int. J. Immunopathol. Pharmacol.* **2021**, *35*, 20587384211002621. [CrossRef]
3. Kwok, M.; Lee, C.; Li, H.S.; Deng, R.; Tsoi, C.; Ding, Q.; Poon, E.N. Remdesivir induces persistent mitochondrial and structural damage in human induced pluripotent stem cell-derived cardiomyocytes. *Cardiovasc. Res.* **2022**, *118*, 2652–2664. [CrossRef]
4. Wen, W.; Chen, C.; Tang, J.; Wang, C.; Zhou, M.; Cheng, Y.; Mao, Q. Efficacy and safety of three new oral antiviral treatment (molnupiravir, fluvoxamine and Paxlovid) for COVID-19: A meta-analysis. *Ann. Med.* **2022**, *54*, 516–523. [CrossRef]
5. Catalano, A.; Iacopetta, D.; Ceramella, J.; Scumaci, D.; Giuzio, F.; Saturnino, C.; Aquaro, S.; Rosano, C.; Sinicropi, M.S. Multidrug Resistance (MDR): A Widespread Phenomenon in Pharmacological Therapies. *Molecules* **2022**, *27*, 616. [CrossRef] [PubMed]
6. Corrêa, R.C.G.; Heleno, A.S.; Alves, M.J.; Isabel, I.C.F.R. Bacterial resistance: Antibiotics of last generation used in clinical practice and the arise of natural products as new therapeutic alternatives. *Curr. Pharm. Des.* **2020**, *26*, 815–837. [CrossRef]
7. Interagency Coordination Group on Antimicrobial Resistance (IACG). No Time to Wait: Securing the Future from Drug Resistant Infections. Available online: https://www.who.int/docs/default-source/documents/no-time-to-wait-securing-the-future-from-drug-resistant-infections-en.pdf (accessed on 21 October 2022).
8. Ayaz, M.; Ullah, F.; Sadiq, A.; Ullah, F.; Ovais, M.; Ahmed, J.; Devkota, H.P. Synergistic interactions of phytochemicals with antimicrobial agents: Potential strategy to counteract drug resistance. *Chem. Biol. Interact.* **2019**, *308*, 294–303. [CrossRef]
9. Huang, Y.; Li, H.; Peng, D.; Wang, Y.; Ren, Q.; Guo, Y. The production and exportation of artemisinin-derived drugs in China: Current status and existing challenges. *Malaria J.* **2016**, *15*, 365. [CrossRef] [PubMed]
10. Tiwari, S.; Dubey, N.K. Traditional medicinal plants as promising source of immunomodulator against COVID-19. *J. Agric. Sci.* **2020**, *8*, 126–138. [CrossRef]
11. Akinbolade, S.; Coughlan, D.; Fairbairn, R.; McConkey, G.; Powell, H.; Ogunbayo, D.; Craig, D. Combination therapies for COVID-19: An overview of the clinical trials landscape. *Br. J. Clin. Pharmacol.* **2022**, *88*, 1590–1597. [CrossRef]

12. Taher, M.; Tik, N.; Susanti, D. Drugs intervention study in COVID-19 management. *Drug. Metab. Pers. Ther.* **2021**. [CrossRef]
13. Berber, E.; Sumbria, D.; Kokkaya, S. A metabolic blueprint of COVID-19 and long-term vaccine efficacy. *Drug Metab. Pers. Ther.* **2022**. [CrossRef]
14. Dhama, K.; Karthik, K.; Khandia, R.; Munjal, A.; Tiwari, R.; Rana, R.; Khurana, S.K.; Sana, U.; Khan, R.; Alagawany, M.; et al. Medicinal and Therapeutic Potential of Herbs and Plant Metabolites/Extracts Countering Viral Pathogens—Current Knowledge and Future Prospects. *Curr. Drug. Metab.* **2018**, *19*, 236–263. [CrossRef]
15. Takahashi, J.A.; Barbosa, B.V.R.; Lima, M.T.N.S.; Cardoso, P.G.; Contigli, C.; Pimenta, L.P.S. Antiviral fungal metabolites and some insights into their contribution to the current COVID-19 pandemic. *Bioorg. Med. Chem.* **2021**, *46*, 116366. [CrossRef] [PubMed]
16. Weng, L.M.; Su, X.; Wang, X.Q. Pain Symptoms in Patients with Coronavirus Disease (COVID-19): A Literature Review. *J. Pain Res.* **2021**, *14*, 147–159. [CrossRef] [PubMed]
17. Molla, M.A.; Yeasmin, M.; Slam, K.; Sharif, M.; Amin, R.; Nafisa, T.; Ghosh, A.K.; Parveen, M.; Arif, M.H.; Alam, J.A.J.; et al. Antibiotic Prescribing Patterns at COVID-19 Dedicated Wards in Bangladesh: Findings from a Single Center Study. *Inf. Prev. Pract.* **2021**, *3*, 100–134. [CrossRef]
18. Mobaraki, P.D.; Zaidi, A.K. Consequences of Intubation in COVID-19 Patients: Are We Ready. *IJOPL* **2020**, *10*, 50–53. [CrossRef]
19. Clancy, J.C.; Buehrle, J.D.; Nguyen, M.H. PRO: The COVID-19 pandemic will result in increased antimicrobial resistance rates. *JAC Antimicrob. Resist.* **2020**, *2*, dlaa049. [CrossRef] [PubMed]
20. Falcone, M.; Tiseo, G.; Giordano, C.; Leonildi, A.; Menichini, M.; Vecchione, A.; Pistello, M.; Guarracino, F.; Ghiadoni, L.; Forfori, F.; et al. Predictors of hospital-acquired bacterial and fungal superinfections in COVID-19: A prospective observational study. *J. Antimicrob. Chemother.* **2020**, *76*, 1078–1084. [CrossRef]
21. Hirschmann, R.; Bortolotto, C.C.; Silva, M.T.; Machado, A.K.F.; Xavier, M.O.; Fernandes, M.P.; Martins, R.C.; Bielemann, R.M.; Rodrigues, L.T.; Wehrmeister, F.C. Simultaneity of risk factors for chronic non-communicable diseases in a rural population of a Southern Brazilian city. *Rev. Bras. Epidemiol.* **2020**, *23*, E200066. [CrossRef]
22. Ye, C.; Zhang, S.; Zhang, X.; Cai, H.; Gu, J.; Lian, J.; Lu, Y.; Jia, H.; Hu, J.; Jin, C.; et al. Impact of comorbidities on patients with COVID-19: A large retrospective study in Zhejiang, China. *J. Med. Virol.* **2020**, *92*, 2821–2829. [CrossRef] [PubMed]
23. Coccia, M. Optimal levels of vaccination to reduce COVID-19 infected individuals and deaths: A global analysis. *Environ. Res.* **2022**, *30*, 10024. [CrossRef]
24. Hoffmann, M.; Krüger, N.; Schulz, S.; Cossmann, A.; Rocha, C.; Kempf, A.; Pöhlmann, S. The Omicron variant is highly resistant against antibody-mediated neutralization: Implications for control of the COVID-19 pandemic. *Cell* **2022**, *185*, 447–456. [CrossRef] [PubMed]
25. Choi, A.; Koch, M.; Wu, K.; Chu, L.; Ma, L.Z.; Hill, A.; Nunna, N.; Huang, W.; Oestreicher, J.; Colpitts, T.; et al. Safety and immunogenicity of SARS-CoV-2 variant mRNA vaccine boosters in healthy adults: An interim analysis. *Nat. Med.* **2021**, *27*, 2025–2031. [CrossRef]
26. Javorac, D.; Grahovac, L.; Manic, L.; Srrojikovic, N.; Andelkovi, A.; Bulat, Z.; Dukic, D.; Curcic, M.; Djordjevic, A.B. An overview of the safety assessment of medicines currently used in the COVID-19 disease treatment. *Food Chem. Toxicol.* **2020**, *144*, 111639. [CrossRef]
27. Salasc, F.; Lahlali, T.; Laurent, E.; Calatrava, M.R.; Pizzorno, A. Treatments for COVID-19: Lessons from 2020 and new therapeutic options. *COPHAR* **2022**, *62*, 43–49. [CrossRef]
28. Shubhangi, K.; Darren, F. Repurposing drugs for treatment of SARS-CoV-2 infection: Computational design insights into mechanisms of action. *J. Biomol. Struct. Dyn.* **2022**, *40*, 1316–1330.
29. Milas, S.; Poncelet, A.; Buttafuoco, F.; Pardo, A.; Lali, S.E.; Cherifi, S. Antibiotic use in patients with Coronavirus disease 2019 (COVID-19): Outcomes and associated factors. *Acta Clin. Belg.* **2022**, *77*, 579–587. [CrossRef] [PubMed]
30. Bakowski, A.M.; Beutler, N.; Chen, E.; Trinh, T.; Nguyen, H.; Kirkpatrick, M.G.; Parren, M.; Yang, L.; Ricketts, J.; Gupta, A.K.; et al. Oral drug repositioning candidates and synergistic remdesivir combinations for the prophylaxis and treatment of COVID-19. *bioRxiv* **2020**. [CrossRef]
31. Khataniar, A.; Pathak, U.; Rajkhowa, S.; Jha, A.N. A Comprehensive Review of Drug Repurposing Strategies against Known Drug Targets of COVID-19. *Covid* **2022**, *2*, 148–167. [CrossRef]
32. Langford, B.J.; So, M.; Raybardhan, S.; Leung, V.; Soucy, R.P.J.; Westwood, D.; Daneman, N.; Macfadden, D.R. Antibiotic prescribing in patients with COVID-19: Rapid review and meta-analysis. *CMI* **2021**, *27*, 520–531. [CrossRef]
33. Miranda, C.; Silva, V.; Capita, R.; Calleja, C.A.; Igrejas, G.; Poeta, P. Implications of antibiotics use during the COVID-19 pandemic: Present and future. *J. Antimicrob. Chemother.* **2020**, *75*, 3413–3416. [CrossRef]
34. Lucien, A.M.B.; Canarie, M.F.; Kilgore, P.E.; Denis, G.J.; Fenelon, N.; Pierre, M.; Cerpa, M.; Joseph, G.; Maki, G.; Zervos, M.J.; et al. Antibiotics and antimicrobial resistance in the COVID-19 era: Perspective from resource-limited settings. *Int. J. Infect. Dis.* **2021**, *104*, 250–254. [CrossRef]
35. Su, L.; Tu, Y.; Kong, D.; Chen, D.; Zhang, C.; Zhang, W.; Zhuang, C.; Wang, Z. Drug repurposing of anti-infective clinical drugs: Discovery of two potential anti-cytokine storm agents. *Biomed. Pharmacother.* **2020**, *131*, 110643. [CrossRef] [PubMed]
36. Chen, X.; Liao, B.; Cheng, L.; Peng, X.; Xu, X.; Li, Y.; Hu, T.; Li, J.; Zhou, X.; Ren, B. The microbial coinfection in COVID-19. *Appl. Microbiol. Biotechnol.* **2020**, *104*, 7777–7785. [CrossRef]
37. Chong, W.H.; Saha, B.K.; Ramani, A.; Chopra, A. State-of-the-art review of secondary pulmonary infections in patients with COVID-19 pneumonia. *J. Infect.* **2021**, *49*, 591–605. [CrossRef] [PubMed]

38. Uchino, K.; Kolikonda, M.K.; Brown, D.; Kovi, S.; Collins, D.; Khawaja, Z.; Buletko, A.B.; Russman, A.N.; Hussain, M.S. Decline in Stroke Presentations During COVID-19 Surge. *Stroke* **2020**, *51*, 2544–2547. [CrossRef] [PubMed]
39. Ma, Y.X.; Wang, C.Y.; Li, Y.Y.; Li, J.; Wan, Q.Q.; Chen, J.H.; Tay, F.R.; Niu, L.N. Considerations and Caveats in Combating ESKAPE Pathogens against Nosocomial Infections. *Adv. Sci.* **2020**, *7*, 1901872. [CrossRef] [PubMed]
40. Nadeem, F.S.; Gohar, U.F.; Tahir, S.F.; Mukhtar, H.; Pornpukdeewattana, S.; Nukthamna, P.; Ali, A.M.M.; Bavisetty, S.C.B.; Massa, S. Antimicrobial resistance: More than 70 years of war between humans and bactéria. *Crit. Rev. Microbiol.* **2020**, *5*, 578–599. [CrossRef]
41. Ilanko, A.; Cock, I.E. The interactive antimicrobial activity of conventional antibiotics and *Petalostigma* spp. extracts against bacterial triggers of some autoimmune inflammatory diseases. *Pharmacogn. J.* **2019**, *11*, 292–309. [CrossRef]
42. Obakiro, S.B.; Kiyimba, K.; Paasi, G.; Napyo, A.; Anthierens, S.; Waako, P.; Royen, P.V.; Iramiot, J.S.; Goossens, H.; Kostyanev, T. Prevalence of antibiotic-resistant bacteria among patients in two tertiary hospitals in Eastern Uganda. *J. Glob. Antimicrob. Resist.* **2021**, *25*, 82–86. [CrossRef] [PubMed]
43. Nelson, R.E.; Hatfield, K.M.; Wolford, H.; Samore, M.H.; Scott, R.D.; Reddy, S.C.; Paul, P.; Jernigan, J.A.; Baggs, L. National Estimates of Healthcare Costs Associated with Multidrug-Resistant Bacterial Infections Among patients hospitalized in the United States. *Clin. Infect. Dis.* **2021**, *72*, 17–26. [CrossRef]
44. Abdulbari, A.; Ali, N.M.; Raghif, A.R.A.; Matloob, A.N. Comparison of Oral Isotretinoin vs Azithromycin in the Treatment of Acne Vulgaris. *Indian J. Med. Forensic Med. Toxicol.* **2021**, *5*, 1485–1489.
45. Salmerón-Manzano, E.; Garrido-Cardenas, J.A.; Manzano-Agugliaro, F. Worldwide research trends on medicinal plants. *Int. J. Environ. Res. Public Health* **2020**, *17*, 3376. [CrossRef]
46. Hao, D.C.; Xiao, P.G. Pharmaceutical resource discovery from traditional medicinal plants: Pharmacophylogeny and pharmacophylogenomics. *Chin. Herb. Med.* **2020**, *12*, 104–117. [CrossRef]
47. Khan, T.; Sankhe, K.; Suvarna, V.; Sherje, A.; Patel, K.; Dravyakar, B. DNA gyrase inhibitors: Progress and synthesis of potent compounds as antibacterial agents. *Biomed. Pharmacother.* **2018**, *103*, 923–938. [CrossRef] [PubMed]
48. Anand, U.; Jacobo-Herrera, N.; Altemimi, A.; Lakhssassi, N. A comprehensive review on medicinal plants as antimicrobial therapeutics: Potential avenues of biocompatible drug discovery. *Metabolites* **2019**, *9*, 258. [CrossRef]
49. Baloyi, I.T.; Idowu, J.A.; Abdullahi, A.Y.; Sekelwa, C. Antibacterial, antiquorum sensing, antibiofilm activities and chemical profiling of selected South African medicinal plants against multi-drug resistant bacteria. *J. Med. Plant Res.* **2022**, *16*, 52–65.
50. Nigussie, D.; Gail, D.; Beyene, T.T.; Malcolm, B.; Adefris, L.B.; Abebaw, F.; Eyasu, M. Antibacterial and Antifungal Activities of Ethiopian Medicinal Plants: A Systematic Review. *Front. Pharmacol.* **2021**, *12*, 1663–9812. [CrossRef]
51. Manandhar, S.; Luitel, S.; Dahal, R.K. In Vitro Antimicrobial Activity of Some Medicinal Plants against Human Pathogenic Bacteria. *J. Trop. Med.* **2019**, *2019*, 1895340. [CrossRef]
52. Choi, J.S.; Jung, H.C.; Baek, Y.J.; Kim, B.Y.; Lee, M.W.; Kim, H.D.; Kim, S.W. Antibacterial Activity of Green-Synthesized Silver Nanoparticles Using *Areca catechu* Extract against Antibiotic-Resistant Bacteria. *Nanomaterials* **2021**, *11*, 205. [CrossRef] [PubMed]
53. Kebede, T.; Gadisa, E.; Tufa, A. Antimicrobial activities evaluation and phytochemical screening of some selected medicinal plants: A possible alternative in the treatment of multidrug-resistant microbes. *PLoS ONE* **2021**, *16*, e0249253. [CrossRef] [PubMed]
54. Cech, N.B.; Junio, H.A.; Ackermann, L.W.; Kavanaugh, J.S.; Horswill, A.R. Quorum quenching and antimicrobial activity of goldenseal (*Hydrastis canadensis*) against methicillin-resistant *Staphylococcus aureus* (MRSA). *Planta Med.* **2012**, *78*, 1556–1561. [CrossRef] [PubMed]
55. Richwagen, N.; Lyles, J.; Dale, B.; Quave, C. Antibacterial Activity of *Kalanchoe mortagei* and *K. fedtschenkoi* against ESKAPE Pathogens. *Front. Pharmacol.* **2019**, *10*, 67–72. [CrossRef] [PubMed]
56. Da Cruz, N.; Ferraz, A.C.; Moraes, T.D.F.S.; Lima, W.G.; Dos Santos, J.P.; Duarte, L.P.; de Magalhães, J.C. Pristimerin isolated from *Salacia crassifolia* (Mart. Ex. Schult.) G. Don. (Celastraceae) roots as a potential antibacterial agent against *Staphylococcus aureus*. *J. Ethnopharmacol.* **2021**, *266*, 113423.
57. Mulat, M.; Khan, F.; Pandita, A. Chemical Composition and Antibacterial, Anti-biofilm and Anti-virulence Activities of Plant Extracts against Human Pathogenic Bacteria. *Nat. Prod. J.* **2022**, *12*, 54–68. [CrossRef]
58. Schultz, F.; Anywar, G.; Tang, H. Targeting ESKAPE pathogens with anti-infective medicinal plants from the Greater Mpigi region in Uganda. *Sci. Rep.* **2020**, *10*, 11935. [CrossRef]
59. Assis, F.V.; Ferreira, J.M.S.; Siqueira, F.L.; Gonçalves, I.E.; Lacerda, R.P.; Nascimento, R.A.; Araújo, S.G.; Trindade, J.T.; Herrera, K.M.S.; Lima, A.R.L.S. Antibacterial activity of *Lamiaceae* plant extracts in clinical isolates of multidrug-resistant bacteria. *Anais da Academia Brasileira de Ciências* **2018**, *90*, 1665–1670. [CrossRef]
60. Al-sa'ady, A. Antibacterial screening for five local medicinal plants against nosocomial pathogens: *Klebsiella pneumoniae* and *Staphylococcus epidermidis*. *Eurasia J. Biosci.* **2020**, *14*, 553–559.
61. Al-Bakri, A.G.; Othman, G.; Afif, F.U. Determination of the antibiofilm, antiadhesive, and anti-MRSA activities of seven *Salvia* species. *Pharmacogn. Mag.* **2010**, *6*, 264–270. [CrossRef]
62. Tsai, C.C.; Lin, C.S.; Hsu, C.R. Using the Chinese herb *Scutellaria barbata* against extensively drug-resistant *Acinetobacter baumannii* infections: In vitro and in vivo studies. *BMC Complement. Altern. Med.* **2018**, *18*, 96. [CrossRef]
63. Bouhrim, M.; Radi, F.Z.; Mechchate, H.; Al-zahrani, M.; Qurtam, A.A.; Aleissa, A.M.; Drioiche, A.; Handaq, N.; Zair, T. Phytochemical Analysis, Antimicrobial and Antioxidant Properties of *Thymus zygis* L. and *Thymus willdenowii* Boiss. Essential Oils. *Plants* **2022**, *11*, 15.

64. Sun, L.; Song, F.; Shi, N.; Liu, F.; Li, S.; Li, P.; Zhang, W.; Jiang, X.; Zhang, Y.; Sun, L.; et al. Combination of four clinical indicators predicts the severe/critical symptom of patients infected COVID-19. *J. Clin. Virol.* **2020**, *128*, 104431. [CrossRef] [PubMed]
65. Chanda, S.; Rakholiya, K. Combination therapy: Synergism between natural plant extracts and antibiotics against infectious diseases. *Microbiol. Book Ser.* **2011**, *1*, 520–529.
66. Hemaiswarya, S.; Kruthiventi, A.K.; Doble, M. Synergism between natural products and antibiotics against infectious diseases. *Phytomedicine* **2008**, *15*, 639–652. [CrossRef]
67. Ju, J.; Xie, Y.; Yu, H.; Guo, Y.; Cheng, Y.; Qian, H.; Yao, W. Synergistic interactions of plant essential oils with antimicrobial agents: A new antimicrobial therapy. *Crit. Rev. Food Sci. Nutr.* **2020**, *62*, 1740–1751. [CrossRef] [PubMed]
68. Saquib, S.A.; AlQahtani, N.A.; Ahmad, I.; Kader, M.A.; Al Shahrani, S.S.; Asiri, E.A. Evaluation and Comparison of Antibacterial Efficacy of Herbal Extracts in Combination with Antibiotics on Periodontal pathobionts: An in vitro Microbiological Study. *Antibiotics* **2019**, *8*, 89. [CrossRef] [PubMed]
69. Embaby, M.A.; El-Raey, M.A.; Zaineldain, M.; Almaghrabi, O.; Marrez, D.A. Synergistic effect and efflux pump inhibitory activity of *Ficus nitida* phenolic extract with tetracycline against some pathogenic bacteria. *Toxin Rev.* **2021**, *40*, 1187–1197. [CrossRef]
70. Khleifat, K.M.; Matar, S.A.; Jaafreh, M.; Qaralleh, H.; Al-limoun, M.O.; Alsharafa, K.Y. Essential Oil of *Centaurea damascena* Aerial Parts, Antibacterial and Synergistic Effect. *J. Essent. Oil-Bear. Plants* **2019**, *22*, 356–367. [CrossRef]
71. Dassanayake, M.K.; Khoo, T.J.; An, J. Antibiotic resistance modifying ability of phytoextracts in anthrax biological agent *Bacillus anthracis* and emerging superbugs: A review of synergistic mechanisms. *Ann. Clin. Microbiol. Antimicrob.* **2021**, *20*, 79. [CrossRef]
72. Bezalwar, P.M.; Charde, V.N. Study on synergistic action of *Coriandrum sativum* seed extracts on antibiotics against multidrug resistant *P. aeruginosa*. *Environ. Conserv. J.* **2019**, *20*, 83–88. [CrossRef]
73. Da Silva, J.B.; de Bessa, M.E.; Mayorga, O.A.S.; Andrade, V.T.; da Costa, Y.F.G.; de Freitas, M.R.; Alves, M.S. A promising antibiotic, synergistic and antibiofilm effects of *Vernonia condensata* Baker (Asteraceae) on *Staphylococcus aureus*. *Microb. Pathog.* **2018**, *123*, 385–392. [CrossRef]
74. Maheshwari, M.; Althubiani, A.S.; Abulreesh, H.H.; Qais, F.A.; Khan, M.S.; Ahmad, I. Bioactive extracts of *Carum copticum* L. enhances efficacy of ciprofloxacin against MDR enteric bacteria. *Saudi J. Biol. Sci.* **2019**, *26*, 1848–1855. [CrossRef]
75. Bahmani, M.; Taherikalani, M.; Khaksarian, M.; Rafieian-Kopaei, M.; Ashrafi, B.; Nazer, M.; Rashidipour, M. The synergistic effect of hydroalcoholic extracts of *Origanum vulgare*, *Hypericum perforatum* and their active components carvacrol and hypericin against *Staphylococcus aureus*. *Future Sci.* **2019**, *5*, FSO371. [CrossRef] [PubMed]
76. Keawchai, K.; Chumkaew, P.; Permpoonpattana, P.; Srisawat, T. Synergistic effect of *Hydnophytum formicarum* tuber and *Vatica diospyroides Symington* cotyledon extracts with ampicillin on pathogenic bacteria. *J. Appl. Biol.* **2022**, *10*, 6–11.
77. Silva, D.M.; Costa, P.A.; Ribon, A.B.; Purgato, G.A.; Gaspar, D.M.; Diaz, M.A. Plant Extracts Display Synergism with Different Classes of Antibiotics. *Anais da Academia Brasileira de Ciências* **2019**, *91*, e20180117. [CrossRef] [PubMed]
78. Melguizo-Rodríguez, L.; García-Recio, E.; Ruiz, C.; De Luna-Bertos, E.; Illescas-Montes, R.; Costela-Ruiz, V.J. Biological properties and therapeutic applications of garlic and its components. *Food Funct.* **2022**, *13*, 2415–2426. [CrossRef] [PubMed]
79. Rana, S.; Chauhan, P. Spices that heal: Review on untapped potential of lesser-known spices as immunity booster during COVID-19 pandemic. *Ann. Phytomedicine* **2022**, *11*, 7–11. [CrossRef]
80. Mei, J.; Zhao, F.; Xu, R.; Huang, Y. A review on the application of spectroscopy to the condiments detection: From safety to authenticity. *Crit. Rev. Food Sci. Nutr.* **2021**, *62*, 6374–6389. [CrossRef] [PubMed]
81. Long, Y.; Zhang, M.; Mujumdar, A.S.; Chen, J. Valorization of turmeric (*Curcuma longa* L.) rhizome: Effect of different drying methods on antioxidant capacity and physical properties. *Dry. Technol.* **2022**, *40*, 1609–1619. [CrossRef]
82. Global Industry Analysts. Condiments Market—Growth, Trends, COVID-19 Impact and Forecasts (2022–2029). Available online: https://www.maximizemarketresearch.com/market-report/condiments-market/148426/#:~{}:text=Condiments%20Market%20Segment%20Analysis%3A&text=The%20Food%20Chain%20Service%20segment,dressing%20markets%20in%20the%20world (accessed on 25 October 2022).
83. Hussain, Y.; Alam, W.; Ullah, H.; Dacrema, M.; Daglia, M.; Khan, H.; Arciola, C.R. Antimicrobial Potential of Curcumin: Therapeutic Potential and Challenges to Clinical Applications. *Antibiotics* **2022**, *11*, 322. [CrossRef]
84. Thimmulappa, R.K.; Mudnakudu-Nagaraju, K.K.; Shivamallu, C.; Subramaniam, K.T.; Radhakrishnan, A.; Bhojraj, S.; Kuppusamy, G. Antiviral and immunomodulatory activity of curcumin: A case for prophylactic therapy for COVID-19. *Heliyon* **2021**, *7*, 6350. [CrossRef]
85. Babaei, F.; Nassiri-asl, M.; Hosseinzadeh, H. Curcumin (a constituent of turmeric): New treatment option against COVID-19. *Food Sci. Nutr.* **2020**, *8*, 5215–5227. [CrossRef] [PubMed]
86. Pawar, K.S.; Mastud, R.N.; Pawar, S.K.; Pawar, S.S.; Bhoite, R.R.; Bhoite, R.R.; Kulkarni, M.V.; Deshpande, A.R. Oral Curcumin with Piperine as Adjuvant Therapy for the Treatment of COVID-19: A Randomized Clinical Trial. *Front. Pharmacol.* **2021**, *12*, 669362. [CrossRef] [PubMed]
87. Ahmadi, R.; Salari, S.; Sharifi, M.D.; Reihani, H.; Rostamiani, M.B.; Behmadi, M.; Elyasi, S. Oral nano-curcumin formulation efficacy in the management of mild to moderate outpatient COVID-19: A randomized triple-blind placebo-controlled clinical trial. *Food Sci. Nutr.* **2021**, *9*, 4068–4075. [CrossRef] [PubMed]
88. Saber-Moghaddam, N.; Salari, S.; Hejazi, S.; Amini, M.; Taherzadeh, Z.; Eslami, S.; Elyasi, S. Oral nano-curcumin formulation efficacy in management of mild to moderate hospitalized coronavirus disease-19 patients: An open label nonrandomized clinical trial. *Phytother Res.* **2022**, *35*, 2616–2623. [CrossRef]

89. Vahedian-Azimi, A.; Abbasifard, M.; Rahimi-Bashar, F.; Guest, P.C.; Majeed, M.; Mohammadi, A.; Sahebkar, A. Effectiveness of Curcumin on Outcomes of Hospitalized COVID-19 Patients: A Systematic Review of Clinical Trials. *Nutrients* **2022**, *14*, 256. [CrossRef] [PubMed]
90. Zhang, X.; Chen, L.; Hu, C.; Fast, D.; Zhang, L.; Yang, B.; Kan, J.; Du, J. Curcumin attenuates poly(I:C)-induced immune and inflammatory responses in mouse macrophages by inhibiting TLR3/TBK1/IFNB cascade. *J. Funct. Foods.* **2022**, *33*, 104949. [CrossRef]
91. Mahmoud, D.B.; Bakr, M.M.; Al-karmalawy, A.A. Scrutinizing the Feasibility of Nonionic Surfactants to Form Isotropic Bicelles of Curcumin: A Potential Antiviral Candidate Against COVID-19. *AAPS PharmSciTech* **2022**, *23*, 44. [CrossRef]
92. Alici, H.; Tahtaci, H.; Demir, K. Design and various in silico studies of the novel curcumin derivatives as potential candidates against COVID-19—Associated main enzymes. *Comput. Biol. Chem.* **2022**, *98*, 107657. [CrossRef]
93. Rampogu, S.; Lee, G.; Park, J.S.; Lee, K.W.; Kim, M.O. Molecular Docking and Molecular Dynamics Simulations Discover Curcumin Analogue as a Plausible Dual Inhibitor for SARS-CoV-2. *Int. J. Mol. Sci.* **2022**, *23*, 1771. [CrossRef] [PubMed]
94. Asha, D.; Kushwaha, K.; Mishra, P.; Mrinalini, S.; Lilly, G.; Deepika, S. Nanocurcumin formulation: A possible therapeutic agent for post COVID inflammatory syndrome. *Immunopharmacol. Immunotoxicol.* **2022**, *44*, 141–146.
95. Soleymani, S.; Naghizadeh, A.; Karimi, M. COVID-19: General strategies for herbal therapies. *J. Evid.-Based Integr. Med.* **2022**, *27*, 1–18. [CrossRef] [PubMed]
96. Kiraç, H.; Dalda, S.A.; Coskun, O.F. Morphological and molecular characterization of garlic (*Allium sativum* L.) genotypes sampled from Turkey. *Genet. Resour. Crop. Evol.* **2022**, *69*, 1833–1841. [CrossRef]
97. Kiymaci, M.E.; Kaskatepe, B. Assessment of Cinnamon as an Antimicrobial Agent. In *Promising Antimicrobials from Natural Products*; Rai, M., Kosalec, I., Eds.; Springer: Cham, Switzerland, 2022; Volume 4, pp. 53–73.
98. Gidwani, B.; Bhattacharya, R.; Shukla, S.S.; Pandey, R.K. Indian spices: Past, present and future challenges as the engine for bio-enhancement of drugs: Impact of COVID-19. *J. Sci. Food Agric.* **2022**, *102*, 3065–3077. [CrossRef] [PubMed]
99. Mir, S.A.; Firoz, A.; Alaidarous, M.; Alshehri, B.; Dukhyil, A.B.; Banawas, S.; Alsagaby, A.S.; Alturaiki, W.; Bhat, A.G.; Kashoo, F.; et al. Identification of SARS-CoV-2 RNA-dependent RNA polymerase inhibitors from the major phytochemicals of *Nigella sativa*: An *in silico* approach. *Saudi J. Biol. Sci.* **2022**, *29*, 394–401. [CrossRef] [PubMed]
100. Bharathi, M.; Sivamaruthi, B.S.; Kesika, P.; Thangaleela, S.; Chaiyasut, C. In Silico Screening of Potential Phytocompounds from Several Herbs against SARS-CoV-2 Indian Delta Variant B.1.617.2 to Inhibit the Spike Glycoprotein Trimer. *Appl. Sci.* **2022**, *12*, 665. [CrossRef]
101. Chinnaiyan, S.K.; Pandiyan, R.; Natesan, S.; Chindam, S.; Gouti, A.K.; Sugumaran, A. Fabrication of basil oil Nanoemulsion loaded gellan gum hydrogel—Evaluation of its antibacterial and anti-biofilm potential. *J. Drug Deliv. Sci. Technol.* **2022**, *68*, 1773–2247. [CrossRef]
102. Zrig, A. The Effect of Phytocompounds of Medicinal Plants on Coronavirus (2019-NCOV) Infection. *Pharm. Chem. J.* **2022**, *55*, 1080–1084. [CrossRef]
103. Santos, S.; Barata, P.; Charmier, A. Cannabidiol and Terpene Formulation Reducing SARS-CoV-2 Infectivity Tackling a Therapeutic Strategy. *Front. Immunol.* **2022**, *13*, 841459. [CrossRef] [PubMed]
104. Ali, S.; Alam, M.; Khatoon, F.; Fatima, U.; Elasbali, A.M.; Adnan, M.; Islam, A.; Hassan, M.I.; Snoussi, M. Natural products can be used in therapeutic management of COVID-19: Probable mechanistic insights. *Biomed. Pharmacother.* **2022**, *147*, 112658. [CrossRef]
105. Bhatwalkar, S.B.; Mondal, R.; Krishna, S.B.N.; Adam, J.K.; Govender, P.; Anupam, R. Antibacterial properties of organosulfur compounds of garlic (*Allium sativum*). *Front. Microbiol.* **2021**, *12*, 613077. [CrossRef]
106. Torres, K.A.D.M.; Lima, S.M.R.R.; Torres, L.M.B.; Gamberini, M.T.; Silva, J.P.I.D. Garlic: An alternative treatment for group B *Streptococcus*. *Microbiol. Spectr.* **2021**, *9*, 121–170. [CrossRef]
107. Rouf, R.; Uddin, S.J.; Sarker, D.K.; Islam, M.T.; Ali, E.S.; Shilpi, J.A.; Sarker, S.D. Antiviral potential of garlic (*Allium sativum*) and its organosulfur compounds: A systematic update of pre-clinical and clinical data. *Trends Food Sci. Technol.* **2020**, *104*, 219–234. [CrossRef]
108. Thuy, B.T.P.; My, T.T.A.; Hai, N.T.T.; Hieu, L.T.; Hoa, T.T.; Thi, P.L.; Nhung, N.T.A. Investigation into SARS-CoV-2 resistance of compounds in garlic essential oil. *ACS Omega* **2020**, *5*, 8312–8320. [CrossRef]
109. Garg, M.; Sharma, A.; Vats, S.; Tiwari, V.; Kumari, A.; Mishra, V.; Krishania, M. Vitamins in Cereals: A Critical Review of Content, Health Effects, Processing Losses, Bioaccessibility, Fortification, and Biofortification Strategies for Their Improvement. *Front. Nutr.* **2021**, *8*, 586815. [CrossRef]
110. Zaer, A.M.; Norouzi, F.; Askari, V.R.; Khazdair, M.R.; Roshan, M.; Boskabady, M.; Hosseini, M.; Boskabady, M.H. The protective effect of *Nigella sativa* extract on lung inflammation and oxidative stress induced by lipopolysaccharide in rats. *J. Ethnopharmacol.* **2020**, *253*, 112653. [CrossRef]
111. Koshak, A.E.; Koshak, E.A. *Nigella sativa* l. as a potential phytotherapy for COVID-19: A mini-review of in-silico studies. *CTR* **2020**, *93*, 100602.
112. Islam, M.N.; Hossain, K.S.; Sarker, P.P.; Ferdous, J.; Hannan, M.A.; Rahman, M.M.; Uddin, M.J. Revisiting pharmacological potentials of *Nigella sativa* seed: A promising option for COVID-19 prevention and cure. *Phytother. Res.* **2021**, *35*, 1329–1344. [CrossRef]

113. Prasanth, D.S.N.B.K.; Murahari, M.; Chandramohan, V.; Panda, S.P.; Atmakuri, L.R.; Guntupalli, C. In silico identification of potential inhibitors from *Cinnamon* against main protease and spike glycoprotein of SARS CoV-2. *J. Biomol. Struct. Dyn.* **2020**, *39*, 4618–4632. [CrossRef]
114. D'aquila, P.; Paparazzo, E.; Crudo, M.; Bonacci, S.; Procopio, A.; Passarino, G.; Bellizzi, D. Antibacterial Activity and Epigenetic Remodeling of Essential Oils from Calabrian Aromatic Plants. *Nutrients* **2022**, *14*, 391. [CrossRef]
115. Qureshi, W.; Saeed, F.; Ajaz, M.; Rasool, S.A. In vitro antimicrobial, antibiofilm and antiphage activity of thyme (*Thymus vulgaris*). *Pak. J. Bot.* **2022**, *22*, 660–666. [CrossRef]
116. Kunnumakkara, A.B.; Rana, V.; Parama, D.; Banik, K.; Girisa, S.; Henamayee, S.; Aggarwal, B.B. COVID-19, cytokines, inflammation, and spices: How are they related? *Life Sci.* **2021**, *284*, 119201. [CrossRef]
117. Nath, M.; Debnath, P. Therapeutic role of traditionally used Indian medicinal plants and spices in combating COVID-19 pandemic situation. *J. Biomol. Struct. Dyn.* **2022**, *20*, 1–20.
118. Singh, N.A.; Kumar, P.; Kumar, N. Spices and herbs: Potential antiviral preventives and immunity boosters during COVID-19. *Phytother. Res.* **2021**, *35*, 2745–2757. [CrossRef]
119. Kumar, B.; Zaidi, S.; Haque, S.; Dasgupta, N.; Hussain, A.; Pramodh, S.; Mishra, B.N. In silico studies reveal antiviral effects of traditional Indian spices on COVID-19. *Curr. Pharm. Des.* **2021**, *27*, 3462–3475. [CrossRef]
120. Devan, A.R.; Nair, B.; Kumar, A.R.; Gorantla, J.N.; Nath, L.R. Unravelling the immune modulatory effect of Indian spices to impede the transmission of COVID-19: A promising approach. *Curr. Pharm. Biotechnol.* **2022**, *23*, 201–220. [CrossRef]
121. Radhika, A.G.; Malik, H. Fight against COVID-19: Survey of Spices & Herbs Used in North India. *Open J. Epidemiol.* **2021**, *11*, 256–266.
122. Sengupta, S.; Bhattacharyya, D.; Kasle, G.; Karmakar, S.; Sahu, O.; Ganguly, A.; Das Sarma, J. Potential Immunomodulatory Properties of Biologically Active Components of Spices Against SARS-CoV-2 and Pan β-Coronaviruses. *Front. Cell. Infect.* **2021**, *11*, 729622. [CrossRef] [PubMed]
123. Khasamwala, R.H.; Ranjani, S.; Nivetha, S.S.; Hemalatha, S. COVID-19: An In Silico Analysis on Potential Therapeutic Uses of Trikadu as Immune System Boosters. *Appl. Biochem. Biotechnol.* **2022**, *194*, 291–301. [CrossRef]
124. Rajamanickam, K.; Rathinavel, T.; Periyannan, V.; Ammashi, S.; Marimuthu, S.; Nasir Iqbal, M. Molecular insight of phytocompounds from Indian spices and its hyaluronic acid conjugates to block SARS-CoV-2 viral entry. *J. Biomol. Struct. Dyn.* **2022**, *12*, 1–20. [CrossRef]
125. Rajan, M.; Gupta, P.; Kumar, A. Promising antiviral molecules from ayurvedic herbs and spices against COVID-19. Chin. *J. Integr. Med.* **2021**, *27*, 243–244.
126. Ratnawatia, S.; Kumbara, A.; Pudentia, M.; Kusuma, W.; Soriente, A. The Indonesian Herbal Heritage Medicine during COVID-19 Pandemic. *Rev. Int. Geogr. Educ.* **2021**, *11*, 1611–1620.
127. Mlozi, S.H. The role of natural products from medicinal plants against COVID-19: Traditional medicine practice in Tanzania. *Heliyon* **2022**, *8*, e09739. [CrossRef]
128. Hajibeygi, R.; Mirghazanfari, S.M.; Pahlavani, N.; Jalil, A.T.; Alshahrani, S.H.; Rizaev, J.A.; Yekta, N.H. Effect of a diet based on Iranian traditional medicine on inflammatory markers and clinical outcomes in COVID-19 patients: A double-blind, randomized, controlled trial. *Eur. J. Integr. Med.* **2022**, *55*, 102179. [CrossRef]
129. Bousquet, J.; Haahtela, T.; Blain, H.; Czarlewski, W.; Zuberbier, T.; Bedbrook, A.; Anto, J.M. Available and affordable complementary treatments for COVID-19: From hypothesis to pilot studies and the need for implementation. *Clin. Transl. Allergy* **2022**, *12*, e12127. [CrossRef]
130. Johnson, T.S.; Narayana, D.A. Role of Spices in Offering Natural Immunity to Fight Various Diseases. *Pharmacogn. Mag.* **2021**, *13*, 600–613. [CrossRef]
131. Farzana, M.; Shahriar, S.; Jeba, F.R.; Tabassum, T.; Araf, Y.; Ullah, M.; Hosen, M.J. Functional food: Complementary to fight against COVID-19. *Be Beni Suef Univ. J. Basic Appl. Sci.* **2022**, *11*, 33. [CrossRef]

molecules

Article

Native Mass Spectrometry Coupled to Spectroscopic Methods to Investigate the Effect of Soybean Isoflavones on Structural Stability and Aggregation of Zinc Deficient and Metal-Free Superoxide Dismutase

Xinyu Bian [1,2], Xiaoyu Zhuang [3], Junpeng Xing [1], Shu Liu [1,2,*], Zhiqiang Liu [1,2] and Fengrui Song [1,2,*]

1 State Key Laboratory of Electroanalytical Chemistry & Jilin Province Key Laboratory of Chinese Medicine Chemistry and Mass Spectrometry, Changchun Institute of Applied Chemistry, Chinese Academy of Sciences, Changchun 130022, China
2 School of Applied Chemistry and Engineering, University of Science and Technology of China, Hefei 230029, China
3 Experiment Center for Science and Technology, Shanghai University of Traditional Chinese Medicine, Shanghai 201203, China
* Correspondence: mslab20@ciac.ac.cn (S.L.); songfr@ciac.ac.cn (F.S.); Tel.: +86-431-85262613 (S.L.); +86-431-85262044 (F.S.)

Abstract: The deficiency or wrong combination of metal ions in Cu, Zn-superoxide dismutase (SOD1), is regarded as one of the main factors causing the aggregation of SOD1 and then inducing amyotrophic lateral sclerosis (ALS). A ligands-targets screening process based on native electrospray ionization ion mobility mass spectrometry (ESI-IMS-MS) was established in this study. Four glycosides including daidzin, sophoricoside, glycitin, and genistin were screened out from seven soybean isoflavone compounds and were found to interact with zinc-deficient or metal-free SOD1. The structure and conformation stability of metal-free and zinc-deficient SOD1 and their complexes with the four glycosides was investigated by collision-induced dissociation (CID) and collision-induced unfolding (CIU). The four glycosides could strongly bind to the metal-free and copper recombined SOD1 and enhance the folding stability of these proteins. Additionally, the ThT fluorescence assay showed that these glycosides could inhibit the toxic aggregation of the zinc-deficient or metal-free SOD1. The competitive interaction experiments together with molecular docking indicate that glycitin, which showed the best stabilizing effects, binds with SOD1 between β-sheet 6 and loop IV. In short, this study provides good insight into the relationship between inhibitors and different SOD1s.

Keywords: superoxide dismutase; isoflavones; ion mobility-mass spectrometry; amyotrophic lateral sclerosis; aggregation

Citation: Bian, X.; Zhuang, X.; Xing, J.; Liu, S.; Liu, Z.; Song, F. Native Mass Spectrometry Coupled to Spectroscopic Methods to Investigate the Effect of Soybean Isoflavones on Structural Stability and Aggregation of Zinc Deficient and Metal-Free Superoxide Dismutase. *Molecules* **2022**, *27*, 7303. https://doi.org/10.3390/molecules27217303

Academic Editors: Raffaele Pezzani and Sara Vitalini

Received: 23 September 2022
Accepted: 24 October 2022
Published: 27 October 2022

Publisher's Note: MDPI stays neutral with regard to jurisdictional claims in published maps and institutional affiliations.

1. Introduction

Cu, Zn-superoxide dismutase (SOD1, a 32 kDa homodimeric protein) was the first protein found to be closely related to amyotrophic lateral sclerosis (ALS) [1]. One of the possible mechanisms of ALS is the death of motor neurons caused by toxic aggregated SOD1. The toxic aggregated SOD1 has been found in both familial and sporadic ALS cases [2]. The misfolding and dissociation of SOD1 due to mutations, wrong metallization, or aberrant post-translational modifications are the main causes of aggregation [3–5].

Although there is still no cure strategy for ALS, researchers never stop making efforts to find effective medicine and treatment methods. Inhibiting the aggregation of SOD1 was thought to be one of the key treatment methods [6]. Some small molecules or peptides that contain high-affinity structures with target proteins may disturb or change the aggregation pathway [7–9]. A variety of such molecules were searched from natural products [10–12].

In our previous work, some natural flavonoids and their derivatives have been studied, and proved to be effective inhibitors of SOD1 misfolding and aggregation [13–16].

Soybean isoflavones including aglycones and glycosides are a series of 3-benzopyranone compounds mainly found in soybeans [17]. Epidemiological investigations and research have connected these compounds to a variety of health factors such as limiting the development of cardiovascular diseases, osteoporosis, type 2 diabetes, and breast cancer [18–21]. The ability of antioxidants, the regulation of hormones, and the cell growth of isoflavones lead to the reduction of disease risk [22,23]. Many studies have proved the antioxidant properties of soybean isoflavones by the expression and activity changes of SOD1 [22]. However, no research has ever been concerned about the possible direct influence on the structure of SOD1 taken by these isoflavones.

Native mass spectrometry (MS) has been proven effective in analyzing the structural dynamics of non-covalent protein interactions such as protein-ligand (P-L) interactions [24]. The binding situation of P-L can be read directly from mass to charge (m/z) changes in the spectra. The signal intensity of the complexes usually provides binding affinity information of ligands. Comparing the structure stability of protein and P-L complexes is a reliable way of screening more targeted drug molecules [25]. Sometimes the research of ligand binding requires denaturing of the protein in order to increase the sensitivity or fragmentation of the protein to study the binding sites. This would reduce the biological significance. Ion mobility spectrometry can show structure information in the gas phase while keeping the native state in the solution during sample pretreatment [26]. Additionally, collision-induced-dissociation and collision-induced-unfolding are two adjunctive methods by giving collision energy (CE) to proteins or P-L complexes. The resulting structural changes bring different drift times (DT) and peak intensity distributions in ion mobility spectrometry (IMS) [27].

Based on our previous work, some kinds of flavonoids [13], stilbenoids [14], catechins [15], and caffeoylquinic acids [16] were found to form stable non-covalent complexes with wild-type SOD1 (Cu_2Zn_2SOD1) or metal-free SOD1 (ApoSOD1). Some of these compounds were proved to inhibit the aggregation of ApoSOD1. As isoflavones contain similar structure as the compounds mentioned above and show multiple biological functions, in this study, we firstly used the ESI-MS method to screen the bioactive ligands, which could effectively interact with different SOD1s from the incubation system of the soybean isoflavone mixture with various SOD1 species and confirm the target SOD1 species of these bioactive ligands (four glycosides). We then investigated the interaction between bioactive isoflavone glycoside and target SOD1s. Partly or wrongly metallated SOD1 has been reported to induce the toxic aggregation linked with ALS [28]. Therefore, two kinds of abnormally metallated SOD1s became our targets for screening, interaction, structure, and aggregation analysis. CID and CIU experiments were performed to show the influence on the target SOD1s' structure taken by the isoflavone glycosides. The method of TFE-induced aggregation was used to investigate the inhibition ability of the isoflavone glycosides to the aggregation of target SOD1s. The binding sites of these bioactive ligands on target SOD1s were predicted by experimental and theoretical methods.

2. Results and Discussion

2.1. Screening of Potential Inhibitors and Targets

Ligands that can bind with SOD1 are firstly screened from a mixture of seven soybean isoflavones, including three aglycones (L_{a1}–L_{a3}) and four glycosides (L_{g1}–L_{g4}). Their structures are shown in Figure 1. ESI–MS experiments of the isoflavone mixture with and without ApoSOD1, Cu_2SOD1, and Cu_2Zn_2 SOD1 (P_1, P_2, and P_3) are performed. Here we used bovine SOD1 instead of human SOD1, as their secondary structure are quite similar (referred from 1SXA and 6FN8 in the PDB database). Observing the MS intensity fading of free components is a usual method in ligand screening [29]. Here, compared with the intensity fading mass spectra (Figure S1, Table S1), it can be found that the intensity of free L_{g1}–L_{g4} decreases after the addition of proteins while the intensity of free L_{a1}–L_{a3} does not change the attenuation of free L_{g1}–L_{g4}, which indicates that part of these glycosides interact with the SOD1s. Thus, the following study focuses on these glycosides.

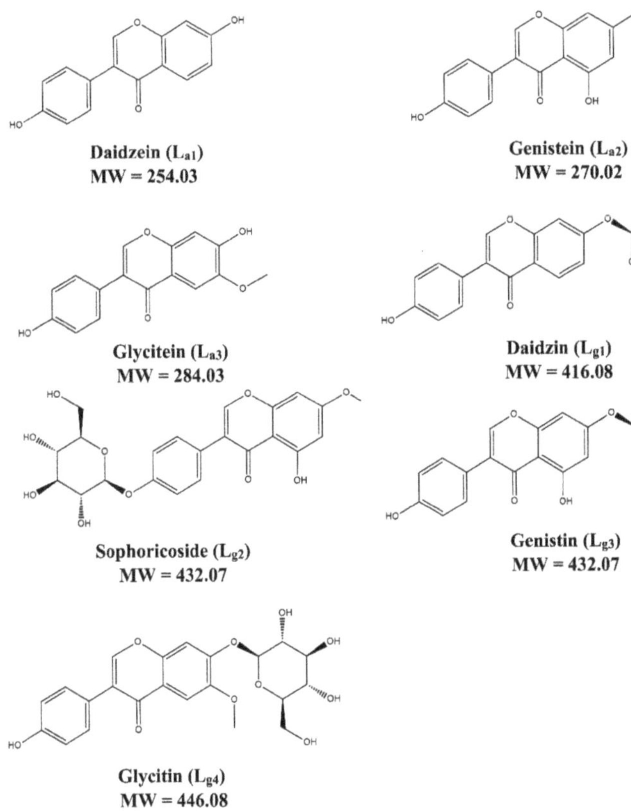

Figure 1. The structure of seven soybean isoflavones.

L_{g2} and L_{g4} are chosen as an example of glycosides interacting with the P_1–P_3 mixture. The possible target proteins for these glycosides include P_1–P_3 dimers (Di) and the monomers (Mo) dissociated from these dimers. Thus, ion mobility spectrometry (IMS) is used together with ESI-MS to confirm the targets. As shown in MS spectra (Figure 2A,A1), (Figure 2B,B1) of glycosides (L_{g2} and L_{g4}) interacted with the P_1-P_3 mixture, and complex ions with different charges were found, such as $[Di\text{-}L]^{11+}$, $[Di\text{-}L]^{12+}$ and $[Di\text{-}2L]^{12+}$ or $[Mo\text{-}L]^{6+}$, etc. It is found that the intensity of all P_3-L complex ions is far less than that of those complexes of P_1-L or P_2-L. Therefore, P_1 and P_2 are regarded as the main binding targets of the glycosides, while P_3 (Cu_2Zn_2SOD1) will no longer be concerned in this study. The 12+ charged dimer ($[Di]^{12+}$) and 6+ charged monomer ($[Mo]^{6+}$) of P_1 or P_2 share the same m/z in ESI-MS spectra (m/z 2599 for P_1 and 2609 for P_2), and can be divided after being extracted into IMS spectra (Figure 2C). According to our previous work, ref. [30] the peak at low drift time corresponds to $[Di]^{12+}$ and the other one corresponds to $[Mo]^{6+}$. Such results also occur for P-L complexes, as shown in Figure 2D,E, which show the extracted IMS spectra from the MS peaks of the complexes including P_1-L_{g2} at m/z 2672, P_2-L_{g2} at m/z 2682, P_1-L_{g4} at m/z 2674, and P_2-L_{g4} at m/z 2684. The IMS spectra profile of the complex ions is similar to that of P_1 and P_2, which means that their complexes also contain $[Di\text{-}2L_g]^{12+}$and $[Mo\text{-}L_g]^{6+}$ forms. Thus, the two peaks in the IMS spectra of each complex can be classified as $[Di\text{-}2L]^{12+}$ and $[Mo\text{-}L]^{6+}$. Therefore, both dimers and monomers of P_1

and P_2 can be regarded as the main binding targets of ligands. A series of calculations reflecting the complex generation quantity is listed in Table S2. According to the results, the dimer of P_1 seems to more easily bind with the four ligands. P_2 complexes also show appreciable percentages. The difference is that the binding rate of the P_2 monomer seems slightly higher than its dimer. P_3 shows the lowest binding ability, which is the same as the results of the visual observation of the spectra.

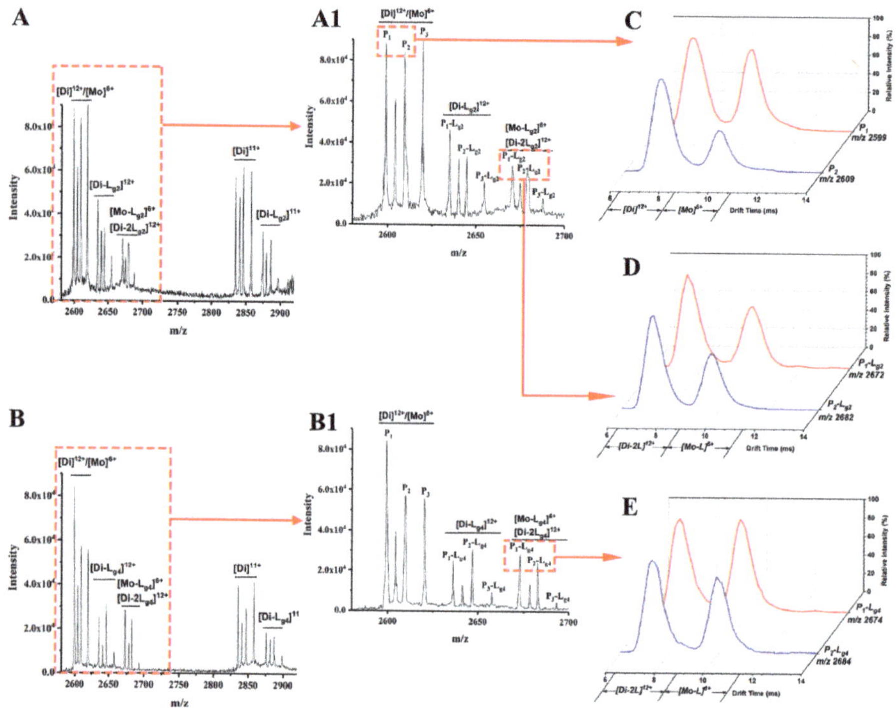

Figure 2. Mass spectra of sophoricoside (L_{g2}) with the SOD1 mixture (**A**) and glycitin (L_{g4}) with the SOD1 mixture (**B**). An enlarged view between *m/z* 2600 and 2700 is shown in (**A1**,**B1**). The IMS spectra of ApoSOD1 (P_1) and Cu_2 SOD1 (P_2) (**C**), P_1-L_{g2} and P_2-L_{g2} complexes (**D**), and P_1-L_{g4} and P_2-L_{g4} complexes (**E**) are extracted from the MS peaks of the *m/z* labeled beside the IMS curves.

2.2. Binding Affinity of ApoSOD1, Cu_2SOD1 with Glycosides

Exerting collision energy has been found to dissociate SOD1 dimer and SOD1-ligand complexes [13,30]. SOD1-ligand complexes may dissociate by losing a ligand or a subunit in CID. A clear dissociation pathway of SOD1 dimer and SOD1-ligand complexes helps explain and compare the binding affinity of ligands and the gas phase stability of different SOD1s and complexes. Thus, we performed CID experiments on the 11+ charged dimer ions of P_1, P_2, and their complexes with L_{g1}–L_{g4}. The ions are given a collision energy between 15 and 40 V in the trap region before being detected. L_{g4} is chosen as the representative compound to compare the binding affinities with different SOD1s. The dissociation behavior of complexes differs as CE increases. Figure 3 shows the change of relative intensity of the $[Di]^{11+}$ and $[Di\text{-}L_{g4}]^{11+}$ ions of proteins (P_1 and P_2) and their Di-L_{g4} (protein dimer-glycitin) complexes with trap CE. In the P_1 group (Figure 3A), the relative intensity of the $[Di]^{11+}$ ion of P_1 dimer is found to decrease as CE grows at a lower CE. This situation is reversed at about 26 V. The relative intensity of $[Di]^{11+}$ ion of the P_1 dimer comes to a minimum value and then increases. This strongly indicates that the complex (P_1-L_{g4}) dissociates into P_1 dimer and ligand (L_{g4}) at high CE because the P_1-L_{g4} complex is the only source of the P_1 dimer that is increasing. If we take the MS spectra of 28 V and 40 V of Trap CE, for example, it can be found that the peaks of the complexes nearly disappear under 40 V but the intensity of $[Di]^{11+}$ is similar as that under 28 V. The IM peaks extracted from the MS peak of the protein (Figure 3D) show that the main products under 28 V or higher CE are mostly at the monomer state. That is to say, the complexes under high Trap CE dissociate into separated proteins and ligands and the protein dimers keep the same charge state as the complexes ($[Di]^{11+}$ for example) or dissociate to monomer ($[Di\text{-}L]^{12+}$ to $[Mo]^{6+}$ and L). However, copper recombinant SOD1 provides very different results (Figure 3B). The downtrend of the dimer only becomes slighter at about 30V for P_2-L_{g4}. The convergence of intensity tendency of protein dimer and protein dimer-ligand complexes means that the ligand may interact more strongly with P_2 than P_1.

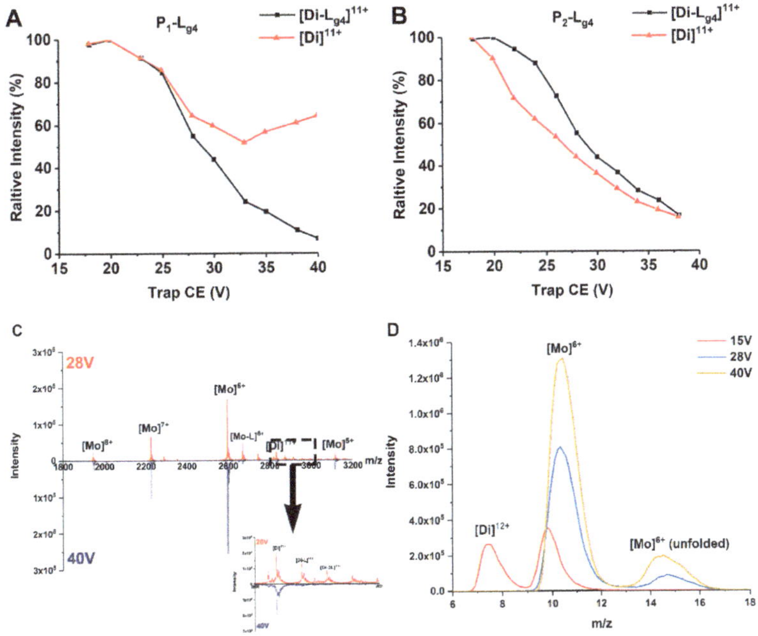

Figure 3. (**A**,**B**) Relative intensity changes of ApoSOD1 and Cu_2SOD1 (P_1 and P_2) and their Di-L_{g4} (protein dimer-glycitin) complexes as trap CE increases. (**C**) The mass spectra of P_1 at 28 V and 40 V of Trap CE with an enlarged view of the mass range between *m/z* 2800 and 3000. (**D**) The IM spectra of *m/z* 2600 extracted from (**C**) with an additional comparison of the same peak under 15 V of Trap CE.

MS/MS experiments of the complexes including P_1-L_{g4} and P_2-L_{g4} provide another comparison of these complexes (Figure S2). Given 30 V of Trap CE, the intensity of the P_1-L_{g4} ion ($[Di\text{-}L_{g4}]^{11+}$) becomes very weak while the P_1 dimer ion ($[Di]^{11+}$) maintains the strongest peak. In the P_2 group, the intensity of the $[Di\text{-}L_{g4}]^{11+}$ ion stays at a high level. The appearance of the $[Di]^{11+}$ ion of the dimer indicates that losing the ligand from the complex is still the main way of dissociation but at a lower extent than P_1-L_{g4}. This result further indicates that the binding of ligands to P_2 is stronger than that to P_1. At the same time, the small intensity of $[Mo\text{-}L_{g4}]^{6+}$ is also found in the P_2-L_{g4} group, indicating that a small part of the $[Di\text{-}L_{g4}]^{11+}$ ion of the P_2-L_{g4} loses one of the subunits before losing a ligand.

Based on the behavior of different SOD1s and their complexes in CID (Figures 3 and S2), it is clear that all the complexes dissociate in the way of losing a ligand, and the recombination of copper considerably enhances the interaction between the four glycosides and SOD1s.

With the increasing collision energy, P_1-L_g loses a ligand so easily that the abundance of the free P_1 dimer can even be increased, as shown in Figure 3A. More affinity of ligand binding taken by copper recombination makes the dissociation slower. As a result, the abundance of the $[Di\text{-}L_{g4}]^{11+}$ ion of P_2-L_{g4} can show parallel reduction with the free P_2 dimer. Figure 4 shows the possible dissociation pathway of the complex ions of different SOD1s (ApoSOD1, Cu_2SOD1) with glycoside in CID.

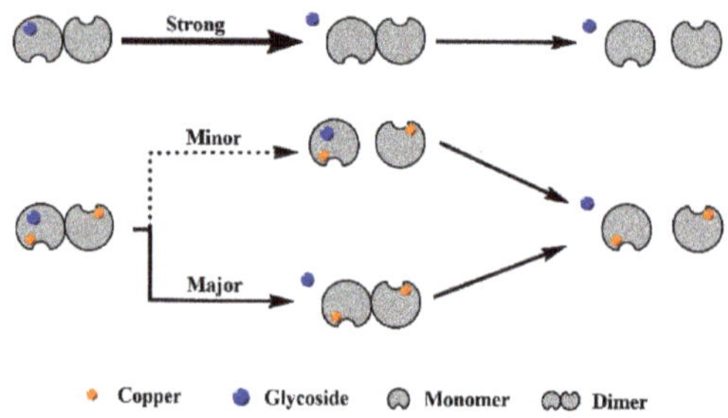

Figure 4. Possible dissociation pathway of the complex ions of different SOD1 (ApoSOD1, Cu_2SOD1) with glycoside in CID.

2.3. *Conformation Stability of the Complexes of ApoSOD1 and Cu_2SOD1 (P_1 and P_2) with Glycosides*

Collision energy not only dissociates SOD1 dimer or bound ligands but also makes SOD1 unfold [30]. Here we performed a collision-induced unfolding (CIU) experiment by IMS-MS to simulate the unfolding process of abnormal SOD1s and evaluate how the binding of soybean isoflavone glycosides influences the conformation stability of ApoSOD1 and Cu_2SOD1 (P_1 and P_2). As the 6+ charged monomer ion in IMS shows the same drift time as partly unfolded 12+ charged dimer ion, we chose the 11+ charged species as our analysis objects.

The unfolding of the SOD1 dimer contains three different states: native state (conformation A or Di-A) at low CE, partially unfolded state (conformation B or Di-B), and wholly unfolded state (conformation C or Di-C) [30]. The unfolded state of protein means more risk of aggregation. The unfolded conformation of protein brings a significant growth in collision cross section (CCS). This means higher drift time in IMS if the charge state does not change. Figure 5 shows the relative contents of three conformations of P_1, P_2, and P_1/P_2-L_{g4} at the CE of 30 V. The binding of the ligand does not bring conformation changes of the three states but rather brings abundance differences. The abundance of compact conformation of the P-L complexes is higher than P and, in turn, P-L complexes unfold less than P. To show the effect on protein conformation from different ligands, the relative abundance of conformation A is read from the IMS spectra and is listed in Table S3. It is found from the P_1 complexes that L_{g4} as a ligand increases the relative abundance of conformation A by about 5%. L_{g2} and L_{g3} behave less effectively in protecting conformation A. L_{g1} is the least effective in preventing the unfolding of P_1. For P_2 complexes, L_{g4} also shows the best in stabilizing the folding structure. L_{g1} still affects little in conformation A.

CIU (collision-induced unfolding) heat maps (Figures 6 and S3), matching drift time (DT), collision energy (CE), and relative intensity are plotted to give a comprehensive analysis of the unfolding difference taken by glycosides. The three independent areas represent the DT-CE distribution of the three conformations. The CE region of high intensity (red area) reflects the folding stability from another aspect. It is found that P_2-L_g needs higher CE to convert conformation A into B or B into C (Table S4), which means the glycosides can stabilize the conformation of P_2 at all unfolding stages. The P_1-L_g only shows a little difference with P_1.

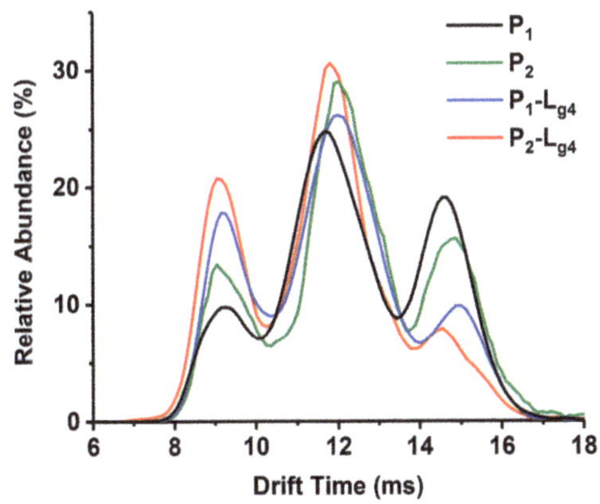

Figure 5. IMS spectra of Apo/Cu_2SOD1 (P_1/P_2) and their complex with glycitin (P_1/P_2-L_{g4}) at a collision energy of 30 V.

The results from the heat maps and IMS spectra at the CE of 30 V are unified. It is indicated from both the content of unfolded proteins and conformation conversion CE that the action of L_{g2}, L_{g3}, and L_{g4} are effective in inhibiting unfolding but different in some aspects. Specifically, the totally unfolded form of the proteins can surely be reduced in varying degrees, decided mainly by the ratio of P-L. L_{g4} shows the strongest ability to keep compact conformation.

2.4. The Effects of Soybean Isoflavone Glycosides on the Aggregation of Different SOD1s

ApoSOD1 can form toxic amyloid-like aggregates rapidly while being induced by TFE [31,32]. Under the induction of TFE, ApoSOD1 can partly unfold (which occurs at structures such as β-5, β-6) and quickly self-assemble. ThT is a fluorescent indicator which can specifically bind with aggregated β sheets. In this work, the ThT fluorescence intensity of TFE-induced samples was measured continuously to show the kinetics of aggregation. The influence of ligands and copper ions could be evaluated by the change in the aggregation process. The kinetics of P_1, P_2, and P_3 are shown in Figure S4. It can be found that the aggregating speed and extent of P_1 (ApoSOD1) are much higher than copper recombined SOD1 (Cu_2SOD1). The results shown in Figure 7 indicate that the four glycosides can inhibit the aggregation of P_1 and P_2. Considering the stochastic formation of aggregation [33], L_{g2}, L_{g3}, and L_{g4} show similar inhibition effects on aggregation and the effects are better than L_{g1}. This is reasonable because L_{g2}, L_{g3}, and L_{g4} were found more effective in stabilizing the folding structure of the target SOD1s than L_{g1} in the CIU results.

2.5. Binding Competition of Glycitin and Two Other Small Molecules to Cu_2SOD1 (P_2)

The binding sites of some small molecules such as isoproterenol and hesperidin on SOD1 have been predicted experimentally or theoretically [34,35]. Competitive binding between such molecules and other molecules showing binding affinity with unknown sites helps in judging the binding sites [14,36]. 5-Furd (L_{c1}) and naringin (L_{c2}) were reported as two stabilizers of SOD1 previously [8,13]. L_{c1} was found to bind with SOD1 on the surface of the β-barrel by XRD measurement, which contributes a lot to fibrous aggregation. The results of molecular docking and molecular dynamics simulation in our previous study showed that L_{c2} combines with SOD1 at the dimer interface. In this work, to study whether the soybean isoflavone glycosides would competitively bind with SOD1 against other ligands due to similar binding sites, we used L_{c1} and L_{c2} as control compounds to investigate the competitive binding of glycitin (L_{g4}) and control compounds to P_2. Different ligands were added into samples containing P_2 in a different order and then analyzed by ESI-MS. Figure 8 shows the ESI-MS spectra of competitive binding of L_{g4} with control compounds (L_{c1} or L_{c2}) to P_2. In the experiments, L_{c1} or L_{g4} was first incubated with P_2 for 1 h followed by the addition of the other ligand and incubation for another 1 h. Then the samples were subjected to MS measurement. Figure 8A,B shows that the addition of L_{g4} in a different order does not affect the intensity of $[Di\text{-}L_{c1}]^{11+}$ and $[Di\text{-}L_{g4}]^{11+}$, while a complex ($[Di\text{-}L_{c1}\text{-}L_{g4}]^{11+}$) of two ligands ($L_{c1}$ and L_{g4}) bound to P_2 is found in similar intensity. This result implies that L_{g4} may not bind at the same site as L_{c1}. The results of L_{c2} as a control compound are similar to that of L_{c1} above. The spectra of L_{g4} added respectively as the first or second ligand are shown in Figure 8C,D. The intensities of the complexes in Figure 8C,D are similar, including $[Di\text{-}L_{c1}]^{11+}$, $[Di\text{-}L_{g4}]^{11+}$, and $[Di\text{-}L_{c1}\text{-}L_{g4}]^{11+}$. The result suggests that L_{g4} and L_{c2} may bond at different sites on P_2. These results indicate that the competition ligands at least do not completely replace the site of the ligand that is added first. Thus, it can be predicted that L_{g4} does not share identical binding sites with either L_{c1} or L_{c2}.

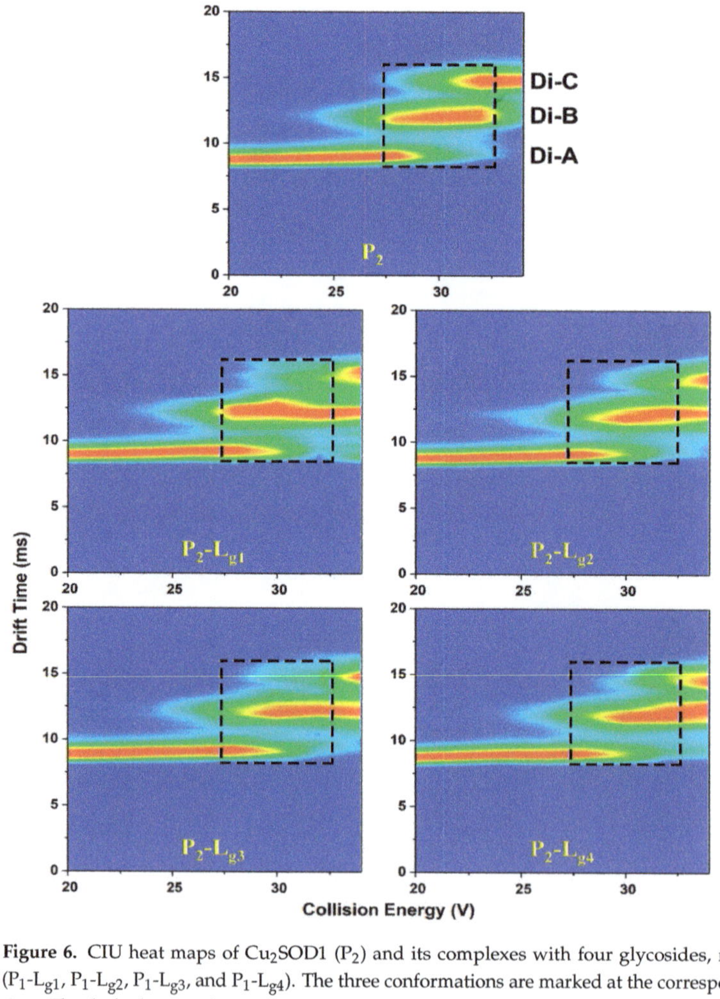

Figure 6. CIU heat maps of Cu_2SOD1 (P_2) and its complexes with four glycosides, respectively ($P_1\text{-}L_{g1}$, $P_1\text{-}L_{g2}$, $P_1\text{-}L_{g3}$, and $P_1\text{-}L_{g4}$). The three conformations are marked at the corresponding drift time. The dashed areas of trap collision voltages (27 V to 32 V) show a comparison between protein and complexes on the boundary voltage of conformation conversion.

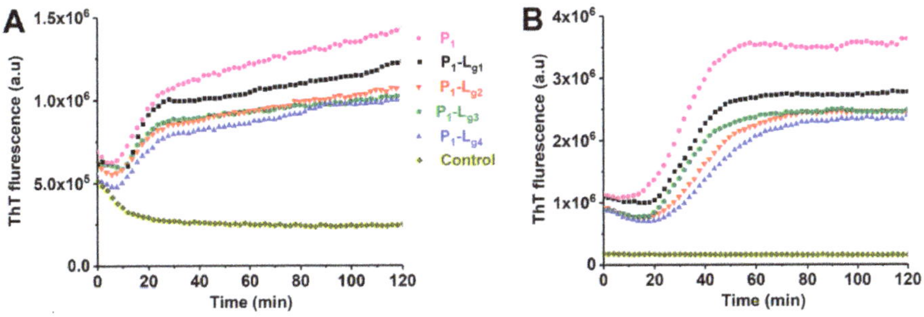

Figure 7. Aggregation kinetics of (**A**) ApoSOD1 (P_1) and (**B**) Cu_2SOD1 (P_2) and their P-L_g complexes. The control groups contain all compounds except for proteins.

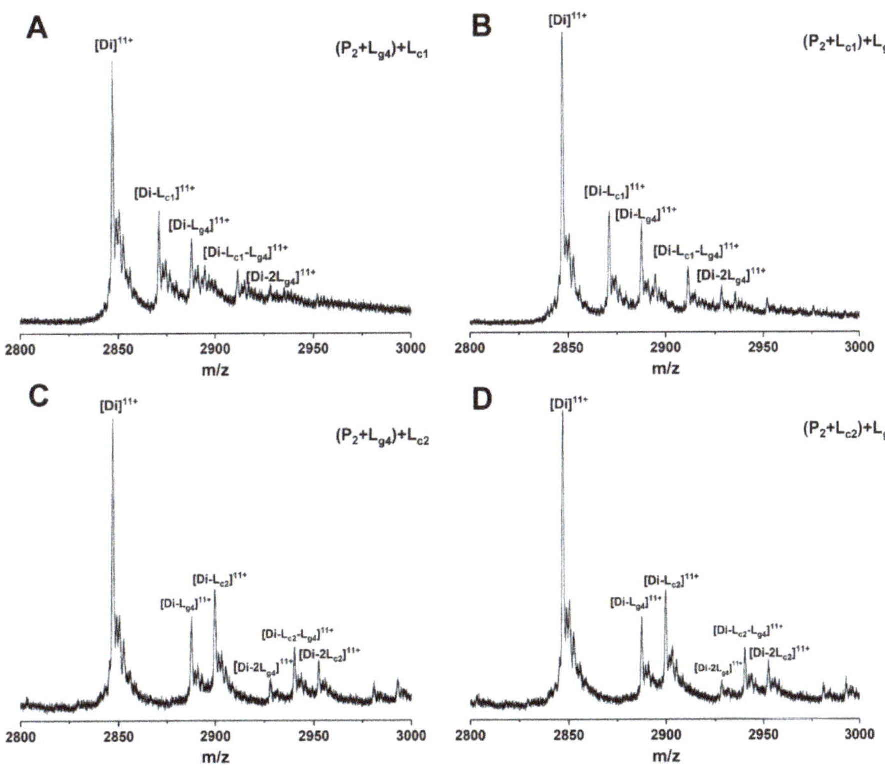

Figure 8. ESI-MS spectra of competition binding of glycitin (L_{g4}) with control compound (L_{c1}:5-Furd or L_{c2}: naringin) to Cu_2SOD1 (P_2). The compounds in the bracket are mixed and incubated for 1 h followed by adding the second ligand. Using 5-Furd (L_{c1}) as the control compound, the spectra of glycitin (L_{g4}) were added, respectively, as the first ligand (**A**) or second ligand (**B**). Using naringin (L_{c2}) as the control compound, the spectra of glycitin (L_{g4}) were added respectively as the first ligand (**C**) or second ligand (**D**).

2.6. Binding Sites of Glycitin Analyzed by Molecular Docking

Molecular docking was performed to predict the binding sites of L_{g4}. The sites (Figure 9) in P_1 and P_2 with low binding energy are placed at similar residues. Specifically, L_{g4} binds between β-sheet 6 (residue 94–100) and part of loop IV (residue 72–77). The ligand is oriented along the β-sheet and builds a connection between the β-sheet and loop IV through H-bonds. As reported, β-sheet 6 is the core region of aggregation in SOD1 [37–39]. Loop IV builds a link between the edge β-sheets and the whole structure and contributes a lot to dimerization and the enzymatic activity of SOD1 [40]. It is quite rigid in Cu_2Zn_2SOD1 but highly flexible in some SOD1 mutants. These fragments in some SOD1 mutants can be more unstable and easily unfold while being induced. Therefore, the binding of glycitin provides additional intermolecular interaction to stabilize the conformation of SOD1 as well as prevent aggregation.

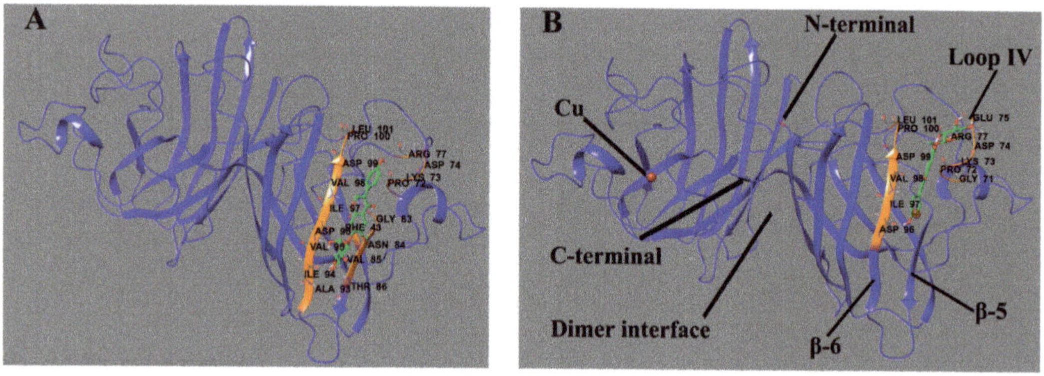

Figure 9. Predicted binding site of glycitin in ApoSOD1 (**A**) (binding energy = −6.33 kcal/mol) and Cu_2 SOD1 (**B**) (binding energy = −5.41 kcal/mol). The residues that interact with glycitin are labeled and shown as secondary structures (colored) as well as primary structures in the form of sticks. Glycitin is also shown in the form of sticks. The H-bonds between the ligand and residues are shown as green spherical dotted lines.

2.7. Calculation and Comparison of Collision Cross Section

Collision cross section (CCS) is usually used to evaluate the stretch degree of proteins. It can be measured by IMS and calculated through a theoretical model. In this study, the CCS of P_1 and P_2 and their L_{g4}-bound complexes are given by two methods: measured CCS from standard lines (Figure S5) and drift times of target components and theoretical CCS calculated through the computational method [41]. All results are listed in Tables 1 and 2. As the CCS of all species has little difference, the recombination of metal ions and glycitin makes little effect on the size of the overall structure of SOD1. It can be inferred from these results that losing metal ions or other stabilizers may not make SOD1 unfold immediately. However, SOD1 without metal ions or stabilizers can hardly stay stable in dynamic processes or external induction such as the collision energy in MS or TFE-induced aggregation.

Table 1. IMS measured CCS (nm^2) of ApoSOD1 and Cu_2 SOD1 (P_1, P_2) and complexes with glycitin (P_1-L_{g4}, P_2-L_{g4}) under six IM wave velocities (WV) and wave heights (WH).

WV(m/s)/WH(V)	430/40	350/40	400/40	430/35	400/37	360/37
P_1	23.82	24.01	23.22	24.62	24.16	23.77
P_1-L_{g4}	24.02	24.23	23.55	25.89	24.33	23.95
P_2	23.50	25.19	23.64	22.97	24.43	23.11
P_2-L_{g4}	23.68	25.35	23.81	23.04	24.57	23.28

Table 2. Average measured CCS (nm^2) and theoretically calculated CCS (nm^2) of ApoSOD1(P_1) and Cu_2SOD1(P_2) and their complexes (P_1-L_{g4}, P_2-L_{g4}). The error of measured CCS is the standard deviation of the CCS measured under six parameters (see Table 1). The error of calculated CCS is the standard deviation of the CCS calculated under different energy states (n = 4).

	IMS Measured	Theoretically Calculated
P_1	23.93 ± 0.46	26.48 ± 0.11
P_1-L_{g4}	24.33 ± 0.81	26.61 ± 0.10
P_2	23.81 ± 0.85	26.84 ± 0.16
P_2-L_{g4}	23.96 ± 0.86	26.86 ± 0.18

3. Materials and Methods

3.1. Materials

Cu_2Zn_2SOD1 from bovine erythrocytes was purchased from Beyotime Biotechnology (Shanghai, China). Daidzein, genistein, sophricoside, genistin, and (purity ≥98% for all) were provided from Nature Institutes for Food and Drug Control(Beijing, China). Glycitein, daidzin, and flycitin were purchased from Chengdu Herbpurify Co., LTD (Chengdu, China). Horse cytochrome c, horse myoglobin, naringin, Thioflavin-T (ThT), ammonium acetate, and 5-furd were obtained from Sigma-Aldrich (St. Louis, MO, USA). Trifluoroethanol (TFE) was purchased from J&K Scientific Ltd. (Beijing, China). Ethylenediaminetetraacetic acid (EDTA) and copric chloride dihydrate ($CuCl_2.2H_2O$) were obtained from Beijing Chemical Works (Beijing, China). Methanol and formic acid were supplied by TEDIA Company (Fairfield, OH, USA). The ultrapure water used in all experiments was prepared by a Milli-Q water purification system (Milford, MA, USA). The dialysis devices (Micro Float-A-Lyzer, MW-cut: 10 kDa) were bought from Spectrum Laboratories (Rancho Dominguez, CA, USA). The isoflavones, 5-furd, and naringin were dissolved at 1 mM in methanol as stock solutions and stored at 4 °C before experiments.

3.2. Preparation of ApoSOD1 and Cu Recombined SOD1

Step 1: Cu_2Zn_2SOD1 was first dialyzed by Float-A-Lyzer G2 (10-kDa MW-cut) in a 20 mM ammonium acetate buffer (pH = 3.2) containing 5 mM EDTA for 24 h.

Step 2: The buffer was changed into 20 mM ammonium acetate (pH = 3.2) to remove EDTA.

Step 3: The buffer differs for ApoSOD1 and Cu_2SOD1. A 20 mM ammonium acetate buffer (pH = 6.8) was used to refold ApoSOD1 while a 20 mM ammonium acetate buffer (pH = 3.2) containing 5 mM $CuCl_2$ for Cu_2SOD1. ApoSOD1 was prepared and collected after 24 h of dialysis followed by ultrafiltration. For Cu_2SOD1, the time of dialysis was usually about 4–6 h.

Steps 4 and 5 are only for Cu_2SOD1.

Step 4: The buffer was changed into 20 mM ammonium acetate (pH = 3.2). The dialysis lasted for 6 h.

Step 5: The buffer was changed into 20 mM ammonium acetate (pH = 6.8). The dialysis lasted for more than 12 h followed by ultrafiltration and product collection.

Step 6: For both products, 10 μL of the product was diluted at 10 mM ammonium acetate (pH = 6.8) to 200 μL. Then the samples were analyzed by ESI-MS after ultrafiltration to confirm the metal content. The products with unexpected metal content would be re-dialyzed from the first step.

The mass spectra of products suitable for other experiments are shown in Figure S6.

3.3. Mass Spectrometry Experiments

For all MS samples, the proteins and ligands were mixed and diluted into 5 μM and 30 μM by 10 mM ammonium acetate (pH = 6.8). The samples were incubated at 37 °C for 1 h before analysis.

The MS experiments were all performed on a quadrupole ion-mobility time-of-flight (Q-IM-TOF) mass spectrometer (Synapt G2-S, Waters Corp., Manchester, UK). Samples were directly infused at a flow rate of 10 μL/min and detected under positive-ion mode. If not mentioned specifically, all the conditions were set at optimized values. The capillary voltage was set at 2.40 kV and the cone voltage at 40 V. The source temperature was kept at 80 °C. As the desolvation gas, we used nitrogen at a flow rate of 400 L/h under 150 °C. Nitrogen was also the IMS gas in the IMS experiments and its flow rate was 90.00 mL/min. At the same time, manual control of IMS parameters was switched on for optimization. As a result, IMS wave velocity was set to 600 m/s and IMS wave height to 33.0 V. All data were acquired and analyzed by Masslynx 4.1 software (Waters Corp., Manchester, UK).

3.4. Analysis of Aggregation Kinetics

To prepare the samples of ThT fluorescence assays, the protein with or without a ligand was diluted with an ammonium acetate buffer (pH = 6.8) at 50 mM. The concentration of protein and ligand was 15 μM and 150 μM. ThT was added to be 25 μM at the same time. After incubating the samples at 37 °C for 1 h, TFE was added to induce the aggregation at the final concentration of 12%. The samples were quickly mixed well by vortex and transferred into a black 96-well plate. The fluorescence was measured every 2 min by a Molecular Devices Spectra Max i3x instrument. The excitation and emission wavelengths were set at 440 and 485 nm.

3.5. Competition Binding Experiments of Glycitin with Control Compounds to Cu_2SOD1

The protein and the first ligand were mixed and diluted into 5 μM and 30 μM by 10 mM ammonium acetate (pH = 6.8) and incubated at 37 °C for 1 h. Then the second ligand was added, and the concentration was also 30 μM. After incubation for another 1 h, the samples were analyzed by ESI-MS. The method and instrument parameters are the same as in other ESI-MS experiments.

3.6. Molecular Docking

The structure of bovine SOD1 (PDB code 1SXA [42,43]) was prepared using Autodock 4.2.6 program [44]. Water and metal ions (all ions for ApoSOD1 and zinc ions for Cu_2 SOD1) were removed from the models and saved as rigid macromolecules for docking. The force field parameters of copper ions were manually into the configuration files of the program for distinguishing the copper ions in Cu_2SOD1. The 3D structure of glycitin was downloaded from the PubChem website and optimized in the Autodock program. For docking parameters, glycitin was set as a flexible ligand for semiflexible docking. The grid box was set as 80 × 108 × 70 and was centered in one of the subunits of SOD1 to include the whole subunit and dimer interface. The binding models were output using the Lamarckian Genetic Algorithm and sorted by binding energy. Reasonable docked structures with the lowest binding energy were chosen as the final results.

3.7. Measurements and Calculation of Collision Cross Section

Theoretically, the calibration drift time (t'_d) and CCS (Ω_c) have a relationship, as Formula (1) shows. The calibration drift time and CCS can be calculated by Formulas (2) and (3) [45–47]. Here, t_d represents IMS measured drift time and Ω means the actual CCS. M_G means the relative molecular mass of IMS gas. The drift time (t_d) of cytochrome c and myoglobin under six different IMS parameters was measured first. The CCS (Ω) of these two standard proteins was provided by the CCS database. Then, linear fitted curves were plotted according to Formula (1). The drift time of SOD1-involved samples was measured under the same IMS parameters. The results were substituted into the formulas to get the measured CCS.

$$\ln(\Omega_c) = X\ln(t'_d) + A \tag{1}$$

$$t'_d = t_d - 1.57 \times 10^{-3} \times (m/z)^{0.5} \tag{2}$$

$$\Omega_c = \Omega/z[(m + M_G)/mM_G]^{0.5} \quad (3)$$

For the calculation of theoretical CCS, Collidoscope [41], an open source program, was employed to calculate the CCS of SOD1 or ligand-bound SOD1. Necessary modification for molecule models was performed before calculation. The calculation was processed using a charge placement algorithm. The net charge was set to be 11 and the collision gas was set as spherical N_2. As there is a lack of Lennard-Jones parameters of Cu in the default files, all Cu atoms in Cu_2SOD1-involved molecules were deleted as the last modification step for the models.

4. Conclusions

In summary, we screened four glycosides from seven soybean isoflavones showing stable non-covalent binding affinity with SOD1 at three different metallization states. The results of CID-MS and MS/MS indicated that the recombination of copper played an important role in increasing the binding affinity of the four glycosides. This may be the result from the stabilization effect of copper according to other experiments and calculations. CIU-IMS-MS showed that ApoSOD1 and Cu_2SOD1 need higher collision energy to unfold after binding with the glycosides. TFE-induced aggregation kinetics experiments showed that both Cu^{2+} and the four glycosides had significant effects on the formation of aggregates. As seen from the results of MS and aggregation kinetics, glycitin showed the best effect in stabilizing the SOD1 structure and inhibiting its aggregation, followed by sophoricoside and genistin. Competitive interaction experiments between two small molecules whose binding sites in SOD1 were known and glycitin were performed, and we made sure that these ligands shared different binding sites. Molecular docking also supported this result and showed that glycitin interacts with SOD1 at β-sheet 6 and loop IV. In conclusion, soybean isoflavone glycosides were hopeful small molecules in inhibiting the aggregation of abnormal metalized SOD1.

Supplementary Materials: The following supporting information is available online on the following link https://www.mdpi.com/article/10.3390/molecules27217303/s1, Table S1: The intensity and fading rate of MS peaks of isoflavone aglycones (La1–La3) and glycosides (Lg1–Lg5) before (I) and after (I′) adding SOD1 mixture. All compounds were analyzed under ESI negative mode. Figure S1: Intensity fading mass spectra of seven soybean isoflavones before (A) and after (B) adding P1, P2, and P3 mixture. All peaks represent [M-H]- ions of corresponded compounds. Table S2: The relative abundance of complexes compared with the protein species at the same charge state. The percentages (P) of dimer (Di)-ligand complexes were calculated directly through MS intensity (I) of +11 charged complexes and proteins. The formula is P = I (complex)/I(protein). The percentages (P) of monomer (Mo)-ligand complexes were calculated by the integrated IM peak area (A) of +6 charged monomers. The formula is P = A (complex)/A(protein). Figure S2: MS/MS spectra of Apo/ Cu2 SOD1-glycitin (P1-Lg4 and P2-Lg4) complexes. The Trap collision energy is set as 30 V. The precursor ions are the 11+ charged ions of dimer of the complexes ($[Di\text{-}Lg4]^{11+}$). Table S3. The relative abundance (%) of conformation A of different SOD1s and complexes. Figure S3: CIU heat maps of ApoSOD1 (P1) and complexes with four glycosides (Lg1-Lg 4). The three conformations are marked at corresponding drift time. Dashed areas of trap collision voltages (26 V to 32 V) show comparison between protein and complexes on the boundary voltage of conformation conversion. Table S4. Conversion CE (V) of Apo and Cu2SOD1(P1, P2) and their complexes with four glycosides (Lg1–Lg4). Figure S4. ThT fluorescence of ApoSOD1 (P1, red), Cu2SOD1 (P2, black), and Cu2Zn2SOD1 (P3, blue) induced by TFE. Figure S5: The linear fitted curves under six IM parameters (wave velocity and wave height) according to Formulas (1), standard CCS and measured drift time of cytochrome c and myoglobin. Figure S6: The MS spectra of ApoSOD1 (left) and Cu_2SOD1 (right) products after dialysis.

Author Contributions: Conceptualization, X.B. and F.S.; methodology, X.B., X.Z. and F.S.; validation, X.B. and F.S.; formal analysis, X.B.; investigation, X.B. and X.Z.; resources, J.X.; data curation, X.B. and J.X.; writing—original draft preparation, X.B.; writing—review and editing, F.S. and S.L.; visualization, X.B. and F.S.; supervision, Z.L. and S.L.; project administration, S.L. and F.S.; funding acquisition, F.S. and X.Z. All authors have read and agreed to the published version of the manuscript.

Funding: This research was funded by National Natural Science Foundation of China [number 81873193, 82173965 and 21904022].

Institutional Review Board Statement: Not applicable.

Informed Consent Statement: Not applicable.

Data Availability Statement: The data presented in this study are available in this article and Supplementary Materials.

Conflicts of Interest: The authors declare no conflict of interest.

Sample Availability: Not available.

References

1. Rosen, D.R.; Siddique, T.; Patterson, D.; Figlewicz, D.A.; Sapp, P.; Hentati, A.; Donaldson, D.; Goto, J.; Oregan, J.P.; Deng, H.X.; et al. Mutations in CU/ZN Superoxide-Dismutase Gene Are Associated with Familial Amyotrophic-Lateral-Sclerosis. *Nature* **1993**, *362*, 59–62. [CrossRef]
2. Gruzman, A.; Wood, W.L.; Alpert, E.; Prasad, M.D.; Miller, R.G.; Rothstein, J.D.; Bowser, R.; Hamilton, R.; Wood, T.D.; Cleveland, D.W.; et al. Common molecular signature in SOD1 for both sporadic and familial amyotrophic lateral sclerosis. *Proc. Natl. Acad. Sci. USA* **2007**, *104*, 12524–12529. [CrossRef]
3. Soto, C.; Pritzkow, S. Protein misfolding, aggregation, and conformational strains in neurodegenerative diseases. *Nat. Neurosci.* **2018**, *21*, 1332–1340. [CrossRef]
4. Trist, B.G.; Hilton, J.B.; Hare, D.J.; Crouch, P.J.; Double, K.L. Superoxide Dismutase 1 in Health and Disease: How a Frontline Antioxidant Becomes Neurotoxic. *Angew. Chem. Int. Ed. Engl.* **2021**, *60*, 9215–9246. [CrossRef] [PubMed]
5. Hilton, J.B.; White, A.R.; Crouch, P.J. Metal-deficient SOD1 in amyotrophic lateral sclerosis. *J. Mol. Med.-Jmm* **2015**, *93*, 481–487. [CrossRef]
6. Auclair, J.R.; Boggio, K.J.; Petsko, G.A.; Ringe, D.; Agar, J.N. Strategies for stabilizing superoxide dismutase (SOD1), the protein destabilized in the most common form of familial amyotrophic lateral sclerosis. *Proc. Natl. Acad. Sci. USA* **2010**, *107*, 21394–21399. [CrossRef] [PubMed]
7. Ray, S.S.; Nowak, R.J.; Brown, R.H.; Lansbury, P.T. Small-molecule-mediated stabilization of familial amyotrophic lateral sclerosis-linked superoxide dismutase mutants against unfolding and aggregation. *Proc. Natl. Acad. Sci. USA* **2005**, *102*, 3639–3644. [CrossRef] [PubMed]
8. Wright, G.S.; Antonyuk, S.V.; Kershaw, N.M.; Strange, R.W.; Samar Hasnain, S. Ligand binding and aggregation of pathogenic SOD1. *Nat. Commun.* **2013**, *4*, 1758. [CrossRef]
9. Goyal, D.; Shuaib, S.; Mann, S.; Goyal, B. Rationally Designed Peptides and Peptidomimetics as Inhibitors of Amyloid-beta (A beta) Aggregation: Potential Therapeutics of Alzheimer's Disease. *ACS Comb. Sci.* **2017**, *19*, 55–80. [CrossRef]
10. Sgarbossa, A. Natural Biomolecules and Protein Aggregation: Emerging Strategies against Amyloidogenesis. *Int. J. Mol. Sci.* **2012**, *13*, 17121–17137. [CrossRef]
11. Bhatia, N.K.; Srivastava, A.; Katyal, N.; Jain, N.; Khan, M.A.I.; Kundu, B.; Deep, S. Curcumin binds to the pre-fibrillar aggregates of Cu/Zn superoxide dismutase (SOD1) and alters its amyloidogenic pathway resulting in reduced cytotoxicity. *Biochim. Biophys. Acta (BBA)-Proteins Proteom.* **2015**, *1854*, 426–436. [CrossRef] [PubMed]
12. Bhatia, N.K.; Modi, P.; Sharma, S.; Deep, S. Quercetin and Baicalein Act as Potent Antiamyloidogenic and Fibril Destabilizing Agents for SOD1 Fibrils. *ACS Chem. Neurosci.* **2020**, *11*, 1129–1138. [CrossRef]
13. Zhuang, X.; Zhao, B.; Liu, S.; Song, F.; Cui, F.; Liu, Z.; Li, Y. Noncovalent Interactions between Superoxide Dismutase and Flavonoids Studied by Native Mass Spectrometry Combined with Molecular Simulations. *Anal. Chem.* **2016**, *88*, 11720–11726. [CrossRef]
14. Zhuang, X.; Li, X.; Zhao, B.; Liu, Z.; Song, F.; Lu, J. Native Mass Spectrometry Based Method for Studying the Interactions between Superoxide Dismutase 1 and Stilbenoids. *ACS Chem. Neurosci.* **2020**, *11*, 184–190. [CrossRef]
15. Zhao, B.; Zhuang, X.; Pi, Z.; Liu, S.; Liu, Z.; Song, F. Determining the Effect of Catechins on SOD1 Conformation and Aggregation by Ion Mobility Mass Spectrometry Combined with Optical Spectroscopy. *J. Am. Soc. Mass Spectrom.* **2018**, *29*, 734–741. [CrossRef]
16. Zhao, B.; Zhuang, X.Y.; Liu, S.; Liu, Z.Q.; Song, F.R.; Liu, S.Y. Investigation of the interaction between superoxide dismutase and caffeoylquinic acids by alkali metal assisted cationization-ion mobility mass spectrometry. *Int. J. Mass Spectrom.* **2018**, *434*, 151–157. [CrossRef]
17. Krizova, L.; Dadakova, K.; Kasparovska, J.; Kasparovsky, T. Isoflavones. *Molecules* **2019**, *24*, 1076. [CrossRef] [PubMed]
18. Liu, X.X.; Li, S.H.; Chen, J.Z.; Sun, K.; Wang, X.J.; Wang, X.G.; Hui, R.T. Effect of soy isoflavones on blood pressure: A meta-analysis of randomized controlled trials. *Nutr. Metab. Cardiovasc. Dis.* **2012**, *22*, 463–470. [CrossRef]
19. Nirmala, F.S.; Lee, H.; Kim, J.S.; Jung, C.H.; Ha, T.Y.; Jang, Y.J.; Ahn, J. Fermentation Improves the Preventive Effect of Soybean Against Bone Loss in Senescence-Accelerated Mouse Prone 6. *J. Food Sci.* **2019**, *84*, 349–357. [CrossRef]
20. Mezei, O.; Banz, W.J.; Steger, R.W.; Peluso, M.R.; Winters, T.A.; Shay, N. Soy isoflavones exert antidiabetic and hypolipidemic effects through the PPAR pathways in obese Zucker rats and murine RAW 264.7 cells. *J. Nutr.* **2003**, *133*, 1238–1243. [CrossRef]

21. Shu, X.O.; Zheng, Y.; Cai, H.; Gu, K.; Chen, Z.; Zheng, W.; Lu, W. Soy Food Intake and Breast Cancer Survival. *JAMA-J. Am. Med. Assoc.* **2009**, *302*, 2437–2443. [CrossRef]
22. Rizzo, G. The Antioxidant Role of Soy and Soy Foods in Human Health. *Antioxidants* **2020**, *9*, 635. [CrossRef]
23. Husain, D.; Khanna, K.; Puri, S.; Haghighizadeh, M. Supplementation of Soy Isoflavones Improved Sex Hormones, Blood Pressure, and Postmenopausal Symptoms. *J. Am. Coll. Nutr.* **2015**, *34*, 42–48. [CrossRef]
24. Sharon, M.; Robinson, C.V. The role of mass Spectrometry in structure elucidation of dynamic protein complexes. *Annu. Rev. Biochem.* **2007**, *76*, 167–193. [CrossRef]
25. Pacholarz, K.J.; Garlish, R.A.; Taylor, R.J.; Barran, P.E. Mass spectrometry based tools to investigate protein-ligand interactions for drug discovery. *Chem. Soc. Rev.* **2012**, *41*, 4335–4355. [CrossRef]
26. Stojko, J.; Fieulaine, S.; Petiot-Becard, S.; Van Dorsselaer, A.; Meinnel, T.; Giglione, C.; Cianferani, S. Ion mobility coupled to native mass spectrometry as a relevant tool to investigate extremely small ligand-induced conformational changes. *Analyst* **2015**, *140*, 7234–7245. [CrossRef] [PubMed]
27. Dupuis, N.F.; Wu, C.; Shea, J.E.; Bowers, M.T. The Amyloid Formation Mechanism in Human IAPP: Dimers Have beta-Strand Monomer-Monomer Interfaces. *J. Am. Chem. Soc.* **2011**, *133*, 7240–7243. [CrossRef] [PubMed]
28. Sahawneh, M.A.; Ricart, K.C.; Roberts, B.R.; Bomben, V.C.; Basso, M.; Ye, Y.; Sahawneh, J.; Franco, M.C.; Beckman, J.S.; Estevez, A.G. Cu,Zn-superoxide dismutase increases toxicity of mutant and zinc-deficient superoxide dismutase by enhancing protein stability. *J. Biol. Chem.* **2010**, *285*, 33885–33897. [CrossRef]
29. Downard, K.M. Indirect study of non-covalent protein complexes by MALDI mass spectrometry: Origins, advantages, and applications of the "intensity-fading" approach. *Mass Spectrom. Rev.* **2016**, *35*, 559–573. [CrossRef] [PubMed]
30. Zhuang, X.; Liu, S.; Zhang, R.; Song, F.; Liu, Z.; Liu, S. Identification of unfolding and dissociation pathways of superoxide dismutase in the gas phase by ion-mobility separation and tandem mass spectrometry. *Anal. Chem.* **2014**, *86*, 11599–11605. [CrossRef]
31. Stathopulos, P.B.; Rumfeldt, J.A.O.; Scholz, G.A.; Irani, R.A.; Frey, H.E.; Hallewell, R.A.; Lepock, J.R.; Meiering, E.M. Cu/Zn superoxide dismutase mutants associated with amyotrophic lateral sclerosis show enhanced formation of aggregates in vitro. *Proc. Natl. Acad. Sci. USA* **2003**, *100*, 7021–7026. [CrossRef]
32. Kumar, V.; Prakash, A.; Pandey, P.; Lynn, A.M.; Hassan, M.I. TFE-induced local unfolding and fibrillation of SOD1: Bridging the experiment and simulation studies. *Biochem. J.* **2018**, *475*, 1701–1719. [CrossRef]
33. Abdolvahabi, A.; Shi, Y.; Chuprin, A.; Rasouli, S.; Shaw, B.F. Stochastic Formation of Fibrillar and Amorphous Superoxide Dismutase Oligomers Linked to Amyotrophic Lateral Sclerosis. *ACS Chem. Neurosci.* **2016**, *7*, 799–810. [CrossRef]
34. Manjula, R.; Wright, G.S.A.; Strange, R.W.; Padmanabhan, B. Assessment of ligand binding at a site relevant to SOD1 oxidation and aggregation. *FEBS Lett.* **2018**, *592*, 1725–1737. [CrossRef] [PubMed]
35. Huang, H.J.; Chang, T.T.; Chen, H.Y.; Chen, C.Y. Finding inhibitors of mutant superoxide dismutase-1 for amyotrophic lateral sclerosis therapy from traditional chinese medicine. *Evid. Based Complement Altern. Med.* **2014**, *2014*, 156276. [CrossRef]
36. Hashempour, S.; Shahabadi, N.; Adewoye, A.; Murphy, B.; Rouse, C.; Salvatore, B.A.; Stratton, C.; Mahdavian, E. Binding Studies of AICAR and Human Serum Albumin by Spectroscopic, Theoretical, and Computational Methodologies. *Molecules* **2020**, *25*, 5410. [CrossRef]
37. Richardson, J.S.; Richardson, D.C. Natural beta-sheet proteins use negative design to avoid edge-to-edge aggregation. *Proc. Natl. Acad. Sci. USA* **2002**, *99*, 2754–2759. [CrossRef]
38. Elam, J.S.; Taylor, A.B.; Strange, R.; Antonyuk, S.; Doucette, P.A.; Rodriguez, J.A.; Hasnain, S.S.; Hayward, L.J.; Valentine, J.S.; Yeates, T.O.; et al. Amyloid-like filaments and water-filled nanotubes formed by SOD1 mutant proteins linked to familial ALS. *Nat. Struct. Biol.* **2003**, *10*, 461–467. [CrossRef] [PubMed]
39. Khare, S.D.; Dokholyan, N.V. Common dynamical signatures of familial amyotrophic lateral sclerosis-associated structurally diverse Cu, Zn superoxide dismutase mutants. *Proc. Natl. Acad. Sci. USA* **2006**, *103*, 3147–3152. [CrossRef]
40. Hornberg, A.; Logan, D.T.; Marklund, S.L.; Oliveberg, M. The coupling between disulphide status, metallation and dimer interface strength in Cu/Zn superoxide dismutase. *J. Mol. Biol.* **2007**, *365*, 333–342. [CrossRef] [PubMed]
41. Ewing, S.A.; Donor, M.T.; Wilson, J.W.; Prell, J.S. Collidoscope: An Improved Tool for Computing Collisional Cross-Sections with the Trajectory Method. *J. Am. Soc. Mass Spectrom.* **2017**, *28*, 587–596. [CrossRef] [PubMed]
42. Berman, H.M.; Westbrook, J.; Feng, Z.; Gilliland, G.; Bhat, T.N.; Weissig, H.; Shindyalov, I.N.; Bourne, P.E. The Protein Data Bank. *Nucleic Acids Res.* **2000**, *28*, 235–242. [CrossRef]
43. Rypniewski, W.R.; Mangani, S.; Bruni, B.; Orioli, P.L.; Casati, M.; Wilson, K.S. Crystal-Structure of Reduced Bovine Erythrocyte Superoxide-Dismutase at 1.9 Angstrom Resolution. *J. Mol. Biol.* **1995**, *251*, 282–296. [CrossRef] [PubMed]
44. Morris, G.M.; Huey, R.; Lindstrom, W.; Sanner, M.F.; Belew, R.K.; Goodsell, D.S.; Olson, A.J. AutoDock4 and AutoDockTools4: Automated Docking with Selective Receptor Flexibility. *J. Comput. Chem.* **2009**, *30*, 2785–2791. [CrossRef]
45. Ruotolo, B.T.; Benesch, J.L.P.; Sandercock, A.M.; Hyung, S.-J.; Robinson, C.V. Ion mobility-mass spectrometry analysis of large protein complexes. *Nat. Protoc.* **2008**, *3*, 1139–1152. [CrossRef] [PubMed]
46. Valentine, S.J.; Counterman, A.E.; Clemmer, D.E. A database of 660 peptide ion cross sections: Use of intrinsic size parameters for bona fide predictions of cross sections. *J. Am. Soc. Mass Spectrom.* **1999**, *10*, 1188–1211. [CrossRef]
47. Bush, M.F.; Hall, Z.; Giles, K.; Hoyes, J.; Robinson, C.V.; Ruotolo, B.T. Collision Cross Sections of Proteins and Their Complexes: A Calibration Framework and Database for Gas-Phase Structural Biology. *Anal. Chem.* **2010**, *82*, 9557–9565. [CrossRef]

molecules

Article

Pharmacokinetic Herb-Drug Interactions of Glipizide with *Andrographis paniculata* (Burm. f.) and Andrographolide in Normal and Diabetic Rats by Validated HPLC Method

Elza Sundhani [1,2], Agung Endro Nugroho [3,*], Arief Nurrochmad [3], Ika Puspitasari [3], Dita Amalia Prihati [4] and Endang Lukitaningsih [5,*]

1 Faculty of Pharmacy, Universitas Gadjah Mada, Sekip Utara, Yogyakarta 55281, Indonesia
2 Department of Pharmacology and Clinical Pharmacy, Faculty of Pharmacy, Universitas Muhammadiyah Purwokerto, Jl. KH. Ahmad Dahlan Dukuhwaluh, Purwokerto 53182, Indonesia
3 Department of Pharmacology and Clinical Pharmacy, Faculty of Pharmacy, Universitas Gadjah Mada, Sekip Utara, Yogyakarta 55281, Indonesia
4 Laboratory of Advanced Pharmaceutical Sciences, Faculty of Pharmacy, Universitas Gadjah Mada, Sekip Utara, Yogyakarta 55281, Indonesia
5 Department of Pharmaceutical Chemistry, Faculty of Pharmacy, Universitas Gadjah Mada, Sekip Utara, Yogyakarta 55281, Indonesia
* Correspondence: nugroho_ae@ugm.ac.id (A.E.N.); lukitaningsih_end@ugm.ac.id (E.L.); Tel.: +62-85643929723 (A.E.N.); +62-81392807988 (E.L.)

Citation: Sundhani, E.; Nugroho, A.E.; Nurrochmad, A.; Puspitasari, I.; Amalia Prihati, D.; Lukitaningsih, E. Pharmacokinetic Herb-Drug Interactions of Glipizide with *Andrographis paniculata* (Burm. f.) and Andrographolide in Normal and Diabetic Rats by Validated HPLC Method. *Molecules* **2022**, 27, 6901. https://doi.org/10.3390/molecules27206901

Academic Editors: Raffaele Pezzani and Sara Vitalini

Received: 13 September 2022
Accepted: 9 October 2022
Published: 14 October 2022

Abstract: Co-administered medicinal herbs can modify a drug's pharmacokinetics (PK), effectiveness, and toxicity. *Andrographis paniculata* (Burm. f.) ethanolic extract (APE) and andrographolide (AND) (a potent CYP2C9 inducer/inhibitor) can alter the pharmacokinetic parameters of glipizide (GLZ). This study aimed to determine the potential pharmacokinetics of herb–drug interactions between GLZ and APE/AND in the plasma of normal and diabetic rats using the HPLC bioanalysis method. The glipizide bioanalytical method established with RP-HPLC/UV instrument was validated following the EMA guidelines. GLZ was administered alone and in combination with APE or AND to normal and diabetic rats. The GLZ pharmacokinetic parameters were estimated according to the correlation between concentration and sampling time using the PK solver program. A simple and rapid GLZ bioanalysis technique with a lower limit of quantitation of 25 ng/mL was developed and presented the following parameters: accuracy (error $\leq$ 15%), precision (CV $\leq$ 15%), selectivity, stability, and linearity (R^2 = 0.998) at concentrations ranging 25–1500 ng/mL. APE administration significantly improved the C_{max} and $AUC_{0-t}/AUC_{0-\infty}$ GLZ values in normal and diabetic rats ($p < 0.05$). AND significantly reduced the bioavailability of GLZ in diabetic rats with small values of T 1/2, C_{max}, and $AUC_{0-t}/AUC_{0-\infty}$ ($p < 0.05$). This combination can be considered in administering medications because it can influence the pharmacological effects of GLZ.

Keywords: herb–drug interactions; *Andrographis paniculata*; andrographolide; glipizide

1. Introduction

Complementary and alternative medicine (CAM) derived from natural products or plant-based substances has been increasingly utilized to treat diseases. Its application is expanding in Asian nations, the United States, Europe, and Australia [1–3]. Diabetes mellitus is a disease that responds favorably to CAM treatments, particularly plant-based natural products [3,4]. Approximately 78% of patients with diabetic mellitus use herbal medicines and supplements as an alternative treatment [5,6].

According to the International Diabetes Federation, diabetes mellitus affects more than 500 million people worldwide (10.5% of the world's adult population) [7]. The prevalence of diabetes mellitus influences the production of many herbal medicines as alternative treatments. The components of herbal medicinal compounds can cause herb–drug interactions

(HDIs) because they can act as reversible inhibitors, irreversible inhibitors, or inductors of CYP450 enzymes or cause synergistic, additive, or antagonistic effects on the pharmacological activity of drugs [8]. The simultaneous usage of medicines and natural compounds can lead to pharmacokinetic (absorption, distribution, metabolism, and elimination) and pharmacodynamic (pharmacological activity) interactions [9–11].

Antidiabetic drugs, including sulfonylureas, meglitinide analogs, thiazolidinediones, and DPP4 inhibitors, are metabolized mainly by the enzymes CYP2C8, CYP2C9, CYP3A4, and CYP2C19 [12]. CYP2C9 is the most common member of the P450 2C (CYP2C) subfamily and has a role in the metabolism of oral hypoglycemic agents. This enzyme is also implicated in the metabolism of numerous anticoagulant drugs, NSAIDs, and antihypertensive drugs [13,14]. Glipizide is a sulfonylurea antidiabetic medication metabolized by CYP2C9 and CYP2C19 [15–17] to inactive metabolites in 3-cis-hydroxyglipizide (15%) and 4-trans-hydroxyglipizide (71%) [18]. Alterations in glipizide plasma concentrations caused by metabolic phase variations can affect the therapy's effectiveness. An example is HDIs that occur when combining drugs with natural substances that have inducing or inhibiting effects on their metabolizing enzymes.

Andrographis paniculata (Burm. f.) or Sambiloto is one of the herbs commonly used as alternative medicine due to its pharmacological activities, including antidiabetic, anti-inflammatory, anti-obesity, antioxidant, and anti-dengue properties [19–22]. In addition to its potential pharmacological efficacy, *A. paniculata* can produce HDIs with several conventional drugs. *A. paniculata*'s secondary metabolites, including andrographolide (AND), 14-deoxy-11,12-didehydroandrographolide, and andrographidine A, exhibit stable affinity and binding to receptors that play a role in the expression of CYP450 metabolizing enzymes, such as constitutive androstane receptor and pregnane X receptor (PXR) [23]. *A. paniculata* and its major secondary metabolite, andrographolide, can inhibit the kinetics of several enzymes, including CYP2E1, and decrease the expression of CYP2C9 and CYP3A proteins on human hepatic cytochrome and Caco-2 cell line [24–27].

A. paniculata ethanolic extract (APE) and AND can alter the pharmacokinetic profile of a drug by increasing the clearance (CL) value and significantly decreasing the AUC of theophylline, etoricoxib, nabumetone, and naproxen [28–31]. APE can also increase the AUC and reduce the CL of gliclazide and midazolam [32,33]. AND, the primary component of APE, can alter the pharmacokinetic parameters of aminophylline, doxofylline, meloxicam, glyburide, glimepiride, metformin, and warfarin by enhancing their AUC and Tmax values and significantly reducing their CL values [34–38].

APE and AND exhibit HDIs with several drugs; however, their pharmacokinetic interactions with the antidiabetic agent glipizide have never been examined. Drug interaction investigations, particularly during the pharmacokinetic phase, require a selective, precise, and efficient method for monitoring plasma glipizide concentration. The chromatographic approach is the most commonly used because it can obtain low detection limits, produce structural information, and analyze analytes with varying polarities [39]. LC–MS/MS and RP-HPLC/UV have been applied extensively to assess glipizide in the plasma matrix. Compared with LC–MS/MS, RP-HPLC/UV is simpler in terms of sample preparation and less expensive. Prior to the analysis of plasma glipizide concentration in normal and diabetic rats, the analytical method was verified following the European Medicines Agency (EMA) guidelines. The best and most valid analytical method was applied to examine the pharmacokinetics of glipizide in plasma from the rats co-administered with APE and AND.

2. Materials and Methods

2.1. Chemicals and Reagents

Glipizide of pharmaceutical secondary standard (Sigma–Aldrich, St. Louis, MO, USA), andrographolide (98%, Sigma–Aldrich), simplicia *Andrographis paniculata* (Burm. f.) collection from B2P2TOOT Tawangmangu, Central Java, Indonesia, acetonitrile of HPLC grade (Smartlab), methanol of HPLC grade, potassium dihydrogen phosphate (KH_2PO_4), distilled water (PT. Brataco, Indonesia), and blank plasma were used in this study.

2.2. HPLC Condition

A set of HPLC instruments (Hitachi UV-vis L-2420 detector [at 233 nm], Hitachi L-2130 HPLC pump, D-2000 HSM elite software, Phenomenex Luna® 5 m C18 100 chromatographic column, LC column 250 mm × 4.6 mm) were employed in this work. The mobile phase consisted of KH_2PO_4 and acetonitrile (57%:43%) and had a pH of 4.25 and a flow rate of 1 mL/min. The mobile phase was filtered using a 0.45 m pore filter and degassed using a bath sonicator before use.

2.3. Preparation of Standard Stock and Working Solutions

In brief, 10 mg of glipizide was weighed and transferred to a 10 mL volumetric flask before methanol was added to the mark (Solution A, glipizide 1000 mg/mL). Afterward, 1 mL of solution A was pipetted into a 10 mL volumetric flask, methanol was added to the calibration mark, and the mixture was homogenized (Solution B, glipizide 100 g/mL). Solution B was pipetted into a 10 mL flask, and methanol was added to the mark to create a series of working standard solutions (0.25–15 µg/mL).

2.4. Preparation of Spiked Plasma

In brief, 200 µL of plasma was administered with a standard solution of glipizide in a centrifuge tube. Deprotonation was conducted by adding 1000 µL of acetonitrile and centrifuging at 14,000 rpm at 40 °C for 10 min. The supernatant was extracted, transferred to a vial (twofold replication), and dried. After drying, 1 mL of mobile phase was added to the residue, which was then vortexed for 1 min, filtered over a 0.45 µm nylon/PVDF membrane, and introduced into the HPLC instrument.

2.5. System Suitability Test (SST)

SST was conducted to measure the amount of glipizide in plasma using a standard solution of glipizide with a concentration of 1500 ng/mL. The SST criteria for the percentage RSD area of glipizide in standard solutions for six injection replications (≤2.0%, tailings factor [asymmetry] ≤ 2.0, and theoretical plate [N] ≥ 2000) were satisfied.

2.6. Method Validation

The bioanalytical method for measuring glipizide in plasma was validated following the EMA guidelines. The validation procedure consisted of selectivity, calibration curve, lower limit of quantitation (LLOQ), accuracy, and precision, carryover, dilution integrity, and stability.

2.6.1. Selectivity and Carryover

Selectivity test was conducted by injecting plasma for interference with glipizide peaks. Selectivity is achieved when the peak interference in the retention time of glipizide in blank plasma is ≤20% of the LLOQ of glipizide. For carryover, blank plasma was injected into the HPLC system after the injection of the highest concentration plasma spike.

2.6.2. Calibration Curve and LLOQ

Calibration standards were prepared by adding standard glipizide solution to 200 µL (spike) of blank plasma to obtain the final serial concentrations of 25, 75, 150, 300, 550, 700, 950, 1150, 1250, and 1500 ng/mL. The values of linear regression, slope, intercept, and percentage recovery were calculated from the concentration versus area correlation. The calibration standard satisfies the parameters when the difference between the observed concentration of the standard solution and its theoretical concentration is less than ±15%, except for LLOQ for which the difference cannot exceed ±20%. A minimum of six series of standard concentrations must comprise the calibration curve, and 75% of the standard solutions must meet the criterion. LLOQ was determined by injecting standard glipizide solution into plasma at concentrations of 15, 25, and 30 ng/mL to obtain six replicates. LLOQ was determined at the measured concentration with a percentage difference of ±20%.

2.6.3. Accuracy and Precision

Glipizide standard solution was added to the plasma at four concentration levels within the calibration curve range: LLOQ, 3 × LLOQ (low QC), 30% to 50% of the calibration curve range (medium QC), and at least 75% of the highest calibration curve (high QC). Accuracy is satisfied when the difference between the average value of the measured QC sample concentration and the theoretical concentration is less than ±15%, except for LLOQ for which the difference must not exceed ± 20%. Precision is satisfied when the standard deviation value of the QC sample concentration measured from six replications did not exceed 15%, except for LLOQ for which the difference must not exceed 20%.

2.6.4. Dilution Integrity

The integrity of dilution was determined by adding analyte at a concentration above the ULOQ to blank plasma, which was subsequently diluted with blank plasma at least five times for each dilution factor (2×, 5×, and 10×). The integrity of dilution must match the criteria for accuracy and precision.

2.6.5. Stability

The glipizide bioanalytical stability test was conducted using low (75 ng/mL) and high QC (1250 ng/mL) samples that were tested immediately after preparation and using low (75 ng/mL) and high QC (1250 ng/mL) samples stored under specific conditions and then examined. This stability test includes a short-term stability test, in which the QC samples of glipizide were prepared at room temperature (T0) and frozen in a freezer (−80 °C) for 4 (T4) and 24 h (T24). The QC glipizide samples were stored in a freezer (−80 °C) and thawed at room temperature three times to determine their liquid-freezing stability. Stable QC samples of glipizide ready for injection were stored in the autosampler for 24 h.

2.7. Experimental Design for In Vivo Studies

2.7.1. Animals Used in the Study

The UGM Integrated Research and Testing Laboratory provided male Wistar rats weighing between 180 and 250 g and aged 8–10 weeks for the experiments. The rats were maintained in cages (50 cm × 45 cm × 15 cm) with controlled temperature (24 ± 1 °C) and humidity (40–70%) place in a room with an auto-adjusted light cycle of 12 h of bright and 12 h of dark conditions. The feed was 5–10 g/day ABS2, and drinking water (groundwater) was administered ad libitum. The Animal Ethics Committee of Integrated Research and Testing Laboratory UGM, Indonesia authorized the protocol for maintenance and pharmacokinetic study (approval number: 00055/04/LPPT/XII/2021; Yogyakarta, Indonesia).

2.7.2. Sample Preparation

Andrographis paniculata (Burm. f.) extract (APE) was obtained using the maceration method with 96% ethanol solvent for 24 h (remaceration two times). Using a rotary evaporator at 50 °C, the filtrate from the maceration was concentrated to get a thick extract. Animals were administered glipizide (GLZ), APE, and andrographolide (AND) standards suspended in 5% sodium carboxy methyl cellulose (CMC-Na).

2.7.3. Pharmacokinetic Interaction Study in Normal and Diabetic Rats

The changes in the pharmacokinetic parameters of glipizide in normal rats were evaluated using pharmacokinetic assays to determine the effects of a 7 day treatment with APE and AND. Twenty animals were used in this pharmacokinetic test and were divided into four groups: CMC-Na or glipizide (5 mg/kg BW) on day 7, a combination of APE (300 mg/kg BW) for 7 days and glipizide (5 mg/kg BW) on day 7, and a combination of AND (15 mg/kg BW) for 7 days and glipizide (5 mg/kg BW) on day 7. The same number of rats in the treatment groups were utilized for the pharmacokinetic evaluation of diabetic rats. The rats were verified to be diabetic (blood glucose values > 200 mg/dL) after being intraperitoneally induced with streptozotocin (STZ) 65 mg/kg BW and nicotinamide 110 mg/kg BW. CMC-Na alone, glipizide (5 mg/kg BW) alone, and combination of APE (300 mg/kg BW) and AND (15 mg/kg BW) with glipizide (5 mg/kg BW) were administered to rats for 28 days. The doses of APE and AND selected relate to optimizing antidiabetic activity, while the dose of glipizide is derived from previous research [40,41].

The blood was collected through the orbital sinus of the rat's eye (250–300 μL) at 0.25, 0.5, 2, 4, 6, 8, 12, and 14 h after treatment administration to measure the pharmacokinetic parameters. The blood was centrifuged at 4000 rpm for 10 min, and the plasma was kept at −80 °C before analysis. In brief, 200 μL of plasma samples were deprotonated with 1000 μL of acetonitrile. The filtrate and residue were separated by centrifugation at 14,000 rpm and 4 °C for 10 min. The filtrate was successfully treated by deprotonation twice and was dried until no acetonitrile residue remained. The residue was then dissolved in the mobile phase in the HPLC instrument. For the subsequent analysis, 1 mL of the sample solution was placed in an HPLC vial, and the injection volume was set at 20 μL. Glipizide concentrations were determined using an optimized and validated HPLC system. Plasma concentrations of glipizide were measured at each collection time to determine the pharmacokinetic parameters.

2.8. Data Analysis

The estimated pharmacokinetic parameters were examined by noncompartmental analysis (NCA) with PKSolver 2.0 USA software. One-way ANOVA was conducted using GraphPad Prism 8 software, and significance was determined at $p < 0.05$ and $p < 0.01$.

3. Results

3.1. SST

SST parameters such as glipizide bioanalysis retention time, area, asymmetry, and theoretical plate conformed with the EMA guidelines (Figure 1 and Table 1). On the basis of the glipizide SST results in spike plasma, the optimal parameters for the HPLC analytical procedure were obtained. The proposed method can reliably and efficiently quantify glipizide in plasma.

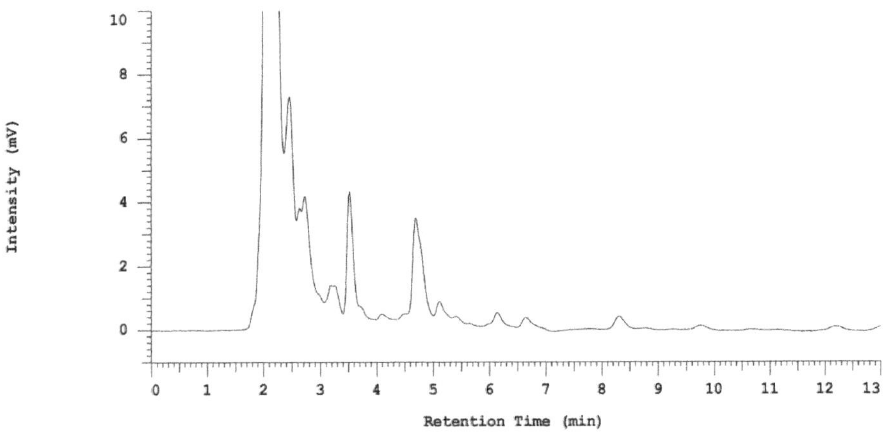

(a)

Figure 1. *Cont.*

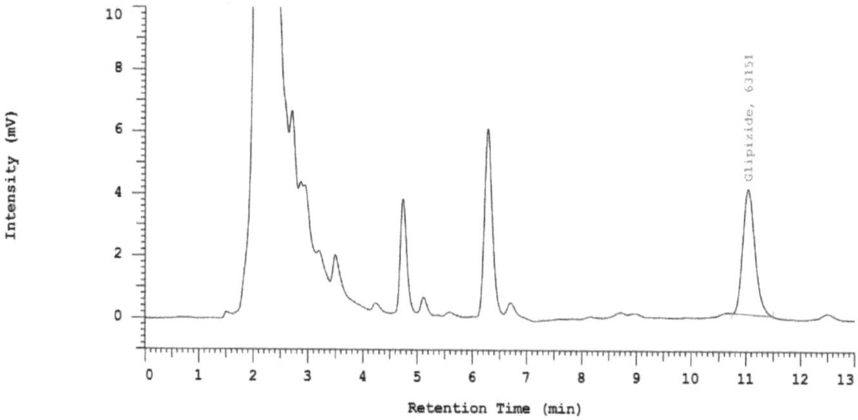

(b)

Figure 1. HPLC chromatogram of blank plasma (**a**) and glipizide (retention time 11.07 min) in plasma (**b**) using mobile phase KH_2PO_4, pH 4.25, acetonitrile (57.43%), and flow rate 1 mL/min.

Table 1. System suitability test (SST) results of glipizide in spike plasma.

Parameters	Result	Acceptance Criteria
Retention time	0.30	RSD ≤ 2.0%
Area	0.36	RSD ≤ 2.0%
Tailing factor (asymmetry)	1.23	≤2.0
Number of theoretical plates	11,336	≥2000

3.2. *Validation Method*

3.2.1. Selectivity

The results for selectivity from six separate samples met the criteria established by the EMA. The interference area on retention time was equivalent to the peak glipizide retention time in blank plasma (≤20% of the glipizide area on the LLOQ). These results demonstrate that this method is selective for analyzing the plasma samples of glipizide.

3.2.2. Calibration Curve and LLOQ

The calibration curve was determined on the basis of the correlation between area (y-axis) and glipizide concentration (x-axis) in the spiked plasma concentration range of 25–1500 ng/mL. Linear regression equations (y = bx + a) were generated using Microsoft Excel 2019. The linearity of the calibration curve was determined using the r-value (correlation coefficient) of 0.999 (R^2 = 0.998) (Figure 2). Linearity results indicated that the analytical method exhibited a proportionate relationship between the detector response and variations in the analyte concentration. The proposed approach produced a LLOQ of 25 ng/mL. The concentration met the LLOQ criterion, i.e., the acquisition of a percent differentiation value of 20%.

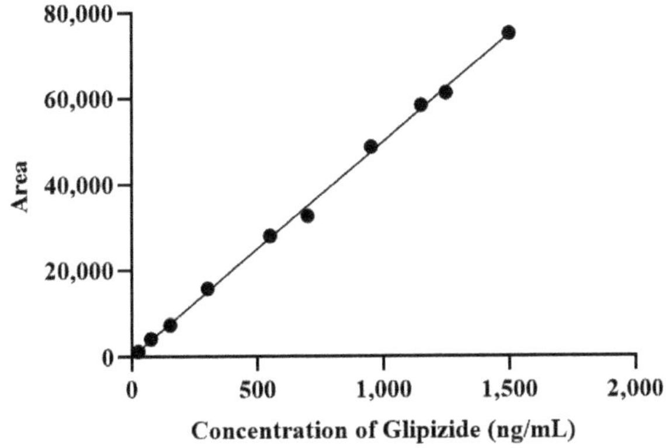

Figure 2. Calibration curves for glipizide using HPLC to determine the relationship between concentration and peak area. The equation obtained was y = 49.858x + 143.83 (R^2 = 0.9983), F value = 4812.72.

3.2.3. Accuracy and Precision

Accuracy and precision were determined using four analyte concentrations in various plasma spikes, namely, concentrations at LLOQ (25 ng/mL), low QC (75 ng/mL), medium QC (700 ng/mL), and high QC (1250 ng/mL). The results of within-run and between-run accuracy tests matched the requirement: a 15% difference between the average measured concentration of the QC sample and the theoretical concentration (%error), except for the LLOQ at 20% difference. The obtained precision values also met the criteria with a %CV of 15%, except for the LLOQ of 20% (Table 2) and recovery data for LLOQ (80–120%), LQC, medium QC, and high QC (85–115%) (Table 3). On the basis of these two validation parameters, the proposed method can be used for glipizide analysis in plasma samples to obtain a good value for the degree of similarity between the analytical results and the actual analyte concentration, repeatability, and reproducibility.

Table 2. Results of glipizide accuracy and precision in spike plasma.

Concentration (ng/mL)	% Error				% CV			
	Within (*n* = 6)			Between (*n* = 18)	Within (*n* = 6)			Between (*n* = 18)
	Day 1	Day 2	Day 3		Day 1	Day 2	Day 3	
25	−2.89	0.54	10.64	2.76	10.82	5.69	3.89	7.36
75	−1.53	3.95	8.84	3.75	6.43	5.08	3.84	5.12
700	−1.65	9.56	−4.17	1.25	5.55	3.92	3.82	4.43
1250	9.20	13.30	5.34	9.28	5.13	0.34	6.08	3.85

Table 3. Recovery data for glipizide in spiked plasma (*n* = 18).

Concentration (ng/mL)	Measured Concentration (ng/mL) (Mean ± SD)	% Recovery (±CV)
25	25.69 ± 1.76	102.76 ± 0.07
75	77.81 ± 3.89	103.42 ± 0.05
700	708.72 ± 5.14	101.29 ± 0.07
1250	1366 ± 49.75	109.28 ± 0.04

3.2.4. Dilution Integrity

The dilution integrity test employed the same accurate and precise criteria parameters (%error ≤ 15%, % CV ≤ 15%) to assess the bioanalytical method's accuracy, precision, and dependability. The findings of dilution integrity test indicated that the analytical method could examine diluted samples with precision and accuracy (Table 4).

Table 4. Uji Dilution integrity glipizide in spike plasma.

Dilution Factor	Mean ± SD Measurable Conc. (ng/mL)	Accuracy (% Error)	Precision (% CV)
2×	199.04 ± 10.94	−0.48	5.50
5×	423.20 ± 24.54	5.80	5.80
10×	1054.29 ± 74.57	5.43	7.07

3.2.5. Stability

Stability tests were conducted to verify whether the analyte in plasma remains stable and does not degrade throughout bioanalysis and storage. The freeze–thaw stability and short-term stability of glipizide stored at room temperature and −80 °C (4 and 24 h) and in the autosampler (24 h) indicated its good stability results. No significant degradation was observed for the glipizide stored under a variety of conditions (%error ≤ 15 %; % CV ≤ 15%) (Table 5).

Table 5. Stability test of glipizide in spike plasma.

Stability Study		Mean ± SD Measurable Conc. (ng/mL)	Accuracy (% Error)	Precision (% CV)
		Short-Term Stability		
T0	Low QC 75 μ/mL High QC 1250 μ/mL	81.10 ± 2.06 1239.79 ± 68.22	8.13 −0.82	2.54 5.50
T4		69.99 ± 1.61 1285.25 ± 10.84	−6.68 2.82	2.30 0.84
T24		70.30 ± 2.90 1311.63 ± 30.52	−6.27 4.93	4.12 2.33
Freeze–thaw stability		78.65 ± 4.84 1230.17 ± 112.55	4.86 −1.59	6.15 9.15
Autosampler		76.37 ± 2.27 1233.47 ± 71.06	1.83 −1.32	2.98 5.76

3.3. *Study of Pharmacokinetic Interaction in Normal and Diabetic Rats*

Bioanalytical validation using HPLC was successfully applied to determine the several pharmacokinetic parameters of glipizide administered alone and in combination with APE and AND in rat plasma. Figure 3 represents the correlation between glipizide concentration and plasma uptake in normal and diabetic rats for up to 14 h following oral administration. The concentration of glipizide significantly increased in the diabetic rats compared with that in the normal rats. In diabetic rats administered with glipizide in combination with APE, the concentration of glipizide increased. In contrast, when combined with AND, it decreased, altering its pharmacokinetic parameters. Table 6 displays the NCA results for the pharmacokinetic parameters of glipizide.

Co-administration of APE and AND reduced the pharmacokinetic parameters of glipizide such as half-life (T 1/2), area under the first moment curve from zero to infinity ($AUMC_{0-\infty}$), and mean residence time ($MRT_{0-\infty}$) in the normal group compared with the control (single glipizide administration). In addition, the maximum observed concentration (C_{max}) and area under the curve (AUC_{0-t}) increased, but the difference was not significant compared with the single glipizide group ($p > 0.05$). Changes in pharmacokinetic parameters were also observed in diabetic rats administered glipizide with APE and AND for 28 days. The C_{max} value increased 2.5 times ($p < 0.01$), and the AUC_{0-t} value increased 1.7 times ($p < 0.05$) in the diabetic rats receiving APE in combination with glipizide compared with those in the diabetic rats receiving glipizide alone. By contrast, the diabetic rats administered with AND exhibited a substantial decrease in various pharmacokinetic parameters, including C_{max} (9 times), AUC_{0-t} (9.6 times), and $AUC_{0-\infty}$ (13.6 times) with p-values of 0.01.

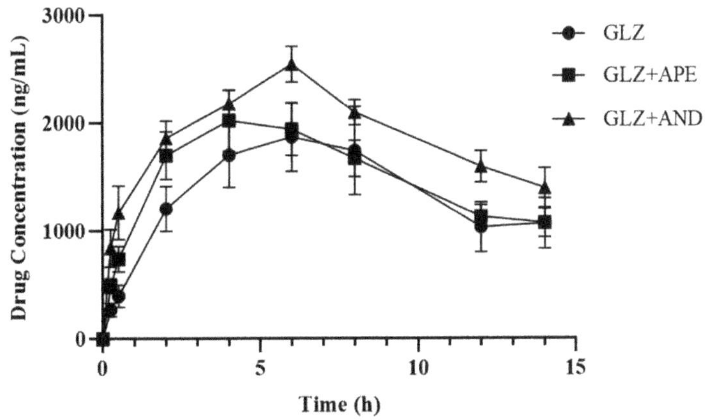

(a)

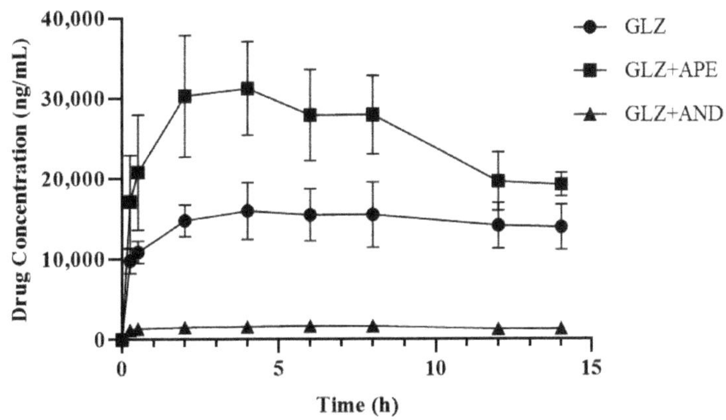

(b)

Figure 3. Concentrations of glipizide in the plasma (ng/mL ± S.E.M.) of normal (**a**) and diabetic (**b**) rats treated with glipizide (GLZ 5 mg/kg BW) alone or in combination with *Andrographis paniculata* extract (APE 300 mg/kg BW) and andrographolide (AND 15 mg/kg BW).

Table 6. Pharmacokinetic parameters of glipizide in the plasma of normal and diabetic rats.

Parameters	Unit	Value (±SEM)					
		Normal Rats			Diabetic Rats		
		GLZ	GLZ + APE	GLZ + AND	GLZ	GLZ + APE	GLZ + AND
T 1/2	h	11.7 ± 4.19	7.07 ± 0.49	6.8 ± 0.6	39.06 ± 12.4	13.5 ± 3.9	23.59 ± 8.3
Tmax	h	5.6 ± 0.7	4.4 ± 0.4	6 ± 0	5.2 ± 1.3	3.7 ± 1.2	6.4 ± 0.75
Cmax	ng/mL	2038.3 ± 359.4	2134 ± 233.2	2545.6 ± 165	17,049.6 ± 3407	44,634.7 ± 5.471 **	1876.4 ± 133.2 *
$AUC_{0\text{–}t}$	ng/mL·h	25,130.3 ± 5.228	28,231.4 ± 2978	35,893 ± 3048	204,126.9 ± 41,879	356,362.8 ± 37.946 *	21,052.9 ± 1.292 **
$AUC_{0\text{–}\infty}$	ng/mL·h	34.266 ± 3379	31,729.2 ± 3120	41,127.7 ± 4339	978,757.3 ± 279,871	734,414.7 ± 123.596 *	71.719 ± 22.203 **
$AUMC_{0\text{–}\infty}$	ng/mL·h^2	637,882.6 ± 162,429	378,090.9 ± 29,818	522,329 ± 89,356	74,019,560 ± 3,708,6791	17,673,567.5 ± 7,373,051	3,632,224 ± 2,286,205
$MRT_{0\text{–}\infty}$	h	19.3 ± 6.1	12 ± 0.4	12.4 ± 0.89	57.8 ± 17.8	20.9 ± 5	35.8 ± 12.04
Cl/F	(ng)/(ng/mL)	2.662 ± 1.1381	1.713 ± 0.30	1.22 ± 0.09	0.3067 ± 0.059	0.1263 ± 0.02006	2.272 ± 0.179
Vd/F	(ng)/(ng/mL)/h	0.152 ± 0.0159	0.165 ± 0.02	0.13 ± 0.02	0.0075 ± 0.002	0.0076 ± 0.00118	0.094 ± 0.023

All values are expressed as the mean ± SEM. GLZ = glipizide; APE = *Andrographis paniculata* extract; AND = andrographolide. * Significant at $p < 0.05$; ** significant at $p < 0.01$, when compared to a single glipizide treatment. T 1/2 = half-life; Tmax = Maximum observed time; Cmax = maximum observed concentration; $AUC_{0\text{–}t}$ = area under the curve from zero to time t; $AUC_{0\text{–}\infty}$ = area under the curve from zero to infinity; $AUMC_{0\text{–}\infty}$ = area under the first moment curve from zero to infinity; MRT = mean residence time from zero to infinity; Cl/F = clearance; Vd/F = volume of distribution.

4. Discussion

The glipizide bioanalytical method was validated to establish an appropriate, stable, and dependable quantitative analytical approach for plasma analysis. RP-HPLC/UV was chosen for the analysis of glipizide concentrations in the plasma because of its accuracy and selectivity for bioanalytical studies [42,43]. Bioanalysis necessitates plasma sample pretreatment by protein precipitation to eliminate interfering substances and optimize the sensitivity of the analytical procedure [44,45]. On the basis of the optimization results of the protein precipitation method, acetonitrile was selected because it creates the best plasma matrix devoid of interfering chemicals compared with methanol. Acetonitrile is suitable for usage as a mobile phase in HPLC systems and can attract plasma proteins. Variations in the ratio of mobile phase composition to pH also have an essential impact on determining the analysis method of glipizide in plasma.

The parameters of optimum drug resolution value of endogenous biologic substances, best peak shape, and reasonable retention time were used to determine the composition of the mobile phase. Validation of the percentage differentiation parameter of $\leq$20% revealed that the developed analytical method acquired a LLOQ of 25 ng/mL, which is comparable with the values in earlier investigations at a concentration of 50 ng/mL [46]. Therefore, the proposed bioanalytic approach can determine analyte concentrations in plasma with accuracy and precision at the LLOQ [47].

Glipizide is a sulfonylurea drug that is more susceptible to thermal degradation than other drugs, such as gliclazide [48,49]. The stability test conducted as part of the validation procedure demonstrated that the changes in glipizide concentrations did not reach statistical significance. Therefore, the proposed technique continued to meet the EMA guidelines. The validated bioanalytical method for glipizide that is selective, accurate, precise, and stable in plasma spikes was then used to analyze the concentration of glipizide co-administered with APE and AND in the plasma of normal and diabetic rats.

Similar to glibenclamide, the elevated concentrations of glipizide in the diabetic group was related to their elevated blood glucose levels [36]. The plasma concentration of glipizide in the diabetic rats was 10 times higher than that in the normal rats. The effects of streptozotocin induction on the gene expression of arachidonic and drug-metabolizing enzymes in the liver of diabetic rats resulted in the elevated concentrations of glipizide [50,51]. These diabetic rats can develop liver diseases that reduce the production and activity of drug-metabolizing enzymes, potentially resulting in pharmacokinetic parameter alterations [52]. Due to micro- and macrovascular alterations, the pathological condition of diabetic rats has been shown to affect the absorption and distribution of a drug. In animal models with diabetes, biotransformation/metabolism and drug excretion are also susceptible to changes [53]. In this study, it was found that differences in the pathological conditions of normal and type 2 diabetic rats (insulin deficiency model) affected the plasma glipizide concentration, thus affecting changes in pharmacokinetic parameters. This insulin-deficient rat model exhibits characteristics of moderated hyperglycemia and is associated with a loss of 60% of the function of β-cells. As a result, the condition of diabetes is reasonably stable [54].

The enhanced bioavailability of glipizide in the diabetic rats considerably increased the values of its pharmacokinetic parameters (T 1/2, C_{max}, AUC_{0-t}, $AUC_{0-\infty}$, $AUMC_{0-\infty}$, and $MRT_{0-\infty}$). The prolonged T 1/2 value caused glipizide to persist longer in the diabetic rats than in the normal rats. Owing to the decreased ability of the CYP2C9 enzyme to generate the inactive metabolites of glipizide, the elevation in glipizide's pharmacokinetic parameters could affect the potential pharmacological activity of this drug. When glipizide was combined with natural compounds that induce or inhibit the activity of metabolizing enzymes, the changes in enzyme activity resulted in pharmacokinetic HDIs in the diabetic rats. APE enhanced the pharmacokinetic parameters (C_{max} and AUC_{0-t}) of glipizide in normal and diabetic rats. The increase in these two pharmacokinetic parameters indicated an increase in plasma glipizide concentration (bioavailability) [55]. In the diabetic rats, the increased bioavailability of glipizide due to its combination with APE had a beneficial effect, i.e., pharmacological enhancement. APE exhibits glucose-lowering efficacy in rats by increasing the mRNA and protein expression of GLUT-4 and insulin expression in pancreatic beta cells [56,57]. A study reported evidence of an increase in the pharmacological action of the medicine (synergistic effect) when the anti-inflammatory agents etoricoxib and naproxen were combined with APE [58].

The additional adjustment in the bioavailability of glipizide in diabetic rats was hypothesized to be the result of APE's metabolizing enzyme's inhibitory impact. In vivo studies and human/rat hepatocyte culture experiments revealed that APE can significantly inhibit CYP2C9 and CYP3A4 expression [25]. Similar to antidiabetic drug sulfonylureas, the APE-induced increase in bioavailability also occurred for gliclazide, which is also metabolized by CYP2C9 [32]. AND, the major secondary metabolite of *A. paniculata*, dramatically reduced the bioavailability of glipizide in diabetic rats (decreased parameters C_{max}, $AUC_{0–t}$, $AUC_{0–\infty}$, and $AUMC_{0–\infty}$) but showed the opposite effect on normal mice (nonsignificant increase in bioavailability). The decrease in the plasma concentration of glipizide might be related to its slow absorption due to co-administration with AND (marked by a long T_{max} value). Tolbutamide, an antidiabetic, similarly experiences a reduction in pharmacokinetic drug bioavailability when combined with AND. AND enhances the gene transcription and enzyme activity of CYP1A1/2, CYP2C6/11, and CYP3A1/2 by activating AhR and binding to PXR in the cell nucleus and by significantly increasing CYP2C activity. In addition, this compound induces the expression and activity of enzymes to increase the metabolism of tolbutamide and inhibit its deposition to the target of action [59]. Owing to its interaction with AND, the bioavailability of nabumetone decreases and negatively affects its pharmacological activity. A significant reduction in the pharmacokinetic parameters (C_{max}, T_{max}, and $AUC_{0–t}$) and antiarthritic efficacy of nabumetone was observed when it was administered with AND [28].

Owing to the inconsistency between the mechanism of HDIs during the pharmacokinetic phase and the pharmacological activity, additional interaction studies are warranted to determine the effect of APE and AND on the antidiabetic pharmacological activity of glipizide. Different mechanism pathways exist for HDIs in the pharmacokinetic and pharmacodynamic phases; pharmacokinetics focuses on interactions in the ADME phase of pharmaceuticals, and pharmacology is mainly concerned with the synergistic, additive, or antagonistic effects of HDIs [11,60]. Changes in drug bioavailability in plasma affect the binding activity of the medication to its target of the action, although the two mechanisms utilize distinct pathways. Therefore, the results of this study can be applied in the therapeutic evaluation of HDIs affecting the efficacy of diabetes treatment. The limitation of this study is that the pharmacokinetic data in the animal model are insufficient to confirm the existence of herb–drug interactions; thus, it is necessary to extrapolate the data in humans to assess clinical significance.

5. Conclusions

A straightforward and rapid bioanalysis method for measuring glipizide in plasma was developed. Validation revealed that the technique exhibits accuracy, precision, selectivity, and sensitivity, and linearity ($r = 0.999$) within the concentration range of 25–1500 ng/mL. Glipizide remained stable in the plasma of normal and diabetic rats. The analytical approach was then utilized to examine the pharmacokinetics of HDIs. APE administration significantly altered the pharmacokinetic parameters (C_{max} and AUC) of glipizide ($p < 0.05$), thus increasing the bioavailability of glipizide. AND administration significantly decreased ($p < 0.05$) the parameters (T 1/2, C_{max}, and AUC) of glipizide in diabetic rats. APE and AND that are co-administered with glipizide are a source of potential herb–drug interactions. Although the effect on antidiabetic activity needs to be studied further, this research can reflect the concern in the combination of herbal use for diabetes therapy.

Author Contributions: Conceptualization, E.S., A.E.N., A.N. and E.L.; methodology, E.S., D.A.P., A.N., E.L. and I.P.; writing—draft preparation, E.S., A.E.N., A.N., I.P. and E.L.; writing—review and editing, E.S. and E.L.; supervision, A.E.N., A.N. and E.L. All authors have read and agreed to the published version of the manuscript.

Funding: This study was financially supported by Program Rekognisi Tugas Akhir (RTA) UGM (No. 3350/UN1.P.III/Dit-lit/PT.01.05/2022).

Institutional Review Board Statement: All animal experiments were approved by the Integrated Research and Testing Laboratory UGM (registration number: 00055/04/LPPT/XII/2021; Yogyakarta, Indonesia) and followed Animal Facility rules. Animal studies were conducted according to UGM's ethical guidelines.

Informed Consent Statement: Not applicable.

Data Availability Statement: The data described in this study are accessible from the author by request.

Conflicts of Interest: The authors declare no conflict of interest.

Sample Availability: Samples of the compounds are not available from the authors.

References

1. Barnes, P.M.; Bloom, B.; Nahin, R.L. Complementary and Alternative Medicine Use Among Adults and Children: United States, 2007. *Natl. Health Stat. Rep.* **2008**, *12*, 1–23. [CrossRef]
2. Fjær, E.L.; Landet, E.R.; McNamara, C.L.; Eikemo, T.A. The use of complementary and alternative medicine (CAM) in Europe. *BMC Complement. Med. Ther.* **2020**, *20*, 108. [CrossRef] [PubMed]
3. Shi, J.; Hu, H.; Harnett, J.; Zheng, X.; Liang, Z.; Wang, Y.-T.; Ung, C.O.L. An evaluation of randomized controlled trials on nutraceuticals containing traditional Chinese medicines for diabetes management: A systematic review. *Chin. Med.* **2019**, *14*, 54. [CrossRef]
4. Manya, K.; Champion, B.; Dunning, T. The use of complementary and alternative medicine among people living with diabetes in Sydney. *BMC Complement. Altern. Med.* **2012**, *12*, 2. [CrossRef] [PubMed]
5. Chang, H.-Y.; Wallis, M.; Tiralongo, E. Use of complementary and alternative medicine among people living with diabetes: Literature review. *J. Adv. Nurs.* **2007**, *58*, 307–319. [CrossRef] [PubMed]
6. Phutrakool, P.; Pongpirul, K. Acceptance and use of complementary and alternative medicine among medical specialists: A 15-year systematic review and data synthesis. *Syst. Rev.* **2022**, *11*, 10. [CrossRef]
7. Sun, H.; Saeedi, P.; Karuranga, S.; Pinkepank, M.; Ogurtsova, K.; Duncan, B.B.; Stein, C.; Basit, A.; Chan, J.C.; Mbanya, J.C.; et al. IDF Diabetes Atlas: Global, regional and country-level diabetes prevalence estimates for 2021 and projections for 2045. *Diabetes Res. Clin. Pract.* **2022**, *183*, 109119. [CrossRef]
8. Bo, L.; Baosheng, Z.; Yang, L.; Mingmin, T.; Beiran, L.; Zhiqiang, L.; Huaqiang, Z. Herb-drug enzyme-mediated interactions and the associated experimental methods: A review. *J. Tradit. Chin. Med.* **2016**, *36*, 392–408. [CrossRef]
9. Babos, M.; Heinan, M.; Redmond, L.; Moiz, F.; Souza-Peres, J.; Samuels, V.; Masimukku, T.; Hamilton, D.; Khalid, M.; Herscu, P. Herb–Drug Interactions: Worlds Intersect with the Patient at the Center. *Medicines* **2021**, *8*, 44. [CrossRef]
10. Brantley, S.J.; Argikar, A.A.; Lin, Y.S.; Nagar, S.; Paine, M.F. Herb–Drug Interactions: Challenges and Opportunities for Improved Predictions. *Drug Metab. Dispos.* **2014**, *42*, 301–317. [CrossRef]
11. Fasinu, P.S.; Bouic, P.J.; Rosenkranz, B. An Overview of the Evidence and Mechanisms of Herb–Drug Interactions. *Front. Pharmacol.* **2012**, *3*, 69. [CrossRef] [PubMed]
12. Tornio, A.; Niemi, M.; Neuvonen, P.J.; Backman, J.T. Drug interactions with oral antidiabetic agents: Pharmacokinetic mechanisms and clinical implications. *Trends Pharmacol. Sci.* **2012**, *33*, 312–322. [CrossRef] [PubMed]
13. Daly, A.K.; Rettie, A.E.; Fowler, D.M.; Miners, J.O. Pharmacogenomics of CYP2C9: Functional and Clinical Considerations. *J. Pers Med.* **2017**, *8*, 1. [CrossRef] [PubMed]
14. Van Booven, D.; Marsh, S.; McLeod, H.; Whirl-Carrillo, M.; Sangkuhl, K.; Klein, T.E.; Altman, R.B. Cytochrome P450 2C9-CYP2C9. *Pharm. Genom.* **2010**, *20*, 277–281. [CrossRef] [PubMed]
15. Desta, Z.; Zhao, X.; Shin, J.G.; Flockhart, D.A. Clinical significance of the cytochrome P450 2C19 genetic polymorphism. *Clin. Pharmacokinet.* **2002**, *41*, 913–958. [CrossRef] [PubMed]
16. Holstein, A.; Beil, W. Oral antidiabetic drug metabolism: Pharmacogenomics and drug interactions. *Expert Opin. Drug Metab. Toxicol.* **2009**, *5*, 225–241. [CrossRef] [PubMed]
17. Tirkkonen, T.; Heikkilä, P.; Huupponen, R.; Laine, K. Potential CYP2C9-mediated drug–drug interactions in hospitalized type 2 diabetes mellitus patients treated with the sulphonylureas glibenclamide, glimepiride or glipizide. *J. Intern. Med.* **2010**, *268*, 359–366. [CrossRef]
18. Tan, B.; Zhang, Y.F.; Chen, X.Y.; Zhao, X.H.; Li, G.X.; Zhong, D.F. The effects of CYP2C9 and CYP2C19 genetic polymorphisms on the pharmacokinetics and pharmacodynamics of glipizide in Chinese subjects. *Eur. J. Clin. Pharmacol.* **2010**, *66*, 145–151. [CrossRef]

19. Arifullah, M.; Namsa, N.D.; Mandal, M.; Chiruvella, K.K.; Vikrama, P.; Gopal, G.R. Evaluation of anti–bacterial and anti–oxidant potential of andrographolide and echiodinin isolated from callus culture of *Andrographis paniculata* Nees. *Asian Pac. J. Trop. Biomed.* **2013**, *3*, 604–610. [CrossRef]
20. Dai, Y.; Chen, S.R.; Chai, L.; Zhao, J.; Wang, Y.; Wang, Y. Overview of pharmacological activities of *Andrographis paniculata* and its major compound andrographolide. *Crit. Rev. Food Sci. Nutr.* **2019**, *59* (Suppl. S1), S17–S29. [CrossRef]
21. Edwin, E.-S.; Vasantha-Srinivasan, P.; Senthil-Nathan, S.; Thanigaivel, A.; Ponsankar, A.; Pradeepa, V.; Selin-Rani, S.; Kalaivani, K.; Hunter, W.B.; Abdel-Megeed, A.; et al. Anti-dengue efficacy of bioactive andrographolide from *Andrographis paniculata* (Lamiales: Acanthaceae) against the primary dengue vector *Aedes aegypti* (Diptera: Culicidae). *Acta Trop.* **2016**, *163*, 167–178. [CrossRef] [PubMed]
22. Fitrawan, L.; Ariastuti, R.; Tjandrawinata, R.R.; Nugroho, A.E.; Pramono, S. Antidiabetic effect of combination of fractionated-extracts of *Andrographis paniculata* and *Centella asiatica*: In vitro study. *Asian Pac. J. Trop. Biomed.* **2018**, *8*, 527. [CrossRef]
23. Sundhani, E.; Nugroho, A.E.; Nurrochmad, A.; Lukitaningsih, E. Molecular interactions of *Andrographis paniculata* Burm. f. Active Compound with Nuclear Receptor (CAR and PXR): An In Silico Assessment Approach. *Indones J. Chem.* **2022**, *22*, 126. [CrossRef]
24. Pan, Y.; Abd-Rashid, B.A.; Ismail, Z.; Ismail, R.; Mak, J.W.; Pook, P.C.K.; Er, H.M.; Ong, C.E. In vitro determination of the effect of *Andrographis paniculata* extracts and andrographolide on human hepatic cytochrome P450 activities. *J. Nat. Med.* **2011**, *65*, 440–447. [CrossRef]
25. Pekthong, D.; Blanchard, N.; Abadie, C.; Bonet, A.; Heyd, B.; Mantion, G.; Berthelot, A.; Richert, L.; Martin, H. Effects of *Andrographis paniculata* extract and Andrographolide on hepatic cytochrome P450 mRNA expression and monooxygenase activities after in vivo administration to rats and in vitro in rat and human hepatocyte cultures. *Chem. Biol. Interact.* **2009**, *179*, 247–255. [CrossRef]
26. Pekthong, D.; Martin, H.; Abadie, C.; Bonet, A.; Heyd, B.; Mantion, G.; Richert, L. Differential inhibition of rat and human hepatic cytochrome P450 by *Andrographis paniculata* extract and andrographolide. *J. Ethnopharmacol.* **2008**, *115*, 432–440. [CrossRef] [PubMed]
27. Qiu, F.; Hou, X.L.; Takahashi, K.; Chen, L.X.; Azuma, J.; Kang, N. Andrographolide inhibits the expression and metabolic activity of cytochrome P450 3A4 in the modified Caco-2 cells. *J. Ethnopharmacol.* **2012**, *141*, 709–713. [CrossRef]
28. Balap, A.; Lohidasan, S.; Sinnathambi, A.; Mahadik, K. Pharmacokinetic and Pharmacodynamic Interaction of Andrographolide and Standardized Extract of *Andrographis paniculata* (Nees) with Nabumetone in Wistar Rats. *Phytother. Res. PTR* **2017**, *31*, 75–80. [CrossRef]
29. Balap, A.; Lohidasan, S.; Sinnathambi, A.; Mahadik, K. Herb-drug interaction of *Andrographis paniculata* (Nees) extract and andrographolide on pharmacokinetic and pharmacodynamic of naproxen in rats. *J. Ethnopharmacol.* **2017**, *195*, 214–221. [CrossRef]
30. Balap, A.; Atre, B.; Lohidasan, S.; Sinnathambi, A.; Mahadik, K. Pharmacokinetic and pharmacodynamic herb–drug interaction of *Andrographis paniculata* (Nees) extract and andrographolide with etoricoxib after oral administration in rats. *J. Ethnopharmacol.* **2016**, *183*, 9–17. [CrossRef]
31. Chien, C.F.; Wu, Y.T.; Lee, W.C.; Lin, L.C.; Tsai, T.H. Herb–drug interaction of *Andrographis paniculata* extract and andrographolide on the pharmacokinetics of theophylline in rats. *Chem. Biol. Interact.* **2010**, *184*, 458–465. [CrossRef] [PubMed]
32. Mouid, M.G. Effect of Ethanolic Extract of Aerial Parts of *Andrographis paniculata* on the Pharmacokinetics of Gliclazide in Rats. *Asian J. Biomed. Pharm. Sci.* **2015**, *5*, 21–24. [CrossRef]
33. Wongnawa, M.; Soontaro, P.; Riditid, W.; Wongpoowarak, P.; Ruengkittisaku, S. The effects of *Andrographis paniculata* (Burm. f.) Nees on the pharmacokinetics and pharmacodynamics of midazolam in healthy volunteers. *Songklanakarin J. Sci. Technol.* **2012**, *34*, 533–539.
34. Vaishali, V.G.; Patel, J.; Varia, R.; Bhavsar, S.; Priti, P.V.; Falguni, F.M.; Tamanna, T.S. Effect of Andrographolide Co-Administration on Pharmacokinetics of Meloxicam in Rats. *Int. J. Curr. Microbiol. Appl. Sci.* **2017**, *6*, 2147–2153. [CrossRef]
35. Li, X.P.; Zhang, C.L.; Gao, P.; Gao, J.; Liu, D. Effects of andrographolide on the pharmacokinetics of aminophylline and doxofylline in rats. *Drug Res.* **2013**, *63*, 258–262. [CrossRef]
36. Samala, S.; Veeresham, C. Pharmacokinetic and Pharmacodynamic Interaction of Boswellic Acids and Andrographolide with Glyburide in Diabetic Rats: Including Its PK/PD Modeling: PK/PD Interactions with Glyburide. *Phytother. Res.* **2016**, *30*, 496–502. [CrossRef]
37. Samala, S.; Veeresham, C. Andrographolide Pretreatment Enhances the Bioavailability and Hypoglycemic Action of Glimepiride and Metformin. *Int. J. Phytomedicine.* **2015**, *7*, 254–264.
38. Zhang, X.; Zhang, X.; Wang, X.; Zhao, M. Influence of andrographolide on the pharmacokinetics of warfarin in rats. *Pharm. Biol.* **2018**, *56*, 351–356. [CrossRef]
39. Kirthi, A.; Shanmugam, R.; Prathyusha, M.S.; Basha, D.J. A Review on Bioanalytical Method Development and Validation by RP-HPLC. *J. Glob. Trends Pharm. Sci.* **2014**, *5*, 2265–2271.
40. Anusha, A.; Narendar, D.; Krishna Murthy, B.; Goverdhan, P. Influence of Single and Multi Dose Treatment of Glipizide on Pharmacokinetics and Pharmacodynamics of Irbesartan in Normal and Hypertensive Rats. *High Blood Press Cardiovasc. Prev. Off J. Ital. Soc. Hypertens.* **2017**, *24*, 179–185. [CrossRef]
41. Chaudhari, S.; Zambad, S.; Ali, M. Effect of Aqueous Extract of Azadirachta indica Leaves on Pharmacokineics and Pharmacodynamics of Glipizide. *Drug Metab. Lett.* **2019**, *13*, 19–24. [CrossRef] [PubMed]

42. Nikolin, B.; Imamović, B.; Medanhodzić-Vuk, S.; Sober, M. High perfomance liquid chromatography in pharmaceutical analyses. *Bosn. J. Basic Med. Sci.* **2004**, *4*, 5–9. [CrossRef] [PubMed]
43. Rezazadeh, M.; Emami, J. A simple and sensitive HPLC method for analysis of imipramine in human plasma with UV detection and liquid-liquid extraction: Application in bioequivalence studies. *Res. Pharm. Sci.* **2016**, *11*, 168–176. [PubMed]
44. Deshpande, M.M.; Kasture, V.S.; Mohan, M.; Chavan, M.M.A.M.J. *Bioanalytical Method Development and Validation: A Review*; IntechOpen: London, UK, 2019. [CrossRef]
45. Sebaiy, M.M.; Abdellatef, H.E.; Elmosallamy, M.A.; Alshuwaili, M.K. Identification of Antioxidative Ingredients from Feverfew (Tanacetum Parthenium) Extract Substantially free of Parthenolide and other Alpha-Unsaturated Gamma-Lactones. *Open J. Anal. Bioanal. Chem.* **2020**, *4*, 001–006. [CrossRef]
46. Atif, M.; Khalid, S.H.; Onn Kit, G.L.; Sulaiman, S.A.S.; Asif, M.; Chandersekaran, A. Development and validation of RP-HPLC-UV method for the determination of Glipizide in human plasma. *J. Young Pharm. JYP* **2013**, *5*, 26–29. [CrossRef]
47. Thippani, R.; Pothuraju, N.R.; Ramisetti, N.R.; Shaik, S. Optimization and validation of a fast RP–HPLC method for the determination of dobutamine in rat plasma: Pharmacokinetic studies in healthy rat subjects. *J. Pharm. Anal.* **2013**, *3*, 434–439. [CrossRef]
48. Gedawy, A.; Al-Salami, H.; Dass, C.R. Advanced and multifaceted stability profiling of the first-line antidiabetic drugs metformin, gliclazide and glipizide under various controlled stress conditions. *Saudi Pharm. J.* **2020**, *28*, 362–368. [CrossRef]
49. Gupta, S.; Bansal, G. Validated Stability-Indicating HPLC-UV Method for Simultaneous Determination of Glipizide and Four Impurities. *J. AOAC Int.* **2011**, *94*, 523–530. [CrossRef]
50. Jarrar, Y.B.; Al-Essa, L.; Kilani, A.; Hasan, M.; Al-Qerem, W. Alterations in the gene expression of drug and arachidonic acid-metabolizing Cyp450 in the livers of controlled and uncontrolled insulin-dependent diabetic mice. *Diabetes Metab. Syndr. Obes. Targets Ther.* **2018**, *11*, 483–492. [CrossRef]
51. Neyshaburinezhad, N.; Rouini, M.; Shirzad, N.; Esteghamati, A.; Nakhjavani, M.; Namazi, S.; Ardakani, Y.H. Evaluating the effect of type 2 diabetes mellitus on CYP450 enzymes and P-gp activities, before and after glycemic control: A protocol for a case–control pharmacokinetic study. *MethodsX* **2020**, *7*, 100853. [CrossRef]
52. Palatini, P.; De Martin, S. Pharmacokinetic drug interactions in liver disease: An update. *World J. Gastroenterol.* **2016**, *22*, 1260–1278. [CrossRef] [PubMed]
53. Dostalek, M.; Akhlaghi, F.; Puzanovova, M. Effect of diabetes mellitus on pharmacokinetic and pharmacodynamic properties of drugs. *Clin. Pharmacokinet.* **2012**, *51*, 481–499. [CrossRef] [PubMed]
54. Furman, B.L. Streptozotocin-Induced Diabetic Models in Mice and Rats. *Curr. Protoc. Pharmacol.* **2015**, *70*, 5–47. [CrossRef]
55. Price, G.; Patel, D.A. *Drug Bioavailability*; StatPearls Publishing: St. Petersburg, FL, USA, 2022.
56. Nugroho, A.E.; Rais, I.R.; Setiawan, I.; Pratiwi, P.Y.; Hadibarata, T.; Tegar, M.; Pramono, S. Pancreatic effect of andrographolide isolated from *Andrographis paniculata* (Burm. f.) Nees. *Pak. J. Biol. Sci. PJBS* **2014**, *17*, 22–31. [CrossRef]
57. Nugroho, A.E.; Lindawati, N.Y.; Herlyanti, K.; Widyastuti, L.; Pramono, S. Anti-diabetic effect of a combination of andrographolide-enriched extract of *Andrographis paniculata* (Burm. f.) Nees and asiaticoside-enriched extract of *Centella asiatica* L. in high fructose-fat fed rats. *Indian J. Exp. Biol.* **2013**, *51*, 1101–1108. [PubMed]
58. Sundhani, E.; Lukitaningsih, E.; Nurrochmad, A.; Nugroho, A.E. Potential pharmacokinetic and pharmacodynamic herb-drug interactions of *Andrographis paniculata* (Burm. f.) and andrographolide: A systematic review. *J. Herbmed. Pharmacol.* **2022**, *11*, 154–165. [CrossRef]
59. Chen, H.-W.; Huang, C.-S.; Liu, P.-F.; Li, C.-C.; Chen, C.-T.; Liu, C.-T.; Chiang, J.-R.; Yao, H.-T.; Lii, C.-K. *Andrographis paniculata* Extract and Andrographolide Modulate the Hepatic Drug Metabolism System and Plasma Tolbutamide Concentrations in Rats. *Evid. Based Complement. Altern. Med.* **2013**, *2013*, 982689. [CrossRef]
60. Niu, J.; Straubinger, R.M.; Mager, D.E. Pharmacodynamic Drug-Drug Interactions. *Clin. Pharmacol. Ther.* **2019**, *105*, 1395–1406. [CrossRef]

molecules

MDPI

Review

β-Cyclocitral: Emerging Bioactive Compound in Plants

Mohammad Faizan [1,*,†], Sadia Haque Tonny [2], Shadma Afzal [3], Zeba Farooqui [4], Pravej Alam [5,*,†], S. Maqbool Ahmed [1], Fangyuan Yu [6] and Shamsul Hayat [7]

1 Botany Section, School of Sciences, Maulana Azad National Urdu University, Hyderabad 500032, India
2 Faculty of Agriculture, Bangladesh Agricultural University, Mymensingh 2202, Bangladesh
3 Department of Biotechnology, Motilal Nehru National Institute of Technology Allahabad, Prayagraj 211004, India
4 College of Pharmacy, University of Houston, Houston, TX 77204, USA
5 Department of Biology, College of Science and Humanities, Prince Sattam Bin Abdulaziz University, Alkharj 11942, Saudi Arabia
6 Collaborative Innovation Center of Sustainable Forestry in Southern China, College of Forest Science, Nanjing Forestry University, Nanjing 210037, China
7 Department of Botany, Faculty of Life Sciences, Aligarh Muslim University, Aligarh 202002, India
* Correspondence: faizanetawah8@gmail.com (M.F.); alamprez@gmail.com (P.A.)
† These authors contributed equally to this work.

Abstract: β-cyclocitral (βCC), a main apocarotenoid of β-carotene, increases plants' resistance against stresses. It has recently appeared as a novel bioactive composite in a variety of organisms from plants to animals. In plants, βCC marked as stress signals that accrue under adverse ecological conditions. βCC regulates nuclear gene expression through several signaling pathways, leading to stress tolerance. In this review, an attempt has been made to summarize the recent findings of the potential role of βCC. We emphasize the βCC biosynthesis, signaling, and involvement in the regulation of abiotic stresses. From this review, it is clear that discussing compound has great potential against abiotic stress tolerance and be used as photosynthetic rate enhancer. In conclusion, this review establishes a significant reference base for future research.

Keywords: apocarotenoid; abiotic stress; β-carotene; gene expression

Citation: Faizan, M.; Tonny, S.H.; Afzal, S.; Farooqui, Z.; Alam, P.; Ahmed, S.M.; Yu, F.; Hayat, S. β-Cyclocitral: Emerging Bioactive Compound in Plants. *Molecules* **2022**, *27*, 6845. https://doi.org/10.3390/molecules27206845

Academic Editor: Raffaele Pezzani

Received: 7 September 2022
Accepted: 6 October 2022
Published: 13 October 2022

Publisher's Note: MDPI stays neutral with regard to jurisdictional claims in published maps and institutional affiliations.

1. Introduction

Plants have evolved several types of secondary metabolites as a defensive shield to protect themselves from phytophagous herbivores [1]. Miscellaneous bioactive compounds induce fuel in various beneficial activities such as wound healing and antifungal, anti-inflammatory, and antimicrobial effects. For eco-friendly and substantial environments, the use of biomolecules is incrementally increasing [2]. β-cyclocitral (βCC) 2,6,6-trimethyl-1-cyclohexene-1-carboxaldehyde is an endogenous volatile compound which is derived from the carotenoid β-carotene [3]. Plant protection proficiency from injury of plants due to free radicals, stimulation of enzymes, and extinguishing singlet oxygen function is controlled by bioactive compounds [4]. The functional characterization of carotenoid compounds like hormones, signals, and biosynthesis through the non-enzymatic method is observed in apocarotenoid [5].

Carotenoids are tetraterpene pigments, which extensively distributed in foodstuffs that have always been part of the human diet. Few carotenoids can be converted into retinoids exhibiting vitamin A properties, which is important for humans. Moreover, they are very versatile as they are present in food not only as a vitamin A component, but also as natural pigment, antioxidant, and health curing compound [6]. They play an important role in lethal reactive oxygen species (ROS) scavenging and also actively participate in light absorption and the protection of photosynthetic supplies and functions [7]. The β-ring holding carotenoids like apocarotenoid obtained from carotenoid cleavage dioxygenases-4 (CCD-4 clade), which generate volatile apocarotenoids are known as βCC. It is derived from singlet oxygen $^1O_2$ invasion, and is present in higher amounts during the chemical reactions in photosynthesis performing cells. Poly-unsaturated fatty acid oxidation is found to function as a co-substrate for performing enzymatic activities [8,9]. Interestingly, βCC contain vital apocarotenoids of β-carotene [10] that play important roles under abiotic stress conditions [11,12]. Plant genomic effect, apoptosis, gene activation processes, and transcription activation are encountered due to low RES and excess ROS oxidation [13–15]. Carotenoid degradation involves βCC, β-ionone, β-ionone rendered in water odor. Organoleptic properties (flowery flavor or fragrance) of βCC in beverages, pharmacy and industries are also economically beneficial [16,17]. In many plant species, βCC has been found in leaves, flowers, fruits and roots [18,19]. Moreover, a substantial source of βCC is found in lichens and mosses [20,21].

Although it acts as a powerful repellent and a signal of poor quality food to grazers, such as Daphnia [22] and the cell dissolution of *Microcystis* [23], βCC is beneficial to vascular plant growth. One of the major benefits of βCC is that it plays an important role in growth regulation, enhancing branching, the emergence of lateral roots, and cell division [24]. The production of provitamin carotenoid is highly dependent on βCC, which helps to improve the yield of the bioreactor and growth index [25]. The quantity of βCC is proportional to the amount of LOX 13S- lipoxygenase process which influences the decrease of apocarotenoid in *Solanum lycopersicum* and *A. thaliana* mutants in comparison with the wild type [26]. The deterioration of aquatic and marine organisms in freshwater due to carotenoid compound degeneration lead to the formation of malodor. One of the major pretenses of this odor and taste compound is βCC along with dimethyl sulfide, β-ionone, etc [27]. A number of vegetables and fruits are the sources of violet and raspberry type fragrance in carotenoid compounds.

βCC is capable of eliciting multiple stress signals and gives strength to survive under unfavorable conditions. Stress signaling of βCC is induced by oxidation and degradation of chloroplasts which enhances the production of ROS and the activity of auxin and brassinosteroids. βCC is able to promote root growth in rice, reducing the toxicity of salinity to rice seedlings [24]. Recent studies have found that βCC can enhance the tolerance of *Arabidopsis thaliana* to high light stress [10]. The acclimatization of photo-oxidative stress causes the inhibition of signal modulation of $^1O_2$ through methylene blue sensitivity, which acts as a zinc finger protein. This function was directed by transcriptome reprogramming [28]. Detoxification activities for resisting oxidative cell damage are increased by βCC [10]. The βCC exhaustion is influenced by foliose lichen during heat and wounding stress [29]. It acts as a secondary precursor for stress signal relocation in *A.* thaliana. Analogous activities of βCC were also found to occur in dihydroactinidiolide during $^1O_2$ oxidation [10]. The excess amount of light interference persuades the increasing amount of glycosylated βCC, which immobilizes the signaling molecule [30,31]. As a consequence, drought tolerance is achieved by signaling and pathway activation by using βCC exploitation. The exogenous application of βCC also ameliorates the drought and light stress in *A. thaliana*, Viola tricolor and Capsicum sp. [31]. Antioxidant signaling and crosstalk are imitated during stress, which causes the upregulation of various enzymes like superoxide dismutases (SOD), catalase (CAT) and peroxidase (POD) [32]. Inhibiting traits of βCC on hydroxylase enzymes causes multiple loop-hole creations for establishing a homeostasis pathway. The inter-connectedness between metabolic activities in primary and secondary metabolism is observed which agitate the consistent state of miscellaneous metabolites [33]. Previous studies revealed that βCC is a volatile compound derived from β-carotene oxidation, which mediates the response of cells to singlet oxygen stress. Beside these well-known examples, the latest research unraveled novel apocarotenoid growth regulators and suggested the presence of yet unidentified ones. However, knowledge of βCC involvement in the complex stress signaling network is very limited. This review highlights the structure and functions of βCC in plants. It presents the βCC mediated stress mitigation, as well as signaling cascades in plants. The authors also demonstrated the βCC-mediated involvement in the regulation of the stress response.

2. Biosynthesis of βCC

The formation of βCC occurs either by direct oxidation of β-carotene through ROS ($^1O_2$) or by an enzymatic pathway. A family of non-heme iron-dependent enzymes in plants catalyzes the carotenoids by an enzymatic cleavage via 9-cis-epoxycarotenoid cleavage dioxygenases (NCEDs) and carotenoid cleavage dioxygenases (CCDs), resulting in apocarotenoids, an oxidation product [34,35]. The first step in abscisic acid (ABA) production is catalyzed by NCED enzymes cleaving the 11, 12 (11′, 12′) double bond of 9-cis-violaxanthin or 9-cisneoxanthin [36]. Furthermore, CCD enzymes and NCED enzymes do not share cleavage specificities. In *Arabidopsis*, there are four CCDs (CCD1, CCD4, CCD7, and CCD8). It is unknown whether one of these CCDs creates βCC from carotene in plant leaves. In each of the four CCDs in *Arabidopsis* deficient mutants, the accumulation of βCC was not affected, which suggests that β-carotene oxidation mediated by CCD in this species is not a major source of this apocarotenoid [37], despite the fact that between 4 CCDs functional redundancy cannot be ruled out. This is similar in cyanobacteria, where βCC formation aided by CCD was not found [38]. Unlike CCDs that are plastidial, cytosolic CCD1 cleaves the double bonds of 9, 10 (9′, 10′) to produce varying volatiles and apocarotenoids of extensive acyclic or monocyclic apocarotenoids and carotenoids.

The strigolactones biosynthesis is dependent on CCD8 and CCD7 [39]. Since CCD4 has a specific cleavage activity at 9, 10 (9′, 10′) and 5, 6 (5′, 6′) double bond, it does not generate βCC [34,35]. Furthermore, in high light conditions, CCD4 is highly downregulated, which activates the accumulation of βCC [37]. However, the cleavage of β-carotene in citrus from the location 7, 8 (7′, 8′), CCD4b is reported under CCD4 enzyme, which results in the production of βCC [8]. Similarly, another CCD4c in the *Crocus stigma* from CCD4 can cleave carotenoids at 9-10 (9′, 10′), resulting into β-ionone and produces βCC with low efficiency at 7 and 8 (7′, 8′) [40]. For the production of βCC, CCD4b gene in *Vitis vinifera* in the carotenoid-accumulating yeast strain is also reported [41]. Another way for the oxidation of carotenoids can be provided by lipoxygenase [9]. Similarly, in leaves of *Solanum lycopersicum* and *Arabidopsis*, knockout mutants for 13-lipoxygenase LOX2 were reported to have low levels of βCC [26]. On the other hand, in the βCC accumulation under high light and $^1O_2$ stresses, it is unknown if this enzyme is involved despite the fact that LOX2 is induced under these circumstances [37]. Eventually, from the fungus *Lepista irina*, extracellular fluid purified a peroxidase which produces βCC and other unstable apocarotenoids from the cleavage of β-carotene [42].

When compared to photosystem II, it is thought that photosystem I does not produce considerable amounts of $^1O_2$. Auto-oxidation of β-carotene can also produce βCC, especially when attacked by the reactive specie $^1O_2$ [31]. Carotenoids quench $^1O_2$ through a physical mechanism that involves energy transfer from $^1O_2$ to the carotenoid, followed by the excited quencher's thermal decay [43]. However, carotene can be oxidised by $^1O_2$, allowing $^1O_2$ to be chemically quenched. $^1O_2$ is an electrophilic molecule that has a strong affinity for double bonds in carotenoid molecules, oxidizing them and creating a range of apocarotenoids, including βCC [10]. In microalgae, the principal oxidation products of β-carotene are β-ionone and βCC, which release large amounts of these chemicals during summer blooms [44].

3. Derivatives of βCC

Recently, a new bioactive compound, βCC, has emerged in a variety of living organisms varying from plants and cyanobacteria to animals and fungi. It is a volatile compound consisting of a short chain of apocarotenoids produced by the non-enzymatic and enzymatic oxidation of the β-carotene. Derivatives of βCC such as lactones, β-cyclocitric acid, glycosylated βCC, 2,2,6-trimethyl-cyclohexanone and 2,6-dimethyl-cyclohexanol are briefly discussed. Lactone, being a potent biologically active compound, occurs naturally and possesses anticancer, antiplasmodial, antifungal and antimicrobial effects. Figure 1 showed the schematic synthesis of lactone. In a previous study, lactone synthesis stated by the addition of $NaBH_4$ in the presence of H_2O/CH_3OH in βCC (10 g, 0.06 mol) which yielded an alcoholic compound βCC (9.65 g, yield 97%) [45]. This alcohol was treated with $CH_3C(OC_2H_5)_3$, CH_3CH_2COOH at 137 °C and resulted in (1,3,3-Trimethyl-2-methylene-cyclohex-1-yl) acetic acid ethyl ester. In the presence of KOH/C_2H_5OH, this acetic acid ethyl ester was transformed into (1,3,3-Trimethyl-2-methylene-cyclohexyl) acetic acid [46]. Furthermore, this acetic acid was converted to either 1-Bromomethyl-2,2,6-trimethyl-9-oxabicyclo [4.3.0] nonan-8-one with the reagent NBS/THF, CH_3COOH or to 1-Chloromethyl-2,2,6-trimethyl-9-oxabicyclo [4.3.0] nonan-8-one with the help of NCS/THF, CH_3COOH. It can also be converted to 1-Iodomethyl-2,2,6-trimethyl-9-oxabicyclo [4.3.0] nonan-8-one with the help of I_2/KI, $NaHCO_3$ which can be further converted to 1,2,2,6-Tetramethyl-9-oxabicyclo [4.3.0] nonan-8-one in the presence of n-Bu_3SnH. The product 1,2,2,6-Tetramethyl-9-oxabicyclo [4.3.0] nonan-8-one, 1-Iodomethyl-2,2,6-trimethyl-9-oxabicyclo [4.3.0] nonan-8-one, 1-Bromomethyl-2,2,6-trimethyl-9-oxabicyclo [4.3.0] nonan-8-one and 1-Chloromethyl-2,2,6-trimethyl-9-oxabicyclo [4.3.0] nonan-8-one are all lactones derived from βCC [46].

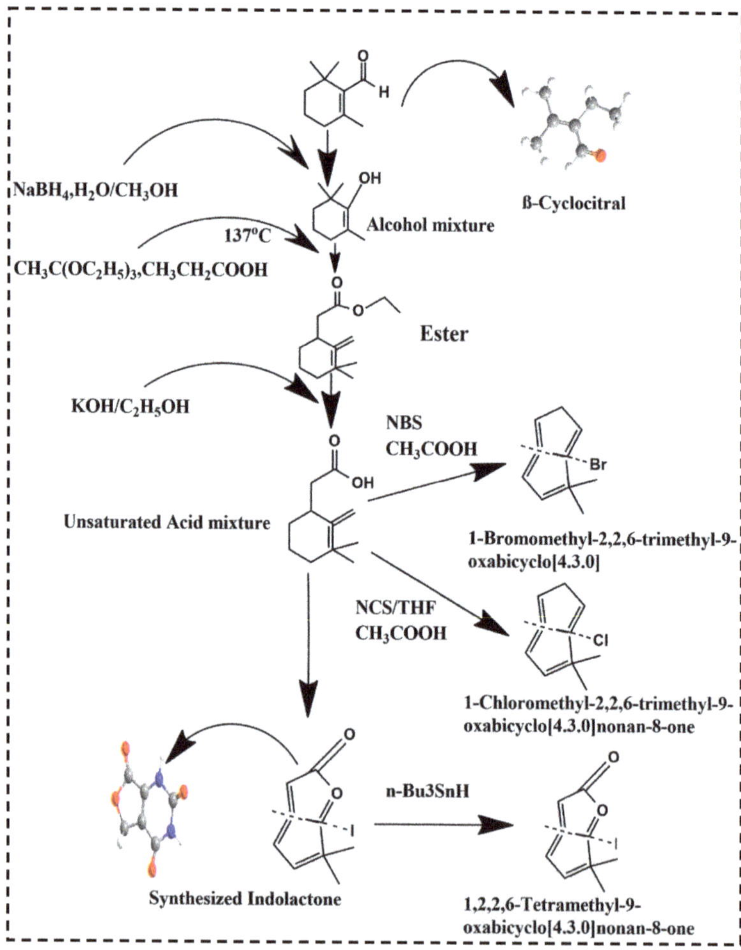

Figure 1. Lactone synthesis from βCC.

In the leaves of *Arabidopsis*, the oxidation of βCC results in β-cyclocitric acid (2,2,6-trimethyl cyclohexene-1-carboxylic acid) which is responsible for the accumulation under strain conditions [31]. The level of β-cyclocitric acid in plants under drought stress increased by an aspect of 15 and doubled the content of βCC only. An accumulation of β-cyclocitric acid in plants exposed to volatile βCC was observed [31]. In the leaves and fruits of tomato, the levels of β-cyclocitric acid are remarkably more than that of βCC. Unlike in water, the conversion of βCC to β-cyclocitric acid was much faster in plants, suggesting the role of enzymatic catalysis [47]. As it remains undiscovered, there might be a connection of a Baeyer-Villiger monooxygenases that yields esters from carboxylic acids and ketones from aldehydes [48]. In brassinosteroid biosynthesis, this type of enzyme responsible for the oxidation of castasterone to brassinolide has been reported [49]. A regulatory mechanism that modulates the βCC-mediated signaling could be represented by the process of glycosylation of βCC. However, from the total βCC pool, glycosylated βCC was found to represent only a small portion (<2%) [30]. The βCC converts to various compounds like 2,2,6-trimethyl-cyclohexanone and 2,6-dimethyl-cyclohexanol after UV-light exposure [50]. In glycosylated form, βCC can occur in plants, and several glycosyl transferases are induced by the βCC treatment [10].

4. Signalling of βCC in Plants

The βCC has emerged as a new dimension for acting as a stress tolerant molecule in adverse conditions. The signaling pathway has been disclosed along with the transportation function within plants. The βCC performs in a hormone-induced marker line and the corresponding mutant responds to phytohormone pathway signals like auxin and brassinosteroids (BRs) and eventually results in cell enlargement [51]. Enzymatic action produces CCD4b from the genetic variance of *Crocus sativus* through β-carotene cleavage in the model plant *Arabidopsis*, which hastens to reduce the dehydration, salinity and oxidation rate [40,52]. The endo-metabolic substances in vascular plants took part in the xenobiotic response with diverse detoxifying agents, such as SCL14, ANAC102, ANAC001 and ANAC031 for oxidation resistance [11]. The βCC induced plants build interdependence with PAP signaling and down-regulate carotenoid substances while ST2A acts as a sulfate donor, and SAL1 has a deleterious effect during the plethora of light and drought stress [53,54]. Along with PAP, Methylerythritol cyclodiphosphate (MEcPP) substrates are also redox regularized and trigger the augmentation rate of the ROS level [55–57]. Sustaining photosensitivity during oxidative stress environment in mbs1 mutant crops for signaling pathway, procurement of protein and partial replacement in the nuclei occurs [28].

5. Functions of βCC in Plants

The βCC is a volatile organic compound that has been reported to have multiple functions in non-vascular plants (Figure 2). Microalgae discharge βCC, which is responsible for transferring stress signals to homogeneous algae and inducing defence. The former compound plays an allelopathic function on heterogeneous algae and aquatic macrophyte for opposite nutrients, as well as providing protection in opposition to predators [58]. The βCC has been reported to make cell rupture in *Nitzschia palea*, a diatom [59]. Ikawa et al. (2001) [60] reported that in cyanobacterium *Microcystis*, βCC is one of the main emitted volatile organic compounds. Sun et al. (2020) [61] suggested that the toxicity of βCC to cells might be associated with nuclear variation, DNA laddering, caspase 9- and caspase 3-like performance, signifying the initiation of a programmed cell death mechanism. In the case of cyanobacterial bloom, β-cyclocitric acid is produced by the oxidation of the βCC compound by *Microcystis*, indirectly causing toxicity. The production of this acidic compound leads to water acidification, causing chlorophyll loss, cell lysis, and phycocyanin pigment release, resulting in a characteristic blue colour [21]. These studies suggest that βCC and other volatile apocarotenoids are the principal allelopathic agents in cyanobacterial volatile organic compounds, but that at high concentrations, these compounds may be harmful to the emitters. However, no evidence has been found that low levels of βCC can elicit defence mechanisms in photosynthetic bacteria, such as those found in vascular plants. Mosses have also been observed to release volatile chemical compounds that could be used in interspecies communication [62]. Experimental evidence exhibited that photosynthetic activators and enzymatic variance treated with βCC in plants increased the photosynthetic rate, root-shoot expansion and carbon assimilation [63].

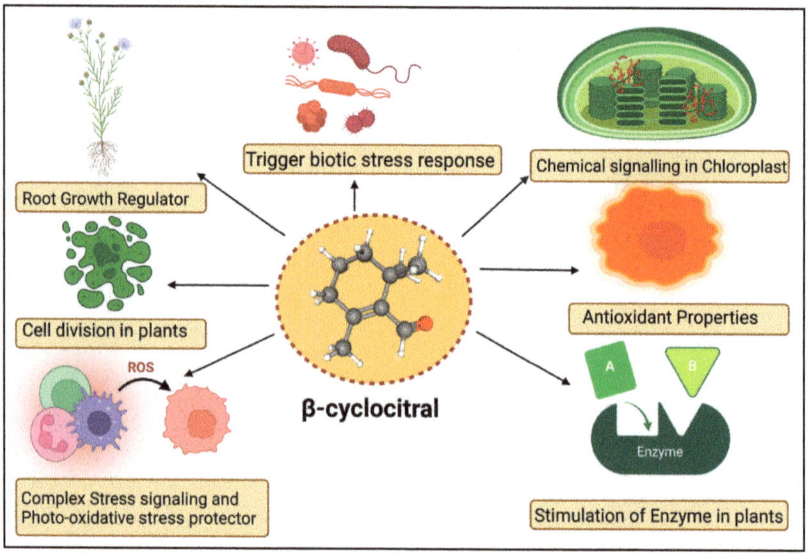

Figure 2. Diagrammatic sketch describing different functions of β-cyclocitral in plants.

The mosses *Hamatocaulis vernicosus* and *Sphagnum fexuosum* compete for the release of volatile organic compounds via increasing emission of a molecule called methyl 2,6, 6-trimethyl-1-cyclohexene-1-carboxylate, which is chemically linked to βCC. As a result, an alarmed mechanism could be set off, signalling the competitive strength of their neighboring moss species. The former is an enzyme that can convert carotene to βCC. The βCC is an intermediary in the $^1O_2$ signalling pathway, which controls gene expression reprogramming. It eventually causes plant cells to shift from active growth to cellular defence, resulting in stress and adaptation. The bulk of the downregulated gene encoded proteins involved in the development, growth, and biogenesis of cellular components [10]. Upregulated genes, on the other hand, were linked to environmental interactions, stress responses, and cellular mobility. Under normal or light stress conditions, βCC produces a tiny zinc finger protein (MBS1; methylene blue sensitivity 1) that is needed for the proper expression of $^1O_2$-responsive genes [64]. The βCC is said to have increased the former protein levels while also causing the protein to relocalize to the nucleus [28]. Further research revealed that the *Arabidopsis* mbs1 mutant (deficient in MBS1) was insensitive to βCC and therefore lacked an increase in photo-tolerance after treatment with βCC [28]. MBS1 is thought to be downstream of βCC in the $^1O_2$ signalling pathway, although it's precise function remains uncertain. Exogenous reactive substances are inactivated by typical detoxifying enzymes in vascular plants, which remove these molecules in three phases. In the first stage, side groups are introduced or modified in harmful substances such as herbicides, pollutants, and so on. The modified molecule is conjugated to sugar moieties or glutathione in the second stage. Finally, inactivated chemical compounds are compartmentalized [65]. The transcriptome of βCC-treated *Arabidopsis* plants showed activation of detoxification pathways [66]. Several glutathione-S-transferases (GST) and UDP-glycosyltransferases (UDP-glycosyltransferases) were involved in the xenobiotic detoxification process. The GRAS protein (SCL14; SCARECROW LIKE 14) and the glutaredoxin (GRX480/ROXY19) fight for interaction with TGAII transcription factors and mediate the activation/inhibition of a large number of detoxifying enzymes during the modification phase [67]. SCL14-controlled xenobiotic detoxification was induced by βCC and photooxidative stress conditions. Similarly, the scl14 knockout mutant did not respond to βCC and remained susceptible to high light stress even after treatment with βCC [68]. A few signalling cascade components downstream of βCC have been identified. The SCL14/TGA II complex, in particular, modulates the transcriptional levels of a transcription factor that regulates other downstream transcription regulators and, eventually, the redox enzymes of the first phase of the detoxification response [68], thus improving cellular detoxification capabilities. Surprisingly, the loss of MBS1 had no effect on βCC-induced cellular detoxification, indicating that there are two pathways in βCC signalling, one regulated by SCL14 and the other dependent on MBS1 [68]. Lipid peroxidation produces the chemicals that are characteristic of photooxidation and contributes unwaveringly to its toxicity. They decompose into reactive aldehydes (acrolein or 4-hydroxynonenal), which obstruct macromolecule function and cause cell death [69]. Diversified functions of βCC in plants are given in Table 1.

In plants including *Solanum lycopersicum*, *Piper nigrum*, and *Arabidopsis thaliana*, chemical βCC can cause changes in gene expression and promote drought tolerance [70]. The available literature on βCC shows that it may activate a signalling cascade that has yet to be fully described. Dickinson et al. (2019) [24] found that *Arabidopsis* seedlings grown in Petri plates treated with βCC stimulated the growth of primary roots. Increased root lengths may be beneficial under salt and water stress conditions, allowing for better soil exploration and water uptake. βCC has the ability to influence root development in *Solanum lycopersicum* and *Oryza sativa* without relying on auxin or brassinosteroid signalling. However, it is unclear whether βCC causes root growth directly through cell division and elongation or indirectly through the activation of cellular detoxification and resistance to oxidative stress. The molecular processes underlying βCC control of root development will need to be clarified in future research [31].

Table 1. Various functions of βCC in Plants.

Plant Species	Functions	References
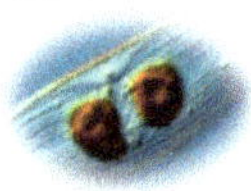 *Nitzschia palea*	Cell rupture at 0.1–0.5 mg mL^{-1} dose, cyanobacterial cell degradation, change in water color	[59]
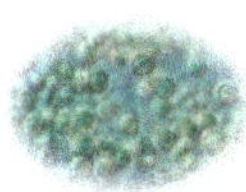 *Cyanidium caldarium*	Unpalatable water odor	[71]
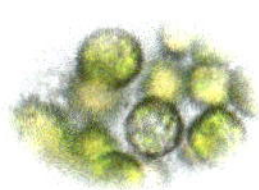 *Chlorella pyrenoidosa*	Inhibition of cell growth and development	[60]
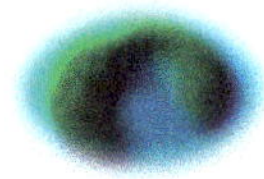 *Chlamydomonas reinhardtii*	Induce programmed cell death, cause poison to other algae	[61]

Table 1. *Cont.*

Plant Species	Functions	References
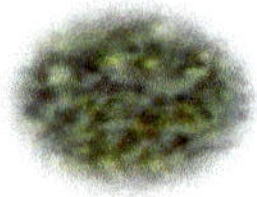 *Microcystis aeruginosa*	Increase βCC emission, expose high ion concentration	[72]
 *Solanum lycopersicum*	Retro nasal olfactory (smell) add to flavor to the fruit, volatile compound induces taste	[73]
 *Oryza sativa*	Scented rice varieties have aroma, more leaves present in vegetative stage	[74]
 *Petroselinum crispum*	Helps to produce essential oil and contribute in anti-fungal activity	[75]
 *Camellia sinensis*	Improve odorant properties and structural functions	[76]
 *Grapevines*	Inhibit infestation of spider mite and reduce symptoms	[77]

6. βCC Involved in the Regulation of Stress Response

The βCC has recently emerged as a unique plant signal in vascular and non-vascular plants that triggers stress tolerance (Figure 3). Ramel et al. (2012) [10] found that βCC increased tolerance to drought stress and high light-induced oxidative stress in *Arabidopsis thaliana* by altering the expression of several nuclear-encoded genes. Overexpression of the Crocus CCD4b gene in *Arabidopsis* confers tolerance to environmental and oxidative stressors, according to Baba et al. (2015) [52]. βCC plays important role in drought stress tolerance. Drought stress hampers photosynthesis, stomatal conductance, and the respiration and transduction system. In plants, leaf senescence, cell expansion, and yield significantly decreased under drought stress [78,79]. ROS provokes drought stress which puts the oxidation process, assimilation and antioxidant systems at risk [80]. The correlation between βCC and ROS during drought merely demonstrated that increase the rate of βCC aid the reduction of β-carotene in *Solanum lycopersicum*. Simulating stress tolerance, SOD and CAT upregulation altered transcription levels in plants during drought stress (Figure 3). The $^1O_2$ simulate as a signaling particle, attacking lipid peroxidation within stressed plants [81,82]. In *Arabidopsis thaliana*, elevated levels of singlet oxygen-regulated genes (SORGs) are a crucial indicator in the early stages of stress. Dispensation of βCC or dhA protein aid in lowering photoinhibition in comparison with plants grown in a control or natural stress environment βCC acts as a secondary messenger imparting towards the nucleus through methylene blue sensitive [28]. The uncoupling of chlorophyll compound produces ROS accordingly. The plethora of light stress in the environment decrement the pace of the photosynthetic electron transport chain (PETC) immobilizes the photosystem and inhibits photosynthesis [83,84]. Photoinhibition, incompatibility of photochemical reaction and the D1 protein in photodamage system degrade [85,86]. A copious amount of light energy is mechanized with a non-photochemical extinguishing method that stimulates $^1O_2$ production in stress conditions and $^1O_2$ mitigation in the dark light [87]. The excessive re-occurrence of mutagenesis took place in the chlorophyll content [88]. Signaling molecules helps to overcome photodegradation in the light stress condition using a wide range of carotenoid content. In high light stress, carotenoid oxidation took place in PS–II which alters the nuclear gene expression [89]. Moreover, low temperature caused the progressive mitigation of βCC content and the substantial increase in H_2O_2 in *Solanum lycopersicum* [90,91]. Plant reaction after treatment with βCC under stress condition is described in Table 2.

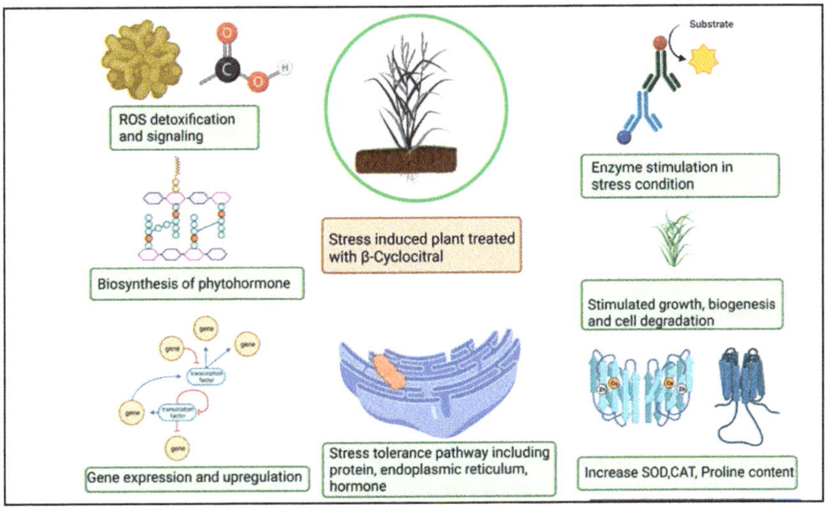

Figure 3. Schematic representation of β-cyclocitral effects on plants under abiotic stress.

Salt stress hampers plant growth amplification, ionic balance, nutritional symmetry and stomatal function in shoots [92–94]. Water absorption blockage, water potential alleviation, oxidative stress occurrence and programmed cell death (PCD) strike the plants functioning during salinity stress [95]. Executer 1 and 2 proteins are accountable for cell death in each route [82]. Enhanced root ontogenesis illustrates a favorable impact on salinity stress via βCC application. Rice and *Arabidopsis* roots in considerable depth stimulate plant vigor and genomic functions which help to defeat stress through nuclear transportation and ion activation [24]. In the presence of carotenoid compounds, the *S. lycopersicum* crop demonstrated enhanced oxidation and biochemical synthesis [31]. The inhibition of curly leaves, prevention of leaves wilting, ameliorating relative water content (RWC) and the stomatal opening is accelerated in βCC treated plants [7]. However, the functioning mechanism of βCC in plants has yet to be extensively studied in both long and short-duration experiments. Photo-oxidative stress is a consecutive production of ROS in a pattern which is harmful to the antioxidant defense system. As a result, the photo-oxidation process continues to develop toward a pessimistic function, hazardous chemicals get acclimated, and chloroplast damage happens and eventually yields losses [96]. Photo-oxidation took place in leaves, ontogeny led to senescence and oxidative cascade occurred under stressed conditions [97,98]. The abundance of stress-tolerant genes becomes active in the raised ROS [99–101]. In the presence of elevated ROS, singlet oxygen oxidizes promptly in the environment and occurs in PS-II complexes, enabling photochemical quenching of $^{3}Chl^{*}$ transfer [102]. ROS is responsible for leaf damage, mitigating photosynthesis competence, photoproduction and photo-oxidative injury [6,70,103–105]. Lichens and cyanobacterial compounds contain reactive radicals in the physiologically active stress condition [29].

Table 2. βCC mediated abiotic alteration in plants.

Plants Name	Stress	Effect in Plants	References
Arabidopsis thaliana	Light stress	βCC acts as a secondary messenger, alters transcription of singlet oxygen via MBS protein	[89]
Solanum lycopersicum	Drought stress	Increases the osmolyte accumulation, root and branching, superoxide exclusion and improves stress tolerance	[7]
Arabidopsis thaliana	Photo-oxidative stress	ROS production, changes gene expression	[10]
Viola tricolor	Drought stress	Elicits drought stress, no wilting occurs in water lack condition	[31]
Arabidopsis thaliana	Oxidative stress	Influences gene expression, suppression in transgenic plants, Cu/Zn SOD is induced	[106]
Oryza sativa	Salt stress	Tolerates the adverse salinity stress condition and increased vigor of plants	[24]
Arabidopsis thaliana	Water stress	Overlaps genetic responses, showing the impact of singlet oxygen	[107]
Microcystis cyanobacteria	Water stress	Oxidation and acidification of βCC; Blue color forms, pH is reduced	[47]
Arabidopsis thaliana	Light stress	SCL14 enhances genetic responses, upregulation of lipid peroxidation	[68]
Oryza sativa	Salt stress	Promotes cell division in root meristems and stimulates lateral root branching	[24]

7. Conclusions and Future Perspective

As an emerging molecule, βCC has been gaining increasing interest to provide stress tolerance due to their excellent features resulting from the altering of the expression of several nuclear-encoded genes. In this work, novel apocarotenoids of β-carotene and βCC applications for stress tolerance were systematically reviewed. Their biosynthesis, derivatives and signalling were discussed. Furthermore, the promising functions of βCC were also discussed and potential directions for future work were suggested.

Author Contributions: Conceptualization, M.F. and P.A.; resources, M.F., S.H.T. and S.A.; writing-original draft preparation, M.F., S.H.T. and S.A.; writing-review and editing, Z.F., S.M.A., F.Y. and S.H.; visualization, P.A. and S.H., supervision, S.H. and S.M.A.; funding acquisition, P.A. All authors have read and agreed to the published version of the manuscript.

Funding: This research received no external funding.

Institutional Review Board Statement: Not applicable.

Informed Consent Statement: Not applicable.

Data Availability Statement: Not applicable.

Conflicts of Interest: The authors declare that they have no conflict of interest.

References

1. Bernhoft, A. A brief review on bioactive compounds in plants. In *Bioactive Compounds in Plants—Benefits and Risks for Man and Animals*; The Norwegian Academy of Science and Letters: Oslo, Norway, 2010; Volume 50, pp. 11–17.
2. Sasidharan, S.; Chen, Y.; Saravanan, D.; Sundram, K.M.; Latha, L.Y. Extraction, isolation and characterization of bioactive compounds from plants' extracts. *Afr. J. Tradit. Complement. Altern. Med.* **2011**, *8*, 1–10. [CrossRef] [PubMed]
3. Tiwari, S. Plants: A rich source of herbal medicine. *J. Nat. Prod.* **2008**, *1*, 27–35.
4. Couladis, M.; Tzakou, O.; Verykokidou, E.; Harvala, C. Screening of some Greek aromatic plants for antioxidant activity. *Phyther. Res.* **2003**, *17*, 194–195. [CrossRef] [PubMed]
5. Havaux, M. Carotenoid oxidation products as stress signals in plants. *Plant J.* **2014**, *79*, 597–606. [CrossRef] [PubMed]
6. Galasso, C.; Corinaldesi, C.; Sansone, C. Carotenoids from marine organisms: Biological functions and inductrial applications. *Antioxidants* **2017**, *6*, 96. [CrossRef]
7. Deshpande, S.; Manoharan, R.; Mitra, S. Exogenous β-cyclocitral treatment primes tomato plants against drought by inducing tolerance traits, independent of abscisic acid. *Plant Biol.* **2021**, *23*, 170–180. [CrossRef]
8. Rodrigo, M.J.; Alquézar, B.; Alós, E.; Medina, V.; Carmona, L.; Bruno, M.; Al-Babili, S.; Zacarías, L. A novel carotenoid cleavage activity involved in the biosynthesis of Citrus fruit-specific apocarotenoid pigments. *J. Exp. Bot.* **2013**, *64*, 4461–4478. [CrossRef]
9. Hayward, S.; Cilliers, T.; Swart, P. Lipoxygenases: From isolation to application. *Compr. Rev. Food Sci. Food Saf.* **2017**, *16*, 199–211. [CrossRef]
10. Ramel, F.; Birtic, S.; Ginies, C.; Soubigou-Taconnat, L.; Triantaphylidès, C.; Havaux, M. Carotenoid oxidation products are stress signals that mediate gene responses to singlet oxygen in plants. *Proc. Natl. Acad. Sci. USA* **2012**, *109*, 5535–5540. [CrossRef]
11. Moretto, J.A.S.; de Freitas, P.N.N.; de Almeida, E.C.; Altarugio, L.M.; da Silva, S.V.; Fiore, M.F.; Pinto, E. Effects of different cultivation conditions on the production of β-cyclocitral and β-ionone in Microcystis aeruginosa. *BMC Microbiol.* **2022**, *22*, 78. [CrossRef]
12. Mitra, S.; Estrada-Tejedor, R.; Volke, D.C.; Phillips, M.A.; Gershenzon, J.; Wright, L.P. Negative regulation of plastidial isoprenoid pathway by herbivore-induced β-cyclocitral in Arabidopsis thaliana. *Proc. Natl. Acad. Sci. USA* **2021**, *118*, e2008747118. [CrossRef] [PubMed]
13. Sharoni, Y.; Danilenko, M.; Dubi, N.; Ben-Dor, A.; Levy, J. Carotenoids and transcription. *Arch. Biochem. Biophys.* **2004**, *430*, 89–96. [CrossRef] [PubMed]
14. Kuntz, E.; Hoeller, U.; Greatrix, B.; Lankin, C.; Seifert, N.; Acharya, S.; Riss, G.; Buchwald-Hunziker, P.; Hunziker, W.; Goralczyk, R. β-Carotene and apocarotenals promote retinoid signaling in BEAS-2B human bronchioepithelial cells. *Arch. Biochem. Biophys.* **2006**, *455*, 48–60. [CrossRef]
15. Liu, J.; Sun, X.; Dong, H.; Sun, C.; Sun, W.; Chen, B.; Song, Y.; Yang, B. β-Ionone suppresses mammary carcinogenesis, proliferative activity and induces apoptosis in the mammary gland of the Sprague-Dawley rat. *Int. J. Cancer* **2008**, *122*, 2689–2698. [CrossRef] [PubMed]
16. Mahattanatawee, K.; Rouseff, R.; Valim, M.F.; Naim, M. Identification and aroma impact of norisoprenoids in orange juice. *J. Agric. Food Chem.* **2005**, *53*, 393–397. [CrossRef] [PubMed]
17. Lalko, J.; Lapczynski, A.; McGinty, D.; Bhatia, S.; Letizia, C.S.; Api, A.M. Fragrance material review on ionone. *Food Chem. Toxicol.* **2007**, *45*, S251–S257. [CrossRef] [PubMed]

18. Yu, C.; Shi, C.; Ji, M.; Xu, X.; Zhang, Z.; Ma, J.; Wang, G. Taste and odor compounds associated with aquatic plants in Taihu Lake: Distribution and producing potential. *Environ. Sci. Pollut. Res.* **2019**, *26*, 34510–34520. [CrossRef]
19. Moraga, Á.R.; Rambla, J.L.; Ahrazem, O.; Granell, A.; Gómez-Gómez, L. Metabolite and target transcript analyses during Crocus sativus stigma development. *Phytochemistry* **2009**, *70*, 1009–1016. [CrossRef]
20. Lewinsohn, E.; Sitrit, Y.; Bar, E.; Azulay, Y.; Meir, A.; Zamir, D.; Tadmor, Y. Carotenoid pigmentation affects the volatile composition of tomato and watermelon fruits, as revealed by comparative genetic analyses. *J. Agric. Food Chem.* **2005**, *53*, 3142–3148. [CrossRef]
21. Arii, S.; Tsuji, K.; Tomita, K.; Hasegawa, M.; Bober, B.; Harada, K.-I. Cyanobacterial blue color formation during lysis under natural conditions. *Appl. Environ. Microbiol.* **2015**, *81*, 2667–2675. [CrossRef]
22. Juttner, F.; Watson, S.B.; von Elert, E.; Koster, O.J. β-cyclocitral, a grazer defense signal unique to the cyanobacterium Microcystis. *Chem. Ecol.* **2010**, *36*, 1387–1397. [CrossRef] [PubMed]
23. Ozaki, K.; Ohta, A.; Iwata, C.; Horikawa, A.; Tsuji, K.; Ito, E.; Ikai, Y.; Harada, K.I. Lysis of cyanobacteria with volatile organic compounds. *Chemosphere* **2008**, *71*, 1531–1538. [CrossRef] [PubMed]
24. Dickinson, A.J.; Lehner, K.; Mi, J.; Jia, K.-P.; Mijar, M.; Dinneny, J.; Al-Babili, S.; Benfey, P.N. β-Cyclocitral is a conserved root growth regulator. *Proc. Natl. Acad. Sci. USA* **2019**, *116*, 10563–10567. [CrossRef] [PubMed]
25. Zheng, X.; Zhu, K.; Ye, J.; Price, E.J.; Deng, X.; Fraser, P.D. The effect of β-cyclocitral treatment on the carotenoid content of transgenic Marsh grapefruit (*Citrus paradisi* Macf.) suspension-cultured cells. *Phytochemistry* **2020**, *180*, 112509. [CrossRef] [PubMed]
26. Miras-Moreno, B.; Pedreño, M.A.; Fraser, P.D.; Sabater-Jara, A.B.; Almagro, L. Effect of diflufenican on total carotenoid and phytoene production in carrot suspension-cultured cells. *Planta* **2019**, *249*, 113–122. [CrossRef] [PubMed]
27. Havaux, M. β-Cyclocitral and Derivatives: Emerging Molecular Signals Serving Multiple Biological Functions. *Plant Physiol. Biochem.* **2020**, *155*, 35–41. [CrossRef]
28. Shumbe, L.; d'Alessandro, S.; Shao, N.; Chevalier, A.; Ksas, B.; Bock, R.; Havaux, M. Methylene Blue Sensitivity 1 (MBS1) is required for acclimation of Arabidopsis to singlet oxygen and acts downstream of β-cyclocitral. *Plant Cell Environ.* **2017**, *40*, 216–226. [CrossRef]
29. García-Plazaola, J.I.; Portillo-Estrada, M.; Fernández-Marín, B.; Kännaste, A.; Niinemets, Ü. Emissions of carotenoid cleavage products upon heat shock and mechanical wounding from a foliose lichen. *Environ. Exp. Bot.* **2017**, *133*, 87–97. [CrossRef]
30. Swapnil, P.; Meena, M.; Singh, S.K.; Dhuldhaj, U.P.; Harish Marwal, A. Vital roles of carotenoids in plants and humans to deteriorate stress with its structure, biosynthesis, metabolic engineering and functional aspects. *Curr. Plant Biol.* **2021**, *26*, 100203. [CrossRef]
31. D'Alessandro, S.; Havaux, M. Sensing β-carotene oxidation in photosystem II to master plant stress tolerance. *New Phytol.* **2019**, *223*, 1776–1783. [CrossRef]
32. Noctor, G.; Arisi, A.-C.M.; Jouanin, L.; Kunert, K.J.; Rennenberg, H.; Foyer, C.H. Glutathione: Biosynthesis, metabolism and relationship to stress tolerance explored in transformed plants. *J. Exp. Bot.* **1998**, *49*, 623–647. [CrossRef]
33. Jamaluddin, N.D.; Rohani, E.R.; Mohd Noor, N.; Goh, H.-H. Transcriptome-wide effect of DE-ETIOLATED1 (DET1) suppression in embryogenic callus of *Carica papaya*. *J. Plant Res.* **2019**, *132*, 181–195. [CrossRef] [PubMed]
34. Giuliano, G.; Al-Babili, S.; Von Lintig, J. Carotenoid oxygenases: Cleave it or leave it. *Trends Plant Sci.* **2003**, *8*, 145–149. [CrossRef]
35. Hou, X.; Rivers, J.; León, P.; McQuinn, R.P.; Pogson, B.J. Synthesis and function of apocarotenoid signals in plants. *Trends Plant Sci.* **2016**, *21*, 792–803. [CrossRef] [PubMed]
36. Nambara, E.; Marion-Poll, A. Abscisic acid biosynthesis and catabolism. *Annu. Rev. Plant Biol.* **2005**, *56*, 165. [CrossRef]
37. Ramel, F.; Mialoundama, A.S.; Havaux, M. Nonenzymic carotenoid oxidation and photooxidative stress signalling in plants. *J. Exp. Bot.* **2013**, *64*, 799–805. [CrossRef]
38. Cui, H.; Wang, Y.; Qin, S. Genomewide analysis of carotenoid cleavage dioxygenases in unicellular and filamentous cyanobacteria. *Comp. Funct. Genom.* **2012**, *2012*, 164690. [CrossRef]
39. Al-Babili, S.; Bouwmeester, H.J. Strigolactones, a novel carotenoid-derived plant hormone. *Annu. Rev. Plant Biol.* **2015**, *66*, 161–186. [CrossRef]
40. Rubio-Moraga, A.; Rambla, J.L.; Fernández-de-Carmen, A.; Trapero-Mozos, A.; Ahrazem, O.; Orzáez, D.; Granell, A.; Gómez-Gómez, L. New target carotenoids for CCD4 enzymes are revealed with the characterization of a novel stress-induced carotenoid cleavage dioxygenase gene from Crocus sativus. *Plant Mol. Biol.* **2014**, *86*, 555–569. [CrossRef]
41. Meng, N.; Yan, G.-L.; Zhang, D.; Li, X.-Y.; Duan, C.-Q.; Pan, Q.-H. Characterization of two *Vitis vinifera* carotenoid cleavage dioxygenases by heterologous expression in *Saccharomyces cerevisiae*. *Mol. Biol. Rep.* **2019**, *46*, 6311–6323. [CrossRef]
42. Zorn, H.; Langhoff, S.; Scheibner, M.; Nimtz, M.; Berger, R.G. A peroxidase from *Lepista irina* cleaves β, β-carotene to flavor compounds. *Biol. Chem.* **2003**, *384*, 1049–1056. [CrossRef] [PubMed]
43. Edge, R.; McGarvey, D.J.; Truscott, T.G. The carotenoids as anti-oxidants—A review. *J. Photochem. Photobiol. B Biol.* **1997**, *41*, 189–200. [CrossRef]
44. Ma, Z.; Niu, Y.; Xie, P.; Chen, J.; Tao, M.; Deng, X. Off-flavor compounds from decaying cyanobacterial blooms of Lake Taihu. *J. Environ. Sci.* **2013**, *25*, 495–501. [CrossRef]
45. Crombie, B.S.; Smith, C.; Varnavas, C.Z.; Wallace, T.W. A conjugate addition–radical cyclisation approach to sesquiterpene-phenol natural products. *J. Chem. Soc. Perkin Trans.* **2001**, *1*, 206–215. [CrossRef]

46. Mazur, M.; Gładkowski, W.; Podkowik, M.; Bania, J.; Nawrot, J.; Białońska, A.; Wawrzeńczyk, C. Lactones 43. New biologically active lactones: β-cyclocitral derivatives. *Pest Manag. Sci.* **2014**, *70*, 286–294. [CrossRef]
47. Tomita, K.; Hasegawa, M.; Arii, S.; Tsuji, K.; Bober, B.; Harada, K. Characteristic oxidation behavior of β-cyclocitral from the cyanobacterium Microcystis. *Environ. Sci. Pollut. Res.* **2016**, *23*, 11998–12006. [CrossRef]
48. Van Berkel, W.J.H.; Kamerbeek, N.M.; Fraaije, M. Flavoprotein monooxygenases, a diverse class of oxidative biocatalysts. *J. Biotechnol.* **2006**, *124*, 670–689. [CrossRef]
49. Kim, T.-W.; Hwang, J.-Y.; Kim, Y.-S.; Joo, S.-H.; Chang, S.C.; Lee, J.S.; Takatsuto, S.; Kim, S.-K. Arabidopsis CYP85A2, a cytochrome P450, mediates the Baeyer-Villiger oxidation of castasterone to brassinolide in brassinosteroid biosynthesis. *Plant Cell* **2005**, *17*, 2397–2412. [CrossRef]
50. Kim, T.; Kim, T.-K.; Zoh, K.-D. Degradation kinetics and pathways of β-cyclocitral and β-ionone during UV photolysis and UV/chlorination reactions. *J. Environ. Manag.* **2019**, *239*, 8–16. [CrossRef]
51. Felemban, A.; Braguy, J.; Zurbriggen, M.D.; Al-Babili, S. Apocarotenoids involved in plant development and stress response. *Front. Plant Sci.* **2019**, *10*, 1168. [CrossRef]
52. Baba, S.A.; Jain, D.; Abbas, N.; Ashraf, N. Overexpression of Crocus carotenoid cleavage dioxygenase, CsCCD4b, in Arabidopsis imparts tolerance to dehydration, salt and oxidative stresses by modulating ROS machinery. *J. Plant Physiol.* **2015**, *189*, 114–125. [CrossRef] [PubMed]
53. Estavillo, G.M.; Crisp, P.A.; Pornsiriwong, W.; Wirtz, M.; Collinge, D.; Carrie, C.; Giraud, E.; Whelan, J.; David, P.; Javot, H. Evidence for a SAL1-PAP chloroplast retrograde pathway that functions in drought and high light signaling in Arabidopsis. *Plant Cell* **2011**, *23*, 3992–4012. [CrossRef] [PubMed]
54. Wilson, P.B.; Estavillo, G.M.; Field, K.J.; Pornsiriwong, W.; Carroll, A.J.; Howell, K.A.; Woo, N.S.; Lake, J.A.; Smith, S.M.; Harvey Millar, A. The nucleotidase/phosphatase SAL1 is a negative regulator of drought tolerance in Arabidopsis. *Plant J.* **2009**, *58*, 299–317. [CrossRef] [PubMed]
55. Rossel, J.B.; Walter, P.B.; Hendrickson, L.; Chow, W.S.; Poole, A.; Mullineaux, P.M.; Pogson, B.J. A mutation affecting ASCORBATE PEROXIDASE 2 gene expression reveals a link between responses to high light and drought tolerance. *Plant Cell Environ.* **2006**, *29*, 269–281. [CrossRef] [PubMed]
56. Kim, B.; Von Arnim, A.G. FIERY1 regulates light-mediated repression of cell elongation and flowering time via its 3′(2′), 5′-bisphosphate nucleotidase activity. *Plant J.* **2009**, *58*, 208–219. [CrossRef] [PubMed]
57. Li, Z.; Sharkey, T.D. Metabolic profiling of the methylerythritol phosphate pathway reveals the source of post-illumination isoprene burst from leaves. *Plant Cell Environ.* **2013**, *36*, 429–437. [CrossRef]
58. Zuo, Z. Why algae release volatile organic compounds—The emission and roles. *Front. Microbiol.* **2019**, *10*, 491. [CrossRef]
59. Chang, D.W.; Hsieh, M.-L.; Chen, Y.-M.; Lin, T.-F.; Chang, J.-S. Kinetics of cell lysis for Microcystis aeruginosa and *Nitzschia palea* in the exposure to β-cyclocitral. *J. Hazard. Mater.* **2011**, *185*, 1214–1220. [CrossRef]
60. Ikawa, M.; Sasner, J.J.; Haney, J.F. Activity of cyanobacterial and algal odor compounds found in lake waters on green alga *Chlorella pyrenoidosa* growth. *Hydrobiologia* **2001**, *443*, 19–22. [CrossRef]
61. Sun, Q.; Zhou, M.; Zuo, Z. Toxic mechanism of eucalyptol and β-cyclocitral on *Chlamydomonas reinhardtii* by inducing programmed cell death. *J. Hazard. Mater.* **2020**, *389*, 121910. [CrossRef]
62. Saritas, Y.; Sonwa, M.M.; Iznaguen, H.; König, W.A.; Muhle, H.; Mues, R. Volatile constituents in mosses (Musci). *Phytochemistry* **2001**, *57*, 443–457. [CrossRef]
63. Schwachtje, J.; Minchin, P.E.H.; Jahnke, S.; van Dongen, J.T.; Schittko, U.; Baldwin, I.T. SNF1-related kinases allow plants to tolerate herbivory by allocating carbon to roots. *Proc. Natl. Acad. Sci. USA* **2006**, *103*, 12935–12940. [CrossRef] [PubMed]
64. Shao, N.; Duan, G.Y.; Bock, R. A mediator of singlet oxygen responses in Chlamydomonas reinhardtii and Arabidopsis identified by a luciferase-based genetic screen in algal cells. *Plant Cell* **2013**, *25*, 4209–4226. [CrossRef] [PubMed]
65. Riechers, D.E.; Kreuz, K.; Zhang, Q. Detoxification without intoxication: Herbicide safeners activate plant defense gene expression. *Plant Physiol.* **2010**, *153*, 3–13. [CrossRef]
66. Sandermann, H., Jr. Plant metabolism of xenobiotics. *Trends Biochem. Sci.* **1992**, *17*, 82–84. [CrossRef]
67. Huang, L.J.; Li, N.; Thurow, C.; Wirtz, M.; Hell, R.; Gatz, C. Ectopically expressed glutaredoxin ROXY19 negatively regulates the detoxification pathway in Arabidopsis thaliana. *BMC Plant Biol.* **2016**, *16*, 200. [CrossRef] [PubMed]
68. D'alessandro, S.; Ksas, B.; Havaux, M. Decoding β-cyclocitral-mediated retrograde signaling reveals the role of a detoxification response in plant tolerance to photooxidative stress. *Plant Cell* **2018**, *30*, 2495–2511. [CrossRef]
69. Mano, J.; Biswas, M.S.; Sugimoto, K. Reactive carbonyl species: A missing link in ROS signaling. *Plants* **2019**, *8*, 391. [CrossRef]
70. D'Alessandro, S.; Mizokami, Y.; Légeret, B.; Havaux, M. The apocarotenoid β-cyclocitric acid elicits drought tolerance in plants. *iScience* **2019**, *19*, 461–473. [CrossRef]
71. Xu, Q.; Yang, L.; Yang, W.; Bai, Y.; Hou, P.; Zhao, J.; Zhou, L.; Zuo, Z. Volatile organic compounds released from *Microcystis flos-aquae* under nitrogen sources and their toxic effects on *Chlorella vulgaris*. *Ecotoxicol. Environ. Saf.* **2017**, *135*, 191–200. [CrossRef]
72. Hasegawa, M.; Nishizawa, A.; Tsuji, K.; Kimura, S.; Harada, K. Volatile organic compounds derived from 2-keto-acid decarboxylase in *Microcystis aeruginosa*. *Microbes Environ.* **2012**, *27*, 525–528. [CrossRef] [PubMed]
73. Tieman, D.; Bliss, P.; McIntyre, L.M.; Blandon-Ubeda, A.; Bies, D.; Odabasi, A.Z.; Rodríguez, G.R.; van der Knaap, E.; Taylor, M.G.; Goulet, C. The chemical interactions underlying tomato flavor preferences. *Curr. Biol.* **2012**, *22*, 1035–1039. [CrossRef] [PubMed]

74. Hinge, V.; Patil, H.; Nadaf, A. Comparative characterization of aroma volatiles and related gene expression analysis at vegetative and mature stages in basmati and non-basmati rice (*Oryza sativa* L.) cultivars. *Appl. Biochem. Biotechnol.* **2016**, *178*, 619–639. [CrossRef] [PubMed]
75. Linde, G.A.; Gazim, Z.C.; Cardoso, B.K.; Jorge, L.F.; Tešević, V.; Glamočlija, J.; Soković, M.; Colauto, N.B. Antifungal and antibacterial activities of *Petroselinum crispum* essential oil. *Genet. Mol. Res.* **2016**, *15*, gmr.15038538. [CrossRef] [PubMed]
76. Ojha, P.K.; Roy, K. PLS regression-based chemometric modeling of odorant properties of diverse chemical constituents of black tea and coffee. *RSC Adv.* **2018**, *8*, 2293–2304. [CrossRef] [PubMed]
77. Lazazzara, V.; Bueschl, C.; Parich, A.; Pertot, I.; Schuhmacher, R.; Perazzolli, M. Downy mildew symptoms on grapevines can be reduced by volatile organic compounds of resistant genotypes. *Sci. Rep.* **2018**, *8*, 1618. [CrossRef] [PubMed]
78. Misra, V.; Solomon, S.; Mall, A.K.; Prajapati, C.P.; Hashem, A.; Abd_Allah, E.F.; Ansari, M.I. Morphological assessment of water stressed sugarcane: A comparison of waterlogged and drought affected crop. *Saudi J. Biol. Sci.* **2020**, *27*, 1228–1236. [CrossRef]
79. Anjum, S.A.; Ashraf, U.; Tanveer, M.; Khan, I.; Hussain, S.; Shahzad, B.; Zohaib, A.; Abbas, F.; Saleem, M.F.; Ali, I. Drought induced changes in growth, osmolyte accumulation and antioxidant metabolism of three maize hybrids. *Front. Plant Sci.* **2017**, *8*, 69. [CrossRef]
80. Tanveer, M.; Shahzad, B.; Sharma, A.; Khan, E.A. 24-Epibrassinolide application in plants: An implication for improving drought stress tolerance in plants. *Plant Physiol. Biochem.* **2019**, *135*, 295–303. [CrossRef]
81. Laloi, C.; Havaux, M. Key players of singlet oxygen-induced cell death in plants. *Front. Plant Sci.* **2015**, *6*, 39. [CrossRef]
82. Dogra, V.; Rochaix, J.; Kim, C. Singlet oxygen-triggered chloroplast-to-nucleus retrograde signalling pathways: An emerging perspective. *Plant. Cell Environ.* **2018**, *41*, 1727–1738. [CrossRef] [PubMed]
83. Murata, N.; Takahashi, S.; Nishiyama, Y.; Allakhverdiev, S.I. Photoinhibition of photosystem II under environmental stress. *Biochim. Biophys. Acta (BBA)-Bioenerg.* **2007**, *1767*, 414–421. [CrossRef] [PubMed]
84. Roach, T.; Krieger-Liszkay, A. Regulation of photosynthetic electron transport and photoinhibition. *Curr. Protein Pept. Sci.* **2014**, *15*, 351–362. [CrossRef] [PubMed]
85. Aro, E.M.; Suorsa, M.; Rokka, A.; Allahverdiyeva, Y.; Paakkarinen, V.; Saleem, A.; Battchikova, N.; Rintamäki, E. Dynamics of photosystem II: A proteomic approach to thylakoid protein complexes. *J. Exp. Bot.* **2005**, *56*, 347–356. [CrossRef] [PubMed]
86. Takahashi, S.; Murata, N. How do environmental stresses accelerate photoinhibition? *Trends Plant Sci.* **2008**, *13*, 178–182. [CrossRef]
87. Demmig-Adams, B.; Garab, G.; Adams III, W.; Govindjee, U. *Non-Photochemical Quenching and Energy Dissipation in Plants, Algae and Cyanobacteria*; Springer: Dordrecht, The Netherlands, 2014.
88. Ouchane, S.; Picaud, M.; Vernotte, C.; Astier, C. Photooxidative stress stimulates illegitimate recombination and mutability in carotenoid-less mutants of *Rubrivivax gelatinosus*. *EMBO J.* **1997**, *16*, 4777–4787. [CrossRef]
89. Shumbe, L.; Bott, R.; Havaux, M. Dihydroactinidiolide, a high light-induced β-carotene derivative that can regulate gene expression and photoacclimation in Arabidopsis. *Mol. Plant* **2014**, *7*, 1248–1251. [CrossRef]
90. Liu, T.; Ye, X.; Li, M.; Li, J.; Qi, H.; Hu, X. H_2O_2 and NO are involved in trehalose-regulated oxidative stress tolerance in cold-stressed tomato plants. *Environ. Exp. Bot.* **2020**, *171*, 103961. [CrossRef]
91. Khan, T.A.; Fariduddin, Q.; Yusuf, M. Lycopersicon esculentum under low temperature stress: An approach toward enhanced antioxidants and yield. *Environ. Sci. Pollut. Res.* **2015**, *22*, 14178–14188. [CrossRef]
92. Hernández, J.A.; Ferrer, M.A.; Jiménez, A.; Barceló, A.R.; Sevilla, F. Antioxidant systems and O_2^-/H_2O_2 production in the apoplast of pea leaves. Its relation with salt-induced necrotic lesions in minor veins. *Plant Physiol.* **2001**, *127*, 817–831. [CrossRef]
93. Isayenkov, S.V. Physiological and molecular aspects of salt stress in plants. *Cytol. Genet.* **2012**, *46*, 302–318. [CrossRef]
94. Rajendran, K.; Tester, M.; Roy, S.J. Quantifying the three main components of salinity tolerance in cereals. *Plant. Cell Environ.* **2009**, *32*, 237–249. [CrossRef]
95. Demidchik, V.; Cuin, T.A.; Svistunenko, D.; Smith, S.J.; Miller, A.J.; Shabala, S.; Sokolik, A.; Yurin, V. Arabidopsis root K^+-efflux conductance activated by hydroxyl radicals: Single-channel properties, genetic basis and involvement in stress-induced cell death. *J. Cell Sci.* **2010**, *123*, 1468–1479. [CrossRef] [PubMed]
96. Borisova, M.M.M.; Kozuleva, M.A.; Rudenko, N.N.; Naydov, I.A.; Klenina, I.B.; Ivanov, B.N. Photosynthetic electron flow to oxygen and diffusion of hydrogen peroxide through the chloroplast envelope via aquaporins. *Biochim. Biophys. Acta (BBA)-Bioenerg.* **2012**, *1817*, 1314–1321. [CrossRef] [PubMed]
97. Juvany, M.; Müller, M.; Munné-Bosch, S. Photo-oxidative stress in emerging and senescing leaves: A mirror image? *J. Exp. Bot.* **2013**, *64*, 3087–3098. [CrossRef] [PubMed]
98. Ahmad, P.; Jaleel, C.A.; Salem, M.A.; Nabi, G.; Sharma, S. Roles of enzymatic and nonenzymatic antioxidants in plants during abiotic stress. *Crit. Rev. Biotechnol.* **2010**, *30*, 161–175. [CrossRef] [PubMed]
99. Apel, K.; Hirt, H. Reactive oxygen species: Metabolism, oxidative stress, and signal transduction. *Annu. Rev. Plant Biol.* **2004**, *55*, 373–399. [CrossRef]
100. Queval, G.; Foyer, C.H. Redox regulation of photosynthetic gene expression. *Philos. Trans. R. Soc. B Biol. Sci.* **2012**, *367*, 3475–3485. [CrossRef]
101. Munné-Bosch, S.; Queval, G.; Foyer, C.H. The impact of global change factors on redox signaling underpinning stress tolerance. *Plant Physiol.* **2013**, *161*, 5–19. [CrossRef]

102. Mozzo, M.; Dall'Osto, L.; Hienerwadel, R.; Bassi, R.; Croce, R. Photoprotection in the antenna complexes of photosystem II: Role of individual xanthophylls in chlorophyll triplet quenching. *J. Biol. Chem.* **2008**, *283*, 6184–6192. [CrossRef]
103. Fischer, B.B.; Hideg, E.; Krieger-Liszkay, A. Production, detection, and signaling of singlet oxygen in photosynthetic organisms. *Antioxid. Redox Signal.* **2013**, *18*, 2145–2162. [CrossRef] [PubMed]
104. Deshpande, S.; Purkar, V.; Mitra, S. β-Cyclocitral, a Master Regulator of Multiple Stress-Responsive Genes in *Solanum lycopersicum* L. Plants. *Plants* **2021**, *10*, 2465. [CrossRef] [PubMed]
105. Roach, T.; Baur, T.; Kranner, I. β-Cyclocitral Does Not Contribute to Singlet Oxygen-Signalling in Algae, but May Down-Regulate Chlorophyll Synthesis. *Plants* **2022**, *11*, 2155. [CrossRef] [PubMed]
106. Sunkar, R.; Kapoor, A.; Zhu, J.-K. Posttranscriptional induction of two Cu/Zn superoxide dismutase genes in Arabidopsis is mediated by downregulation of miR398 and important for oxidative stress tolerance. *Plant Cell* **2006**, *18*, 2051–2065. [CrossRef]
107. Koh, E.; Carmieli, R.; Mor, A.; Fluhr, R. Singlet oxygen-induced membrane disruption and serpin-protease balance in vacuolar-driven cell death. *Plant Physiol.* **2016**, *171*, 1616–1625. [CrossRef]

molecules

Article

Evaluation of Major Constituents of Medicinally Important Plants for Anti-Inflammatory, Antidiabetic and AGEs Inhibiting Properties: In Vitro and Simulatory Evidence

Abdul Rafey [1], Adnan Amin [1,*], Muhammad Kamran [1,2], Muhammad Imran Aziz [1], Varda Athar [1,2], Shah Iram Niaz [3] and Luc Pieters [4]

1 Natural Products Research Lab (NPRL), Gomal Centre of Pharmaceutical Sciences, Faculty of Pharmacy, Gomal University, Dera Ismail Khan 29050, Pakistan
2 Gomal Centre of Biochemistry and Biotechnology (GCBB), Gomal University, Dera Ismail Khan 29050, Pakistan
3 Institute of Chemical Sciences (ICS), Gomal University, Dera Ismail Khan 29050, Pakistan
4 Natural Products & Food Research and Analysis (NatuRA), Department of Pharmaceutical Sciences, University of Antwerp, Universiteitsplein 1, 2610 Antwerp, Belgium
* Correspondence: adnan.amin@gu.edu.pk

Abstract: Diabetes mellitus (DM) is a global health concern that is associated with several micro- and macrovascular complications. We evaluated several important medicinal plant constituents, including polyphenols and flavonoids, for α-glucosidase inhibition, AGEs' inhibitory activities using oxidative and no-oxidative assays, the inhibition of protein cross link formation, 15-lipoxydenase inhibition and molecular docking. The molecular docking studies showed high binding energies of flavonoids for transcriptional regulars 1IK3, 3TOP and 4F5S. In the α-glucosidase inhibition assay, a significant inhibition was noted for quercitrin (IC_{50} 7.6 μg/mL) and gallic acid (IC_{50} 8.2 μg/mL). In the AGEs inhibition assays, quercetin showed significant results in both non-oxidative and (IC_{50} 0.04 mg/mL) and oxidative assays (IC_{50} 0.051 mg/mL). Furthermore, quercitrin showed inhibitory activity in the non-oxidative (IC_{50} 0.05 mg/mL) and oxidative assays (IC_{50} 0.34 mg/mL). A significant inhibition of protein cross link formation was observed by SDS-PAGE analysis. Quercitrin (65%) and quercetin (62%) showed significant inhibition of 15-lipoxygenase. It was thus concluded that flavonoids and other polyphenols present in plant extracts can be effective in management of diabetes and allied co-morbidities.

Keywords: diabetes mellitus; advanced glycation end products (AGEs); protein cross link formation; *Punica granatum* peel; methylglyoxal

Citation: Rafey, A.; Amin, A.; Kamran, M.; Aziz, M.I.; Athar, V.; Niaz, S.I.; Pieters, L. Evaluation of Major Constituents of Medicinally Important Plants for Anti-Inflammatory, Antidiabetic and AGEs Inhibiting Properties: In Vitro and Simulatory Evidence. *Molecules* **2022**, 27, 6715. https://doi.org/10.3390/molecules27196715

Academic Editors: Raffaele Pezzani and Sara Vitalini

Received: 10 August 2022
Accepted: 6 October 2022
Published: 9 October 2022

1. Introduction

Diabetes mellitus (DM) is an ailment of the endocrine system that is associated with chronic insulin resistance and progressive exhaustion and death of β-cells in the pancreas that leads to hyperglycemia [1]. At present, 382 million people are currently diagnosed with diabetes globally with an expected massive increase to 600 million by the year 2035 [2]. About 4.9 million people die every year as a result of diabetes and 50% of this death toll is a consequence of diabetic complications [3]. In diabetes, several mechanisms initiate and impart injury to different vascular structures and repair mechanisms in such patients [4]. Thus, patients suffering from diabetes are susceptible to intense long-standing impediments like atherosclerosis, impaired wound healing, neuropathy, retinopathy, periodontitis, cataracts and nephropathy [5–7].

Advanced Glycation End products (AGE)s are the products produced via a chain of non-enzymatic reactions between reducing sugars, such as glucose, and the amino functionalities of proteins [8]. This reaction, also known as the Maillard reaction, starts when reducing sugars bind with the lysine side chain of a protein [9]. It primarily initiates the formation of a Schiff base that reshuffles into an intermediate product called an Amadori product. This is a crucial feature that permits the development of more complex and irreversible products called AGEs [10]. AGEs are yellow to brown, can show fluorescence, and can form unsolvable adducts with long-lived proteins, impairing their normal biological functions [11]. In this way, higher or uncontrolled glucose levels in diabetic patients contribute to the formation of AGEs and diabetic complications.

An association between inflammation, vascular complications and hyperglycemia has been observed in diabetic persons [12]. The pathogenicity of diabetes also involves the immune system. At the time a diabetes type 2 patient gets diagnosed, 50% of his pancreatic β-cells have already been destroyed [13]. Gradually, the pancreatic β-cells dysfunction increases and persistent hyperglycemia activates the immune system, thus leading to an increase in the inflammatory response. Thus, inflammation contributes to the pathogenesis of diabetes type 2 [14]. Moreover, in diabetes high blood plasma glucose and increased levels of free fatty acids could arouse inflammation processes that further increase glucose consumption through variations in the oxidative phosphorylation pathway [15]. It is also believed that oligomers of polypeptides (amyloid) of the pancreas might trigger inflammation through excitation of the NLRP3 inflammasome as well as production of IL-1β in type 2 diabetes [16].

Ayurvedic medicines are mainly based on natural products and may include plants, animals or minerals. They have centuries-old evidence of safety and efficacy [17]. Based on a survey of the Ayurvedic literature, some traditional medicinal plants including *Juglans regia, Salvadora persica, Syzygium aromaticum, Myristica fragrans, Punica granatum* and *Azadirachta indica* were selected for this investigation.

Juglans regia L. (Juglandaceae) is an important medicinal plant that is used for treatment of diabetes mellitus, inflammation and cancer [18–20]. It mainly contains ellagitannins, juglone and polyphenolic compounds. *Salvadora persica* L. (Salvadoraceae) is mainly linked with several biological activities including antimicrobial, antidiabetic, antirheumatic, anti-asthmatic, and anti-gonorrhea disorders [21]. This plant is a rich source of several classes of compounds including alkaloids, flavonoids, essential oils, sterols and terpenes [22,23]. *Syzygium aromaticum* (L.) Merr. & L.M.Perry (Myrtaceae) is a common Indian spice and is reported to contain several classes of compounds including polyphenols, triterpenes and an essential oil [24,25]. The plant is associated with several traditional medicinal claims including antioxidant, antimicrobial, antidiabetic and anticancer activities [26,27]. Likewise, *Myristica fragrans* Houtt. (Myristicaceae) is a traditional medicinal plant that has two main parts including mace and nutmeg [28]. Both parts of this plant are a rich source of polyphenols and essential oils containing myristicin, eugenol, and elemicin [29]. Traditional uses of nutmeg include antimicrobial, antidiabetic, carminative, hepatoprotective, anti-inflammatory and anti-rheumatic activities [30,31]. *Punica granatum* L. (Punicaceae) is an important medicinal plant that has many medicinal and nutritional benefits [32]. The peel part is a rich source of polyphenols, flavonoids and tannins including gallic acid, granatin A, granatin B, punicalin and punicalagin; quercetin, apigenin and anthocyanins such as pelargonidin [33]. Due to the presence of several important polyphenolic compounds, the peel is used for treatment of several diseases related to inflammation, diabetes, allergic reactions, and as an antioxidant [34,35]. Lastly, *Azadirachta indica* A. Juss (Meliaceae) is a commonly available plant in the sub-continent including India and Pakistan [36]. This plant is commonly used in Ayurvedic medicine for treatment of several ailments including bacterial and helminthic infections and diabetes mellitus [37]. Major chemical constituents include nimbidine, azadirachtin (azadirachtin A), salannol and salannin [38].

Keeping in mind the importance of these medicinal plants, we investigated the plant extracts and their major constituents for antidiabetic, antiglycation (oxidative and non-oxidative modes), and anti-inflammatory potentials using both computational and in vitro models. The study is important since it provides evidence-based information regarding designing a herbal formulation that can be helpful in the management of diabetes and associated health concerns including AGEs and inflammation.

2. Results

2.1. Molecular Docking Results

In the case of transcriptional regulator 3TOP (α-glucosidase), quercetin showed the best fit in the active pocket with pose no. 4 (Figure 1) (−7.9 ΔG (kJ mol^{-1}). The number of H-bonding interactions was seven, including Pro686, Trp685, Glu682, Asn681, Gln665, Gln689, Met688 and the neighboring amino acid were Lys687, Lys680, Trp668, Lys669, involved in hydrophobic interactions (Table 1). Similarly, quercitrin presented a good fitting with pose no. 4 (−8.2 ΔG (kJ mol^{-1}) with amino acids Ala163, Lys545, His548, Tyr544, Asp261, Asn146 (Table 1, Figure S3). Hydrophobic interactions were reported with Phe 264, Phe161, Cys145, Val539, Val544, Pro549, Asn788, and Val144 (Figure 1). Juglone and gallic acid also presented strong H-bonding interactions (Table 1, Figure 1), whereas caryophyllene-oxide α-humulene and *trans*-caryophyllene did not show any fitting within the active pocket of the transcriptional regulator (Table 1, Figures S1–S3). Docking images of all other compounds with transcriptional regulator 3TOP are shown in Supplementary Materials (Figures S1–S3).

Table 1. Docking score, H and non-H bonding interactions of test compounds for transcriptional regulator 3TOP (α-glucosidase).

	Compound	Binding Free Energy ΔG (kJ mol^{-1})	Pose No	H Bond	H Bond Interaction Residues	Neighbor Interacting Residues
			3TOP			
1	Eugenol	−5.3	3	3	Ser129, Gln96, Leu92	Thr91, Thr15, Leu89, Thr131, Val17, His142
2	Caryophyllene	−7.6	1	0	0	Lys545, Val539, Cys145, Asn146, Val144, His548, Trp791, Pro789, Asn788, Pro549, Tyr544
3	Caryophyllene oxide	−6.6	1	0	-	Leu154, Glu 199, Trp198, Arg200, Asn186, Ser157, Phe155,
4	α-Humulene	−6.4	1	0	-	Lys388, Val589, Asp592, Tyr512, Pro432, Val588, Asp431, Lys388
5	Juglone	−6.3	3	3	Asn146, Val539, Lys545	Cys145, Trp791, Val144
6	2-Phenylethyl-isothiocyanate	−5.5	1	1	Lys545	Pro549, Trp791, Asp 787, Val144, Val539
7	Quercetin	−7.9	4	7	Pro686, Trp685, Glu682, Asn681, Gln665, Gln689, Met688.	Lys687, Lys680, Trp668, Lys669
8	Quercitrin	−8.2	4	6	Ala163, Lys545, His548, Tyr544, Asp261, Asn146	Phe 264, Phe161, Cys145, Val539, Val544, Pro549 Asn788, Val144,
9	Apigenin	−7.4	5	4	Glu197, Asn556, Thr274, Lys 278	Tyr257, Arg260, Val256, Leu258, Tyr275, Ala263

Molecular docking of all pure compounds on the active pockets of transcriptional regulator 1IK3 (15-lipoxygenase) was performed, and it was evident that flavonoids including quercitrin, quercetin and apigenin showed fitting within the active pocket with high binding free energy (ΔG (kJ mol^{-1}) (Table 2). In the case of quercetin, the best fitting was noticed in pose 1 (−8.9 ΔG (kJ mol^{-1}), where 5 H- bonding interactions were recorded with His1584, Asp1279, Thr1586, Asp1157 and Asp1526 (Table 2). The non-H bonding interactions/hydrophobic interactions were seen with neighboring amino acids Phe1559, Trp1418, Trp1355, Phe1427, Trp1369, Phe1560 and Tyr1251 (Figure 2). Juglone and gallic acid also presented strong H-bonding interactions (Table 2, Figure 2). Quercitrin showed 7 H bonding interactions (−8.2 ΔG (kJ mol^{-1}) with Glu970, Asp969, Tyr967, Asp965, Trp985, Cys996 and Ser990 (Table 2). The other interactions including Vander Waal's and *pi* sigma were also recorded with neighboring amino acids Gly992, Val993, Ser991, Phe995, Ala973 and Pro968 (Figure S6). 2-Phenyl-isothiocyanate, α-humulene and caryophyllene did not show any fitting within the active pocket of transcriptional regulator 3TOP. Docking images of all other compounds with transcriptional regulator 1IK3 are shown in Supplementary Materials (Figures S4–S6).

(**A**)

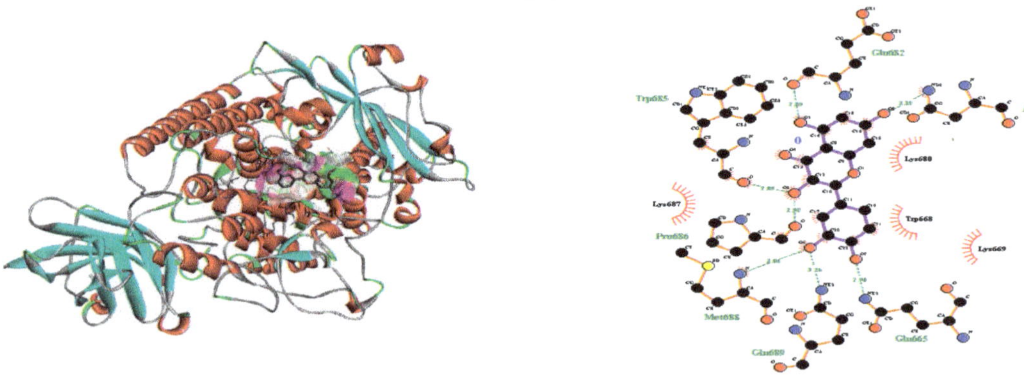

(**B**)

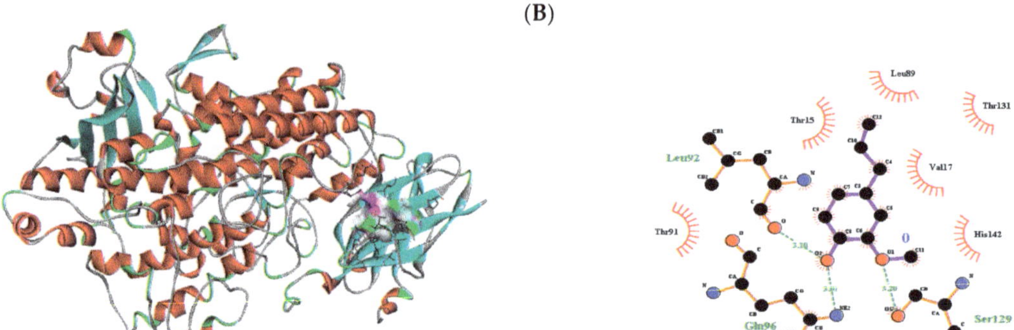

(**C**)

Figure 1. 3D H-bonding interactions of juglone pose no. 3 (**A**), quercetin pose no, 4 (**B**); gallic acid pose no. 2 (**C**) with binding sites of transcriptional regulator 3TOP.

Furthermore, the docking of ligands with the transcriptional regulator "4F5S" (BSA) yielded interesting results. Amongst flavonoids, quercetin showed the best fit in the active pocket of 4F5S with pose 4 (-7.3 ΔG (kJ mol^{-1}) (Table 3). The number of contributing H-bonding interactions was seven, including Tyr340, Val342, Gln220, Lys221, Lys294, Ala290, Pro338, and the neighboring amino acids were Ala341, Arg217, and Glu291 that were involved in hydrophobic interactions (Figure 3). Juglone and gallic acid also presented strong H-bonding interactions (Table 3, Figure 3). Similarly, quercitrin presented good fitting by pose 5 (-7.2 ΔG (kJ mol^{-1}) with amino acids Glu339, Lys294, Lys221, Ala290,

Lys439, Pro338 (Table 3). The hydrophobic interactions were reported with Pro446, Arg217, Asp450, Tyr451 and Cys447 (Figure S9). Likewise, juglone showed the best fitting in the active pocket with pose 5 (−6.0 ΔG (kJ mol^{-1}) with amino acids Thr421, Ser418 and Lys465 (Table 3). Terpenes including caryophyllene, caryophyllene oxide, α-humulene and 2-phenyl-isothiocyanate did not present any H-bonding affinity with BSA. Docking images of all other compounds with transcriptional regulator 4F5S are shown in Supplementary Materials (Figures S7–S9).

Table 2. Docking score, H- and non-H bonding interactions of test compounds for transcriptional regulator 1IK3 (15-lipoxygenase).

	Compound	Binding Free Energy ΔG (kJ mol^{-1})	Pose No	H Bond	H Bond Interaction Residues	Neighbor Interacting Residues
			1IK3			
1	Eugenol	−6.1	2	2	Glu1400, Glu1397	Phe1289, Pro1405, Asn1404, Leu1412, Leu1401,
2	Caryophyllene	−6.3	1	0	0	Asp1157, Trp1355 Asp1526, Phe1560, Tyr1251, Thr1586 Trp1369, Phe1427,
3	Caryophyllene oxide	−6.1	1	2	Gln1372, Arg1377	Ile1587, Tyr1251, Ile1280, Asp1281, Gln 1286, Asp1357
4	α-Humulene	−6.1	1	0	0	Pro1159, Phe1427, Asp1157, Asp1526 Phe1560, Trp1355 Trp1369
5	Juglone	−5.2	8	2	Asp1157, Lys1460	Phe1427, Trp1355, Trp1369
6	2-Phenylethyl-isothiocyanate	−4.4	1	0	0	Pro1327, Pro1329, Glu1397, Glu1400, Leu1401, Phe1289,
7	Quercetin	−8.9	1	5	His1584, Asp1279, Thr1586, Asp1157 Asp1526	Phe1559, Trp1418, Trp1355, Phe1427 Trp1369, Phe1560, Tyr1251,
8	Quercitrin	−8.2	4	7	Glu970, Asp969, Tyr967, Asp965 Trp985, Cys996 Ser990	Gly992, Val993, Ser991 Phe995, Ala973, Pro968
9	Apigenin	−7.6	7	5	Glu1284, Glu1397, Arg1333, Glu1400, Leu1291	Phe1289, Pro1239, Thr1290,

Table 3. Docking score, H- and non-H bonding interactions of test compounds with transcriptional regulator 4F5S (BSA).

	Compound	Binding Free Energy ΔG (kJ mol^{-1})	Pose No	H Bond	H Bond Interaction Residues	Neighbor Interacting Residues
			4F5S			
1	Eugenol	−5.3	1	1	Glu125	Phe133, Tyr137, Tyr160, Leu115, Leu122,
2	Caryophyllene	−7.6	1	0	0	Val481, Leu346, Ala212, Ala209, Arg208, Phe205, Leu480,
3	Caryophyllene oxide	−5.7	1	0	0	Gln203, Ile202, Cys245, His246, Lys242, Leu112, Glu125, Lys136, Phe133,
4	α-Humulene	−6.9	1	0	0	Tyr160, Tyr137, Glu140, Leu115, Leu122
5	Juglone	−6.0	5	3	Thr421, Ser418, Lys465	Pro420, Ser109, Thr466, Leu462,
6	2-Phenylethyl-isothiocyanate	−4.7	1	0	0	Lys132, Lys131, Leu24, Gly21, Val43, Lys20
7	Quercetin	−7.3	4	7	Tyr340, Val342, Gln220, Lys221, Lys294, Ala290, Pro338,	Ala341, Arg217, Glu291,
8	Quercitrin	−7.2	5	6	Glu339, Lys294, Lys221, Ala290, Lys439, Pro338,	Pro446, Arg217, Asp450, Tyr451, Cys447,
9	Apigenin	−7.8	3	4	Lys431, Arg458, Asn457, Leu454,	Arg435, Leu189, His145, Arg196, Ala193,

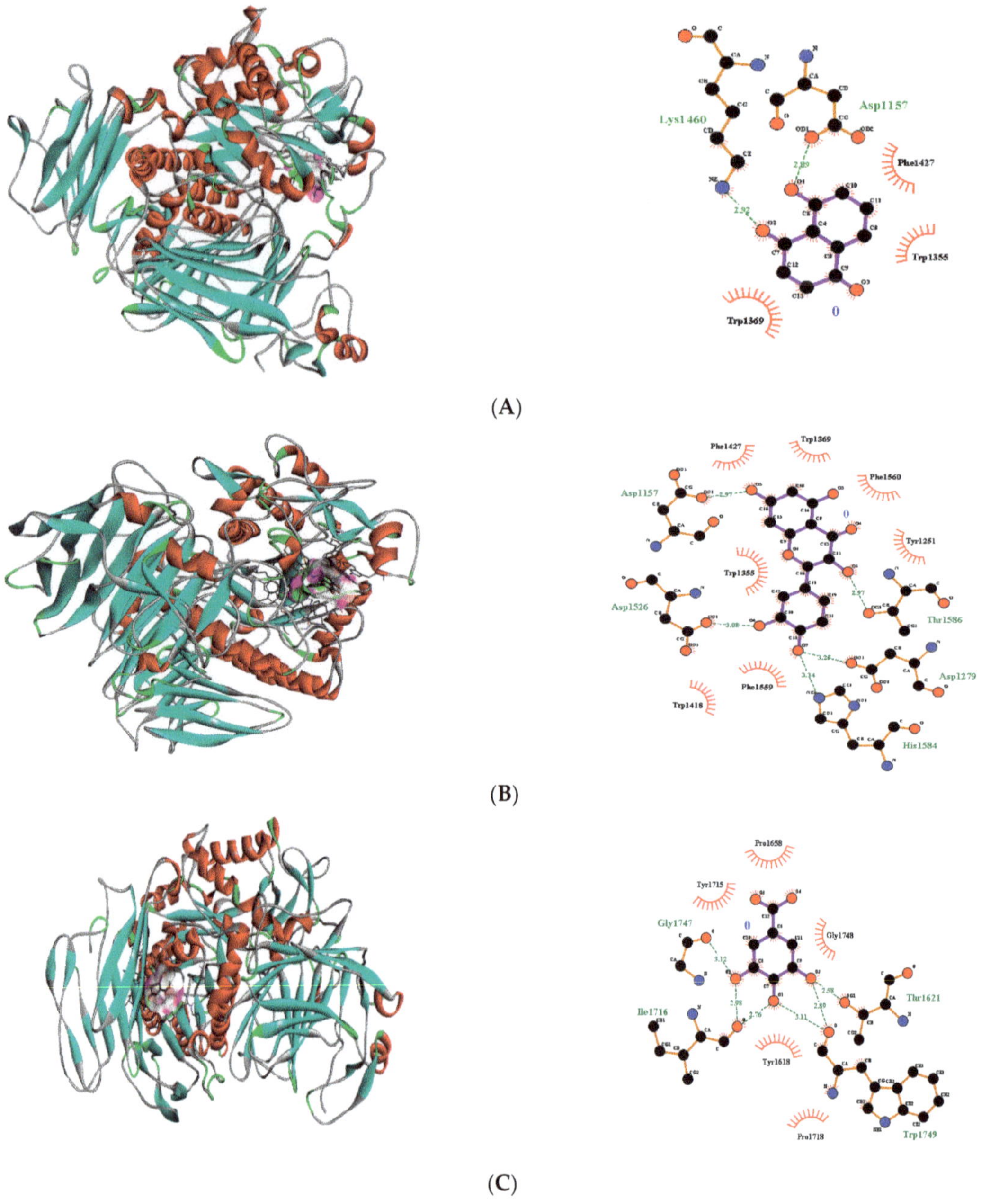

Figure 2. 3D H-bonding Interactions of juglone pose no. 8 (**A**), quercetin pose no. 1 (**B**) and gallic acid pose no. 3 (**C**) with binding sites of transcriptional regulator 1IK3.

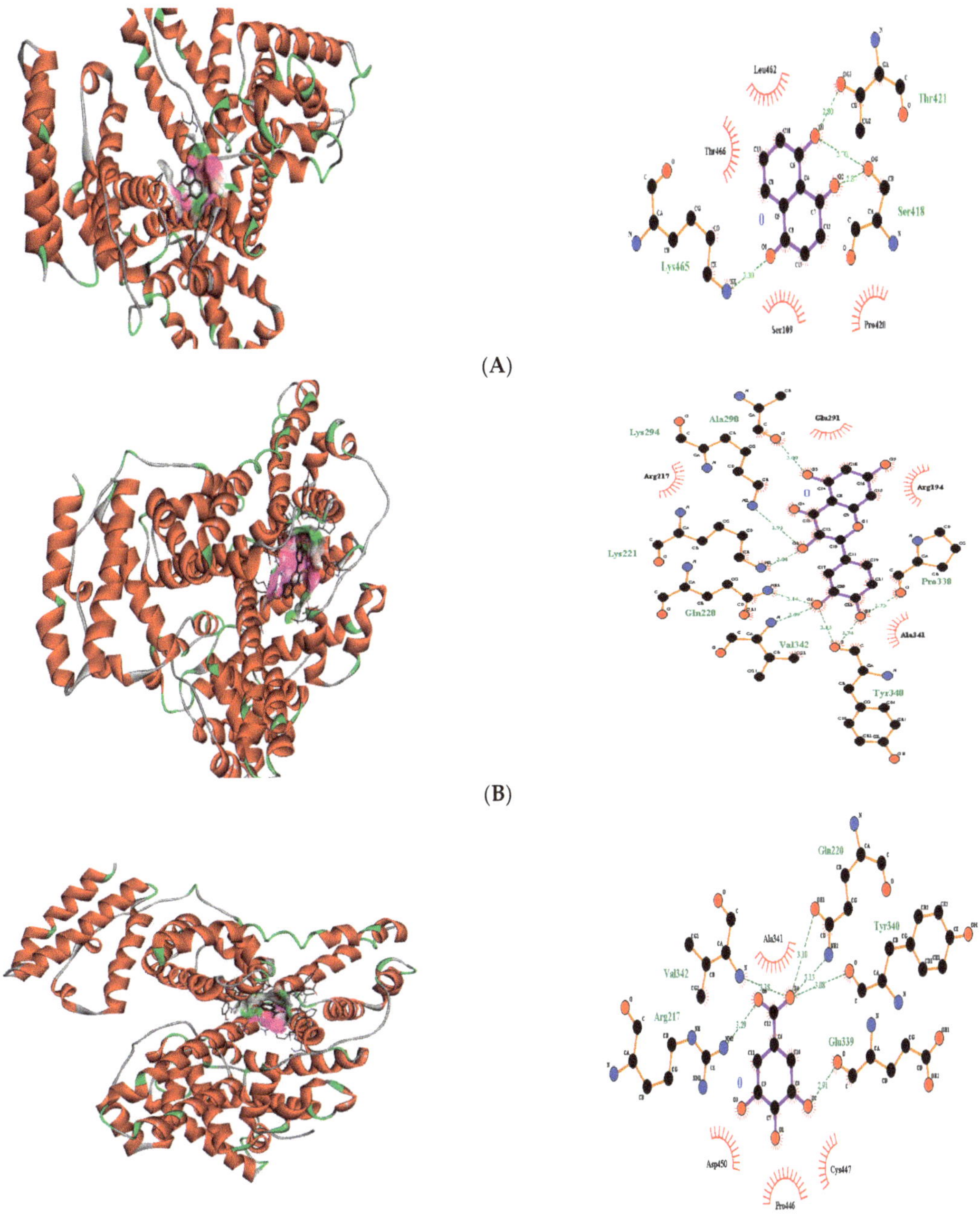

Figure 3. 3D H-bonding interactions of juglone pose no. 5 (**A**), quercetin pose no. 4 (**B**) and gallic acid pose no. 1 (**C**) with binding sites of transcriptional regulator 4F5S (BSA).

2.2. α-Glucosidase Assay

The extracts and pure compounds were assessed for their potential antidiabetic activities using the α-glucosidase assay. Amongst all tested extracts, only *Juglans regia* (61%) and the peel of *Punica granatum* (52%) showed the inhibition of α-glucosidase (Table 4). All other plants were considered inactive, since only mild inhibition was seen. In the case of 15-lipoxygenase activity, *Syzygium aromaticum* showed significant inhibition (70%), followed by *Myristica fragrans* (62%). A mild to moderate inhibition was seen for the other plant extracts.

Table 4. α-Glucosidase and 15-Lipoxygenase assay of selected medicinal plants.

	Name	Assay Type	
		α-Glucosidase *	**15-Lox Assay ***
1	*Juglans regia*	61.0 ± 0.1%	14.0 ± 0.2%
2	*Syzygium aromaticum*	Inactive	70.0 ± 0.1%
3	*Eruca sativa*	Inactive	45.0 ± 0.1%
4	*Myristica fragrans*	Inactive	62.00 ± 0.04%
5	*Punica granatum*	52.0 ± 0.1%	10.00 ± 0.04%
6	*Azadirachta indica*	Inactive	45.0 ± 0.1%
	Standard	72.0 [a] ± 0.1%	62.0 [b] ± 0.1%

* 60 μg/mL; [a] Acarbose (0.01 mM), [b] Rutin (12.5 μg/mL).

With regard to the pure compounds, juglone (IC_{50} 7.6 μg/mL), quercitrin (IC_{50} 7.6 μg/mL) and gallic acid (IC_{50} 8.2 μg/mL) showed the inhibition of α-glucosidase. With regard to 15-lipoxygenase activity, quercitrin (65%), quercetin (62%) and α-humulene (60%) showed inhibition >50%. Inhibition results for other compounds are presented (Table 5).

Table 5. α-Glucosidase and 15-LOX inhibitory activity of test compounds.

	Compound	Assay Type	
		α-Glucosidase (IC_{50}) (μg/mL)	**15-LOX Assay (% Inhibition)**
1	Eugenol	Inactive	51.0 ± 0.2% [4]
2	Caryophyllene	Inactive	Inactive [6]
3	Caryophyllene oxide	Inactive	50.0± 0.1% [2]
4	α-Humulene	Inactive	60.0 ± 0.1% [3]
5	Juglone	5.7 ± 0.1	57.0 ± 0.1% [1]
6	2-Phenylethylisothiocyanate	28.9 ± 0.1	Inactive [5]
7	Quercetin	21.2 ± 0.1	62.0 ± 0.1% [7]
8	Quercitrin	7.6 ± 0.1	65.0± 0.1% [9]
9	Apigenin	24.4 ± 0.1	53.0 ± 0.1% [8]
10	Standard	6.49 ± [a] 0.02	62.0 ± 0.1% [b]

[1] 2.5 μg/mL, [2] 12.5 μg/mL, [3] 13.0 μg/mL, [4] 83.0 μg/mL, [5] 25 μg/mL, [6] 25 μg/mL, [7] 25 μg/mL, [8] 14.1 μg/mL, [9] 16.2 μg/mL. [a] Acarbose (μg/mL), [b] Rutin (12.5 μg/mL).

2.3. Antiglycation Assays

In the AGEs inhibition assays, both oxidative (BSA-MGO) and non-oxidative (BSA-glucose) modes of inhibition were analyzed (Table 6). Among flavonoids, quercetin presented significant results in both the non-oxidative and (IC_{50} 0.04 mg/mL) oxidative assay (IC_{50} 0.051 mg/mL). Quercitrin showed a better inhibition in the non-oxidative (IC_{50} 0.09 mg/mL) than in the oxidative assay (IC_{50} 0.34 mg/mL) (Table 6). Apigenin (IC_{50} 0.45 mg/mL) was only active in the non-oxidative mode. Juglone has shown its activity in both the oxidative (IC_{50} 0.11 mg/mL) and non-oxidative (IC_{50} 0.06 mg/mL) assays, which reflects its potential use for diabetes and its complications (Table 6). All other tested compounds were recorded as inactive.

Table 6. Antiglycation assay of compounds.

	Compound	Protein Glycation	
		BSA-Glucose (IC_{50}) (μg/mL)	**BSA-MGO (IC_{50}) (μg/mL)**
1	Eugenol	0.040 ± 0.007	Inactive
2	Caryophyllene	Inactive	Inactive
3	Caryophyllene oxide	Inactive	Inactive
4	α-Humulene	Inactive	Inactive
5	Juglone	0.060 ± 0.004	0.11 ± 0.04
6	2-Phenylethylisothiocyanate	Inactive	Inactive
7	Quercetin	0.040 ± 0.004	0.050 ± 0.005
8	Quercitrin	0.090 ± 0.008	0.34 ± 0.06
9	Apigenin	0.45 ± 0.02	Inactive
10	Standard [1]	0.030 ± 0.001	1.02 ± 0.21

[1] Rutin.

2.4. Protein Cross-Linking Assay

The compounds were further tested for the inhibition of cross-link formation using both the non-oxidative (BSA-glucose) and oxidative (BSA-MGO) modes. SDS-PAGE was employed for measuring cross-linking. BSA incubation with either glucose or MGO produces small amounts of dimerization that is not visible with incubation of BSA alone. All tested compounds showed inhibition of cross-linked AGEs causing a decrease in intensity of dimerized (cross-linked) bands (Figure 4).

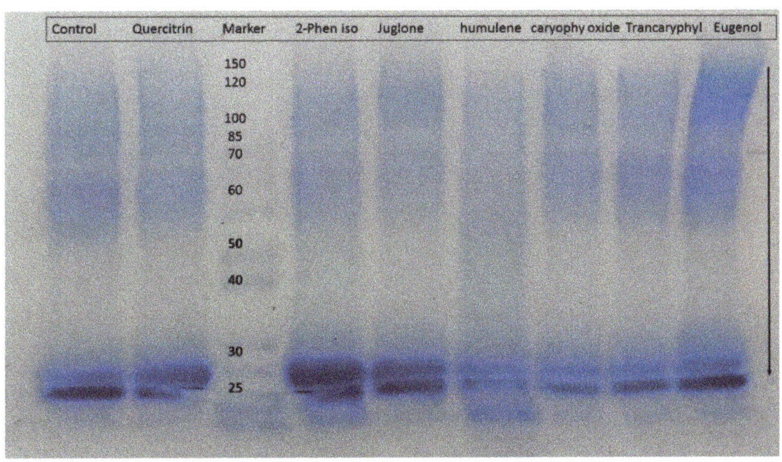

Figure 4. Gel showing BSA-MGO glycation with test compound (Control = BSA + MGO); Quercitrin = Quercitrin + MGO; Marker = Protein ladder; 2-Phen iso = 2-Phenylethylisothiocyanate + MGO; Juglone = Juglone + MGO; Humulene = α-Humulene + MGO; Caryophy Oxide = Caryophyllene oxide; Transcaryophyl = caryophyllene oxide + MGO; Eugenol = Eugenol + MGO.

In the BSA-MGO assay, amongst all tested compounds, 2-phenylethylisothiocyanate presented the highest inhibition (48%) followed by α-humulene (40%), juglone (39%) and caryophyllene oxide (35%). The flavonoids showed a mild inhibition ranging from 21–34% (Figure 5). It was thus concluded that all tested compounds showed a mild inhibition of protein cross-link formation.

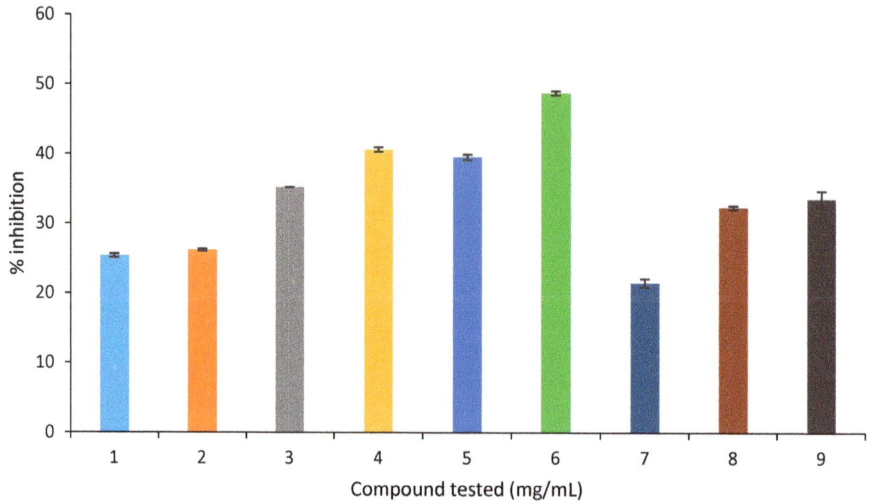

Figure 5. Inhibition of protein cross-link formation in the BSA-MGO assay (1 = eugenol, 2 = caryophyllene, 3 = caryophyllene oxide, 4 = α-humulene, 5 = juglone, 6 = 2-phenylethylisothiocyanate, 7 = quercetin, 8 = quercitrin, 9 = apigenin).

In the BSA-glucose assay (Figure 6), amongst all tested compounds, quercitrin presented the highest inhibition (55%), followed by quercetin (53%), and apigenin (51%). The flavonoids showed a mild inhibition ranging from 36–46% (Figure 7). It was thus concluded that all tested compounds showed a mild inhibition of protein cross-link formation.

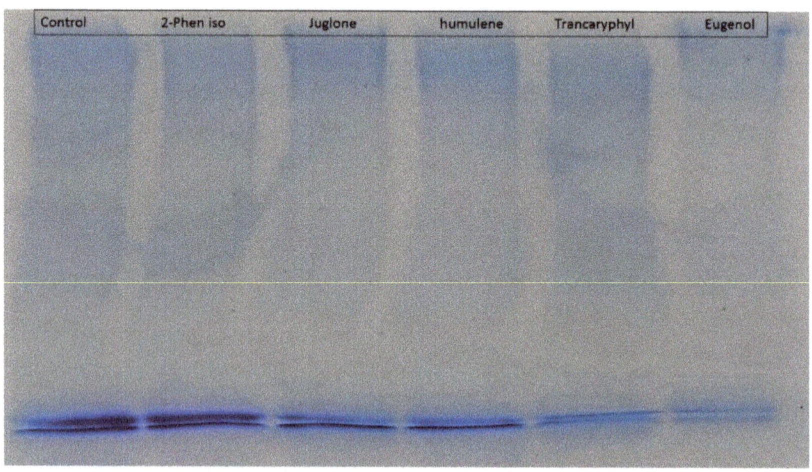

Figure 6. Gel showing BSA-Glucose glycation with tested compounds.

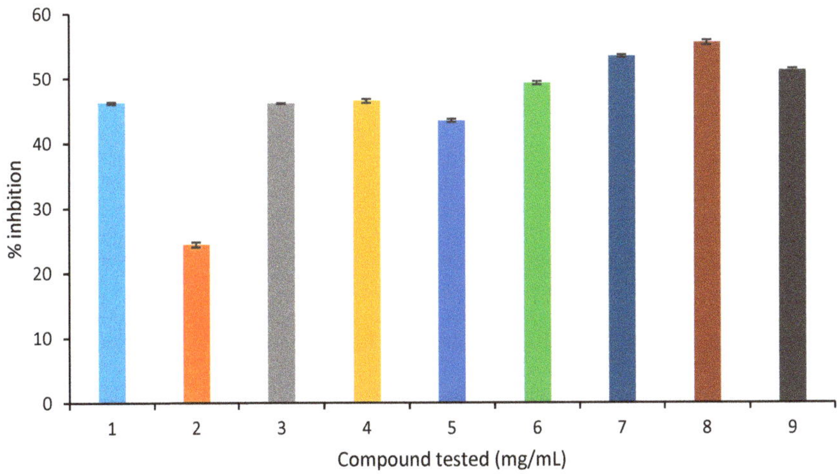

Figure 7. Inhibition of protein cross-link formation in the BSA-glucose assay (1 = eugenol, 2 = caryophyllene, 3 = caryophyllene oxide, 4 = α-humulene, 5 = juglone, 6 = 2-phenylethylisothiocyanate, 7 = quercetin, 8 = quercitrin, 9 = apigenin).

3. Discussion

In this study, the effect of medicinal plant extracts and their major components on diabetes, glycation and inflammation was demonstrated using in vitro and in silico methods. Structural diversification of natural compounds enables multiple biological activities due to diverse mechanisms of action. Polyphenols and flavonoids possess several health benefits on account of their antioxidant, anti-inflammatory, and enzyme inhibiting properties [39]. Generally, the strong antioxidant potential of such compounds may contribute towards antidiabetic, antiglycation and anti-inflammatory activities [40,41].

α-Glucosidases are enzymes located at the intestinal lumen brush border that catalyze the hydrolysis of terminal, non-reducing α-1-4-linked glucose residues of disaccharides or oligosaccharides [42]. These enzymes are therefore helpful to facilitating carbohydrate absorption. Inhibitors of α-glycosidase can interrupt the digestion of carbohydrates to glucose and therefore they can be used for the treatment of type 2 diabetes [43]. In this study, flavonoids and polyphenols exhibited significant inhibition of α-glucosidase, which may be attributed to the presence of OH groups in the C-ring (flavonoids). It has been reported earlier that 3-hydroxylation of the C-ring facilitates the inhibitory activity against α-glucosidase [44]. The data is also in agreement with the molecular docking assessment, where strong H-bonding and hydrophobic interactions were observed for the transcriptional regulator gene for 3TOP (α-glucosidase).

Tissue inflammation is primarily related to immune system activation in diabetic patients [45]. Further phenotype conversion of macrophages from M2-type to M1-type is also very important in inflammatory conditions [46]. Lipoxygenase activation (12, 15-lipoxygenase) also plays a key role in the development of diabetes, and evidence has suggested that Lox inhibitors can greatly protect against diabetes [47]. Our results indicated that flavonoids and polyphenolic compounds possess significant inhibition of 15-lipopxygenase. This is further supported by molecular docking results, which indicate strong H-bonding, and hydrophobic interactions with transcription regulator 3TOP. This could be due to *pi*-interaction of the bond linking the B and C rings [40] that gives a near planar region of these two rings. Such structures (like flavonoids) easily enter the hydrophobic pockets in enzymes and can subsequently increase their inhibitory effect. Thus the C2–C3 double bond of flavonoids is crucial for their anti-inflammatory activity [48].

AGEs build-up in the body can activate several signaling pathways through receptors and thus interrupt various biological activities and cellular functions that finally lead to cell death [49]. Various mechanisms have been documented to explain AGEs synthesis in the human body including oxidative and non-oxidative modes [50]. In this investigation, flavonoids are reported with antiglycation activities in both oxidative and non-oxidative models that may be due to their radical scavenging properties, thus a delay in the progression of glycation is expected [51,52]. The cross-linking AGEs possess strong affinity towards diverse proteins and therefore are resistant to degradation. This leads to further toxicity to the human body [53]. Furthermore, flavonoids also showed inhibition of protein cross-link formation. It was evident from earlier findings that flavonoids and polyphenols inhibit cross-link formation because of antioxidant properties and some other mechanisms [54]. Thus, radical scavenging and inhibition of cross-link formation may provide a protective effect against hyperglycemia-mediated damaging effects on proteins [55]. Our in silico results also supported this fact by indicating strong H-bonding interactions of flavonoids and polyphenols with transcriptional regulator 4F5S (BSA).

4. Materials and Methods

4.1. Plant Materials

A detailed literature review of Ayurvedic medicinal plants and their formulations was performed and medicinal plants were purchased from local herbal markets and identified by a taxonomist in the Institute of Biological Sciences, Gomal University, Pakistan. The plants used during this investigation are shown in the Supplementary Materials (Table S1). The plants were analyzed as reported before using HPLC-DAD, GC-MS, LC-QTOF-MS [56], and major components are highlighted in the Supplementary Materials (Figure S1).

4.2. Extraction and Drying

Plant material was dried in an oven below 40 °C. Next, the plant material was thoroughly grinded followed by cold maceration in 90% methanol. Solvent evaporation was accomplished by a rotary evaporator (Büchi, Flawil, Switzerland), and stored at 4 °C until use. The essential oils were obtained using a Clevenger apparatus (hydrodistillation) and fixed oils were obtained using cold pressing [56].

4.3. Chemicals Reagents and Solvents

15-LOX (lipoxygenase), α-glucosidase, DPPH (2,2-diphenyl-1-picrylhydrazyl radical), bovine serum albumin (BSA), trichloroacetic acid (TCA), methylglyoxal (MGO) and D-glucose were purchased from Merck, (Dorset, UK), and Oxoid (Hampshire, UK), whereas NaN_3 was purchased from DaeJung (Siheung-si, Korea). Chemical for Gel analysis including Coomassie blue, Tris–HCl, sodium dodecyl sulphate, 2-mercaptoethanol, Glycerine, bromophenol blue were purchased from Sigma Aldrich, St. Louis, MO, USA. The standard compounds included aminoguanidine (≥98.5%, Sigma Aldrich, St. Louis, MO USA), eugenol (≥99%, Fluka, Riedstr, Germany), 2-phenylethylisothiocyanate (>99%, Sigma Aldrich, St. Louis, MO, USA), juglone (≥98.5%, Santa Cruz Biotechnology, Santa Cruz, CA, USA), quercetin (≥99%, Sigma Aldrich, St. Louis, MO, USA), quercitrin (≥85%, Sigma Aldrich, St. Louis, MO, USA), *trans*-caryophyllene (≥98.5%, Fluka, , Riedstr, Germany), α-humulene (>98%, Extrasynthese, Genay, France), caryophyllene-oxide (≥99%, Fluka Honeywell, Seelze, Germany) and apigenin (≥95.5% Sigma Aldrich, St. Louis, MO, USA) (structures are shown in the Supplementary Materials).

4.4. Molecular Docking

For molecular docking studies, the X-ray crystallographic structures of the transcriptional regulators 3TOP [57] and 1IK3 [58] were taken from the Protein Data Bank (PDB) and active pocket dimensions for each protein were checked using the CAST*p* 3.0 online tool. The optimization of transcriptional regulators was performed using DS Visualizer 2.0 [59]. Furthermore, the structures of all the phytoconstituents were downloaded from the Pubchem database and PDB files were generated in DS Visualizer 2.0 [59]. The molecular docking was performed using Lamarckian Genetic Algorithm embedded in AutoDock v 4.2 [60]. A total number of nine poses were generated (for each target) and grouped according to their RMSD values. Every set was prudently checked in Discovery Studio Visualizer and presumed binding modes were highlighted for further analysis. Best docked structures based on the binding energy scores (ΔG) and H- binding were chosen for further analysis. The hydrogen bonding and hydrophobic interactions between ligand and protein were calculated by Ligplot$^+$ and DS Visualizer 2.0.

4.5. *Antidiabetic Assays*

4.5.1. α-Glucosidase Inhibition Assay

The α-glucosidase inhibition experiment was accomplished by using a modified method [61]. Initially, the enzyme solution (from *Saccharomyces cerevisiae*) (0.2 units/mL dissolved in 0.1 M phosphate buffer; pH 6.8) was mixed with the test sample (1 to 0.039 mg/mL) and incubated in an oven (37 °C for 10 min). After incubation, the substrate (*p*-nitrophenyl-α-D-glucopyranoside; 0.29 mM) was added to the enzyme and test sample solution and incubated for another 30 min (37 °C). The reaction was halted by adding Na_2CO_3 (100 µL, 200 mM stock) to this mixture and absorbance was noted at 400 nm. Acarbose was used as positive control.

The percentage of inhibition was determined using the following formula:

$$\% \text{ Inhibition} = [1 - \text{absorbance of test sample/absorbance of control}] \times 100$$

4.5.2. AGEs Assay (BSA-Glucose Assay)

The AGEs assay was accomplished by using a standard protocol [62,63]. The protein source (bovine serum albumin, BSA) (10 mg/mL; 135 µL) was mixed with D-glucose solution (500 mM, 135 µL in phosphate buffer; 50 mM, pH 7.4), NaN_3 (sodium azide; 0.02%) and test samples (various concentrations). The reaction mixtures were kept at 60 °C for 1 week to facilitate glycation. Finally, trichloroacetic acid (10 µL, 100%) was added to it to stop the reaction and precipitate the unbound material. The supernatant was removed and the pellet was dissolved in alkaline phosphate buffer saline (137 mM NaCl, 8.1 mM Na2HPO4, 2.68 mM KCl, 1.47 mM KH2PO4, pH10). Finally the fluorescence intensity (ex 370/emis 440; ex 335/emis 385 nm) was recorded on a spectrofluorometer (FLx800, BioTek, Winooski, USA). Aminoguanidine was used as positive control. The control samples used for the experiment were prepared using the same protocol without the test sample.

The AGEs inhibition was calculated as

$$\% \text{ inhibition} = \{1 - [\text{Fluo(BSA + glucose + test substance)} - \text{Fluo (BSA + test substance)}]/[\text{Fluo (BSA + glucose)} - \text{Fluo(BSA)}]\} \times 100$$

where Fluo is the fluorescence intensity. The IC_{50} was calculated using MS Excel.

4.5.3. BSA-MGO Assay

In this oxidative glycation assay, the protein source (BSA) was mixed (10 mg/mL, 135 µL) with methylglyoxal (5.75 mM, 135 µL) and liquefied in phosphate buffer (50 mM, pH 7.4) and NaN_3 (sodium azide; 0.02%). The test samples were added to this reaction mixture and stored at 37 °C for a week. The change in fluorescence intensity (ex 370/emis 440; ex 335/emis 385 nm) was recorded on a spectrofluorometer (FLUOstar Omega®, BMG Lab Tech, Aylesbury, UK). Aminoguanidine and quercetin were used as positive controls. The control samples used for the experiment were prepared using the same protocol without a test sample.

The AGEs inhibition was determined as

$$\% \text{ inhibition} = \{1 - [(\text{Fluo (BSA + MGO + test substance)} - \text{Fluo (BSA + test substance)}]/[(\text{Fluo (BSA + MGO)} - \text{Fluo (BSA)}]\} \times 100$$

where Fluo is the fluorescence intensity. The IC_{50} was calculated using MS Excel.

4.5.4. Protein Cross-Linking Assay

Protein cross-linked by AGEs were observed by SDS-PAGE (sodium dodecyl sulphate-polyacrylamide gel electrophoresis) [64]. In this experiment, gels (10%) were made and stained with dye (0.25% Coomassie blue). The protein samples (6 μL) were diluted with Tris–HCl buffer (3.75 mL, 0.05 M; pH 6.8) also comprising sodium dodecyl sulphate (150 μL, 10% w/v), 2-mercaptoethanol 1% (v/v) and glycerin 20% v/v), followed by boiling for 5 min. The ladder (12 μL) was loaded into the well on gel preceded by the loading of bromophenol blue (2 μL). After loading, the electrophoresis was accomplished by using the Mini-Protean® Tetra Cell apparatus (Bio-Rad, UK). Afterwards, the gels were stained in a solution containing Coomassie blue [0.25% (w/v)], methanol [50% (v/v)] and acetic acid [10% (v/v)].

4.5.5. SDS-PAGE Gels Image Analysis

The developed gel was stained with Coomassie blue and images were obtained using GelDoc. Finally, the Image J tool was used for the determination of integrated density (IntDen). The integrated density was employed further for the determination of percentage inhibition. The integrated density (ID) was determined as follows:

$$\text{Integrated density (ID)} = \text{N} \times (\text{mean} - \text{background})$$

where N is the number of pixels in the selection and the background is the modal grey value (most common pixel value) after smoothing the histogram.

The percentage inhibition of cross-linked AGEs was determined using the formula:

$$\%\ \text{inhibition} = 100 \times (\text{ID without inhibitor} - \text{ID with inhibitor})/\text{ID without inhibitor}$$

4.6. 15-Lipoxygenase Assay

The anti-inflammatory activity of test samples was evaluated using a standard protocol [65]. To the enzyme solution (200 units/mL; 487.5 μL), different concentrations of test sample (12.5 μL (2–0.062 mM) in DMSO were added and incubated for 5 min at 37 °C. The absorbance was recorded immediately after the addition of substrate (500 μL of substrate (250 mM linoleic acid in 0.2 M borate buffer, pH 9) and after every min up to 5 min at 234 nm by using a UV spectrophotometer (UV-1601, SHEMADZU, Kyoto, Japan).

The percentage inhibition of enzyme activity was calculated as follows:

$$\%\ \text{inhibition} = ([\Delta A1/\Delta t] - [\Delta A2/\Delta t]/(\Delta A1/\Delta t) \times 100$$

where $\Delta A1/\Delta t$ and $\Delta A2/\Delta t$ are the increase rate in absorbance at 234 nm for a sample without test substance and with test substance, respectively.

4.7. Statistical Analysis

All experiments were performed in triplicate and results were expressed as mean ± SD.

5. Conclusions

Diabetes, a very prevalent metabolic disorder, is an important health problem worldwide. Persistent hyperglycemia results in the development of inflammation and several life threatening complications due to the production of AGEs. Various strategies are being used by researchers to introduce new treatment options that have a dual effect i.e., both blood glucose lowering and anti-AGEs potential. Traditional medicinal plants are considered as effective and reliable alternatives of conventional medical therapy due to proven safety and efficacy. In this investigation, selected medical plants and their major constituents were analyzed using in silico and in vitro models. Interactions of flavonoids and polyphenols were observed with transcription regulators 1IK3, 3TOP and 4F5S. Further in vitro assays presented anti diabetic, antiglycation, and anti-inflammatory activities of constituents including juglone, quercetin, quercitrin, apigenin and 2-phenylethylisothiocyanate from *Juglans regia*, *Punica granatum* and *Myristica fragrans*. Thus, it was concluded that these plant species may be considered as candidates for the management of diabetes mellitus and co-morbidities occurring due to AGEs.

Supplementary Materials: The following Tables and Figures are available online at https://www.mdpi.com/article/10.3390/molecules27196715/s1, Table S1. Ayurvedic medical plants and their major constituents. Figure S1. Structure of test compounds. Figure S2. 3D H-bonding interactions of 2-phenyl isothiocyanate pose no. 1 [1], Apigenin pose no. 5 [2] caryophyllene oxide pose no. 1 [3] with binding sites of transcriptional regulator 1IK3. Figure S3. 3D H-bonding interactions of eugenol pose no. 3 [4], α-humulene pose no. 1 [5] with binding sites of transcriptional regulator 1IK3. Figure S4. 3D H-bonding interactions of quercitrin pose no. 4 [6] and caryophyllene pose no. 1 [7] with binding sites of transcriptional regulator 1IK3. Figure S5. 3D H-bonding Interactions of 2-phenylethylisothiocyanate pose no. 1 [8], apigenin pose no. 1 [9]; caryophyllene oxide pose no. 1 [10] with binding sites of transcriptional regulator 3TOP. Figure S6. 3D H-bonding interactions of eugenol pose no. 2 [11] and α-humulene pose no. 1 [12] with binding sites of transcriptional regulator 3TOP. Figure S6. 3D H-bonding interactions of quercitrin pose no. 4 [13] and caryophyllene pose no. 1 [14] with binding sites of transcriptional regulator 3TOP. Figure S7. 3D H-bonding interactions of 2-phenylethylisothiocyanate pose no. 1 [15], apigenin pose no. 3 [16] and caryophyllene oxide pose no. 1 [17] with binding sites of transcriptional regulator 4F5S. Figure S9. 3D H-bonding interactions of quercitrin pose no. 5 [20] and caryophyllene pose no. 1 [21] with binding sites of transcriptional regulator 4F5S [66–69].

Author Contributions: A.R. and M.K. and M.I.A. performed analysis and interpretation of enzyme inhibitory data. V.A. performed molecular docking. M.K. and S.I.N. performed SDS-PAGE. L.P. and A.A. designed the method and the main project, edited the manuscript and gave the final approval of the version to be submitted. All authors have read and agreed to the published version of the manuscript.

Funding: This research was funded by GA Society for Medicinal Plant and Natural Products Research and the Foundation "Plants for Health".

Institutional Review Board Statement: The study was conducted according to the guidelines of the declaration of Helsinki and approved by the institutional Review Board and Ethics Committee of Gomal University (2019).

Informed Consent Statement: Not applicable.

Data Availability Statement: The data presented in this study are available in this article and also as Supplementary Materials.

Acknowledgments: GA Society for Medicinal Plant and Natural Products Research and the Foundation "Plants for Health" are acknowledged for supporting this project.

Conflicts of Interest: The authors declare that they have no conflict of interest.

Sample Availability: Samples of the compounds are not available from the authors.

References

1. Lin, Y.; Sun, Z. Current views on type 2 diabetes. *J. Endocrinol.* **2010**, *204*, 1. [CrossRef]
2. Kharroubi, A.T.; Darwish, H.M. Diabetes mellitus: The epidemic of the century. *World J. Diabetes* **2015**, *6*, 850–867. [CrossRef]
3. Seaquist, E.R.; Anderson, J.; Childs, B.; Cryer, P.; Dagogo-Jack, S.; Fish, L.; Heller, S.; Rodriguez, H.; Rosenzweig, J.; Vigersky, R. Hypoglycemia and diabetes: A report of a workgroup of the American Diabetes Association and the Endocrine Society. *Diabetes Care* **2013**, *36*, 1384–1395. [CrossRef] [PubMed]
4. Tamarat, R.; Silvestre, J.S.; Huijberts, M.; Benessiano, J.; Ebrahimian, T.G.; Duriez, M.; Lévy, B.I. Blockade of advanced glycation end-product formation restores ischemia-induced angiogenesis in diabetic mice. *Proc. Natl. Acad. Sci. USA* **2003**, *100*, 8555–8560. [CrossRef] [PubMed]
5. Negi, G.; Kumar, A.; Joshi, R.P.; Ruby, P.K.; Sharma, S.S. Oxidative stress and diabetic neuropathy: Current status of antioxidants. *Inst. Integr. Omics Appl. Biotechnol. J.* **2011**, *2*, 71–78.
6. Colosia, A.D.; Palencia, R.; Khan, S. Prevalence of hypertension and obesity in patients with type 2 diabetes mellitus in observational studies: A systematic literature review. *Diabetes Metab. Syndr. Obes. Targets Ther.* **2013**, *6*, 327. [CrossRef]
7. Chan, G.C.; Tang, S.C. Diabetic nephropathy: Landmark clinical trials and tribulations. *Nephrol. Dial. Transpl.* **2013**, *31*, 359–368. [CrossRef]
8. Brownlee, M. The pathobiology of diabetic complications: A unifying mechanism. *Diabetes* **2005**, *54*, 1615–1625. [CrossRef]
9. Nursten, H.E. *The Maillard Reaction. Chemistry, Biochemistry, and Implications*; Royal Society of Chemistry: Cambridge, UK, 2005.
10. Sharma, C.; Kaur, A.; Thind, S.S.; Singh, B.; Raina, S. Advanced glycation End-products (AGEs): An emerging concern for processed food industries. *J. Food Sci. Technol.* **2015**, *52*, 7561–7576. [CrossRef]

11. Lapolla, A.; Traldi, P.; Fedele, D. Importance of measuring products of non-enzymatic glycation of proteins. *Clin. Biochem.* **2005**, *38*, 103–115. [CrossRef]
12. Pollack, R.M.; Marc, Y.; Donath, R.; LeRoith, D.; Leibowitz, G. Anti-inflammatory Agents in the Treatment of Diabetes and Its Vascular Complications. *Diabetes Care* **2016**, *39*, S244–S252. [CrossRef] [PubMed]
13. Nichols, C.G.; Remedi, M.S. The diabetic β-cell: Hyperstimulated vs. hyperexcited. *Diabetes Obes. Metab.* **2012**, *14*, 129–135. [CrossRef]
14. Zhou, R.; Tardivel, A.; Thorens, B.; Choi, I.; Tschopp, J. Thioredoxin-interacting protein links oxidative stress to inflammasome activation. *Nat. Immunol.* **2010**, *11*, 136–140. [CrossRef] [PubMed]
15. Böni-Schnetzler, M.; Boller, S.; Debray, S.; Bouzakri, K.; Meier, D.T.; Prazak, R.; Kerr-Conte, J.; Pattou, F.; Ehses, J.A.; Schuit, F.C.; et al. Free fatty acids induce a proinflammatory response in islets via the abundantly expressed interleukin-1 receptor I. *Endocrinology* **2009**, *150*, 5218–5229. [CrossRef]
16. Masters, S.L.; Dunne, A.; Subramanian, S.L.; Hull, R.L.; Tannahill, G.M.; Sharp, F.A.; Becker, C.; Franchi, L.; Yoshihara, E.; Chen, Z.; et al. Activation of the NLRP3 inflammasome by islet amyloid polypeptide provides a mechanism for enhanced IL-1b in type 2 diabetes. *Nat. Immunol.* **2010**, *11*, 897–904. [CrossRef] [PubMed]
17. Kumar, S.; Dobos, G.J.; Rampp, T. The Significance of Ayurvedic Medicinal Plants. *J. Evid. Based Complement. Altern. Med.* **2017**, *22*, 494–501. [CrossRef]
18. Hosseini, S.; Huseini, H.F.; Larijani, B.; Mohammad, K.; Najmizadeh, A.; Nourijelyani, K.; Jamshidi, L. The hypoglycemic effect of *Juglans regia* leaves aqueous extract in diabetic patients: A first human trial. *Daru J. Pharm. Sci.* **2014**, *22*, 19. [CrossRef]
19. Rahman, M.A.; Mossa, J.S.; Al-Said, M.S.; Al-Yahya, M.A. Medicinal plant diversity in the flora of Saudi Arabia 1: A report on seven plant families. *Fitoterapia* **2004**, *75*, 149–161. [CrossRef]
20. Amaral, J.S.; Seabra, R.M.; Andrade, P.B.; Valent, A.O.P.; Pereira, J.A.; Ferreres, F. Phenolic profile in the quality control of walnut (*Juglans regia* L.) leaves. *J. Food Chem.* **2004**, *88*, 373–379. [CrossRef]
21. Farag, M.; Abdel-Mageed, M.; El Gamal, A.A.; Basudan, O.A. *Salvadora persica* L.: Toothbrush tree with health benefits and industrial applications—An updated evidence-based review. *Saudi Pharm. J.* **2021**, *29*, 751–763. [CrossRef]
22. Hooda, M.S.; Pal, R.; Bhandari, A.; Singh, J. Antihyperglycemic and antihyperlipidemic effects of *Salvadora persica* in streptozotocin-induced diabetic rats. *Pharm. Biol.* **2014**, *52*, 745–749. [CrossRef] [PubMed]
23. Bader, A.; Flamini, G. The composition of the root oil of *Salvadora persica* Linn. *J. Essent. Oil Res.* **2002**, *14*, 128–129. [CrossRef]
24. Adefegha, S.A.; Oboh, G. *In vitro* inhibition activity of polyphenol-rich extracts from *Syzygium aromaticum* (L.) Merr. and Perry (Clove) buds against carbohydrate hydrolyzing enzymes linked to type 2 diabetes and Fe($^{2+}$)-induced lipid peroxidation in rat pancreas. *Asian Pac. J. Trop. Biomed.* **2012**, *2*, 774–781. [CrossRef]
25. *Syzygium aromaticum* (L.) Merr. and L.M.Perry. Available online: http://powo.science.kew.org/taxon/urn:lsid:ipni.org:names:601421-1 (accessed on 6 June 2022).
26. Vijayasteltar, L.; Nair, G.G.; Maliakel, B.; Kuttan, R.M.K. Safety assessment of a standardized polyphenolic extract of clove buds: Subchronic toxicity and mutagenicity studies. *Toxicol. Rep.* **2016**, *3*, 439–449. [CrossRef]
27. Zulcafli, A.S.; Lim, C.; Ling, A.P.; Chye, S.; Koh, R. Antidiabetic Potential of *Syzygium* sp.: An Overview. *Yale J. Biol. Med.* **2020**, *93*, 307–325.
28. Nikolic, V.; Nikolic, L.; Dinic, A.; Gajic, I.; Urosevic, M.; Stanojevic, L.; Stanojevic, J.; Danilovic, B. Chemical Composition, Antioxidant and Antimicrobial Activity of Nutmeg (*Myristica Fragrans* Houtt.) Seed Essential Oil. *J. Essent. Oil Bear. Plants* **2021**, *24*, 218–227. [CrossRef]
29. Du, S.S.; Yang, K.; Wang, C.F.; You, C.X.; Geng, Z.F.; Guo, S.S.; Deng, Z.W.; Liu, Z.L. Chemical constituents and activities of the essential oil from *Myristica fragrans* against cigarette beetle *Lasioderma serricorne*. *Chem. Biodivers.* **2014**, *11*, 1449–1456. [CrossRef]
30. Naeem, N.; Rehman, R.; Mushtaq, A.; Ghania, J. Ben Nutmeg: A Review on Uses and Biological Properties. *Int. J. Chem. Biochem. Sci.* **2016**, *9*, 107–110.
31. Matulyte, I.; Jekabsone, A.; Jankauskaite, L.; Zavistanaviciute, P.; Sakiene, V.; Bartkiene, E.; Ruzauskas, M.; Kopustinskiene, D.M.; Santini, A.; Bernatoniene, J. The Essential Oil and Hydrolats from *Myristica fragrans* Seeds with Magnesium Aluminometasilicate as Excipient: Antioxidant, Antibacterial, and Anti-Inflammatory Activity. *Foods* **2020**, *9*, 37. [CrossRef]
32. Karimi, M.; Sadeghi, R.; Kokini, J. Pomegranate as a promising opportunity in medicine and nanotechnology. *Trends Food Sci. Technol.* **2017**, *69*, 59–73. [CrossRef]
33. Singh, B.; Singh, J.P.; Kaur, A.; Singh, N. Phenolic compounds as beneficial phytochemicals in pomegranate (*Punica granatum* L.) peel: A review. *Food Chem.* **2018**, *261*, 75–86. [CrossRef] [PubMed]
34. Xiao, J. Dietary flavonoid aglycones and their glycosides: Which show better biological significance? *Crit. Rev. Food Sci. Nutr.* **2017**, *57*, 1874–1905. [CrossRef] [PubMed]
35. Khan, H.; Jawad, M.; Kamal, M.A.; Baldi, A.; Xiao, J.; Nabavi, S.M.; Daglia, M. Evidence and prospective of plant derived flavonoids as antiplatelet agents: Strong candidates to be drugs of future. *Food Chem. Toxicol.* **2018**, *119*, 355–367. [CrossRef] [PubMed]
36. Subapriya, R.; Nagini, S. Medicnal properties of neem leaves: A review. *Curr. Med. Chem. Anticancer Agents* **2005**, *5*, 146–149. [CrossRef]

37. Quelemes, P.V.; Perfeito, M.L.; Guimarães, M.A.; dos Santos, R.C.; Lima, D.F.; Nascimento, C.; Silva, M.P.; Soares, M.J.; Ropke, C.D.; Eaton, P.; et al. Effect of neem (*Azadirachta indica* A. Juss) leaf extract on resistant *Staphylococcus aureus* biofilm formation and *Schistosoma mansoni* worms. *J. Ethnopharmacol.* **2015**, *175*, 287–294. [CrossRef]
38. Biswas, K.; Chattopadhyay, R.K.; Banerjee, R.K.; Bandyopadhyay, U. Biological activities and medicinal properties of neem (*Azadirachta indica*). *Curr. Sci.* **2002**, *82*, 1336–1345.
39. Pandey, K.B.; Rizvi, S.I. Plant polyphenols as dietary antioxidants in human health and disease. *Oxid. Med. Cell. Longev.* **2009**, *2*, 270–278. [CrossRef]
40. Xiao, J.B.; Hogger, P. Dietary polyphenols and type 2 diabetes: Current insights and future perspectives. *Curr. Med. Chem.* **2015**, *22*, 23–38. [CrossRef]
41. Hussain, T.; Tan, B.; Yin, Y.; Blachier, F.; Tossou, M.C.; Rahu, N. Oxidative Stress and Inflammation: What Polyphenols Can Do for Us? *Oxid. Med. Cell. Longev.* **2016**, *2016*, 7432797. [CrossRef]
42. Terra, W.R.; Ferreira, C. Biochemistry of digestion. In *Comprehensive Molecular Insect Science*; Gilbert, L.I., Iatrov, K., Gill, S., Eds.; Elsevier: Oxford, UK, 2005; Volume 4, pp. 171–224.
43. Liu, S.K.; Hao, H.; Bian, Y.; Ge, Y.X.; Lu, S.; Xie, H.X.; Wang, K.M.; Tao, H.; Yuan, C.; Zhang, J.; et al. Discovery of New α-Glucosidase Inhibitors: Structure-Based Virtual Screening and Biological Evaluation. *Front. Chem.* **2021**, *9*, 639279. [CrossRef]
44. Proença, C.; Freitas, M.; Ribeiro, D.; Oliveira, E.; Sousa, J.; Tomé, S.M.; Ramos, M.J.; Silva, A.; Fernandes, P.A.; Fernandes, E. α-Glucosidase inhibition by flavonoids: An in vitro and in silico structure-activity relationship study. *J. Enzym. Inhib. Med. Chem.* **2017**, *32*, 1216–1228. [CrossRef] [PubMed]
45. Sell, H.C.; Habich, C.; Eckel, J. Adaptive immunity in obesity and insulin resistance. *Nat. Rev. Endocrinol.* **2012**, *8*, 709–716. [CrossRef] [PubMed]
46. Nikolajczyk, B.S.; Jagannathan-Bogdan, M.; Shin, H.; Gyurko, R. State of the union between metabolism and the immune system in type 2 diabetes. *Genes Immunol.* **2011**, *12*, 239–250. [CrossRef] [PubMed]
47. Ma, K.; Nunemaker, C.S.; Wu, R.; Chakrabarti, S.K.; Taylor-Fishwick, D.A.; Nadler, J.L. 12-Lipoxygenase Products Reduce Insulin Secretion and {beta}-Cell Viability in Human Islets. *J. Clin. Endocrinol. Metab.* **2010**, *95*, 887–893. [CrossRef] [PubMed]
48. Ribeiro, D.; Freitas, M.; Lima, J.L.; Fernandes, E. Pro inflammatory pathways: The modulation by flavonoids. *Med. Res. Rev.* **2015**, *35*, 877–936. [CrossRef]
49. Song, Q.; Liu, J.; Dong, L.; Wang, X.; Zgang, X. Novel advances in inhibiting advanced glycation end product formation using natural. *Biomed. Pharmacother.* **2021**, *140*, 111750. [CrossRef]
50. Kosmopoulos, M.; Drekolias, D.; Zavras, P.D.; Piperi, C.; Papavassiliou, A.G. Impact of advanced glycation end products (AGEs) signaling in coronary artery disease. *Biochim. Biophys. Acta (BBA) Mol. Basis Dis.* **2019**, *1865*, 611–619. [CrossRef]
51. Wu, G. Amino acids: Metabolism, functions, and nutrition. *Amino Acids* **2009**, *37*, 1–17. [CrossRef]
52. Deetae, P.; Parichanon, P.; Trakunleewatthana, P.; Chanseetis, C.; Lertsiri, S. Antioxidant and anti-glycation properties of Thai herbal teas in comparison with conventional teas. *Food Chem.* **2012**, *133*, 953–959. [CrossRef]
53. Deluyker, D.; Evens, L.; Bito, V. Advanced glycation end products (AGEs) and cardiovascular dysfunction: Focus on high molecular weight AGEs. *Amino Acids* **2017**, *49*, 1535–1541. [CrossRef]
54. Wu, C.H.; Yen, G.C. Inhibitory Effect of Naturally Occurring Flavonoids on the Formation of Advanced Glycation Endproducts. *J. Agric. Food Chem.* **2005**, *53*, 3167–3173. [CrossRef] [PubMed]
55. Matough, F.A.; Budin, S.B.; Hamid, Z.A.; Alwahaibi, N.; Mohamed, J. The role of oxidative stress and antioxidants in diabetic complications. *Sultan Qaboos Univ. Med. J.* **2012**, *12*, 5–18. [CrossRef] [PubMed]
56. Rafey, A.; Amin, A.; Kamran, M.; Haroon, U.; Farooq, K.; Foubert, K.; Pieters, L. Analysis of Plant Origin Antibiotics against Oral Bacterial Infections Using In Vitro and In Silico Techniques and Characterization of Active Constituents. *Antibiotics* **2021**, *10*, 1504. [CrossRef] [PubMed]
57. Ren, L.; Qin, X.; Cao, X.; Wang, L.; Bai, F.; Bai, G.; Shen, Y. Structural insight into substrate specificity of human intestinal maltase-glucoamylase. *Protein Cell* **2011**, *2*, 827–836. [CrossRef]
58. Serzypczak-Jankun, E.; Zhou, K.; Jankun, J. Inhibition of lipoxygenase by (-)-epigallocatechin gallate: X-ray analysis at 2.1 A reveals degradation of EGCG and shows soybean LOX-3 complex with EGC instead. *Int. J. Mol. Med.* **2003**, *12*, 415–420.
59. Accelrys Software Inc. *Discovery Studio Modelling Environment, Release 3.5*; Accelrys Discovery Studio (Accelrys Software Inc.): San Diego, CA, USA, 2002.
60. Trott, O.; Olson, A.J. AutoDock Vina: Improving the speed and accuracy of docking with a new scoring function, efficient optimization, and multithreading. *J. Comput. Chem.* **2010**, *31*, 455–461. [CrossRef]
61. Choudhary, M.I.; Shah, S.A.; Atta-ur-Rahman; Khan, S.N.; Khan, M.T. α-glucosidase and tyrosinase inhibitors from fungal hydroxylation of tibolone and hydroxytibolones. *Steroids* **2010**, *75*, 956–966. [CrossRef]
62. Matsuura, N.; Aradate, T.; Sasaki, C.; Kojima, H.; Ohara, M.; Hasegawa, J. Screening system for the Maillard reaction inhibitor from natural product extracts. *J. Health Sci.* **2002**, *48*, 520–526. [CrossRef]
63. Harris, C.S.; Beaulieu, L.P.; Fraser, M.H.; Mcintyre, K.L.; Owen, P.L.; Martineau, L.C. Inhibition of advanced glycation end product formation by medicinal plant extracts correlates with phenolic metabolites and antioxidant activity. *Planta Med.* **2011**, *77*, 196–204. [CrossRef]
64. Elosta, A.; Slevin, M.; Rahman, K.; Ahmed, N. Aged garlic has more potent antiglycation and antioxidant properties compared to fresh garlic extract in vitro. *Sci. Rep.* **2017**, *7*, 39613. [CrossRef]

65. Malterud, K.E.; Rydland, K.M. Inhibitors of 15-lipoxygenase from orange peel. *J. Agric. Food Chem.* **2000**, *48*, 5576–5580. [CrossRef] [PubMed]
66. Raafa, K. Phytochemical analysis of *Juglans regia* oil and kernel exploring their antinociceptive and anti-inflammatory potentials utilizing combined bio-guided GC–FID, GC–MS and HPLC analyses. *Rev. Bras. Farmacogn.* **2018**, *3*, 358–368. [CrossRef]
67. Alma, M.H.; Ertas, M.; Nitz, S.; Kollmannsberger, H. Chemical composition and content of essential oil from the bud of cultivated Turkish clove (*Syzygium aromaticum* L.). *Bio Resour.* **2007**, *2*, 265–269. [CrossRef]
68. Khoobchandani, M.; Ojeswi, B.K.; Ganesh, N.; Srivastava, M.M.; Gabbanini, S.; Matera, R.; Iori, R.; Valgimigli, R. Antimicrobial properties and analytical profile of traditional *Eruca sativa* seed oil: Comparison with various aerial and root plant extracts. *Food Chem.* **2010**, *120*, 217–224. [CrossRef]
69. Shafiq, M.I.; Ahmed, M.; Rasul, A.; Samra, Z.Q.; Qadir, M.A.; Mazhar, S.; Ali, A. Chemical Composition of the Essential Oils of Nutmeg and Mace by GC-FID/MS Indigenous to Pakistan and Evaluation of their Biological Activities. *Lat. Am. J. Pharm.* **2016**, *35*, 2176–2184.

molecules

MDPI

Review

Quercetin: Its Antioxidant Mechanism, Antibacterial Properties and Potential Application in Prevention and Control of Toxipathy

Weidong Qi †, Wanxiang Qi †, Dongwei Xiong † and Miao Long *

Key Laboratory of Livestock Infectious Diseases, Ministry of Education, College of Animal Science and Veterinary Medicine, Shenyang Agricultural University, Shenyang 110866, China
* Correspondence: longmiao@syau.edu.cn
† These authors contributed equally to this work and should be considered co-first authors.

Abstract: Quercetin, as a flavonol compound found in plants, has a variety of biological activities. It is widely present in nature and the human diet, with powerful oxidative properties and biological activities. In this review, the antioxidant mechanism and broad-spectrum antibacterial properties of quercetin are revealed; the intervention effects of quercetin on pesticide poisoning and the pathway of action are investigated; the toxic effects of main mycotoxins on the collection and the detoxification process of quercetin are summarized; whether it is able to reduce the toxicity of mycotoxins is proved; and the harmful effects of heavy metal poisoning on the collection, the prevention, and control of quercetin are evaluated. This review is expected to enrich the understanding of the properties of quercetin and promote its better application in clinical practice.

Keywords: quercetin; antioxidant; toxipathy

Citation: Qi, W.; Qi, W.; Xiong, D.; Long, M. Quercetin: Its Antioxidant Mechanism, Antibacterial Properties and Potential Application in Prevention and Control of Toxipathy. *Molecules* **2022**, *27*, 6545. https://doi.org/10.3390/molecules27196545

Academic Editors: Raffaele Pezzani and Sara Vitalini

Received: 3 September 2022
Accepted: 28 September 2022
Published: 3 October 2022

1. Introduction

During the last few decades, medicinal plants have gained wide popularity due to their low incidence, mildness of side effects, low price, and natural origin, among which quercetin is one of the well-known types of plant metabolites [1]. Quercetin is a flavonoid widely found in vegetables and fruits. Its name comes from Quercetum (oak forest), used since 1857, its molecular formula is $C_{15}H_{10}O_7$; its chemical structure formula (Figure 1) has unique biological properties, can improve physical and mental status, and reduce viral infection [2,3]. It is a naturally occurring acute coenzyme transport inhibitor [4]. Quercetin exists in its various glycoside forms and the five yellow compounds isolated from dietary quercetin (e.g., quercetin-3-glucoside or isoquercitrin, quercetin-4′-glucoside, quercetin-3,4′-diglucoside [5]). These five compounds are quercetin 3-O-beta-D-glucopyranoside (DA), kaempferol 3-O-(6″-trans-coumarin)-beta-D-glucopyranoside (D1), kaempferol 3-O-(2″,4″-diacetyl-p-coumaroyl-6″-trans-coumaroyl)-beta-D-glucopyranoside (A), kaempferol 3-O-(2″-6″-di-trans-p-coumaroyl)-beta-D-glucopyranoside (D7), and kaempferol 3-O-beta-D-glucopyranoside (B). Quercetin contains multiple hydroxyl groups and its molecular structure comprises four reactive groups, i.e., dihydroxy group between the A ring; o-dihydroxy group B; and C2 and C3 double bonds of the C ring, and the 4-carbonyl group has physiological activities that readily undergo esterification with carboxyl groups and anti-cardiovascular disease [6]. The antioxidant and anti-inflammatory properties of quercetin are closely related to the prevention and treatment of pesticide poisoning and heavy metal poisoning. In addition, quercetin plays an essential role in reducing mycotoxins and protecting cells from damage [3]. However, there are few studies related to the blocking effects of quercetin on various toxic diseases. This review aims to summarize and analyze the preventative effects of quercetin on heavy metal poisoning, pesticide poisoning, mycotoxin poisoning, and inflammation caused by toxic diseases so as to provide a scientific basis for its better clinical application.

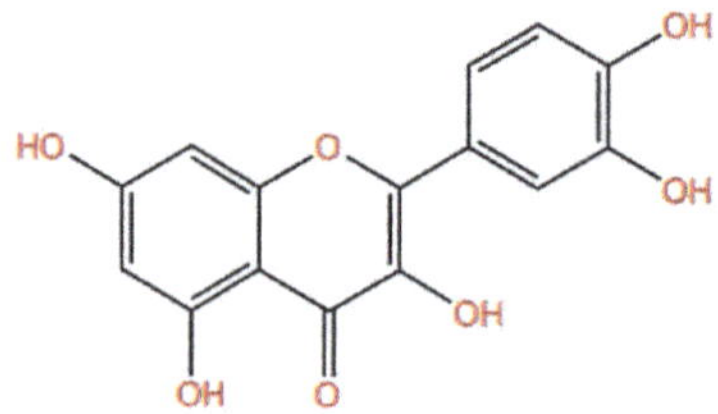

Figure 1. Structural formula of quercetin.

2. Main Pharmacological Activities of Quercetin

2.1. Antioxidants

By studying the chemical structure of quercetin, the International Union of Pure and Applied Chemistry named quercetin as 3,3′,4′,5,7-pentahydroxyflavone, which indicates that quercetin has an OH group attached at positions 3, 5, 7, 3′, and 4′ [2]. The antioxidant mechanism of quercetin in vivo is mainly reflected in its effects on glutathione (GSH), signal transduction pathways, reactive oxygen species (ROS), and enzyme activities. The antioxidant properties of quercetin show a concentration dependence in the low dose range but too much of the antioxidant brings about the opposite result [7].

2.1.1. Quercetin Achieves Antioxidant Effects by Affecting GSH as a Reactive Hydrogen Donor

Quercetin can enhance the antioxidant capacity of the body by regulating the level of GSH. This is because free radicals are produced by the body during metabolic processes, which cause genetic mutations and cell membrane damage; induce various diseases, such as heart disease, liver disease, and diabetes; and accelerates aging of the body [8,9]. Once the body produces free radicals, superoxide dismutase (SOD) will quickly convert O_2^- to H_2O_2 and will be further catalyzed into non-toxic H_2O and GSH. As the reaction hydrogen donor, the level of GSH determines the rate of the reaction [10]. In addition, one study found that the application of quercetin treatment in renal ischemia/reperfusion (I/R) increased the level of GSH [11]. The ability of quercetin to induce GSH synthesis was also confirmed in subsequent studies. Gao et al. [12] reported the inhibitory effect of quercetin on GSH at the 0.5% level and found that GSH reductase can catalyze the reduction reaction of GSSG in erythrocytes and liver, leading to the formation of GSH, whereas while high doses of quercetin can be used to regulate GSH, dynamic balance is affected by the action of peroxidase; H_2O_2 is converted to H_2O and GSH is oxidized to oxidized glutathione disulfide, due to the aforementioned reactions resulting in a dynamic balance of GSH. This may lead to the suppression of the level of GSH at low doses.

2.1.2. Quercetin Achieves Antioxidant Effects through Positive Effects on Various Signal Transduction Pathways

It was found that quercetin counteracts atherosclerosis by reversing the increased expression of NADPH oxidase in oxLDL and the oxLDL-induced decreases in activation of AMP-activated protein kinase, thereby inhibiting NF-κB signaling and holding AKT/eNOS function [13]. In addition, Granado-Serrano et al. [10] found that quercetin improves the antioxidant capacity of cells by activating the intracellular p38 MAPK pathway, increasing intracellular GSH levels and providing a source of hydrogen donors in the scavenging of free radical reactions. It is known that adverse environmental factors increase the production of ROS. These factors increase the activity of mitochondrial electron transport chains, which is an important source of intracellular ROS production [14]. The body fights against free radicals through two main defense systems: non-enzymatic antioxidants represented by vitamins and trace elements (such as vitamin C, vitamin E, selenium, copper, manganese, etc.) and enzymatic antioxidants represented by SOD, including catalase, glutathionase, etc. Quercetin can promote the antioxidant defense system and maintain oxidative homeostasis not only by regulating the non-enzyme-dependent antioxidant defense system and enzyme-mediated antioxidant defense system but also by regulating MAPK, NRFB, AMPK, and other signaling pathways induced by ROS [15–19].

By affecting signal transduction pathways, quercetin can modulate enzymes or antioxidant substances and enhance antioxidant properties, thereby preventing disease progression. In psoriasis, quercetin was found to promote the disease recovery by up-regulating the expression of TNF receptor-associated factor 3 and down-regulating the expression of NF-κB, inducing kinase. In addition, quercetin achieves protection against acute spinal cord injury by up-regulating the activity of SOD, down-regulating the level of malondialdehyde (MDA), and inhibiting the p38MAPK/iNOS signaling pathway [20]. In addition, quercetin reduces imiquimod (IMQ)-induced MDA levels in skin tissues and enhances catalase, SOD, and GSH activities, which together improve the antioxidant properties of the body [21].

It was found that quercetin improves antioxidant capacity by modulating signaling pathways. For example, quercetin promotes the functional recovery of mobile mediators after cerebral ischemia by promoting antioxidant signaling, translating the TGFβ-2/PI3K/AKT pathway, and increasing resistance to apoptosis. It also controls the development of atherosclerosis induced by high fructose diet by enhancing PI3K/AKT and inhibiting ROS [22–24].

2.1.3. Quercetin Prevents Antioxidant Damage by Eliminating ROS

It is known that quercetin can scavenge ROS, and most of the oxidative damage in vivo is attributed to ROS, so quercetin can resist oxidative damage, such as radiation-induced ultraviolet radiation B (UVB) skin lesions, respiratory damage, and other oxidative damage diseases [25]. Human skin is highly functional and can withstand many types of environmental damage; however, UVB induces an imbalance of endogenous antioxidant systems and a transient increase in ROS, which increases the level of inflammation and free radicals and affects cellular processes. Studies have shown that quercetin not only prevents UVB radiation damage by reducing ROS-induced damage to mitochondria but also by scavenging ROS, in addition to inhibiting mitochondrial membrane depolarization and cell membrane movement. Thus, it seems that quercetin can prevent UVB-induced skin damage by suppressing this imbalance [26,27].

In addition, exposure to ambient fine particulate matter (PM2.5) was found to cause respiratory disease, resulting in respiratory damage and a range of adverse changes, such as increased ROS production, suppressed mitochondrial expression, and weakened 16HBE cell activity. Quercetin may stimulate 16HBE cells to repair oxidative damage after PM2.5 exposure through the anti-inflammatory process and the production of ROS [28].

Quercetin enhances antioxidant activity and inhibits lipid cultivation, and it is effective in the treatment of oxidative liver damage [29]. Quercetin was found to restore endogenous redox homeostasis not only by scavenging free radicals and increasing the level of GSH but also by directly scavenging ROS and hydroxyl radicals under hypoxia. The reduction of oxidative stress-induced neurodegeneration in the hippocampal region thereby reversed hypoxia-induced memory impairment [30].

It was found that ionizing radiation induces this type of damage through increased cellular damage or increased free radical formation and cell death due to ROS, whereas bioflavonoids, as a redox agent, can inhibit the toxicity of free radicals and enhance antioxidant properties in vivo. Quercetin protects cells from radiation and genotoxicity-induced damage by increasing endogenous antioxidant and scavenging free radical levels [31–33].

On the other hand, quercetin can prevent oxidant damage by inhibiting oxidative stress. Oxidative stress is caused by an imbalance between antioxidants and oxidants in the body, and the reaction process tends to be oxidized, which, once oxidized, leads to the secretion of high protein enzymes and inflammatory infiltration of neutrophils. In contrast, quercetin can regulate the balance between antioxidants and oxidants to suppress oxidative stress. Various experimental studies demonstrated that quercetin inhibited acrylamide-induced oxidative damage in rats, radiation-induced brain damage in rats, cadmium fluoride-induced neurodegenerative disease and oxidative stress, and diabetes-induced nerve damage in the rat retina. It protects nerves, the brain, or other cells of the body from damage caused by oxidation by regulating the level of antioxidants [34–36].

It was found that quercetin can modulate the expression of antioxidant-related genes in A549 cells to alleviate oxidative stress. In addition, quercetin can increase the level of GSH and decrease the level of ROS to prevent paraquat-induced oxidative damage [37]. Due to the strong inhibitory and scavenging effects of quercetin on ROS, it also protects sperm from the adverse effects of ROS and maintains the function of male germ cells [38]. As quercetin exerts protective effects on gastric epithelial GES-1 cells, it was able to inhibit ROS-induced damage to gastric epithelial cells caused by oxidants such as H_2O_2. In gastric epithelial cells, quercetin can prevent oxidative damage and inhibit ROS production during acute gastric mucosal injury in mice [39].

2.1.4. Effect of Quercetin on Enzyme Activity

Previous research has shown that two enzymes, acetylcholinesterase and butyrylcholinesterase, are associated with oxidative properties, while the -OH group on the lateral benzene ring of quercetin binds to important amino acid residues in the active sites of both enzymes and inhibits their oxidative effects [40]. In addition, Odbayar et al. found that quercetin increases the activity of antioxidant enzymes, such as GSH transferase and the aldo-keto reductase [41].

Pre-treatment with quercetin was reported to significantly increase the levels of expression of endogenous antioxidant enzymes, such as GSH peroxidase, Cu/Zn SOD, Mn SOD, and peroxidase, in CA1 vertebral neurons of the hippocampus of animals suffering ischemic injury. This suggests that quercetin may be a potential neuroprotective agent against ischemia, which protects CA1 vertebral neurons from I/R injury in the hippocampal region of animals [42].

Quercetin has also been shown to prevent heart damage by scavenging anaerobic free radicals induced by lipopolysaccharide-induced endotoxemia. Lipopolysaccharide has been reported to induce biochemical and histopathological damage to the myocardium in the models of endotoxemia. In rat model experiments, rats treated with lipopolysaccharide exhibited decreases in catalase and SOD activities and a significant increase in MDA levels in heart tissue. In contrast, treatment with quercetin significantly reduced MDA levels and increased SOD and catalase levels. This finding indicated that quercetin enhanced the antioxidant defense system [43].

2.2. Antibacterial Properties

The antibacterial mechanisms of quercetin have been reported to include: altering the permeability of bacterial cells; disrupting bacterial cell walls; inhibiting the synthesis of nucleic acids, thereby affecting the synthesis and expression of proproteins; and reducing enzyme activity. The results indicated that quercetin has broad-spectrum antibacterial properties, and it exerts a good inhibitory effect not only on bacteria but also on fungi to a significant extent. Quercetin also shows a good inhibitory effect on the growth of pathogenic bacteria, such as *Salmonella enterica, Escherichia coli, Pseudomonas aeruginosa, Staphylococcus aureus,* and *Aspergillus* [44]. In addition, quercetin affects the growth of *E. coli* by altering the activity of adenosine triphosphate [45].

Furthermore, TEM images showed that the treatment with quercetin (50× minimum inhibitory concentration (MIC)) eventually caused *E. coli* to be cavitated and killed, and such treatment can disrupt the cell membrane and cell wall of *S. aureus* (10× MIC). In the same study by Hossion et al. [44], a novel artificially designed synthetic quercetin acyl glucoside was found to significantly inhibit the growth of *Pseudomonas aeruginosa, S. aureus,* and *E. coli.* In addition, the extract of poplar plum (containing quercetin) showed significant antibacterial activity against Listeria monocytogenes, *Salmonella* spp. and *Shigella* spp. with MIC values ranging from 2.07 to 8.28 mg/mL. Wang et al. [46] found that quercetin could protect rats from catheter-associated *S. aureus* infection by inhibiting the activity of thrombin.

In addition, sugarcane bagasse (with 470 mg quercetin/g polyphenol) extract was found to inhibit the growth of *L. monocytogenes*, *S. aureus*, *E. coli*, and *Salmonella typhi* [47].

In addition, quercetin was found to suppress abiotic surface colonization genes of *L. monocytogenes* at concentrations below the MIC [48]. Qayyum et al. [49] found that quercetin ($\frac{1}{2}$ × MIC) remained inhibitory to *E. faecalis* MTCC 2729 at MIC (256 g/mL) under scanning electron microscopy and confocal laser scanning microscopy. A study by Wang et al. confirmed that quercetin can inhibit the formation of *Streptococcus hepatitis* biofilm. In addition, quercetin prevents bacterial adhesion, inhibits population intervention pathways, alters or disrupts plasma membranes, and inhibits efflux pumps, thereby preventing nucleic acid synthesis. A study by Lee et al. [50] discovered that quercetin has an inhibitory effect on genes related to bacterial adhesion.

3. Preventing Poisoning as an Application of Quercetin

3.1. Preventing of Mycotoxin Poisoning

Mycotoxins are toxic substances produced in many foods and are secondary metabolites synthesized by molds, which are one of the major causes of food spoilage and associated issues in various foodstuffs and fodder and act as a ubiquitous environmental pollutant [51–53]. As shown in Table 1, mycotoxins have significant toxic effects on tissues and organs of humans and animals.

Table 1. Toxic effects of different fungal toxins and protective effects and mechanisms of quercetin.

Fungi	*Fusarium*		*Aspergillus flavus, Aspergillus parasiticus*	*Aspergillus ochratoxin, Aspergillus sulphureus*
Mycotoxins	Deoxynivalenol (DON)	Zearalenone (ZEN)	Aflatoxin B1 (AFB1)	Ochratoxin A (OTA)
Structure	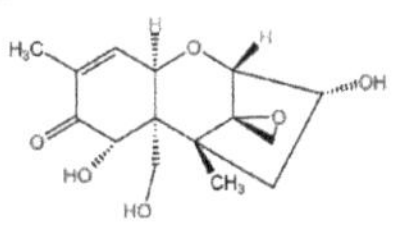	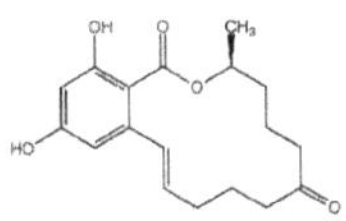	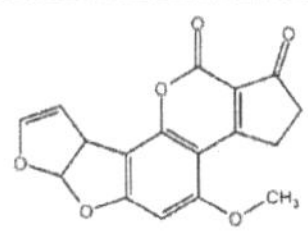	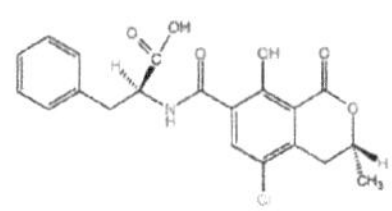
Main area of action	Gastrointestinal and immune system	Reproductive organs	Liver, spleen, kidney	Liver, kidney
Main toxic effects	Fine cell and body fluid mediated, inhibition of protein synthesis	Lipid peroxidation, mitochondrial pathway, and DNA damage	Mutagenic, deformogenic, and carcinogenic	Hepatotoxicity, nephrotoxicity, teratogenicity, and immunosuppression
Protective effect of quercetin	Protects caco-2 cells from damage	Protects HEK293 and HCT116 cells and inhibits apoptosis	Improves the brain, enhances learning and memory, inhibits biotransformation of AFB1, and delays degenerative neurological diseases	Protects cells from damage
Protective mechanism of quercetin	Inhibits the production of ROS and increases cellular activity	Inhibits ROS production, antioxidant activity, and reduces ER256 levels	Increases GSH levels, competes with AFB1 for binding sites, increases glutathione peroxidase levels, increases oxidative dismutase activity, and reduces lipid peroxidation reaction	Activation of PI3K/AKT signaling pathway and reduction of ROS levels
References	[24,25,35]	[54–57]	[37,39,40,42,44]	[50,58]

3.1.1. Prevention of Deoxynivalenol (DON) Poisoning

Fusarium spp. can cause significant yield losses, of which deterioration in quality through contamination of grains with fungal toxins, such as trichothecenes, is an important category of fungal toxin contamination and *Fusarium* spp. includes a wide variety of fungal pathogens [59]. Among the Fusarium toxins, DON poses the greatest risk to humans and livestock and is generally referred to as a vomitoxin [59,60]. The toxic effects of DON arise mainly via interference with the gastrointestinal and immune systems. Studies have shown that low doses of DON can only cause the irritation of the gastrointestinal tract and do not bring about significant clinical symptoms, whereas high doses of DON cause vomiting and esophageal gastric ulcers [61]. The toxicity of DON goes beyond this, as it also inhibits protein synthesis, and reduces tryptophan uptake by the brain and the neurotoxicity of DON in microglia [62,63]. Currently, the main method used to inhibit Fusarium is mainly through azole insecticides, but the disadvantages of this insecticide are also obvious; not only is the effect not very significant because of drug resistance and other reasons but also due to the long-term repeated use of dispersion and persistence in the environment [64–66]. Therefore, quercetin, which is not only novel and environmentally friendly but also biologically active, came to our awareness on account of its mechanisms of defense against abiotic and biotic stresses and environmental interactions [54,67,68]. It was shown that cells pretreated with quercetin (1 mM) showed better resistance to DON-induced lipid peroxidation, loss of mitochondrial membrane potential, ROS generation, cell cycle arrest, down-regulation of neuronal biomarkers, and DNA damage [55]. In a study by Yang et al. [56], it was shown that 15-acetyl deoxynivalenol (15ADON) accounts for a large proportion of the DON family and co-exists with the prototype DON. The results indicated that quercetin treatment largely restored the increased ROS levels and inhibited their growth rate. The results showed that quercetin treatment largely reduced ROS levels and inhibited their growth. Furthermore, Pomothy et al. studied the effects of quercetin on DON-exposed porcine intestinal epithelial cell lines [69]. Metabolomic analysis showed that quercetin plays an important role in the cytotoxic effects and protective functions induced by DON and 15ADON.

3.1.2. Aflatoxin (AFT) Poisoning

Aspergillus flavus is a saprophytic filamentous fungus that contaminates both pre-harvest and post-harvest seed crops, producing the carcinogenic secondary metabolite aflatoxin [70], which is a mutagenic, teratogenic, and carcinogenic toxin in animals and humans [57,71,72]. To prevent the adverse effects produced by *Aspergillus flavus* and its metabolites, the addition of mold inhibitors is one of the important measures used, among which quercetin, a natural mold inhibitor, has become the primary choice due to its low impact on the environment and animal organism [73]. It has been shown that although quercetin does not affect phosphatidylserine externalization and nucleation in *Aspergillus flavus*, it reduces ROS levels and regulates the expression of development-related genes and AFT production-related genes to inhibit proliferation and AFT biosynthesis [74]. In a study by Siess et al. [75], the mechanism of anti-AFB1 initiation effect of quercetin was expounded, which included the enhancement of enzymes involved in AFB1 detoxification (GSH S-transferase, UDP-glucuronosyltransferase), the increased formation of AFB1-GSH conjugates, and the inhibition of AFB1 binding to DNA. Aflatoxin B1 (AFB1) accounts for a high proportion in *A. flavus* and AFB1 is the most toxic, and its toxic effects specifically include the induction of oxidative stress pathways, increase in lipid peroxidation, and decrease in the level of antioxidant enzymes [76]. Not only that, Tan et al. [77] confirmed that AFB1 binds to human serum albumin (HSA) with high affinity, but it was found that quercetin competes with AFB1 to bind HSA, and the binding constant of quercetin-HAS complex is significantly higher than that of the AFB1-HAS complex, and quercetin is able to remove part of HAS from AFB1 and reduce its binding fraction. Numerous studies have investigated the toxicity of AFB1 exposure in animals during embryonic life to their central and peripheral nervous systems. Previous research has implied that the exposure of AFB1 to rat offspring prenatally delays the development of motor activity, motor coordination, exploratory behavior, reflex responses, and learning abilities [78,79]. With increasing concentrations of exposed AFB1, histopathological alterations in the brain of rats may include the dilatation of the lateral ventricles, the contraction of white and grey matter, and the depletion of nerve fibers in the spinal cord [80]. Furthermore, a study by Gugliandolo et al. demonstrated that oral supplementation with quercetin increased

SOD activity, glutathione peroxidase, and catalase levels in the brain and reduced lipid peroxidation in AFB1-treated mice. This was confirmed in a study by Choi [81], in which the effect of quercetin on AFB1-treated HepG2 cells was evaluated and the results showed that quercetin not only inhibited lipid peroxidation and promoted antioxidant defense systems but also suppressed ROS and cytotoxicity. Furthermore, a study by Ghadiri et al. [82] indicated that quercetin exerts its beneficial effects by inhibiting the biotransformation of AFB1 and counteracting its pro-oxidant effects. In addition, studies by Buening et al. and Guengerich et al. [83,84] suggested that quercetin inhibited cytochrome c (P-450) reductase in human liver microsomes in vitro sex-specific liver experiments.

3.1.3. Preventing of Ochratoxin A (OTA) Poisoning

Ochratoxins are mainly produced by *Aspergillus* and *Penicillium* [85]. Studies have shown that OTA is one of the most important fungal toxin contaminants in a wide range of foods produced by molds of the genera *Penicillium* and *Aspergillus*. It has hepatotoxic, nephrotoxic, teratogenic, and immunosuppressive effects [86]. Studies have found that quercetin has inhibitory and mitigating effects on OTA-induced immunotoxicity [87]. Due to its widespread nature, OTA is present in various animal tissues, even in human blood and breast milk, adversely affecting the safety of humans [88]. OTA has been shown to disrupt the immune system by inhibiting the lipogenic component of bone marrow mesenchymal stem cells [89]. In contrast, the inhibitory effect of quercetin on adipocyte differentiation is dose-dependent and is accompanied by a decrease in ROS production [90]. A study by Abdelrahman et al. similarly confirmed that quercetin can attenuate OTA-induced immunotoxicity by activating the PI3K/AKT signaling pathway and ameliorating oxidative stress [87]. Furthermore, in a study by Romero et al. [91], by adding quercetin at concentrations of 250 mg/L and 500 mg/L sequentially to agar containing OTA, the former showed significant signs of reduction in the growth rate of OTA, while the latter completely inhibited its growth. OTA binds to plasma albumin with equally high affinity [92]. Quercetin is also known to bind to HSA [93] and reducing the binding sites of OTA on HAS accelerates its elimination and potentially decreases toxicity [94]. In the results of Poór et al. [95], among the 13 flavonoids tested in the experiment, galangin and quercetin are the most effective competitors of OTA. Furthermore, research by Ramyaa et al. [96] found that quercetin not only prevented OTA-induced apoptosis but also inhibited the activation of the caspase cascade that leads to DNA breakage.

3.1.4. Prevention of Zearalenone (ZEN) Poisoning

ZEN is a fungal toxin produced by four species of Fusarium, namely *Fusarium graminearum*, *Fusarium* spp, *Fusarium sylvatica*, and *Fusarium cereals* [97,98]. ZEN is widely present in nature and many important crops are contaminated and animal bodies are contaminated by feeding on crops containing ZEN [99]. Different studies have shown that ZEN has toxic effects both in vivo and in vitro, and in vivo the main target organ of action is the reproductive organ, which can induce endoplasmic reticulum (ER) stress-mediated apoptosis in leukemic cells [100]. It also induces cytotoxicity through the production of ROS, which leads to apoptosis by lipid peroxidation, mitochondrial pathways, and DNA damage [101–103]. ER plays a crucial role in multicellular organism protein processes, including protein transport, protein folding, and intracellular calcium regulation. In contrast, ZEN leads to ER stress due to the disruption of the redox state, the accumulation of unfolded proteins and protein transport, which then further triggers the unfolded protein response [104]. The unfolded protein response is an adaptive response that restores ER homeostasis by activating the three proximal sensors TF6 (activating transcription factor 6), PKR-like endoplasmic reticulum kinase, and inositol-requiring enzyme 1α (IRE1α) under short and mild ER stress, but severe and prolonged ER stress activates downstream effectors, including JNK, CHOP (C/EBP homologous protein), and members of the cystathionine and Bcl2 families, leading to apoptosis [105,106]. The antioxidant properties of quercetin were shown to help combat ER stress and reduce ZEN-induced apoptosis [107].

In addition, in a study by Salem et al. [108,109], quercetin had a significant protective effect during the activation of ER stress and apoptosis by α- and β-zearalenol in HCT116 cells. Saffronin and quercetin protect HCT116 and HEK293 cells from ZEN-induced apoptosis by reducing ER stress.

3.2. *Preventing Pesticide Poisoning*

Pesticides are mostly used to control pests that harm plants, animals, and humans, and their application may cause harm to humans through occupational exposure or ingestion of contaminated food and water, which can result in a range of injuries, such as cardiotoxicity, once the pesticide enters the body [110]. Quercetin is one of the most effective drugs to improve cardiotoxic diseases caused by pesticides [110]. The harmful effects of pesticides, especially traditional pesticides, on pollinators and their residues in floral fragrances are of wide concern [111]. Conservative estimates have implied that approximately 258,234 people die globally each year from pesticide self-poisoning [112].

3.2.1. Preventing Imidacloprid Poisoning

Imidacloprid, as a systemic insecticide belonging to the neonicotinoid family of insecticides, has been one of the main products used to control agricultural pests since its introduction in 1991 [113]. Imidacloprid can act as an agonist of acetylcholine receptors interfering with neuronal signaling and can cause paralysis, hyperstimulation, and death [114]. A study by Liu et al. found that the lifespan of *Apis cerana* workers exposed to long-term imidacloprid is significantly shortened, while the treatment of the aforementioned *Apis cerana* workers with appropriate amounts of quercetin prolonged their lifespan [115]. Imidacloprid exposure caused ROS accumulation, blocked the activity of antioxidant enzymes, and enhanced mitochondrial apoptosis, which can stimulate oxidative stress pathways to trigger apoptosis in grass carp hepatocytes. In contrast, quercetin counteracts these effects through the PTEN/PI3K/AKT pathway [116]. Ardalani et al. [117] used quercetin as one of the most abundant phytochemicals in plants to assess the up-regulatory effect of quercetin on the function of the honeybee interpretation system. The results indicated that the intake of quercetin led to a significant decrease in the concentration of imidacloprid in honeybees.

3.2.2. Preventing Organophosphorus Pesticide Poisoning

Because the mammalian gestation process consists of two critical periods, the first of which is the formation of organs, organophosphorus pesticide poisoning is extremely damaging to the mammalian fetal brain. Among other things, the process of rapid brain growth is at this stage [118,119]. As a result, organophosphorus interferes with the growth and differentiation of many tissues (especially the brain) during fetal life, which thus affects cell proliferation and leads to brain dysfunction. Studies have shown that even low doses of organophosphorus can cause biochemical and neurobehavioral abnormalities. In contrast, a combined treatment with quercetin improved organophosphorus-induced developmental neurotoxicity by inhibiting oxidative stress and neurotransmission disturbances that can promote cellular redox status [120].

3.3. *Preventing Heavy Metal Poisoning*

Heavy metals cause the oxidative deterioration of biomolecules by initiating free radical-mediated chain reactions that lead to protein oxidation and the oxidation of DNA and RNA, lipid peroxidation, and quercetin as an antioxidant for heavy metal poisoning diseases intervenes both directly and indirectly. This not only directly chelates toxic metals and quenches free radicals to affect biological systems but also indirectly regenerates endogenous antioxidants to enhance cellular antioxidant defense mechanisms [121].

3.3.1. Preventing Cadmium (Cd)-Induced Toxic Diseases

Cd is accumulative in the animal organism and is difficult to excrete after entering the organism, and its toxic effects are mainly exerted through oxidative stress causing slow growth and liver and kidney dysfunction [122]. The main mechanism of Cd-induced cytotoxicity is the regulation of cellular redox status by briefly increasing ROS levels and mitochondrial damage [123,124]. The application of Cd induces an increase in plasma-labeled enzyme activity, and although quercetin does not inhibit this process, it weakens the oxidative damage induced by Cd [125,126]. Since Cd-induced dysfunction relies mainly on oxidative stress, quercetin is an effective chelator of oxygen radicals and metals. Previous research showed that quercetin treatment can prevent chronic Cd-induced oxidative stress and renal tubular damage. In addition, quercetin has shown its efficient action in the reproductive system of animals affected by the heavy metal Cd, restoring the activity of germ cells and improving the survival of follicles damaged by Cd [127]. The above studies have verified the inhibitory effects of quercetin on Cd-induced oxidative stress and other toxic functions, and further studies have shown that the combined treatment with α-tocopherol (AT) and quercetin provided more significant protection against Cd-induced oxidative stress and lipid metabolism compared with quercetin alone, thereby reducing Cd-induced cardiovascular disease [128].

3.3.2. Preventing Iron-Induced Toxic Diseases by Quercetin

Iron is an essential nutrient for the body and its deficiency affects many normal physiological functions; however, iron overload has been found to be associated with various human diseases, with diabetes being the most typical and the fastest growing worldwide [129,130]. The number of people with diabetes is increasing every year, with type 2 diabetes mellitus accounting for most cases of diabetes (about 90%), which places a heavy burden on the world economy and health systems [131]. Abnormal iron status was observed in many cross-sections in type 2 diabetes [132], and iron deposition in the islets is higher in the presence of type 2 diabetes [133]. In addition, iron sagging is a newly discovered form of regulated cell death characterized by irreparable lipid peroxidation due to overproduction of ROS in an iron-dependent manner, which has been shown to play a fundamental pathological role in various diseases associated with iron overload or dysfunction [134,135]. In contrast, there is a potential beneficial effect of quercetin on iron sagging caused by iron overload and consequently a range of diseases [136]. Lesjak [137] confirmed that quercetin can regulate iron homeostasis in the presence of iron overload. The unique properties of novel iron oxide nanoparticles (IONPs) as targeting carriers have been reported to make them suitable biomaterials for medical applications [138]. Although the advent of IONPs has provided additional convenience to medical technology, in vivo and in vitro studies have provided evidence of the possible neurotoxicity of IONPs due to free iron accumulation, ROS production, and protein aggregation [139–141]. Quercetin counteracts iron loading through iron chelating activity, iron homeostasis gene regulation, the attenuation of the Fenton/Haber–Weiss reaction and free radical scavenging, and the inhibition of protein aggregation [142].

3.3.3. Preventing Lead-Induced Toxic Diseases

When Pb enters the body, it harms several systems, such as neurohematopoietic, digestive and renal, and cardiovascular and endocrine systems, inducing cognitive impairment and neuronal degeneration [143]. Most of the common cases of lead poisoning today involve mild chronic lead poisoning, and the main lesions are the effects of lead on metal ions and enzyme systems in the body, which cause plant nervous disorders, anemia, and immune deficiency. One study administered lead acetate intravenously to mice immediately following daily intravenous injections and proved that quercetin reduced the oxidative burden on the brain induced by lead poisoning and inhibited Pb-induced neurotoxicity in a dose-dependent manner, thereby maintaining normal physiological function in lead-poisoned mice [144,145].

4. Summary and Prospects

Quercetin has shown good therapeutic activity against various diseases in previous research and practice. Its powerful oxidative properties and biological activity hold great promise for clinical applications: quercetin is a natural and safe antioxidant with minimal side effects, which can be widely applied in medicine and animal feed. Quercetin exerts a good blocking effect on various toxic diseases discussed in this review and is expected to be a new drug that can prevent and treat various toxic diseases.

In addition, the therapeutic potential of quercetin against COVID-19 was found in studies by Ruben et al. and Derosa et al. [146,147]. Quercetin has shown broad antiviral properties that interfere with multiple steps of pathogen virulence-viral entry, viral assistance, and protein assembly was found to potentiate these therapeutic effects by combining with vitamin C [146].

For autoimmune diseases, which are inherently incurable systemic diseases such that patients are unable to eradicate the disease despite long-term treatment, quercetin was predicted to afford a potential opportunity and complement to the treatment and prevention of autoimmune diseases in a study by Shen et al. [148].

Among the impeding effects of quercetin on cancers, ovarian cancer is the dominant gynecological tumor due to its serious threat to women with multiple molecular mechanisms of etiology, specifically inflammation, oxidative stress, and DNA damage. A study by Tang and Shafabakhsh et al. [149,150] confirmed that quercetin can exert anti-cancer effects by inhibiting cell proliferation, promoting apoptosis, altering cell cycle progression, and affecting autophagy.

Work related to quercetin and its application in disease treatment or therapy remains sparse, and most research therein focuses on non-modified quercetin (agarose form), but the effects of the basic form of quercetin (dietary quercetin glycosides) and its effects remain to be investigated.

Author Contributions: Writing—original draft preparation, W.Q. (Weidong Qi); writing—review and editing, W.Q. (Wanxiang Qi), D.X. and M.L.; supervision, M.L. All authors have read and agreed to the published version of the manuscript.

Funding: This research was funded by the National Natural Science Foundation of China (Grant Nos.32273074, 31972746, 31772809, and 31872538) and through a Key Grant Project of Liaoning Provincial Department of Education (Grant No. LJKZ0632).

Conflicts of Interest: The authors declare no conflict of interest.

References

1. Jafarinia, M.; Sadat Hosseini, M.; Kasiri, N.; Fazel, N.; Fathi, F.; Ganjalikhani Hakemi, M.; Eskandari, N. Quercetin with the potential effect on allergic diseases. *Allergy Asthma Clin. Immunol.* **2020**, *16*, 1–11. [CrossRef] [PubMed]
2. Li, Y.; Yao, J.; Han, C.; Yang, J.; Chaudhry, M.T.; Wang, S.; Liu, H.; Yin, Y. Quercetin, inflammation and immunity. *Nutrients* **2016**, *8*, 167. [CrossRef] [PubMed]
3. Yang, D.; Wang, T.; Long, M.; Li, P. Quercetin: Its main pharmacological activity and potential application in clinical medicine. *Oxid. Med. Cell. Longev.* **2020**, *2020*, 8825387. [CrossRef] [PubMed]
4. Fischer, C.; Speth, V.; Fleig-Eberenz, S.; Neuhaus, G. Induction of zygotic polyembryos in wheat: Influence of auxin polar transport. *Plant Cell* **1997**, *9*, 1767–1780. [CrossRef]
5. Nakamura, T.; Kinjo, C.; Nakamura, Y.; Kato, Y.; Nishikawa, M.; Hamada, M.; Nakajima, N.; Ikushiro, S.; Murota, K. Lymphatic metabolites of quercetin after intestinal administration of quercetin-3-glucoside and its aglycone in rats. *Arch. Biochem. Biophys.* **2018**, *645*, 126–136. [CrossRef]
6. Lu, C.; Huang, F.; Li, Z.; Ma, J.; Li, H.; Fang, L. Synthesis and bioactivity of quercetin aspirinates. *Bull. Korean Chem. Soc.* **2014**, *35*, 518–520. [CrossRef]
7. Boots, A.W.; Haenen, G.R.; Bast, A. Health effects of quercetin: From antioxidant to nutraceutical. *Eur. J. Pharmacol.* **2008**, *585*, 325–337. [CrossRef]
8. Ullah, F.; Iqbal, N.; Ayaz, M.; Sadiq, A.; Ullah, I.; Ahmad, S.; Imran, M. DPPH, ABTS free radical scavenging, antibacterial and phytochemical evaluation of crude methanolic extract and subsequent fractions of Chenopodium botrys aerial parts. *Pak. J. Pharm. Sci.* **2017**, *30*, 761–766.

9. Ghosh, N.; Chakraborty, T.; Mallick, S.; Mana, S.; Singha, D.; Ghosh, B.; Roy, S. Synthesis, characterization and study of antioxidant activity of quercetin-magnesium complex. *Spectrochim. Acta A Mol. Biomol. Spectrosc.* **2015**, *151*, 807–813. [CrossRef]
10. Granado-Serrano, A.B.; Martin, M.A.; Bravo, L.; Goya, L.; Ramos, S. Quercetin modulates Nrf2 and glutathione-related defenses in HepG2 cells: Involvement of p38. *Chem. Biol. Interact.* **2012**, *195*, 154–164. [CrossRef]
11. Kinaci, M.K.; Erkasap, N.; Kucuk, A.; Koken, T.; Tosun, M. Effects of quercetin on apoptosis, NF-kappaB and NOS gene expression in renal ischemia/reperfusion injury. *Exp. Ther. Med.* **2012**, *3*, 249–254. [CrossRef] [PubMed]
12. Gao, W.; Pu, L.; Chen, M.; Wei, J.; Xin, Z.; Wang, Y.; Yao, Z.; Shi, T.; Guo, C. Glutathione homeostasis is significantly altered by quercetin via the Keap1/Nrf2 and MAPK signaling pathways in rats. *J. Clin. Biochem. Nutr.* **2018**, *62*, 56–62. [CrossRef] [PubMed]
13. Hung, C.H.; Chan, S.H.; Chu, P.M.; Tsai, K.L. Quercetin is a potent anti-atherosclerotic compound by activation of SIRT1 signaling under oxLDL stimulation. *Mol. Nutr. Food Res.* **2015**, *59*, 1905–1917. [CrossRef] [PubMed]
14. Fernandez-Sanchez, A.; Madrigal-Santillan, E.; Bautista, M.; Esquivel-Soto, J.; Morales-Gonzalez, A.; Esquivel-Chirino, C.; Durante-Montiel, I.; Sanchez-Rivera, G.; Valadez-Vega, C.; Morales-Gonzalez, J.A. Inflammation, oxidative stress, and obesity. *Int. J. Mol. Sci.* **2011**, *12*, 3117–3132. [CrossRef]
15. Kawamura, K.; Qi, F.; Kobayashi, J. Potential relationship between the biological effects of low-dose irradiation and mitochondrial ROS production. *J. Radiat. Res.* **2018**, *59*, i91–i97. [CrossRef]
16. Zhao, Y.; Hu, X.; Liu, Y.; Dong, S.; Wen, Z.; He, W.; Zhang, S.; Huang, Q.; Shi, M. ROS signaling under metabolic stress: Cross-talk between AMPK and AKT pathway. *Mol. Cancer* **2017**, *16*, 79. [CrossRef]
17. Jalmi, S.K.; Sinha, A.K. ROS mediated MAPK signaling in abiotic and biotic stress- striking similarities and differences. *Front. Plant Sci.* **2015**, *6*, 769. [CrossRef]
18. Stoiber, W.; Obermayer, A.; Steinbacher, P.; Krautgartner, W.D. The role of reactive oxygen species (ROS) in the formation of extracellular traps (ETs) in humans. *Biomolecules* **2015**, *5*, 702–723. [CrossRef]
19. Vurusaner, B.; Poli, G.; Basaga, H. Tumor suppressor genes and ROS: Complex networks of interactions. *Free Radic. Biol. Med.* **2012**, *52*, 7–18. [CrossRef]
20. Song, Y.; Liu, J.; Zhang, F.; Zhang, J.; Shi, T.; Zeng, Z. Antioxidant effect of quercetin against acute spinal cord injury in rats and its correlation with the p38MAPK/iNOS signaling pathway. *Life Sci.* **2013**, *92*, 1215–1221. [CrossRef]
21. Chen, H.; Lu, C.; Liu, H.; Wang, M.; Zhao, H.; Yan, Y.; Han, L. Quercetin ameliorates imiquimod-induced psoriasis-like skin inflammation in mice via the NF-kappaB pathway. *Int. Immunopharmacol.* **2017**, *48*, 110–117. [CrossRef] [PubMed]
22. Lu, X.L.; Zhao, C.H.; Yao, X.L.; Zhang, H. Quercetin attenuates high fructose feeding-induced atherosclerosis by suppressing inflammation and apoptosis via ROS-regulated PI3K/AKT signaling pathway. *Biomed. Pharmacother.* **2017**, *85*, 658–671. [CrossRef] [PubMed]
23. Chang, H.C.; Yang, Y.R.; Wang, P.S.; Wang, R.Y. Quercetin enhances exercise-mediated neuroprotective effects in brain ischemic rats. *Med. Sci. Sports Exerc.* **2014**, *46*, 1908–1916. [CrossRef] [PubMed]
24. Du, L.; Hao, M.; Li, C.; Wu, W.; Wang, W.; Ma, Z.; Yang, T.; Zhang, N.; Isaac, A.T.; Zhu, X.; et al. Quercetin inhibited epithelial mesenchymal transition in diabetic rats, high-glucose-cultured lens, and SRA01/04 cells through transforming growth factor-beta2/phosphoinositide 3-kinase/Akt pathway. *Mol. Cell. Endocrinol.* **2017**, *452*, 44–56. [CrossRef] [PubMed]
25. Xu, D.; Hu, M.J.; Wang, Y.Q.; Cui, Y.L. Antioxidant activities of quercetin and its complexes for medicinal application. *Molecules* **2019**, *24*, 1123. [CrossRef]
26. Casagrande, R.; Georgetti, S.R.; Verri, W.J.; Dorta, D.J.; Dos, S.A.; Fonseca, M.J. Protective effect of topical formulations containing quercetin against UVB-induced oxidative stress in hairless mice. *J. Photochem. Photobiol. B* **2006**, *84*, 21–27. [CrossRef]
27. Zhu, X.; Li, N.; Wang, Y.; Ding, L.; Chen, H.; Yu, Y.; Shi, X. Protective effects of quercetin on UVB irradiationinduced cytotoxicity through ROS clearance in keratinocyte cells. *Oncol. Rep.* **2017**, *37*, 209–218. [CrossRef]
28. Jin, X.; Su, R.; Li, R.; Song, L.; Chen, M.; Cheng, L.; Li, Z. Amelioration of particulate matter-induced oxidative damage by vitamin c and quercetin in human bronchial epithelial cells. *Chemosphere* **2016**, *144*, 459–466. [CrossRef]
29. Kalantari, H.; Foruozandeh, H.; Khodayar, M.J.; Siahpoosh, A.; Saki, N.; Kheradmand, P. Antioxidant and hepatoprotective effects of Capparis spinosa L. Fractions and Quercetin on tert-butyl hydroperoxide- induced acute liver damage in mice. *J. Tradit. Complement. Med.* **2018**, *8*, 120–127. [CrossRef]
30. Prasad, J.; Baitharu, I.; Sharma, A.K.; Dutta, R.; Prasad, D.; Singh, S.B. Quercetin reverses hypobaric hypoxia-induced hippocampal neurodegeneration and improves memory function in the rat. *High Alt. Med. Biol.* **2013**, *14*, 383–394. [CrossRef]
31. Patil, S.L.; Rao, N.B.; Somashekarappa, H.M.; Rajashekhar, K.P. Antigenotoxic potential of rutin and quercetin in Swiss mice exposed to gamma radiation. *Biomed. J.* **2014**, *37*, 305–313. [CrossRef] [PubMed]
32. Ozyurt, H.; Cevik, O.; Ozgen, Z.; Ozden, A.S.; Cadirci, S.; Elmas, M.A.; Ercan, F.; Goren, M.Z.; Sener, G. Quercetin protects radiation-induced DNA damage and apoptosis in kidney and bladder tissues of rats. *Free Radic. Res.* **2014**, *48*, 1247–1255. [CrossRef] [PubMed]
33. Patil, S.L.; Mallaiah, S.H.; Patil, R.K. Antioxidative and radioprotective potential of rutin and quercetin in Swiss albino mice exposed to gamma radiation. *J. Med. Phys.* **2013**, *38*, 87–92. [CrossRef] [PubMed]
34. Kale, A.; Piskin, O.; Bas, Y.; Aydin, B.G.; Can, M.; Elmas, O.; Buyukuysal, C. Neuroprotective effects of Quercetin on radiation-induced brain injury in rats. *J. Radiat. Res.* **2018**, *59*, 404–410. [CrossRef] [PubMed]
35. Ola, M.S.; Ahmed, M.M.; Shams, S.; Al-Rejaie, S.S. Neuroprotective effects of quercetin in diabetic rat retina. *Saudi J. Biol. Sci.* **2017**, *24*, 1186–1194. [CrossRef] [PubMed]

36. Heo, H.J.; Lee, C.Y. Protective effects of quercetin and vitamin C against oxidative stress-induced neurodegeneration. *J. Agric. Food Chem.* **2004**, *52*, 7514–7517. [CrossRef]
37. Zerin, T.; Kim, Y.S.; Hong, S.Y.; Song, H.Y. Quercetin reduces oxidative damage induced by paraquat via modulating expression of antioxidant genes in A549 cells. *J. Appl. Toxicol.* **2013**, *33*, 1460–1467. [CrossRef]
38. Tvrda, E.; Tusimova, E.; Kovacik, A.; Paal, D.; Libova, L.; Lukac, N. Protective effects of quercetin on selected oxidative biomarkers in bovine spermatozoa subjected to ferrous ascorbate. *Reprod. Domest. Anim.* **2016**, *51*, 524–537. [CrossRef]
39. Hu, X.T.; Ding, C.; Zhou, N.; Xu, C. Quercetin protects gastric epithelial cell from oxidative damage in vitro and in vivo. *Eur. J. Pharmacol.* **2015**, *754*, 115–124. [CrossRef]
40. Ademosun, A.O.; Oboh, G.; Bello, F.; Ayeni, P.O. Antioxidative properties and effect of quercetin and its glycosylated form (Rutin) on acetylcholinesterase and butyrylcholinesterase activities. *J. Evid. Based Complement. Altern. Med.* **2016**, *21*, P11–P17. [CrossRef]
41. Odbayar, T.O.; Kimura, T.; Tsushida, T.; Ide, T. Isoenzyme-specific up-regulation of glutathione transferase and aldo-keto reductase mRNA expression by dietary quercetin in rat liver. *Mol. Cell. Biochem.* **2009**, *325*, 121–130. [CrossRef] [PubMed]
42. Chen, B.H.; Park, J.H.; Ahn, J.H.; Cho, J.H.; Kim, I.H.; Lee, J.C.; Won, M.H.; Lee, C.H.; Hwang, I.K.; Kim, J.D.; et al. Pretreated quercetin protects gerbil hippocampal CA1 pyramidal neurons from transient cerebral ischemic injury by increasing the expression of antioxidant enzymes. *Neural Regen. Res.* **2017**, *12*, 220–227. [PubMed]
43. Liu, C.J.; Yao, L.; Hu, Y.M.; Zhao, B.T. Effect of Quercetin-Loaded mesoporous silica nanoparticles on myocardial Ischemia-Reperfusion injury in rats and its mechanism. *Int. J. Nanomed.* **2021**, *16*, 741–752. [CrossRef] [PubMed]
44. Wang, S.; Yao, J.; Zhou, B.; Yang, J.; Chaudry, M.T.; Wang, M.; Xiao, F.; Li, Y.; Yin, W. Bacteriostatic effect of quercetin as an antibiotic alternative in vivo and its antibacterial mechanism in vitro. *J. Food Prot.* **2018**, *81*, 68–78. [CrossRef] [PubMed]
45. Plaper, A.; Golob, M.; Hafner, I.; Oblak, M.; Solmajer, T.; Jerala, R. Characterization of quercetin binding site on DNA gyrase. *Biochem. Biophys. Res. Commun.* **2003**, *306*, 530–536. [CrossRef]
46. Wang, L.; Li, B.; Si, X.; Liu, X.; Deng, X.; Niu, X.; Jin, Y.; Wang, D.; Wang, J. Quercetin protects rats from catheter-related *Staphylococcus aureus* infections by inhibiting coagulase activity. *J. Cell. Mol. Med.* **2019**, *23*, 4808–4818. [CrossRef]
47. Zhao, Y.; Chen, M.; Zhao, Z.; Yu, S. The antibiotic activity and mechanisms of sugarcane (*Saccharum officinarum* L.) bagasse extract against food-borne pathogens. *Food Chem.* **2015**, *185*, 112–118. [CrossRef]
48. Liang, Y.; Xu, K.; Zhang, P.; Zhang, J.; Chen, P.; He, J.; Fang, Y.; Zhou, Y.; Wang, J.; Bai, J. Quercetin reduces tendon adhesion in rat through suppression of oxidative stress. *BMC Musculoskelet Disord.* **2020**, *21*, 608. [CrossRef]
49. Qayyum, S.; Sharma, D.; Bisht, D.; Khan, A.U. Identification of factors involved in Enterococcus faecalis biofilm under quercetin stress. *Microb. Pathog.* **2019**, *126*, 205–211. [CrossRef]
50. Lee, J.H.; Park, J.H.; Cho, H.S.; Joo, S.W.; Cho, M.H.; Lee, J. Anti-biofilm activities of quercetin and tannic acid against *Staphylococcus aureus*. *Biofouling* **2013**, *29*, 491–499. [CrossRef]
51. Torovic, L.; Dimitrov, N.; Assuncao, R.; Alvito, P. Risk assessment of patulin intake through apple-based food by infants and preschool children in Serbia. *Food Addit. Contam. Part A* **2017**, *34*, 2023–2032. [CrossRef] [PubMed]
52. Jennessen, J.; Nielsen, K.F.; Houbraken, J.; Lyhne, E.K.; Schnurer, J.; Frisvad, J.C.; Samson, R.A. Secondary metabolite and mycotoxin production by the *Rhizopus microsporus* group. *J. Agric. Food Chem.* **2005**, *53*, 1833–1840. [CrossRef] [PubMed]
53. Pinotti, L.; Ottoboni, M.; Giromini, C.; Dell'Orto, V.; Cheli, F. Mycotoxin contamination in the EU feed supply chain: A focus on cereal byproducts. *Toxins* **2016**, *8*, 45. [CrossRef] [PubMed]
54. Tohge, T.; Watanabe, M.; Hoefgen, R.; Fernie, A.R. Shikimate and phenylalanine biosynthesis in the green lineage. *Front. Plant Sci.* **2013**, *4*, 62. [CrossRef]
55. Kalagatur, N.K.; Abd, A.E.; Poda, S.; Kadirvelu, K.; Hashem, A.; Mudili, V.; Siddaiah, C. Quercetin mitigates the deoxynivalenol mycotoxin induced apoptosis in SH-SY5Y cells by modulating the oxidative stress mediators. *Saudi J. Biol. Sci.* **2021**, *28*, 465–477. [CrossRef]
56. Yang, Y.X.; Yu, S.; Jia, B.X.; Liu, N.; Wu, A. Metabolomic profiling reveals similar cytotoxic effects and protective functions of quercetin during deoxynivalenol- and 15-acetyl deoxynivalenol-induced cell apoptosis. *Toxicol. Vitr.* **2020**, *66*, 104838. [CrossRef]
57. Alshannaq, A.F.; Gibbons, J.G.; Lee, M.K.; Han, K.H.; Hong, S.B.; Yu, J.H. Controlling aflatoxin contamination and propagation of *Aspergillus flavus* by a soy-fermenting *Aspergillus oryzae* strain. *Sci. Rep.* **2018**, *8*, 16871. [CrossRef]
58. Pestka, J.J.; Smolinski, A.T. Deoxynivalenol: Toxicology and potential effects on humans. *J. Toxicol. Env. Health B Crit. Rev.* **2005**, *8*, 39–69. [CrossRef]
59. Karlsson, I.; Friberg, H.; Kolseth, A.K.; Steinberg, C.; Persson, P. Agricultural factors affecting *Fusarium* communities in wheat kernels. *Int. J. Food Microbiol.* **2017**, *252*, 53–60. [CrossRef] [PubMed]
60. Ramana, M.V.; Balakrishna, K.; Murali, H.C.; Batra, H.V. Multiplex PCR-based strategy to detect contamination with mycotoxigenic *Fusarium* species in rice and fingermillet collected from southern India. *J. Sci. Food Agric.* **2011**, *91*, 1666–1673. [CrossRef]
61. Pestka, J.J.; Bondy, G.S. Alteration of immune function following dietary mycotoxin exposure. *Can. J. Physiol. Pharm.* **1990**, *68*, 1009–1016. [CrossRef]
62. Wang, X.; Fan, M.; Chu, X.; Zhang, Y.; Rahman, S.U.; Jiang, Y.; Chen, X.; Zhu, D.; Feng, S.; Li, Y.; et al. Deoxynivalenol induces toxicity and apoptosis in piglet hippocampal nerve cells via the MAPK signaling pathway. *Toxicon* **2018**, *155*, 1–8. [CrossRef]
63. Razafimanjato, H.; Benzaria, A.; Taieb, N.; Guo, X.J.; Vidal, N.; Di Scala, C.; Varini, K.; Maresca, M. The ribotoxin deoxynivalenol affects the viability and functions of glial cells. *Glia* **2011**, *59*, 1672–1683. [CrossRef]

64. Kahle, M.; Buerge, I.J.; Hauser, A.; Muller, M.D.; Poiger, T. Azole fungicides: Occurrence and fate in wastewater and surface waters. *Environ. Sci. Technol.* **2008**, *42*, 7193–7200. [CrossRef]
65. Rodrigues, E.T.; Lopes, I.; Pardal, M.A. Occurrence, fate and effects of azoxystrobin in aquatic ecosystems: A review. *Environ. Int.* **2013**, *53*, 18–28. [CrossRef] [PubMed]
66. Bollmann, U.E.; Tang, C.; Eriksson, E.; Jonsson, K.; Vollertsen, J.; Bester, K. Biocides in urban wastewater treatment plant influent at dry and wet weather: Concentrations, mass flows and possible sources. *Water Res.* **2014**, *60*, 64–74. [CrossRef] [PubMed]
67. Boutigny, A.L.; Barreau, C.; Atanasova-Penichon, V.; Verdal-Bonnin, M.N.; Pinson-Gadais, L.; Richard-Forget, F. Ferulic acid, an efficient inhibitor of type B trichothecene biosynthesis and Tri gene expression in Fusarium liquid cultures. *Mycol. Res.* **2009**, *113*, 746–753. [CrossRef] [PubMed]
68. Maeda, H.; Dudareva, N. The shikimate pathway and aromatic amino Acid biosynthesis in plants. *Annu. Rev. Plant Biol.* **2012**, *63*, 73–105. [CrossRef]
69. Pomothy, J.M.; Gatt, K.; Jerzsele, A.; Gere, E.P. The impact of quercetin on a porcine intestinal epithelial cell line exposed to deoxynivalenol. *Acta Vet. Hung.* **2021**, *68*, 380–386. [CrossRef]
70. Amaike, S.; Keller, N.P. Aspergillus flavus. *Annu. Rev. Phytopathol.* **2011**, *49*, 107–133. [CrossRef]
71. Cary, J.W.; Han, Z.; Yin, Y.; Lohmar, J.M.; Shantappa, S.; Harris-Coward, P.Y.; Mack, B.; Ehrlich, K.C.; Wei, Q.; Arroyo-Manzanares, N.; et al. Transcriptome analysis of aspergillus flavus reveals veA-Dependent regulation of secondary metabolite gene clusters, including the novel aflavarin cluster. *Eukaryot. Cell* **2015**, *14*, 983–997. [CrossRef]
72. Caceres, I.; Snini, S.P.; Puel, O.; Mathieu, F. *Streptomyces roseolus*, a Promising Biocontrol Agent Against *Aspergillus flavus*, the Main Aflatoxin B(1) Producer. *Toxins* **2018**, *10*, 442. [CrossRef] [PubMed]
73. Li, X.M.; Liu, J.; Pan, F.F.; Shi, D.D.; Wen, Z.G.; Yang, P.L. Quercetin and aconitine synergistically induces the human cervical carcinoma HeLa cell apoptosis via endoplasmic reticulum (ER) stress pathway. *PLoS ONE* **2018**, *13*, e191062. [CrossRef]
74. Li, X.M.; Li, Z.Y.; Wang, Y.D.; Wang, J.Q.; Yang, P.L. Quercetin Inhibits the Proliferation and Aflatoxins Biosynthesis of *Aspergillus flavus*. *Toxins* **2019**, *11*, 154. [CrossRef] [PubMed]
75. Siess, M.H.; Le Bon, A.M.; Canivenc-Lavier, M.C.; Suschetet, M. Mechanisms involved in the chemoprevention of flavonoids. *Biofactors* **2000**, *12*, 193–199. [CrossRef] [PubMed]
76. Gugliandolo, E.; Peritore, A.F.; D'Amico, R.; Licata, P.; Crupi, R. Evaluation of Neuroprotective Effects of Quercetin against Aflatoxin B1-Intoxicated Mice. *Anim.* **2020**, *10*, 898. [CrossRef]
77. Tan, H.; Chen, L.; Ma, L.; Liu, S.; Zhou, H.; Zhang, Y.; Guo, T.; Liu, W.; Dai, H.; Yu, Y. Fluorescence spectroscopic investigation of competitive interactions between quercetin and aflatoxin b(1) for binding to human serum albumin. *Toxins* **2019**, *11*, 214. [CrossRef]
78. Kihara, T.; Matsuo, T.; Sakamoto, M.; Yasuda, Y.; Yamamoto, Y.; Tanimura, T. Effects of prenatal aflatoxin B1 exposure on behaviors of rat offspring. *Toxicol. Sci.* **2000**, *53*, 392–399. [CrossRef]
79. Supriya, C.; Reddy, P.S. Prenatal exposure to aflatoxin B1: Developmental, behavioral, and reproductive alterations in male rats. *Naturwissenschaften* **2015**, *102*, 26. [CrossRef]
80. Wangikar, P.B.; Dwivedi, P.; Sharma, A.K.; Sinha, N. Effect in rats of simultaneous prenatal exposure to ochratoxin a and aflatoxin B1. II. Histopathological features of teratological anomalies induced in fetuses. *Birth Defects Res B Dev. Reprod. Toxicol.* **2004**, *71*, 352–358. [CrossRef]
81. Choi, K.C.; Chung, W.T.; Kwon, J.K.; Yu, J.Y.; Jang, Y.S.; Park, S.M.; Lee, S.Y.; Lee, J.C. Inhibitory effects of quercetin on aflatoxin B1-induced hepatic damage in mice. *Food Chem. Toxicol.* **2010**, *48*, 2747–2753. [CrossRef] [PubMed]
82. Ghadiri, S.; Spalenza, V.; Dellafiora, L.; Badino, P.; Barbarossa, A.; Dall'Asta, C.; Nebbia, C.; Girolami, F. Modulation of aflatoxin B1 cytotoxicity and aflatoxin M1 synthesis by natural antioxidants in a bovine mammary epithelial cell line. *Toxicol. Vitr.* **2019**, *57*, 174–183. [CrossRef] [PubMed]
83. Buening, M.K.; Chang, R.L.; Huang, M.T.; Fortner, J.G.; Wood, A.W.; Conney, A.H. Activation and inhibition of benzo(a)pyrene and aflatoxin B1 metabolism in human liver microsomes by naturally occurring flavonoids. *Cancer Res.* **1981**, *41*, 67–72.
84. Guengerich, F.P.; Kim, D.H. In vitro inhibition of dihydropyridine oxidation and aflatoxin B1 activation in human liver microsomes by naringenin and other flavonoids. *Carcinogenesis* **1990**, *11*, 2275–2279. [PubMed]
85. Zhai, S.; Zhu, Y.; Feng, P.; Li, M.; Wang, W.; Yang, L.; Yang, Y. Ochratoxin a: Its impact on poultry gut health and microbiota, an overview. *Poult Sci.* **2021**, *100*, 101037. [CrossRef] [PubMed]
86. Wang, Y.; Wang, L.; Liu, F.; Wang, Q.; Selvaraj, J.N.; Xing, F.; Zhao, Y.; Liu, Y. Ochratoxin a producing fungi, biosynthetic pathway and regulatory mechanisms. *Toxins* **2016**, *8*, 83. [CrossRef]
87. Abdelrahman, R.E.; Khalaf, A.; Elhady, M.A.; Ibrahim, M.A.; Hassanen, E.I.; Noshy, P.A. Quercetin ameliorates ochratoxin A-Induced immunotoxicity in broiler chickens by modulation of PI3K/AKT pathway. *Chem. Biol. Interact.* **2022**, *351*, 109720. [CrossRef]
88. Clark, H.A.; Snedeker, S.M. Ochratoxin a: Its cancer risk and potential for exposure. *J. Toxicol. Env. Health B Crit. Rev.* **2006**, *9*, 265–296. [CrossRef]
89. Lim, S.; Jang, H.J.; Kim, J.K.; Kim, J.M.; Park, E.H.; Yang, J.H.; Kim, Y.H.; Yea, K.; Ryu, S.H.; Suh, P.G. Ochratoxin a inhibits adipogenesis through the extracellular signal-related kinases-peroxisome proliferator-activated receptor-gamma pathway in human adipose tissue-derived mesenchymal stem cells. *Stem Cells Dev.* **2011**, *20*, 415–426. [CrossRef]

90. Dobrocsyova, V.; Krskova, K.; Capcarova, M.; Zorad, S. Modulation of Adipogenesis and Oxidative Status by Quercetin and Ochratoxin a: Positive or Negative Impact on Rat Adipocyte Metabolism? *Molecules* **2019**, *24*, 3726. [CrossRef]
91. Romero, S.M.; Alberto, M.R.; Manca, D.N.M.; Vaamonde, G. Inhibition of growth and ochratoxin a biosynthesis in *Aspergillus carbonarius* by flavonoid and nonflavonoid compounds. *Mycotoxin Res.* **2009**, *25*, 165–170. [CrossRef] [PubMed]
92. Sueck, F.; Poor, M.; Faisal, Z.; Gertzen, C.; Cramer, B.; Lemli, B.; Kunsagi-Mate, S.; Gohlke, H.; Humpf, H.U. Interaction of ochratoxin a and its thermal degradation product 2′R-Ochratoxin a with human serum albumin. *Toxins* **2018**, *10*, 256. [CrossRef] [PubMed]
93. Wani, T.A.; Bakheit, A.H.; Zargar, S.; Alanazi, Z.S.; Al-Majed, A.A. Influence of antioxidant flavonoids quercetin and rutin on the in-vitro binding of neratinib to human serum albumin. *Spectrochim. Acta A Mol. Biomol. Spectrosc.* **2021**, *246*, 118977. [CrossRef] [PubMed]
94. Zhao, Q.; Lv, Q.; Wang, H. Identification of allosteric nucleotide sites of tetramethylrhodamine-labeled aptamer for noncompetitive aptamer-based fluorescence anisotropy detection of a small molecule, ochratoxin a. *Anal. Chem.* **2014**, *86*, 1238–1245. [CrossRef]
95. Poor, M.; Kunsagi-Mate, S.; Bencsik, T.; Petrik, J.; Vladimir-Knezevic, S.; Koszegi, T. Flavonoid aglycones can compete with Ochratoxin a for human serum albumin: A new possible mode of action. *Int. J. Biol. Macromol.* **2012**, *51*, 279–283. [CrossRef]
96. Ramyaa, P.; Padma, V.V. Ochratoxin-induced toxicity, oxidative stress and apoptosis ameliorated by quercetin–modulation by Nrf2. *Food Chem. Toxicol.* **2013**, *62*, 205–216. [CrossRef]
97. Habschied, K.; Sarkanj, B.; Klapec, T.; Krstanovic, V. Distribution of zearalenone in malted barley fractions dependent on *Fusarium graminearum* growing conditions. *Food Chem.* **2011**, *129*, 329–332. [CrossRef]
98. Rodrigues, I.; Naehrer, K. A three-year survey on the worldwide occurrence of mycotoxins in feedstuffs and feed. *Toxins* **2012**, *4*, 663–675. [CrossRef]
99. Zinedine, A.; Soriano, J.M.; Molto, J.C.; Manes, J. Review on the toxicity, occurrence, metabolism, detoxification, regulations and intake of zearalenone: An oestrogenic mycotoxin. *Food Chem. Toxicol.* **2007**, *45*, 1–18. [CrossRef]
100. Banjerdpongchai, R.; Kongtawelert, P.; Khantamat, O.; Srisomsap, C.; Chokchaichamnankit, D.; Subhasitanont, P.; Svasti, J. Mitochondrial and endoplasmic reticulum stress pathways cooperate in zearalenone-induced apoptosis of human leukemic cells. *J. Hematol. Oncol.* **2010**, *3*, 50. [CrossRef]
101. Abid-Essefi, S.; Ouanes, Z.; Hassen, W.; Baudrimont, I.; Creppy, E.; Bacha, H. Cytotoxicity, inhibition of DNA and protein syntheses and oxidative damage in cultured cells exposed to zearalenone. *Toxicol. Vitr.* **2004**, *18*, 467–474. [CrossRef] [PubMed]
102. Hassen, W.; Ayed-Boussema, I.; Oscoz, A.A.; Lopez, A.C.; Bacha, H. The role of oxidative stress in zearalenone-mediated toxicity in Hep G2 cells: Oxidative DNA damage, gluthatione depletion and stress proteins induction. *Toxicology* **2007**, *232*, 294–302. [CrossRef] [PubMed]
103. Abbes, S.; Ouanes, Z.; Salah-Abbes, J.B.; Abdel-Wahhab, M.A.; Oueslati, R.; Bacha, H. Preventive role of aluminosilicate clay against induction of micronuclei and chromosome aberrations in bone-marrow cells of Balb/c mice treated with Zearalenone. *Mutat. Res.* **2007**, *631*, 85–92. [CrossRef] [PubMed]
104. Arumugam, S.; Thandavarayan, R.A.; Veeraveedu, P.T.; Ma, M.; Giridharan, V.V.; Arozal, W.; Sari, F.R.; Sukumaran, V.; Lakshmanan, A.; Soetikno, V.; et al. Modulation of endoplasmic reticulum stress and cardiomyocyte apoptosis by mulberry leaf diet in experimental autoimmune myocarditis rats. *J. Clin. Biochem. Nutr.* **2012**, *50*, 139–144. [CrossRef]
105. Ron, D.; Walter, P. Signal integration in the endoplasmic reticulum unfolded protein response. *Nat. Rev. Mol. Cell Biol.* **2007**, *8*, 519–529. [CrossRef]
106. Anjaneyulu, M.; Chopra, K. Quercetin, an anti-oxidant bioflavonoid, attenuates diabetic nephropathy in rats. *Clin. Exp. Pharm. Physiol.* **2004**, *31*, 244–248. [CrossRef]
107. Ben, S.I.; Prola, A.; Boussabbeh, M.; Guilbert, A.; Bacha, H.; Abid-Essefi, S.; Lemaire, C. Crocin and Quercetin protect HCT116 and HEK293 cells from Zearalenone-induced apoptosis by reducing endoplasmic reticulum stress. *Cell Stress Chaperones* **2015**, *20*, 927–938.
108. Ben, S.I.; Boussabbeh, M.; Prola, A.; Guilbert, A.; Bacha, H.; Lemaire, C.; Abid-Essefi, S. Crocin protects human embryonic kidney cells (HEK293) from alpha- and beta-Zearalenol-induced ER stress and apoptosis. *Env. Sci. Pollut. Res. Int.* **2016**, *23*, 15504–15514.
109. Ben, S.I.; Prola, A.; Boussabbeh, M.; Guilbert, A.; Bacha, H.; Lemaire, C.; Abid-Essefi, S. Activation of ER stress and apoptosis by alpha- and beta-zearalenol in HCT116 cells, protective role of Quercetin. *Neurotoxicology* **2016**, *53*, 334–342.
110. El-Nahhal, Y.; El-Nahhal, I. Cardiotoxicity of some pesticides and their amelioration. *Environ. Sci. Pollut. Res. Int.* **2021**, *28*, 44726–44754. [CrossRef]
111. Woodcock, B.A.; Isaac, N.J.; Bullock, J.M.; Roy, D.B.; Garthwaite, D.G.; Crowe, A.; Pywell, R.F. Impacts of neonicotinoid use on long-term population changes in wild bees in England. *Nat. Commun.* **2016**, *7*, 12459. [CrossRef] [PubMed]
112. Gunnell, D.; Eddleston, M.; Phillips, M.R.; Konradsen, F. The global distribution of fatal pesticide self-poisoning: Systematic review. *BMC Public Health* **2007**, *7*, 357. [CrossRef] [PubMed]
113. Elbert, A.; Haas, M.; Springer, B.; Thielert, W.; Nauen, R. Applied aspects of neonicotinoid uses in crop protection. *Pest. Manag. Sci.* **2008**, *64*, 1099–1105. [CrossRef] [PubMed]
114. Matsuda, K.; Buckingham, S.D.; Kleier, D.; Rauh, J.J.; Grauso, M.; Sattelle, D.B. Neonicotinoids: Insecticides acting on insect nicotinic acetylcholine receptors. *Trends Pharmacal. Sci.* **2001**, *22*, 573–580. [CrossRef]
115. Liu, J.; Li, Y.; Zhang, Z.; Luo, W.; Cao, L.; Liu, H. Low Concentration of Quercetin Reduces the Lethal and Sublethal Effects of Imidacloprid on *Apis cerana* (Hymenoptera: Apidae). *J. Econ. Entomol.* **2021**, *114*, 1053–1064. [CrossRef]

116. Miao, Z.; Miao, Z.; Wang, S.; Shi, X.; Xu, S. Quercetin antagonizes imidacloprid-induced mitochondrial apoptosis through PTEN/PI3K/AKT in grass carp hepatocytes. *Environ. Pollut.* **2021**, *290*, 118036. [CrossRef]
117. Ardalani, H.; Vidkjaer, N.H.; Laursen, B.B.; Kryger, P.; Fomsgaard, I.S. Dietary quercetin impacts the concentration of pesticides in honey bees. *Chemosphere* **2021**, *262*, 127848. [CrossRef]
118. Wasterlain, C.G.; Shirasaka, Y. Seizures, brain damage and brain development. *Brain Dev.* **1994**, *16*, 279–295. [CrossRef]
119. Braun, J.M. Early-life exposure to EDCs: Role in childhood obesity and neurodevelopment. *Nat. Rev. Endocrinol.* **2017**, *13*, 161–173. [CrossRef]
120. Ibrahim, K.A.; Eleyan, M.; Abd, E.H.; Khwanes, S.A.; Mohamed, R.A. Quercetin attenuates the oxidative Injury-Mediated upregulation of apoptotic gene expression and catecholaminergic neurotransmitters of the fetal rats' brain following prenatal exposure to fenitrothion insecticide. *Neurotox. Res.* **2020**, *37*, 871–882. [CrossRef]
121. Flora, S.J.; Shrivastava, R.; Mittal, M. Chemistry and pharmacological properties of some natural and synthetic antioxidants for heavy metal toxicity. *Curr. Med. Chem.* **2013**, *20*, 4540–4574. [CrossRef] [PubMed]
122. Thomas, M. A comparative study of the factors affecting uptake and distribution of Cd with Ni in barley. *Plant Physiol. Biochem.* **2021**, *162*, 730–736. [CrossRef]
123. Buha, A.; Matovic, V.; Antonijevic, B.; Bulat, Z.; Curcic, M.; Renieri, E.A.; Tsatsakis, A.M.; Schweitzer, A.; Wallace, D. Overview of cadmium thyroid disrupting effects and mechanisms. *Int. J. Mol. Sci.* **2018**, *19*, 1501. [CrossRef] [PubMed]
124. Teklebrhan, R.B.; Ge, L.; Bhattacharjee, S.; Xu, Z.; Sjoblom, J. Initial partition and aggregation of uncharged polyaromatic molecules at the oil-water interface: A molecular dynamics simulation study. *J. Phys. Chem. B* **2014**, *118*, 1040–1051. [CrossRef] [PubMed]
125. Vicente-Sanchez, C.; Egido, J.; Sanchez-Gonzalez, P.D.; Perez-Barriocanal, F.; Lopez-Novoa, J.M.; Morales, A.I. Effect of the flavonoid quercetin on cadmium-induced hepatotoxicity. *Food Chem. Toxicol.* **2008**, *46*, 2279–2287. [CrossRef] [PubMed]
126. Morales, A.I.; Vicente-Sanchez, C.; Sandoval, J.M.; Egido, J.; Mayoral, P.; Arevalo, M.A.; Fernandez-Tagarro, M.; Lopez-Novoa, J.M.; Perez-Barriocanal, F. Protective effect of quercetin on experimental chronic cadmium nephrotoxicity in rats is based on its antioxidant properties. *Food Chem. Toxicol.* **2006**, *44*, 2092–2100. [CrossRef]
127. Izaguirry, A.P.; Soares, M.B.; Vargas, L.M.; Spiazzi, C.C.; Dos, S.B.D.; Noremberg, S.; Mendez, A.S.; Santos, F.W. Blueberry (Vaccinium ashei Reade) extract ameliorates ovarian damage induced by subchronic cadmium exposure in mice: Potential delta-ALA-D involvement. *Environ. Toxicol.* **2017**, *32*, 188–196. [CrossRef]
128. Prabu, S.M.; Shagirtha, K.; Renugadevi, J. Amelioration of cadmium-induced oxidative stress, impairment in lipids and plasma lipoproteins by the combined treatment with quercetin and alpha-tocopherol in rats. *J. Food Sci.* **2010**, *75*, T132–T140. [CrossRef]
129. Scicchitano, P.; Cortese, F.; Gesualdo, M.; De Palo, M.; Massari, F.; Giordano, P.; Ciccone, M.M. The role of endothelial dysfunction and oxidative stress in cerebrovascular diseases. *Free Radic. Res.* **2019**, *53*, 579–595. [CrossRef]
130. Bardou-Jacquet, E.; Morcet, J.; Manet, G.; Laine, F.; Perrin, M.; Jouanolle, A.M.; Guyader, D.; Moirand, R.; Viel, J.F.; Deugnier, Y. Decreased cardiovascular and extrahepatic cancer-related mortality in treated patients with mild HFE hemochromatosis. *J. Hepatol.* **2015**, *62*, 682–689. [CrossRef]
131. Roden, M.; Shulman, G.I. The integrative biology of type 2 diabetes. *Nature* **2019**, *576*, 51–60. [CrossRef] [PubMed]
132. Zhang, C.; Rawal, S. Dietary iron intake, iron status, and gestational diabetes. *Am. J. Clin. Nutr.* **2017**, *106*, 1672S–1680S. [CrossRef]
133. Coffey, R.; Knutson, M.D. The plasma membrane metal-ion transporter ZIP14 contributes to nontransferrin-bound iron uptake by human beta-cells. *Am. J. Physiol. Cell Physiol.* **2017**, *312*, C169–C175. [CrossRef] [PubMed]
134. Strzyz, P. Iron expulsion by exosomes drives ferroptosis resistance. *Nat. Rev. Mol. Cell Biol.* **2020**, *21*, 4–5. [CrossRef] [PubMed]
135. Doll, S.; Conrad, M. Iron and ferroptosis: A still ill-defined liaison. *IUBMB Life* **2017**, *69*, 423–434. [CrossRef] [PubMed]
136. Li, D.; Jiang, C.; Mei, G.; Zhao, Y.; Chen, L.; Liu, J.; Tang, Y.; Gao, C.; Yao, P. Quercetin Alleviates Ferroptosis of Pancreatic beta Cells in Type 2 Diabetes. *Nutrients* **2020**, *12*, 2954. [CrossRef] [PubMed]
137. Lesjak, M.; Srai, S.K.S. Role of dietary flavonoids in iron homeostasis. *Pharmaceuticals* **2019**, *12*, 119. [CrossRef]
138. Kumar, P.; Agnihotri, S.; Roy, I. Preparation and characterization of superparamagnetic iron oxide nanoparticles for magnetically guided drug delivery. *Int. J. Nanomed.* **2018**, *13*, 43–46. [CrossRef]
139. Yarjanli, Z.; Ghaedi, K.; Esmaeili, A.; Rahgozar, S.; Zarrabi, A. Iron oxide nanoparticles may damage to the neural tissue through iron accumulation, oxidative stress, and protein aggregation. *BMC Neurosci.* **2017**, *18*, 51. [CrossRef]
140. Veiseh, O.; Gunn, J.W.; Zhang, M. Design and fabrication of magnetic nanoparticles for targeted drug delivery and imaging. *Adv. Drug Deliv. Rev.* **2010**, *62*, 284–304. [CrossRef]
141. Wahajuddin; Arora, S. Superparamagnetic iron oxide nanoparticles: Magnetic nanoplatforms as drug carriers. *Int. J. Nanomed.* **2012**, *7*, 3445–3471.
142. Bardestani, A.; Ebrahimpour, S.; Esmaeili, A.; Esmaeili, A. Quercetin attenuates neurotoxicity induced by iron oxide nanoparticles. *J. Nanobiotechnology* **2021**, *19*, 327. [CrossRef] [PubMed]
143. Hanna-Attisha, M.; Lanphear, B.; Landrigan, P. Lead Poisoning in the 21st Century: The Silent Epidemic Continues. *Am. J. Public Health* **2018**, *108*, 1430. [CrossRef] [PubMed]
144. Chander, K.; Vaibhav, K.; Ejaz, A.M.; Javed, H.; Tabassum, R.; Khan, A.; Kumar, M.; Katyal, A.; Islam, F.; Siddiqui, M.S. Quercetin mitigates lead acetate-induced behavioral and histological alterations via suppression of oxidative stress, Hsp-70, Bak and upregulation of Bcl-2. *Food Chem. Toxicol.* **2014**, *68*, 297–306. [CrossRef] [PubMed]

145. Liu, C.M.; Zheng, G.H.; Cheng, C.; Sun, J.M. Quercetin protects mouse brain against lead-induced neurotoxicity. *J. Agric. Food Chem.* **2013**, *61*, 7630–7635. [CrossRef]
146. Colunga, B.R.; Berrill, M.; Catravas, J.D.; Marik, P.E. Quercetin and vitamin c: An experimental, synergistic therapy for the prevention and treatment of SARS-CoV-2 related disease (COVID-19). *Front. Immunol.* **2020**, *11*, 1451. [CrossRef]
147. Derosa, G.; Maffioli, P.; D'Angelo, A.; Di Pierro, F. A role for quercetin in coronavirus disease 2019 (COVID-19). *Phytother. Res.* **2021**, *35*, 1230–1236. [CrossRef]
148. Shen, P.; Lin, W.; Deng, X.; Ba, X.; Han, L.; Chen, Z.; Qin, K.; Huang, Y.; Tu, S. Potential implications of quercetin in autoimmune diseases. *Front. Immunol.* **2021**, *12*, 689044. [CrossRef]
149. Shafabakhsh, R.; Asemi, Z. Quercetin: A natural compound for ovarian cancer treatment. *J. Ovarian Res.* **2019**, *12*, 55. [CrossRef]
150. Tang, S.M.; Deng, X.T.; Zhou, J.; Li, Q.P.; Ge, X.X.; Miao, L. Pharmacological basis and new insights of quercetin action in respect to its anti-cancer effects. *Biomed. Pharmacother.* **2020**, *121*, 109604. [CrossRef]

molecules

Article

Antioxidant and Antimicrobial Activities of Thai Edible Plant Extracts Prepared Using Different Extraction Techniques

Pimmada Junsathian [1], Soichiro Nakamura [2], Shigeru Katayama [2,*] and Saroat Rawdkuen [1,3,*]

[1] Food Science and Technology Program, School of Agro-Industry, Mae Fah Luang University, Chiang Rai 57100, Thailand
[2] Institute for Biomedical Sciences, Interdisciplinary Cluster for Cutting Edge Research, Shinshu University, 8304 Minamiminowa, Nagano 399-4598, Kamiina, Japan
[3] Unit of Innovative Food Packaging and Biomaterials, School of Agro-Industry, Mae Fah Luang University, Chiang Rai 57100, Thailand
* Correspondence: skata@shinshu-u.ac.jp (S.K.); saroat@mfu.ac.th (S.R.)

Abstract: This study investigated the antioxidant and antimicrobial activities of six Thai edible plant leaf extracts, including Cashew (CN), Chamuang (CM), Monpu (MP), Thurianthet (TT), Kradon (KD) and Pakliang (PL), extracted using ethanol extraction (EE), microwave-assisted extraction (MAE), and ultrasonic-assisted extraction (UAE). The leaf extracts were characterized for percentage yield, total phenolic content (TPC), total flavonoid content (TFC), 2,2-diphenyl-1-picrylhydrazyl (DPPH) and-ferric reducing antioxidant power (FRAP) activity, and antimicrobial activity against spoilage. MAE produced the highest percentage yields, among which MAE-extracted MP exhibited the highest yield. Furthermore, the highest TPC and TFC were obtained for MAE, with MAE-extracted KD and CN showing the highest TPC and TFC, respectively, among the samples. The highest DPPH and FRAP values were seen in MAE-processed CN, KD, and MP extracts. The inhibition zone of pathogenic bacteria, minimum inhibitory concentration, and minimum bacterial concentration were determined in all samples except TT. These findings indicate that, compared to EE and UAE, MAE improved the antioxidant and antimicrobial efficacy of the leaf extracts. The aforementioned extracts could be employed as natural food additives to prevent chemical and microbial spoilage of foods.

Keywords: antioxidative activity; antimicrobial activity; Thai edible plant; microwave-assisted extraction; ultrasonic-assisted extraction

Citation: Junsathian, P.; Nakamura, S.; Katayama, S.; Rawdkuen, S. Antioxidant and Antimicrobial Activities of Thai Edible Plant Extracts Prepared Using Different Extraction Techniques. *Molecules* **2022**, *27*, 6489. https://doi.org/10.3390/molecules27196489

Academic Editors: Raffaele Pezzani, Sara Vitalini and Vincenzo De Feo

Received: 31 August 2022
Accepted: 28 September 2022
Published: 1 October 2022

1. Introduction

Plants are widely used for dietary and medicinal purposes in several parts of the world. Bioactive compounds from various plants benefit human health by providing protection and nutrition, as well as protection against microorganisms [1,2]. Recently, many researchers have shown interest in identifying the antioxidative properties and antimicrobial efficacy of plant materials [3,4]. Bioactive compounds found in fruits and vegetables are secondary metabolites classified as a subclass of phytochemicals with biological activities, and plant polyphenols are potential antioxidants for human health [5]. Antioxidants can reduce oxidative damage by neutralizing free radicals and eliminating them from the body [6]. Several epidemiological studies have revealed that plant polyphenols reduce the risk of chronic human disease [7,8].

In addition to antioxidative activities, plant polyphenols serve as preservatives in food processing. Several studies have shown that antioxidant substances can act in various ways, including as free radical scavengers or chelators, preventing lipid oxidation and thereby preventing nutrient losses and inhibiting the potential formation of toxic compounds [9,10]. Food spoilage and food poisoning caused by the growth of pathogenic bacteria are major issues in the food industry. There is a growing interest in using natural active preservatives, particularly plant extracts, to prevent food spoilage and preserve foods without posing health risks [11].

The antioxidant and antimicrobial activities of various plant polyphenols have been studied extensively and are used in various food, cosmetic, and pharmaceutical applications [12,13]. The bioactivity of plant-derived extracts containing high amounts of polyphenols depends on their efficiency of extraction [14]. Novel extraction techniques have several benefits over conventional methods, such as excellent recovery and improved stability and quality of the extract [15]. According to Dahmoune, et al. [16], microwave-assisted extraction (MAE) and ultrasonic-assisted extraction (UAE) can increase plant polyphenol yields under controlled time and temperature conditions. UAE is a novel green chemistry method that allows for the extraction of bioactive compounds with low solvent and energy costs, while preserving the integrity of the extracted molecules [17]. MAE is a process that extracts the target compounds from various matrices using microwave energy and extraction solvents. Aqueous ethanol is more effective than absolute ethanol for the extraction of bioactive compounds because it improves the extractability of the compounds by increasing the polarity of the extraction solvent.

Thailand is home to several edible plant species that can grow and survive in tropical zones. Thai people use these plants in various traditional diets, dishes, and medicinal beverages. Six edible Thai plants, including Cashew (*Anacardium occidentale* L.) (CN), Chamuang (*Garcinia cowa* Roxb.) (CM), Monpu (*Glochidion wallichianum* Muell Arg.) (MP), Thurianthet (*Annona murica*) (TT), Kradon (*Careya sphaerica* Roxb.) (KD), and Pakliang (*Gnetum gnemon* var. Temerum) (PL), were studied to compare the efficacy of MAE, UAE, and the conventional ethanol extraction (EE), particularly in terms of enhancement of extraction yield and bioactive properties. This is the first comparative study of edible Thai plant extracts extracted using different extraction methods. The purpose of this study was to compare the efficacy of two novel extraction techniques (UAE and MAE) with that of the conventional EE technique on the total phenolic and flavonoid yield, as well as on the antioxidant and antimicrobial activities of various Thai leaf extracts.

2. Results and Discussion

2.1. Extraction Efficiency of MAE, UAE, and EE

Figure 1 depicts the extraction yield of six Thai plant extracts prepared using EE, UAE, and MAE, with 60% ethanol. The results clearly demonstrate that MAE produced the highest yield compared to the other extraction methods ($p < 0.05$). This may be because the heat generated within the materials affects the heating kinetics and increases the pressure within the cells. These effects could degrade the plant cell wall structure faster and more easily than the other methods, resulting in a higher diffusion rate of the solvent, and thereby increasing the extraction yield [18]. This indicates that MAE is more efficient than the other two methods in increasing the extraction yield. The highest percentage yield was obtained for MP extracted via the MAE method ($p < 0.05$). UAE showed lower extraction yields ($p < 0.05$) than EE and MAE. Polyphenol extraction from carob (*Ceratonia siliqua*) kibbles using a microwave-assisted technique resulted in a high yield [18].

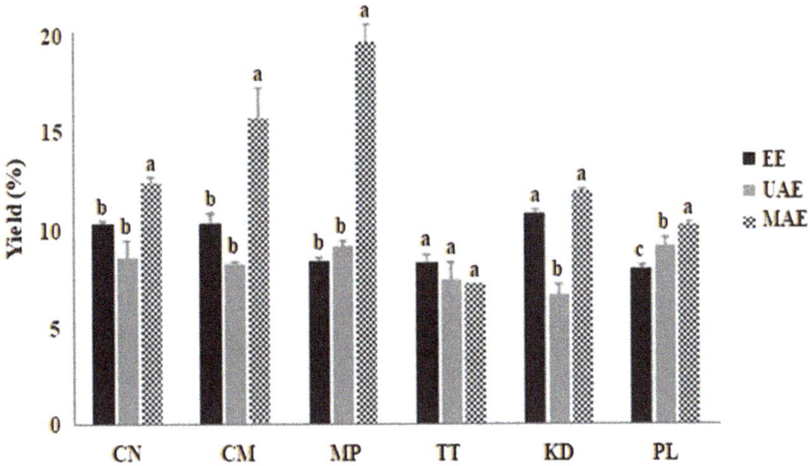

Figure 1. Percentage yield of edible Thai plant extracts by different extraction techniques. Values are presented as the mean ± standard deviation (SD) of triplicate analyses (n = 3). Different letters (a–c) denote significant differences ($p < 0.05$) between mean values of the extraction method for each plant material. CN; Cashew, CM; Chamuang, MP; Monpu, TT; Thurianthet, KD; Kradon, PL; Pakliang, EE; ethanol extraction, UAE; ultrasonic-assisted extraction, MAE; microwave assisted extraction.

2.2. Total Phenolic and Flavonoid Contents of Plant Extracts Prepared by MAE, UAE, and EE Processes

The total phenolic content (TPC) and total flavonoid content (TFC) of the six plant extracts are shown in Figure 2a. MAE produced the highest TPC for each extracted plant material, followed by UAE and EE. MAE and UAE methods were more efficient than the conventional EE ($p < 0.05$). Amongst the extracts, MAE-processed KD showed the highest TPC ($p < 0.05$). Meanwhile, the TPCs obtained by EE were the lowest among the three methods ($p < 0.05$). The CM and PL extracts had the lowest TPC, regardless of the extraction process, compared to the other plant materials. This may be due to the low polyphenol content in their plant cell matrices. MAE has been reported to increase the TPC of plant materials during the extraction process compared with ultrasonic-assisted pressurized liquid extraction and EE [19].

The MAE-extracted samples showed the highest TFC compared with the UAE- and EE-extracted samples ($p < 0.05$) (Figure 2b). Among the MAE-extracted samples, the CN, MP, and KD showed the highest amount of flavonoids (207.7 ± 1.3, 179.1 ± 1.7, and 158.6 ± 1.2 mg QUE/g dry extract, respectively). The highest TFC was observed in MAE-extracted CN, extracted using 60% ethanol. The high TFC obtained via MAE can be associated with the heat generated due to the interaction of microwaves with polar molecules within the plant cells, which creates high pressure to disrupt the cell wall. After cell wall disruption, the diffusion of solvent into plant cells leads to the extraction of higher amounts of phenolic substances [20]. The MAE technique has been documented to obtain higher levels of flavonoids and phenolics from *Herba Epimedii* and ginger [21,22]. UAE works via the cavitation effect, by creating bubbles in the extraction solvent which burst and produce high flashes of energy and thereby disrupt the cell wall of the plant material [23]. Ethanol (60%) also affects the yield of phenolic substances owing to its polarity, as well as water content, which enhances the extraction efficiency by extracting hydrophilic bioactive compounds [24]. In this study, the powder samples of leaf were used to obtain extracts using EE, UAE, and MAE systems; however, the particle size was not measured in these experiments. Further investigations will be needed to assess the extraction efficiency of different particle size samples.

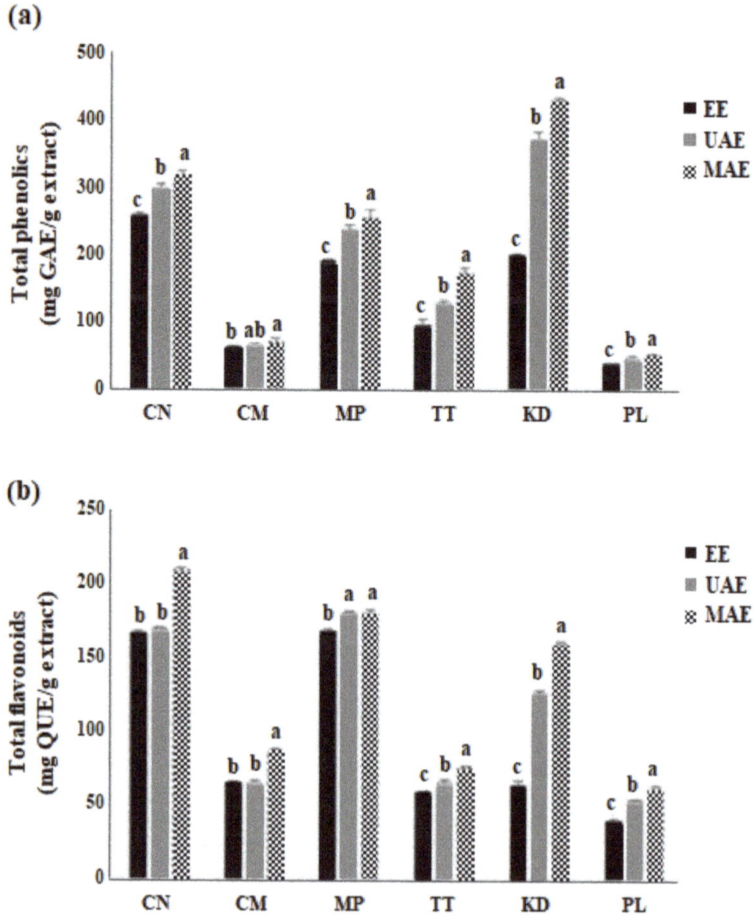

Figure 2. Total phenolic (**a**) and total flavonoid contents (**b**) of six Thai edible plant extracts extracted using EE, UAE, and MAE. The data are presented as the mean ± SD of triplicate analyses (n = 3). Different letters (a–c) denote significant differences (p < 0.05) between extraction methods for each plant. GAE, gallic acid equivalent; QUE, quercetin equivalent.

2.3. *Impact of MAE, UAE, and EE on the Antioxidant Activity of Plant Extracts*

The 2,2-Diphenyl-1-picrylhydrazyl (DPPH) and ferric-reducing antioxidant power (FRAP) activities of the Thai plant extracts were investigated for antioxidant activity, as shown in Figure 3. All samples extracted via the MAE process exhibited higher DPPH radical-scavenging activity (p < 0.05) compared to the samples extracted using UAE and EE. Irrespective of the extraction method, CN, KD, and MP presented the highest values of DPPH activity, with CN extracted via MAE having the highest value (20.6 ± 0.2 mM Trolox equivalent (TE)/g extract), as shown in Figure 3a. MAE-extracted MP showed the second-highest DPPH value (20.23 ± 0.1 mM TE/g extract). The FRAP results were also in line with the DPPH activity, with MAE-extracted KD, CN, and MP exhibiting stronger antioxidant power than the other samples with values of 6.5 ± 0.3, 6.3 ± 0.5 and 6.1 ± 0.1 mM Fe(II)/g extract, respectively (Figure 3b). Based on the above values, the highest FRAP activity was observed for MAE in CN and KD, followed by MP, compared to that obtained by the conventional EE process (p < 0.05). Although lower than the values obtained for MAE, the UAE processed samples also showed higher antioxidant activity when compared to EE ($p < 0.05$), irrespective of the individual activities of the sample extracts. PL showed the lowest DPPH and FRAP activities regardless of the extraction method used.

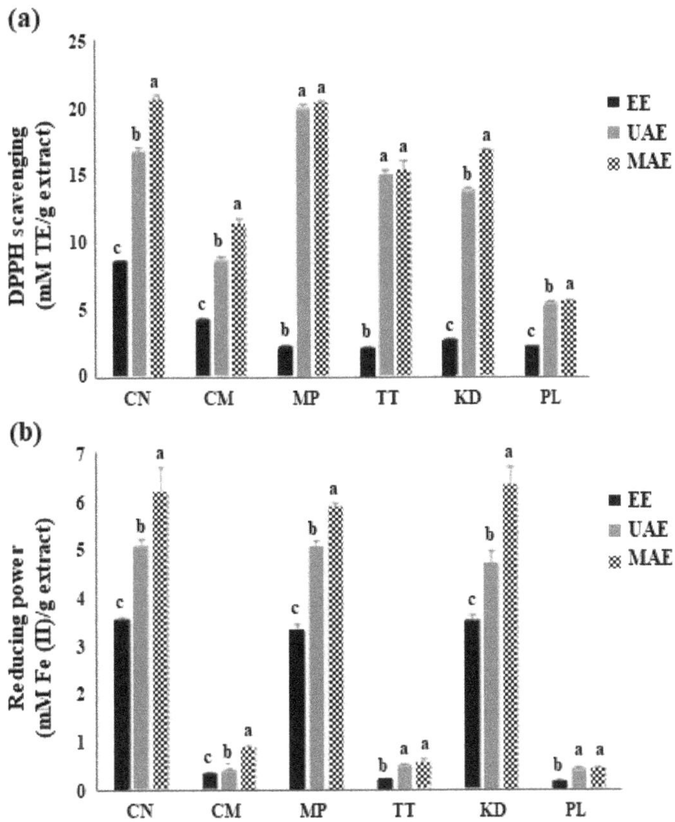

Figure 3. The antioxidant activity of six Thai edible plant extracts extracted by the conventional EE, UAE, and MAE; 2,2-Diphenyl-1-picrylhydrazyl (DPPH) scavenging (**a**) and ferric-reducing antioxidant power (FRAP) (**b**). The data are presented as the mean ± SD of triplicate analyses ($n = 3$). Different letters (a–c) on the bars denote significant difference ($p < 0.05$) of means between extraction methods for each plant. TE; Trolox equivalent.

Zhang, et al. [25] demonstrated that flavonoids extracted from *Epimedium sagittatum* using MAE showed the highest antioxidant activity, yield percentage, and chemical composition. Moreover, UAE was found to optimize the antioxidant activity of phenolic compounds extracted from peaches and pumpkins [26]. It is reasonable to assume that a positive relationship exists between antioxidant assays (DPPH and FRAP) and phenolic compounds, because high amounts of phenolic compounds exhibited strong antioxidant activity by inhibiting free radicals and electron donation from the phenol groups [27]. As expected, the enhanced extraction of phenolic and flavonoid contents by MAE resulted in an increase in antioxidant activity. Therefore, DPPH free radical scavenging and FRAP involving electron transfer reactions are important parameters for predicting the bioactivity of plant extracts [28].

2.4. Disc Diffusion Test, MIC, and Minimum MBC of Thai Edible Plant Extracts Prepared by MAE, UAE, and EE

The inhibition zones of the extracts prepared by MAE, UAE, and EE against the four selected pathogenic bacterial strains are shown in Figure 4 and summarized in Table 1. Most of the extracts showed antimicrobial activity with inhibition zones ranging from 14 to 17 mm for Gram-positive and 9 to 16 mm for Gram-negative bacterial strains, except for the TT extract, which revealed no microbial inhibition. Gram-positive bacteria were more sensitive to the extracts than Gram-negative bacteria, possibly because of the thickness of the cell membrane, which permits the passage of bioactive compounds more easily [29]. Although not observed in our study, a previous study reported that TT has antimicrobial properties. Haro, et al. [30] reported that the methanol extract of soursop (*Annona muricata* L.) leaves showed effective inhibition against *S. aureus* and *E. coli* at concentrations of 150 and 250 mg/mL, respectively.

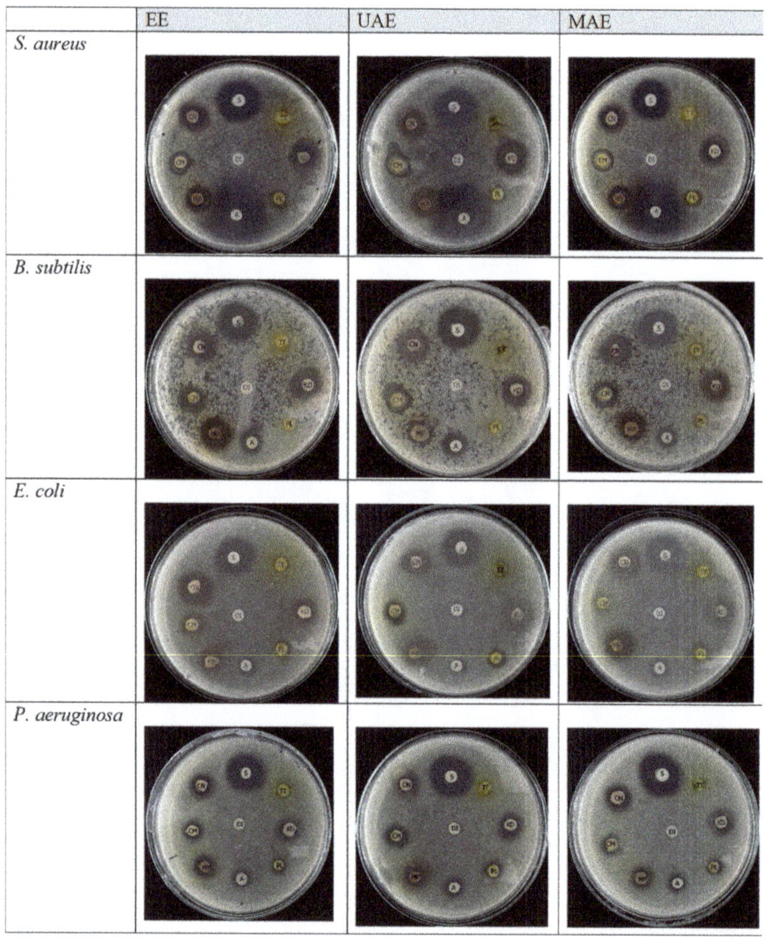

Figure 4. Zone of inhibition of plant extracts prepared via MAE, UAE, and EE against Gram-positive (*Staphylococcus aureus* and *Bacillus subtilis*) and Gram-negative (*Escherichia coli* and *Pseudomonas aeruginosa*) bacteria assessed by the disc-diffusion method. Streptomycin (S), ampicillin (A), and deionized water (DI) (30 μg/mL) were used as positive controls. Paper disc diameter, 6 mm.

Table 1. Inhibition zones of Thai edible plant extracts against pathogenic bacterial strains obtained by the disc-diffusion test.

Bacterial Strains	Samples	Zone of Inhibition (mm)				
		EE	UAE	MAE	Ampicillin	Streptomycin
S. aureus					29	25
	CN	16	14	17		
	CM	14	14	14		
	MP	16	14	14		
	TT	NI	NI	NI		
	KD	17	17	18		
	PL	10	10	11		
B. subtilis					12	23
	CN	15	16	16		
	CM	14	15	15		
	MP	15	16	16		
	TT	NI	NI	NI		
	KD	17	17	17		
	PL	NI	NI	NI		
E. coli					13	22
	CN	12	14	15		
	CM	11	11	11		
	MP	10	15	15		
	TT	NI	NI	NI		
	KD	11	15	16		
	PL	9	10	11		
P. aeruginosa					12	26
	CN	14	15	16		
	CM	12	13	12		
	MP	13	13	14		
	TT	NI	NI	NI		
	KD	13	14	16		
	PL	9	10	10		

The concentrations of all sample extracts were 100 mg/mL and a volume of 20 μL was pipetted onto each paper disc (diameter 6 mm). NI, no inhibition.

The Thai plant extracts were analyzed for MIC and MBC, and the results are presented in Table 2. All plant extracts were effective against Gram-positive bacterial strains, including *S. aureus* and *B. subtilis*, with MIC ranging from 0.78 to 50 mg/mL and MBC ranging from 12.5 to 50 mg/mL. For Gram-negative strains, the MIC values ranged from 1.56 to 50 mg/mL. The lowest MIC values against Gram-positive bacteria were obtained for the MP and KD extracts prepared using MAE and UAE. The KD sample was also effective against Gram-negative strains with the lowest MIC values, along with CN (1.56 mg/mL). Additionally, the MBC values of the selected plant extracts for Gram-positive and Gram-negative strains ranged from 12.5 to 50 mg/mL and 25 to 100 mg/mL, respectively, indicating marked inhibition of the tested bacteria with the lowest bacterial counts. The lowest MBC against *S. aureus* was obtained for the CN sample extracted using MAE (12.5 mg/mL). MAE-extracted CN, MP, and KD showed the lowest MBC values for *B. subtilis* (25 mg/mL). This indicated that the Gram-positive strains were more sensitive to the CN, MP, and KD extracts.

MIC and MBC values were obtained for all plant extracts, except TT. According to Solomon-Wisdom, et al. [31], methanolic and aqueous extracts of TT leaves inhibited *S. aureus*, *E. coli*, and *P. aeruginosa* at a concentration of 200 mg/mL. The extract obtained from KD was found to be a strong antimicrobial agent against both Gram-positive and Gram-negative bacterial strains, with the lowest MIC values compared to other extracts. A similar finding was observed by Daduang, et al. [32] who reported that the extract of the edible part of *Careya sphaerica Roxb.* (Kradon), which had high phenolic and antioxidant contents, inhibited *S. aureus*, *S. typhimurium*, and *E. coli*. Additionally, Chanudom, et al. [33]

reported that the ethanol crude extracts of traditional Thai plants, including CN and MP, possess high phenolic and antioxidant content and are effective against *S. aureus*, *E. coli*, and *P. aeruginosa*.

Table 2. MIC and MBC of edible Thai plant extracts obtained via MAE, UAE, and EE processes.

Bacterial Strains	Plant	MIC (mg/mL)			MBC (mg/mL)		
		EE	UAE	MAE	EE	UAE	MAE
Staphylococcus aureus	CN	1.56	<1.56	0.78	25	25	12.5
	CM	3.12	<1.56	0.78	25	25	25
	MP	<1.56	<1.56	0.78	25	25	25
	TT	N/D	N/D	N/D	N/D	N/D	N/D
	KD	<1.56	<1.56	<1.56	25	25	25
	PL	50	50	12.5	50	50	50
Bacillus subtilis	CN	1.56	<1.56	<1.56	50	50	25
	CM	1.56	<1.56	<1.56	50	50	50
	MP	<1.56	<1.56	<1.56	50	50	25
	TT	N/D	N/D	N/D	N/D	N/D	N/D
	KD	<1.56	<1.56	<1.56	50	50	25
	PL	25	12.5	6.25	50	50	50
Escherichia coli	CN	1.56	1.56	1.56	50	50	25
	CM	12.5	12.5	12.5	100	100	100
	MP	6.25	3.12	3.12	50	50	50
	TT	N/D	N/D	N/D	N/D	N/D	N/D
	KD	3.12	3.12	1.56	50	50	50
	PL	50	50	50	100	100	100
Pseudomonas aeruginosa	CN	1.56	1.56	1.56	50	50	25
	CM	12.5	12.5	12.5	100	100	50
	MP	3.12	6.25	3.12	50	50	25
	TT	N/D	N/D	N/D	N/D	N/D	N/D
	KD	1.56	1.56	1.56	50	25	25
	PL	50	25	25	100	100	100

The concentration of all plant extracts was varied between the range of 100 to 0.78 mg/mL by two-fold dilution in a 96-well plate. N/D, not determined.

There is considerable interest regarding the antioxidant activity of natural plant polyphenols against free radicals. In addition, in some cases, these substances also serve as active compounds against microorganisms such as bacteria, viruses, and fungi. Flavonoids possess antimicrobial activity because of their ability to interact with and disrupt extracellular soluble proteins and bacterial cell wall [1]. Some plant phenolic compounds, such as chlorogenic acid, curcumin, ellagic acid, (−) epicatechin, rutin, and tannic acid, were found to control the growth of foodborne pathogenic strains, including *Bacillus*, *Listeria*, and *Clostridium* species [34]. Jarriyawattanachaikul, et al. [35] reported the antimicrobial activity of Thai herbal plants against various foodborne pathogens (*E. coli*, *S. aureus*, and *Campylobacter jejuni*). Thai herbal extracts have been reported to be effective against both Gram-negative and Gram-positive bacteria. Moreno, et al. [36] reported on the mechanisms associated with the interaction between active phytochemical groups and cellular enzymes. The rate of penetration of bioactive compounds into microbial cells depends on membrane permeability, and disruption of cell membranes may lead to cell death due to the loss of cellular integrity. The water extract of rosella calyx (*Hibiscus sabdariffa*) extracted via the MAE method showed a higher zone of inhibition against *S. aureus* (Gram-positive bacteria) and *E. coli* (Gram-negative bacteria) [37]. MAE provides high extraction efficiency with minimum solvent requirements. Under microwave irradiation, the plant matrix is easily impregnated with solvents with higher heating efficiency. The higher temperature attained by microwave irradiation can hydrolyze ether linkages of cellulose, which are the main constituent of the plant cell wall, and can convert into soluble fractions within short period of time [38].

Costa et al. [39] demonstrated that the leaf extract of CN contained phytochemicals such as gallic acid and derivatives, quercetin derivatives, and luteolin using UPLC-DAD/QTOF-MS analysis. Phukhatmuen et al. [40] reported that 11 compounds including a novel decahydro-1H-xanthene derivative were identified from CM leaf extract, and most compounds were xanthones and benzophenones. In contrast, there are few studies focused on the separation and identification from the leaf extract of MP, TT, KD, and PL. Future studies will be needed to identify active phytoconstituents responsible for antioxidant and antimicrobial activities of these leaf extracts.

3. Materials and Methods

3.1. Materials

Edible Thai plants, including CN, CM, MP, TT, KD, and PL, were collected from Chiang Rai and other Thai provinces. The leaves of these plants were washed with distilled water, ground into fine powder in liquid nitrogen using a stainless blender, and stored in a freezer at −20 °C until sample extraction. Folin–Ciocalteu reagent, Trolox, gallic acid, quercetin, dimethyl sulfoxide (DMSO), and streptomycin sulfate were purchased from Sigma-Aldrich (St. Louis, MO, USA). Nutrient broth and Mueller-Hinton broth (MHB) were purchased from HIMEDIA (Mumbai, India). Ampicillin and sodium chloride were purchased from Bio Basic (Markham, ON, Canada). Bacterial cultures were obtained from the Thailand Institute of Scientific and Technological Research. All the chemicals were of analytical grade.

3.2. Preparation of Thai Edible Plant Extracts Using the Conventional EE, UAE, and MAE

Plant extracts were prepared from six Thai edible leaves using the conventional EE method, as described by Nagappan [41]. In brief, powdered leaf samples were mixed with 60% (*v*/*v*) ethanol at a ratio of 1:30 (g/mL of powder to solvent ratio) and placed in a shaker bath (Memmert, Germany) at 25 °C for 60 min. The UAE method was adapted from a previous study by Kazibwe, et al. [42]. In brief, the powdered leaf samples were mixed with 60% (*v*/*v*) ethanol at a ratio of 1:30 (g/mL of powder to solvent ratio) and sonicated using an ultrasonic bath (DT 255 H, Bandelin Co., Germany) at 25 °C for 60 min at a frequency of 35 kHz. The MAE method was adapted from Dahmoune, Nayak, Moussi, Remini and Madani [16] and Piovesan, et al. [43]. For this, a microwave oven (Samsung, Korea) was used as the heat source to facilitate extraction. Powdered leaf samples from the six Thai edible plants were mixed with 60% (*v*/*v*) ethanol at 1:30 (g/mL of powder to solvent ratio) and subjected to irradiation at 500 W for 60 s. After extraction using the three methods, the extracted samples were collected and filtered using filter paper (Whatman No.4) placed on a Buchner funnel equipped with a vacuum pump and concentrated using a rotary evaporator (IKA/RV 10 basic, Germany) at 45 °C for 10 min. The concentrated extracts were dried in a freeze-dryer (FD 8–55, Australia) at −50 °C for 72 h. The plant extracts were then stored at −20 °C in an airtight container prior to the determination of bioactive chemicals and microbial characteristics.

3.3. Determination of Extraction Yield and Chemical Properties of Thai Edible Plant Extracts

3.3.1. Extraction yield

Percentage yield of the freeze-dried plant extracts was determined using the following equation:

$$\text{Extraction yield (\%)} = \text{gram of freeze-dried plant extract} \times 100/\text{gram of leaf powder}$$

3.3.2. Determination of TPC and TFC

The TPC of Thai edible plant extracts was determined using the Folin–Ciocalteu assay (ISO 14502-1, 2005), and gallic acid was used as a standard. Samples of the extracts (500 μL each) were mixed with 2.5 mL of 10% (*w*/*v*) Folin–Ciocalteu reagent and 2 mL of 7.5% (*w*/*v*) sodium carbonate. The mixture was stirred and incubated in darkness for 1 h at room temperature (25 °C). The absorbance of the samples was measured at 765 nm using a microplate spectrophotometer (Thermo Fisher Scientific, Multiskan GO, USA) [44]. The TPC was expressed as milligrams of gallic acid equivalent (mg GAE)/g of dry extracts [45]. The TFC of the plant extracts was determined using the aluminum trichloride method, as reported by Rebaya, et al. [46], with some modifications. In brief, 36 µL of 5% $NaNO_2$) and 45 µL of diluted sample were transferred to a 96-well microplate and left for 6 min. Next, 36 µL of 10% aluminum chloride hexahydrate was added to each well and the mixtures were incubated for 5 min. Thereafter, 180 µL of 10% sodium hydroxide was added and the final volume was adjusted to 300 µL using distilled water. The mixtures were incubated for another 15 min and the absorbance was measured at 510 nm using a microplate reader (Multiskan GO, USA). The TFC was expressed as milligrams of quercetin equivalent (mg QUE)/g of the extract.

3.3.3. Determination of DPPH and FRAP Activities

The free-radical-scavenging activities of Thai edible plant extracts were analyzed using the DPPH method [47]. First, the DPPH solution (60 mM) was prepared by dissolving 0.00236 g of 2,2-diphenyl-1-picrylhydrazyl in 95% ethanol (v/v). The solution was then mixed with the plant extracts (50 µL). Trolox (10,000 µM) was used as the standard solution and methanol was used as a blank. The mixtures were incubated at room temperature for 30 min. Absorbance was measured at 517 nm using a microplate spectrophotometer (Thermo Fisher Scientific, Multiskan GO, USA) [44]. DPPH activities of the plant extract samples were expressed as µmol TE/g of dry extract [48]. FRAP was evaluated using the method described by Sadeghi, et al. [49], with some modifications. Diluted plant extracts (35 µL) were pipetted into a 96-well microplate, and 265 µL of freshly prepared FRAP reagent (pre-incubated at 37 °C for a few minutes) was added to each well. Ethanol was used as a blank sample. The mixtures containing the plant extracts and FRAP reagent were incubated for 30 min, and the absorbance was measured at 595 nm. The FRAP activities of all samples were calculated based on the standard curve of ferrous sulfate ($FeSO_4{\cdot}7H_2O$) in the range of 100 to 1000 µM. Activity is expressed as mM ferrous sulfate equivalent (mM Fe(II)) per gram of the extract.

3.4. Determination of Antimicrobial Properties of Edible Thai Plant Extracts

The antimicrobial activity of the Thai plant extracts was evaluated using four bacterial strains. The bacterial strains were obtained from the Thailand Institute of Scientific and Technological Research (TISTR). Two strains of Gram-positive bacteria, *Staphylococcus aureus* TISTR 746 and *Bacillus subtilis* TISTR 008, and two strains of Gram-negative bacteria, *Escherichia coli* TISTR 527 and *Pseudomonas aeruginosa* TISTR 1278, were used to prepare the bacterial suspensions.

3.4.1. Bacterial Suspension Preparation and Disk Diffusion Method

All the bacterial strains were activated in MHB for 18–24 h and sub-cultured overnight at 37 °C in MHB medium. The activated bacterial strains were diluted in 5 mL 0.85% sterile saline solution. Microbial suspensions were standardized according to a 0.5 McFarland scale (assumed cell count density 1.5×10^8 CFU/mL) at 625 nm. The zone of inhibition was determined by the disc diffusion method to test the antimicrobial efficacy of the plant extracts, as described by Bauer, et al. [50] with some modifications. Bacterial suspensions of all strains (0.5 mL) were mixed with 25 mL of MHB and bacteriological agar. The agar medium containing the MHB-inoculated bacterial suspensions was poured into sterile Petri dishes and solidified for a few minutes. The disc diffusion method was performed by dropping 20 µL each of the sample extracts prepared in 20% DMSO (100 mg/mL) on a disc of 6 mm diameter and left in a laminar air flow for a few minutes before the incubation process. Distilled water and antibiotics (streptomycin and ampicillin) were used as the negative and positive controls, respectively. The plates containing the plant extracts, control (added with distilled water), and antibiotics were incubated for 12 to 24 h at 37 °C. The inhibition zone was then measured and expressed in millimeters.

3.4.2. Determination of MIC and MBC

The antimicrobial activity of edible Thai plant extracts was studied using a microdilution method adopted from Mosaddik, et al. [51]. Two Gram-positive (*S aureus* and *B. subtilis*) and two Gram-negative bacteria (*E. coli* and *P. aeruginosa*) were tested. The inoculum was prepared from a subculture using Muller-Hinton agar. Plant extracts were dissolved in 20% DMSO (100 mg/mL) and filtered using a 0.45 μm filter. Fifty microliters of each sample was diluted with 50 μL of MHB (50 mg/mL). The samples were then diluted two-fold in a 96-well microplate to attain a concentration range of 1.56 to 50 mg/mL. Bacterial strains without extracts were used as negative controls. Each well was inoculated with 5 μL of bacterial suspension (1×10^7 CFU/mL). Inoculated plates were incubated at 36 °C for 18–24 h. Bacterial growth was then measured using a microplate reader at 625 nm. MIC values were determined as the lowest concentration of each extract which inhibited microbial growth. The results were expressed as milligrams per milliliter. A positive control (only cells plus medium) and negative control (only sample plus medium) were prepared for each tested bacterial strain.

3.5. Statistical Analysis

All experiments were conducted in triplicate (n = 3). Data are expressed as the mean ± standard deviation. The IBM SPSS Statistics software ver. 23 was used for analysis of variance (ANOVA), as well as Duncan's multiple range tests. p-values of less than 0.05 were considered statistically significant.

4. Conclusions

MAE, UAE, and the conventional EE process were employed to prepare phytochemical extracts from six edible Thai plant leaves. CN, CM, MP, TT, KD, and PL samples were processed using 60% ethanol, followed by MAE or UAE. The extraction yield was highest in the MAE-extracted MP leaf extract compared to the UAE and EE methods. TPC and TFC, including DPPH and FRAP values, were higher in the MAE-extracted KD and CN samples. MIC and MBC, including zones of inhibition against Gram-positive and Gram-negative bacterial strains, were highest in all MAE-processed plant samples except TT, which possibly required higher levels of the extract. Therefore, MAE is a promising method for extracting bioactive compounds using aqueous ethanol as an extraction solvent to enhance yield, polyphenol content, and antioxidant properties. MAE of plant extracts will be further investigated in the future for the isolation and characterization of active compounds and their fortification as natural food additives in perishable foods.

Author Contributions: P.J.: Formal analysis, Data curation, Investigation, Methodology, Writing—original draft. S.N.: Supervision. S.K.: Conceptualization, Supervision, Writing—review and editing. S.R.: Conceptualization, Supervision, Formal analysis, Writing-review and editing. All authors have read and agreed to the published version of the manuscript.

Funding: This research was funded by Mae Fah Luang University under the grant number of 5951401001.

Institutional Review Board Statement: Not applicable.

Informed Consent Statement: Not applicable.

Data Availability Statement: Data sharing is not applicable.

Acknowledgments: The authors would like to thank Laboratory of Food Science and Technology, Laboratory of Microbiology, Mae Fah Luang University for providing all the chemicals and instruments. Thailand Institute of Scientific and Technology (TISTR Culture Collection) for providing microbial strains was also acknowledged. Mae Fah Luang University was also specially thanked for providing financial support throughout this project.

Conflicts of Interest: The authors declare no conflict of interest.

Sample Availability: Samples are available from the author now.

References

1. Cowan, M.M. Plant products as antimicrobial agents. *Clin. Microbiol. Rev.* **1999**, *12*, 564–582. [CrossRef] [PubMed]
2. See, J.A.; Gall, M.M. Eating Plants for Health: From Fiber to Phytochemicals. *J. Am. Diet. Assoc.* **1997**, *97* (Suppl. S9), A60. [CrossRef]
3. Lee, J.H.; Cho, S.; Paik, H.D.; Choi, C.W.; Nam, K.T.; Hwang, S.G.; Kim, S.K. Investigation on antibacterial and antioxidant activities, phenolic and flavonoid contents of some thai edible plants as an alternative for antibiotics. *Asian Australas J. Anim. Sci.* **2014**, *27*, 1461–1468. [CrossRef] [PubMed]
4. Sakunpak, A.; Panichayupakaranant, P. Antibacterial activity of Thai edible plants against gastrointestinal pathogenic bacteria and isolation of a new broad spectrum antibacterial polyisoprenylated benzophenone, chamuangone. *Food Chem.* **2012**, *130*, 826–831. [CrossRef]
5. Campos-Vega, R.; Oomah, B.D. Chemistry and classification of phytochemicals. In *Handbook of Plant Food Phytochemicals;* Wiley-Blackwell: Chicester, UK, 2013; pp. 5–48.
6. Pinchuk, I.; Shoval, H.; Dotan, Y.; Lichtenberg, D. Evaluation of antioxidants: Scope, limitations and relevance of assays. *Chem. Phys. Lipids* **2012**, *165*, 638–647. [CrossRef]
7. Fraga, C.G.; Galleano, M.; Verstraeten, S.V.; Oteiza, P.I. Basic biochemical mechanisms behind the health benefits of polyphenols. *Mol. Aspects Med.* **2010**, *31*, 435–445. [CrossRef]
8. Pandey, K.B.; Rizvi, S.I. Plant polyphenols as dietary antioxidants in human health and disease. *Oxid. Med. Cell. Longev.* **2009**, *2*, 270–278. [CrossRef]
9. Finley, J.W.; Kong, A.N.; Hintze, K.J.; Jeffery, E.H.; Ji, L.L.; Lei, X.G. Antioxidants in foods: State of the science important to the food industry. *J. Agric. Food Chem.* **2011**, *59*, 6837–6846. [CrossRef]
10. Poljsak, B.; Kovač, V.; Milisav, I. Antioxidants, Food Processing and Health. *Antioxidants* **2021**, *10*, 433. [CrossRef]
11. Mostafa, A.A.; Al-Askar, A.A.; Almaary, K.S.; Dawoud, T.M.; Sholkamy, E.N.; Bakri, M.M. Antimicrobial activity of some plant extracts against bacterial strains causing food poisoning diseases. *Saudi. J. Biol. Sci.* **2018**, *25*, 361–366. [CrossRef]
12. Santajit, S.; Indrawattana, N. Mechanisms of Antimicrobial Resistance in ESKAPE Pathogens. *Biomed. Res. Int.* **2016**, *2016*, 2475067. [CrossRef] [PubMed]
13. Anand, G.; Ravinanthan, M.; Basaviah, R.; Shetty, A.V. In vitro antimicrobial and cytotoxic effects of *Anacardium occidentale* and *Mangifera indica* in oral care. *J. Pharm. Bioallied. Sci.* **2015**, *7*, 69. [PubMed]
14. Stalikas, C.D. Extraction, separation, and detection methods for phenolic acids and flavonoids. *J. Sep. Sci.* **2007**, *30*, 3268–3295. [CrossRef]
15. Gupta, A.; Naraniwal, M.; Kothari, V. Modern extraction methods for preparation of bioactive plant extracts. *Int. J. Appl. Nat. Sci.* **2012**, *1*, 8–26.
16. Dahmoune, F.; Nayak, B.; Moussi, K.; Remini, H.; Madani, K. Optimization of microwave-assisted extraction of polyphenols from *Myrtus communis* L. leaves. *Food Chem.* **2015**, *166*, 585–595. [CrossRef] [PubMed]
17. Carreira-Casais, A.; Otero, P.; Garcia-Perez, P.; Garcia-Oliveira, P.; Pereira, A.G.; Carpena, M.; Soria-Lopez, A.; Simal-Gandara, J.; Prieto, M.A. Benefits and drawbacks of ultrasound-assisted extraction for the recovery of bioactive compounds from marine algae. *Int. J. Environ. Res. Public Health* **2021**, *18*, 9153. [CrossRef]
18. Huma, Z.-E.; Jayasena, V.; Abbas, S.; Imran, M.; Khan, M. Process optimization of polyphenol extraction from carob (*Ceratonia siliqua*) kibbles using microwave-assisted technique. *J. Food Process. Preserv.* **2017**, *42*, e13450. [CrossRef]
19. Osorio-Tobón, J.F. Recent advances and comparisons of conventional and alternative extraction techniques of phenolic compounds. *J. Food Sci. Technol.* **2020**, *57*, 4299–4315. [CrossRef]
20. Chan, C.-H.; Yeoh, H.K.; Yusoff, R.; Ngoh, G.C. A first-principles model for plant cell rupture in microwave-assisted extraction of bioactive compounds. *J. Food Eng.* **2016**, *188*, 98–107. [CrossRef]
21. Chen, L.; Jin, H.; Ding, L.; Zhang, H.; Li, J.; Qu, C.; Zhang, H. Dynamic microwave-assisted extraction of flavonoids from Herba Epimedii. *Sep. Purif. Technol.* **2008**, *59*, 50–57. [CrossRef]
22. Rahath Kubra, I.; Kumar, D.; Rao, L.J.M. Effect of microwave-assisted extraction on the release of polyphenols from ginger (*Zingiber officinale*). *Int. J. Food Sci. Technol.* **2013**, *48*, 1828–1833. [CrossRef]
23. Vinatoru, M. An overview of the ultrasonically assisted extraction of bioactive principles from herbs. *Ultrason. Sonochem.* **2001**, *8*, 303–313. [CrossRef]
24. Sun, C.; Wu, Z.; Wang, Z.; Zhang, H. Effect of ethanol/water solvents on phenolic profiles and antioxidant properties of Beijing propolis extracts. *Evid. Based Complement. Alternat. Med.* **2015**, *2*, 595393. [CrossRef] [PubMed]
25. Zhang, H.-F.; Zhang, X.; Yang, X.-H.; Qiu, N.-X.; Wang, Y.; Wang, Z.-Z. Microwave assisted extraction of flavonoids from cultivated Epimedium sagittatum: Extraction yield and mechanism, antioxidant activity and chemical composition. *Ind. Crop. Prod.* **2013**, *50*, 857–865. [CrossRef]
26. Altemimi, A.; Watson, D.G.; Choudhary, R.; Dasari, M.R.; Lightfoot, D.A. Ultrasound assisted extraction of phenolic compounds from peaches and pumpkins. *PLoS ONE* **2016**, *11*, e0148758. [CrossRef]

27. Büyüktunce, E.; Porgalı, E.; Çolak, C. Comparison of total phenolic content and total antioxidant activity in local red wines determined by spectrophotometric methods. *Food Nutr. Sci.* **2014**, *2014*, 1660–1667.
28. Moharram, H.; Youssef, M. Methods for determining the antioxidant activity: A review. *Alex. J. Food Sci. Tech.* **2014**, *11*, 31–42.
29. Ahmad Shiekh, K.; Odunayo Olatunde, O.; Zhang, B.; Huda, N.; Benjakul, S. Pulsed electric field assisted process for extraction of bioactive compounds from custard apple (*Annona squamosa*) leaves. *Food Chem.* **2021**, *359*, 129976. [CrossRef]
30. Haro, G.; Utami, N.; Sitompul, E. Study of the antibacterial activities of Soursop (*Annona muricata* L.) leaves. *Int. J. Pharmtech. Res.* **2014**, *6*, 575–581.
31. Solomon-Wisdom, G.; Ugoh, S.; Mohammed, B. Phytochemical screening and antimicrobial activities of *Annona muricata* (*L*) leaf extract. *Am. J. Biol. Chem. Pharm. Sci.* **2014**, *2*, 1–7.
32. Daduang, J.; Daduang, S.; Hongsprabhas, P.; Boonsiri, P. High phenolics and antioxidants of some tropical vegetables related to antibacterial and anticancer activities. *Afr. J. Pharm. Pharmacol.* **2011**, *5*, 608–615. [CrossRef]
33. Chanudom, L.; Bhoopong, P.; Khwanchuea, R.; Tangpong, J. Antioxidant and antimicrobial activities of aqueous & ethanol crude extracts of 13 Thai traditional plants. *Int. J. Curr. Microbiol. Appl. Sci.* **2014**, *3*, 549–558.
34. Cetin-Karaca, H.; Newman, M.C. Antimicrobial efficacy of plant phenolic compounds against Salmonella and Escherichia Coli. *Food Biosci.* **2015**, *11*, 8–16. [CrossRef]
35. Jarriyawattanachaikul, W.; Chaveerach, P.; Chokesajjawatee, N. Antimicrobial activity of Thai-Herbal plants against food-borne pathogens E. Coli, S. Aureus and C. Jejuni. *Agric. Agric. Sci. Procedia* **2016**, *11*, 20–24. [CrossRef]
36. Moreno, S.; Scheyer, T.; Romano, C.S.; Vojnov, A.A. Antioxidant and antimicrobial activities of rosemary extracts linked to their polyphenol composition. *Free Radic. Res.* **2006**, *40*, 223–231. [CrossRef] [PubMed]
37. Alam, G.; Natsir, S.; Alfath, A. Comparison of microwave assisted extraction (MAE) with variations of power and infusion extraction method on antibacterial activity of rosella calyx extract (*Hibiscus sabdariffa*). *J. Phys. Conf. Ser.* **2019**, *1341*, 072002. [CrossRef]
38. Latha, C. Microwave-assisted extraction of embelin from Embelia ribes. *Biotechnol. Lett.* **2007**, *29*, 319–322. [CrossRef]
39. Costa, A.R.; de Lima Silva, J.R.; de Oliveira, T.J.S.; da Silva, T.G.; Pereira, P.S.; de Oliveira Borba, E.F.; de Brito, E.S.; Ribeiro, P.R.V.; Almeida-Bezerra, J.W.; Júnior, J.T.C. Phytochemical profile of Anacardium occidentale L. (cashew tree) and the cytotoxic and toxicological evaluation of its bark and leaf extracts. *S. Afr. J. Bot.* **2020**, *135*, 355–364. [CrossRef]
40. Phukhatmuen, P.; Raksat, A.; Laphookhieo, S.; Charoensup, R.; Duangyod, T.; Maneerat, W. Bioassay-guided isolation and identification of antidiabetic compounds from Garcinia cowa leaf extract. *Heliyon* **2020**, *6*, e03625. [CrossRef]
41. Nagappan, R. Evaluation of aqueous and ethanol extract of bioactive medicinal plant, *Cassia didymobotrya* (Fresenius) Irwin & Barneby against immature stages of filarial vector, *Culex quinquefasciatus* Say (Diptera: Culicidae). *Asian Pac. Trop. Biomed.* **2012**, *2*, 707–711.
42. Kazibwe, Z.; Kim, D.-H.; Chun, S.; Gopal, J. Ultrasonication assisted ultrafast extraction of Tagetes erecta in water: Cannonading antimicrobial, antioxidant components. *J. Mol. Liq.* **2017**, *229*, 453–458. [CrossRef]
43. Piovesan, N.; Viera, V.; Mello, R.d.O.; dos Santos, R.; Vaucher, R.d.A.; Dressler, V.; Bizzi, C.; Fries, L. Microwave-assisted extraction of bioactive compounds from blueberry (*Vaccinium ashei* Reade) and their antioxidant and antimicrobial capacity. *Int. Food Res. J.* **2017**, *24*, 2526–2533.
44. Lu, Q.; Lv, S.; Peng, Y.; Zhu, C.; Pan, S. Characterization of phenolics and antioxidant abilities of red navel orange "Cara Cara" harvested from five regions of China. *Int. J. Food Prop.* **2018**, *21*, 1107–1116. [CrossRef]
45. Olatunde, O.O.; Benjakul, S.; Vongkamjan, K. Antioxidant and antibacterial properties of guava leaf extracts as affected by solvents used for prior dechlorophyllization. *J. Food Biochem.* **2018**, *42*, e12600. [CrossRef]
46. Rebaya, A.; Belghith, S.I.; Baghdikian, B.; Leddet, V.M.; Mabrouki, F.; Olivier, E.; Cherif, J.; Ayadi, M.T. Total phenolic, total flavonoid, tannin content, and antioxidant capacity of *Halimium halimifolium* (Cistaceae). *J. App. Pharm. Sci.* **2014**, *5*, 52–57.
47. Molyneux, P. The use of the stable radical Diphenylpicrylhydrazyl (DPPH) for estimating antioxidant activity. *Songklanakarin J. Sci. Technol.* **2003**, *26*, 211–219.
48. Tagrida, M.; Benjakul, S. Betel (*Piper betle* L.) leaf ethanolic extracts dechlorophyllized using different methods: Antioxidant and antibacterial activities, and application for shelf-life extension of Nile tilapia (*Oreochromis niloticus*) fillets. *RSC Adv.* **2021**, *11*, 17630–17641. [CrossRef] [PubMed]
49. Sadeghi, Z.; Valizadeh, J.; Shermeh, O.A.; Akaberi, M. Antioxidant activity and total phenolic content of *Boerhavia elegans* (choisy) grown in Baluchestan, Iran. *Avicenna J. Phytomed.* **2015**, *5*, 1.
50. Bauer, A.W.; Kirby, W.M.; Sherris, J.C.; Turck, M. Antibiotic susceptibility testing by a standardized single disk method. *Am. J. Clin. Pathol.* **1966**, *45*, 493–496. [CrossRef]
51. Mosaddik, M.; Banbury, L.; Forster, P.; Booth, R.; Markham, J.; Leach, D.; Waterman, P.G. Screening of some Australian Flacourtiaceae species for in vitro antioxidant, cytotoxic and antimicrobial activity. *Phytomedicine* **2004**, *11*, 461–466. [CrossRef]

molecules

MDPI

Article

Antifungal Activity of *Lavandula angustifolia* Essential Oil against *Candida albicans*: Time-Kill Study on Pediatric Sputum Isolates

Stefan Mijatovic [1], Jelena Antic Stankovic [2,*], Ivana Colovic Calovski [1], Eleonora Dubljanin [1], Dejan Pljevljakusic [3], Dubravka Bigovic [3] and Aleksandar Dzamic [1]

1 Institute of Microbiology and Immunology, Faculty of Medicine, University of Belgrade, 11000 Belgrade, Serbia
2 Department of Microbiology, Faculty of Pharmacy, University of Belgrade, 11000 Belgrade, Serbia
3 Institute for Medicinal Plants Research "Dr Josif Pancic", 11000 Belgrade, Serbia
* Correspondence: jelena.stankovic@pharmacy.bg.ac.rs

Abstract: The aim of our study was to determine the susceptibility of 15 *Candida albicans* sputum isolates on fluconazole and caspofungin, as well as the antifungal potential of *Lavandula angustifolia* essential oil (LAEO). The commercial LAEO was analyzed using gas chromatography-mass spectrometry. The antifungal activity was evaluated using EUCAST protocol. A killing assay was performed to evaluate kinetics of 2% LAEO within 30 min treatment. LAEO with major constituents' linalool (33.4%) and linalyl acetate (30.5%) effective inhibited grows of *C. albicans* in concentration range 0.5–2%. Fluconazole activity was noted in 67% of the isolates with MICs in range 0.06–1 μg/mL. Surprisingly, 40% of isolates were non-wild-type (non-WT), while MICs for WT ranged between 0.125–0.25 μg/mL. There were no significant differences in the LAEO MICs among fluconazole-resistant and fluconazole-susceptible sputum strains (p = 0.31) and neither among caspofungin non-WT and WT isolates (p = 0.79). The 2% LAEO rapidly achieved 50% growth reduction in all tested strains between 0.2 and 3.5 min. Within 30 min, the same LAEO concentration exhibited a 99.9% reduction in 27% isolates. This study demonstrated that 2% solution of LAEO showed a significant antifungal activity which is equally effective against fluconazole and caspofungin susceptible and less-susceptible strains.

Keywords: *Candida albicans*; *Lavandula angustifolia*; essential oil; fluconazole; caspofungin; time-kill curve

Citation: Mijatovic, S.; Stankovic, J.A.; Calovski, I.C.; Dubljanin, E.; Pljevljakusic, D.; Bigovic, D.; Dzamic, A. Antifungal Activity of *Lavandula angustifolia* Essential Oil against *Candida albicans*: Time-Kill Study on Pediatric Sputum Isolates. *Molecules* **2022**, *27*, 6300. https://doi.org/10.3390/molecules27196300

Academic Editors: Raffaele Pezzani and Sara Vitalini

Received: 24 August 2022
Accepted: 20 September 2022
Published: 24 September 2022

1. Introduction

Yeasts of the genus *Candida*, as members of the skin and mucosal microbiota, can cause opportunistic fungal infections, both superficial and invasive. For hospitalized pediatric patients, *Candida* is a significant pathogen identified in the setting of bloodstream infections. Children with acute myeloid leukemia (AML), hematopoietic stem cell transplant (HSCT) recipients and patients admitted to intensive care units (ICU) have the highest risk of invasive candidiasis development [1]. With the risky patients, abnormal *Candida* colonization correlates with a higher probability of its invasive infection [2]. According to the origin, the mentioned mycoses are endogenous, so the damage and the increased permeability of the mucous membranes as consequences of the underlying disease, cytostatic or radiation therapy are the entry points for these yeasts. Studies of the pediatric candidiasis showed that *Candida albicans* was the most common isolated species, but collectively *Candida parapsilosis*, *Candida glabrata*, *Candida lusitaniae* and *Candida krusei* predominated [3,4].

Prophylactic and pre-emptive therapies with high-risk patients are very important in prevention of invasive candidiasis. As drugs with low toxicity, azoles and echinocandins are widely used in prevention and treatment of the fore mentioned mycoses [5]. The introduction of echinocandins into standard protocols alone or in combination with other antifungals has greatly improved the clinical outcomes. Although with lower incidence of adverse events than other antifungals, echinocandins are only available as intravenous formulations [6]. *Candida* resistance to echinocandins has emerged and it is most commonly associated with *C. glabrata* [7]. On the other side, azoles have good tolerability, but a narrow activity spectrum. *C. krusei* is intrinsically resistant to fluconazole whereas *C. glabrata* is highly resistant to the mentioned antifungal [8]. However, the increasing resistance of *C. albicans* and non-albicans species to these antifungals, recurrent infections and persistent colonization require a different strategy in prophylaxis and in treatment of the described disease.

Traditionally, essential oils (EOs) and extracts of aromatic plants have been used in the treatments of wound infections and burns [9,10]. Some EOs in certain cultures are used in aromatherapy as a relaxant and as a carminative and a sedative agent [11]. In addition, dried parts of plants are used as repellents for various arthropods [12]. However, in vitro and clinical studies show different and often conflicting results regarding the efficacy of EOs in human medicine. So far, the antibacterial and antifungal effects of aromatic plants' extracts and EOs have been investigated. EOs of oregano, thyme and lavender, cloves, cinnamon and tea tree showed broad and potent anti *Candida* activity in in vitro studies [13–15]. It is assumed that EOs interfere with fungal membranes, but are distinct from currently available polyenes [16]. Furthermore, inhibition of cell wall polysaccharide synthesis could be another method for EOs antifungal activity [17]. EOs show in vitro activity against *Candida* azole-resistant strains, implicating a different mode of action [18]. It is certain that EOs have an antifungal effect, but there are some limitations. In a laboratory, determinations of effective antimicrobial concentration are difficult to compare because their outcome is influenced primarily by the chemical composition of EOs and methodologies of research. Lipophilic character makes them potentially toxic, especially if used at high concentration and should not be administrated orally and parenterally. Due to their broad-spectrum antimicrobial activity, EOs and their components would be very interesting solution for improving prophylactic and therapeutic failures related to the emergence of resistant strains. Aromatherapy could be an additional method for the prevention of massive upper respiratory tract colonization and the consequent bloodstream, but also some systemic *Candida* infections.

The aim of our study was to determine the susceptibility of *Candida albicans* sputum isolates to *Lavandula angustifolia* essential oil (LAEO) as well as to determine its fungicidal potential on tested cultures in a liquid phase. In addition, we compared the antifungal activity of LAEO to a selected azole (fluconazole) and a representative echinocandin (caspofungin).

2. Results

2.1. Microbial Identification

Out of the 15 analyzed *Candida* isolates, 10 strains were recognized as *C. albicans* using chromogenic media. Additional five isolates were identified by matrix-assisted laser desorption in ionization-time of flight mass spectrometry (MALDI-TOF MS). Peptide mass fingerprinting of these isolates as well as *C. albicans* ATCC 10259 performed by MALDI-TOF MS are shown in Figure 1.

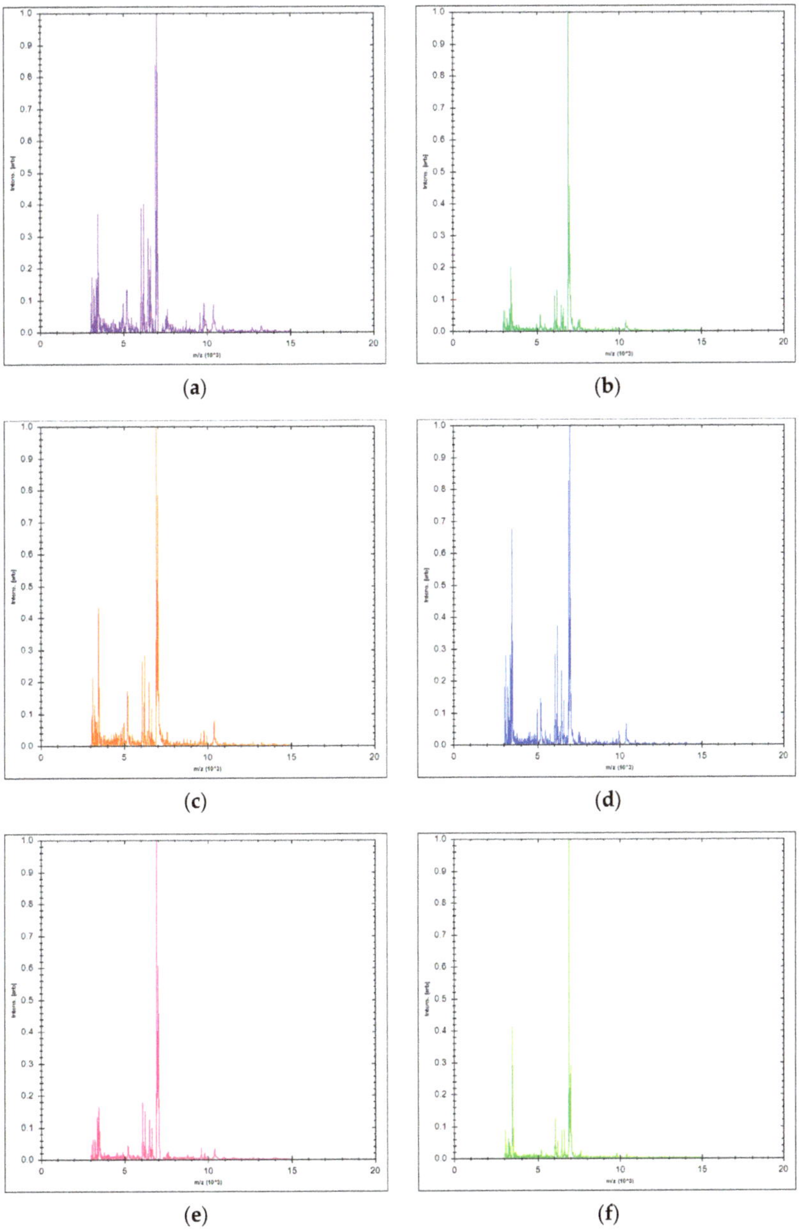

Figure 1. MALDI-TOF mass spectra (m/z 3000 to 15,000) of: (**a**) *C. albicans* PED02; (**b**) *C. albicans* PED06; (**c**) *C. albicans* PED12; (**d**) *C. albicans* PED13; (**e**) *C. albicans* PED15; (**f**) *C. albicans* ATCC 10259.

2.2. Chemical Composition of Essential Oil

A total of 50 compounds were detected in LAEO (Figure 2), of which 30 were identified (97.4% of whole composition) (Table 1). These constituents could be classified in six different groups: monoterpene hydrocarbons, oxygenated monoterpenes, sesquiterpene hydrocarbons, oxygenated sesquiterpenes, diterpene hydrocarbons and oxygenated diterpenes. In Table 1, the 10 main components determined by European Pharmacopoeia representing 84.6% of the whole LAEO composition are bolded [19]. The main constituents of the LAEO were linalool and linalyl acetate, constituting 33.4% and 30.5%, respectively. The contribution of 1,8-cineole (5.0%) and camphor (4.6%) deviated from the ranges determined by European Pharmacopoeia and the ISO 3515 standard, therefore this LAEO can be considered an essential oil of non-standard quality.

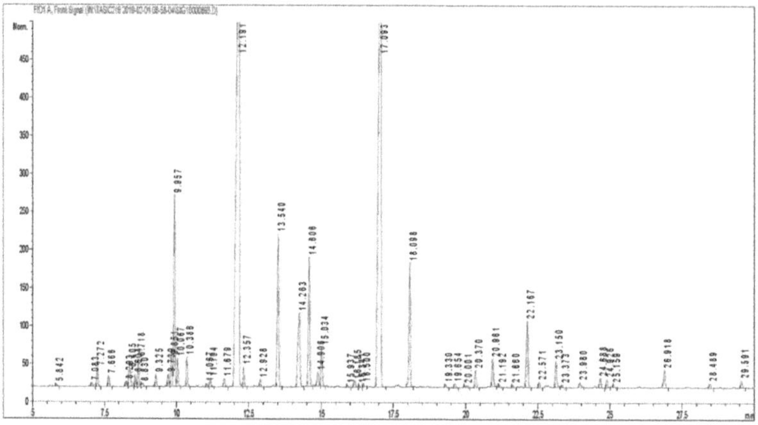

Figure 2. Chromatogram of *L. angustifolia* essential oil showing the separation of 50 chemical components.

Table 1. Composition of the *L. angustifolia* essential oil used in this study.

Components	Contents (%) According to Ph. Eur [1]	Content (%) of LAEO [2] Sample
α-thujene		0.1
α-pinene		0.5
camphene		0.3
octen-3-ol		0.2
octanone-3	0.1–5	0.8
myrcene		0.7
α-phellandrene		0.3
p-cymene		0.3
limonene	Max 1	0.9
1,8-cineole	Max 2.5	5.0
cis-β-ocimene		0.7
trans-β-ocimene		0.7
linalool oxide cis		0.2
linalool oxide trans		0.2
linalool	20–45	33.4
octenyl acetate		0.5
camphor	Max 1.2	4.6
borneol		3.2
terpinene-4-ol	0.1–8	3.9
lavandulol	Min 0.1	0.4
α-terpineol	Max 2	1.2
hexyl isovalerate		0.2
linalyl acetate	25–47	30.5
lavandulyl acetate	Min 0.2	3.5
neryl acetate		0.5
geranyl acetate		0.8
β-caryophyllene		2.3
trans-α-bergamotene		0.2
trans-β-farnesene		0.8
caryophyllene oxide		0.6

[1] European Pharmacopoeia. [2] *L. angustifolia* essential oil.

2.3. Minimum Inhibitory Concentration (MIC) and Minimum Fungicidal Concentration (MFC)

The MICs of LAEO that effectively inhibited grows of *C. albicans* sputum strains ranged from 0.5 to 2% (mean MIC 1.17%; MIC_{50} 1%; MIC_{90} 2%) (Table 2). LAEO MFCs for the tested isolates ranged between 1–4% (mean MFC 1.93%; MFC_{50} 2%; MFC_{90} 2%). The MIC and MFC of LAEO for *C. albicans* ATCC 10259 were in the range observed among clinical isolates (MIC 2%; MFC 2%).

Table 2. MIC and MFC values of *L. angustifolia* essential oil and drugs against *Candida albicans* strains evaluated by the broth microdilution method.

Strains	LAEO [1]		Fluconazole		Caspofungin	
	MIC [2]	MFC [3]	MIC	MFC	MIC	MFC
C. albicans PED01	1	2	64	>64	0.25	2
C. albicans PED02	1	2	0.5	>64	0.25	2
C. albicans PED03	0.5	1	64	>64	4	4
C. albicans PED04	1	2	1	>64	0.25	0.5
C. albicans PED05	1	2	0.5	>64	0.25	0.25
C. albicans PED06	2	2	0.125	>64	0.25	2
C. albicans PED07	1	2	64	>64	0.5	0.5
C. albicans PED08	1	2	64	>64	0.5	2
C. albicans PED09	1	2	64	>64	0.5	2
C. albicans PED10	0.5	1	0.06	>64	0.25	1
C. albicans PED11	0.5	1	0.125	>64	0.25	2
C. albicans PED12	2	4	0.5	>64	0.5	0.5
C. albicans PED13	2	2	0.5	>64	0.25	2
C. albicans PED14	2	2	0.125	>64	0.125	0.5
C. albicans PED15	1	2	0.125	>64	0.5	1
C. albicans ATCC 10259	2	2	0.5	>64	0.5	2

[1] LAEO = *L. angustifolia* essential oil; [2] MIC = minimum inhibitory concentration (values expressed in %vol./vol.); MIC for fluconazole and caspofunin are expressed in µg/mL; [3] MFC = minimum fungicidal concentration (values expressed in %vol./vol.); MFC for fluconazole and caspofunin are expressed in µg/mL.

Five clinical strains (*C. albicans* PED01, PED03, PED07, PED08 and PED09) were resistant to fluconazole with MICs 64 µg/mL, whereas MICs for susceptible isolates ranged from 0.06 to 1 µg/mL (mean MIC 21.57 µg/mL; MIC_{50} 0.5 µg/mL; MIC_{90} 64 µg/mL). Fluconazole MFC_{50} and MFC_{90} were >64 µg/mL. Six sputum isolates (*C. albicans* PED03, PED07, PED08, PED09, PED12 and PED15) were categorized as caspofungin non-WT strains with MICs > 0.25 µg/mL, while MIC for WT isolates ranged between 0.125–0.25 µg/mL (mean MIC 0.57 µg/mL; MIC_{50} 0.25 µg/mL; MIC_{90} 0.5 µg/mL). Caspofungin concentration 2 µg/mL completely kills 50% and 90% of tested isolates. In addition, there were no significant differences in the LAEO MICs among fluconazole-resistant (median 1%) and fluconazole-susceptible (median 1%) sputum strains (Mann–Whitney U test; $p = 0.31$). In addition, no LAEO MICs differences were observed among caspofungin non-WT (median 1%) and WT (median 1%) isolates (Mann–Whitney U test; $p = 0.79$).

2.4. Time-Kill Assay

The time-kill curves of the 2% LAEO against the 16 *Candida* strains are illustrated in Figure 3. Time intervals to reach 50%, 90% and 99.9% reduction in the number of yeast cells from the starting inoculums for each strain are summarized in Table 3. After 30 min exposure to LAEO in 4 strains, the cell count was reduced $\geq$3.0 log_{10} CFU/mL in range from 3.0 to 3.2 log_{10} CFU/mL, whereas in the remaining 12 strains, the reduction ranged only 0.47 to 2.47 log_{10} CFU/mL. This implicates that the mentioned concentration of LAEO showed more fungistatic than fungicidal activity during 30 min exposure.

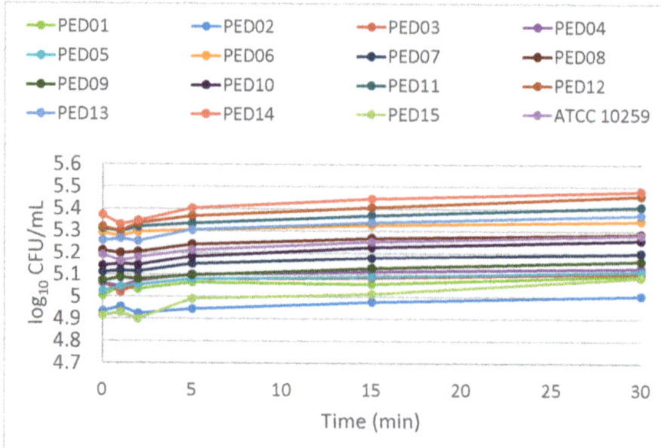

(a)

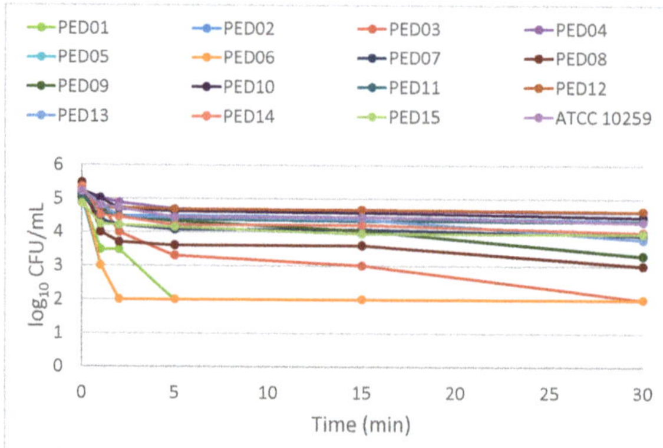

(b)

Figure 3. Time-kill of *C. albicans* sputum strains on different times (0–30 min): (**a**) untreated (grow controls) of *C. albicans* (PED01–PED15) and *C. albicans* ATCC 10259; (**b**) *C. albicans* isolates (PED01–PED15) and *C. albicans* ATCC 10259 treated with 2% concentration of lavender essential oil (LAEO); Within 30 min exposure in 4 strains (PED01, PED03, PED05, PED06) cell count was reduced $\geq$ 3.0 log_{10} CFU/mL, where in rest 12 strains reduction ranged from 0.47 log_{10} CFU/mL (PED04) to 2.47 log_{10} CFU/mL (PED08).

A 50% growth reduction for all sputum strains was rapidly achieved between 0.2 and 3.5 min (approximately 0.83 min). In addition, the time points required to reach 90% and 99.9% were more variable among the tested strains. In fact, within 30 min of exposure to LAEO, 90% reduction was observed with nine strains (*C. albicans* PED01, PED03, PED05, PED06, PED07, PED08, PED09, PED13 and PED14) in time range 30 s–18 min. LAEO exhibited 99.9% reduction in four strains (*C. albicans* PED01, PED03, PED05 and PED06) between 1.8 and 29.7 min. For other isolates, together with *C. albicans* ATCC 10259 no reduction was achieved in the CFU of 90% and 99.9% during 30 min of LAEO exposure.

The number of *C. albicans* cells significantly declined over time in all LAEO treated sputum strains (Friedman test; $p < 0.05$) (Figure 4). Pairwise comparisons using Wilcoxon signed rank test for related samples showed significant differences in all measurement times except 5 min–15 min ($p = 0.14$) and 15 min–30 min ($p = 0.06$) (Figure 4).

Table 3. Times for 2% *L. angustifolia* essential oil to achieve 50%, 90% and 99.9% reductions in growth of starting inoculum during 30 min treatment.

Strains	2% Solution of LAEO [1]		
	50% Reduction [2]	90% Reduction	99.9% Reduction
C. albicans PED01	0.2	0.7	5
C. albicans PED02	1.4	>30	>30
C. albicans PED03	1.1	2	29.7
C. albicans PED04	3.5	>30	>30
C. albicans PED05	0.2	0.6	2
C. albicans PED06	0.2	0.5	1.8
C. albicans PED07	0.4	3.3	>30
C. albicans PED08	0.2	0.7	>30
C. albicans PED09	1.1	15.2	>30
C. albicans PED10	1.3	>30	>30
C. albicans PED11	0.6	>30	>30
C. albicans PED12	0.6	>30	>30
C. albicans PED13	0.7	18	>30
C. albicans PED14	0.4	3.1	>30
C. albicans PED15	0.5	>30	>30
C. albicans ATCC 10259	0.8	>30	>30

[1] LAEO = *L. angustifolia* essential oil; [2] Values are expressed in minutes.

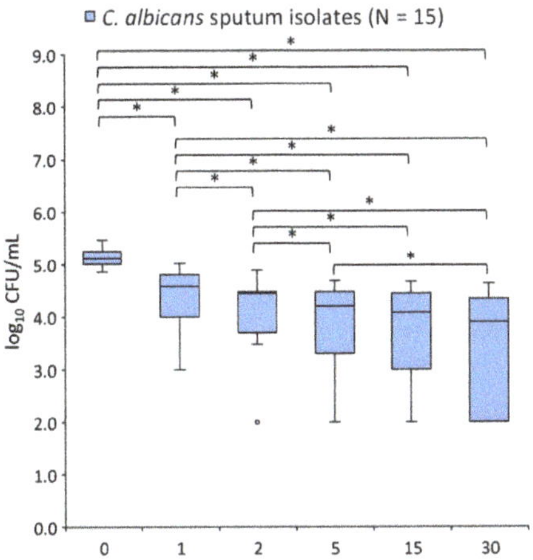

Figure 4. Box plot of cell counts ($\log_{10}$ CFU/mL) of 15 *C. albicans* sputum strains noted from pre-exposure (0 min) and within 30 min exposure to 2% lavender essential oil (LAEO). Friedman test with post hoc pairwise comparison with applied Bonferroni correction showed significant differences among all pairs (* $p < 0.05$) except 5 min–15 min and 15 min–30 min.

Among sputum strains 2% LAEO was reducing the cell counts rapidly after 1 min by approximately 0.49 $\log_{10}$ CFU/mL (68%) (Figure 5). This activity remained, reducing cell viability by approximately 0.73 $\log_{10}$ CFU/mL (81%), 0.85 $\log_{10}$ CFU/mL (86%), 0.91 $\log_{10}$ CFU/mL (87%) and 1.03 $\log_{10}$ CFU/mL (91%) within 2, 5, 15 and 30 min, respectively (Figure 5). In *C. albicans* ATCC 10259 cells reduction calculated at 2, 5, 15 and 30 min were 0.41 $\log_{10}$ CFU/mL (60%), 0.48 $\log_{10}$ CFU/mL (67%), 0.77 $\log_{10}$ CFU/mL (83%), 0.77 $\log_{10}$ CFU/mL (83%) and 0.91 $\log_{10}$ CFU/mL (88%), respectively (Figure 5).

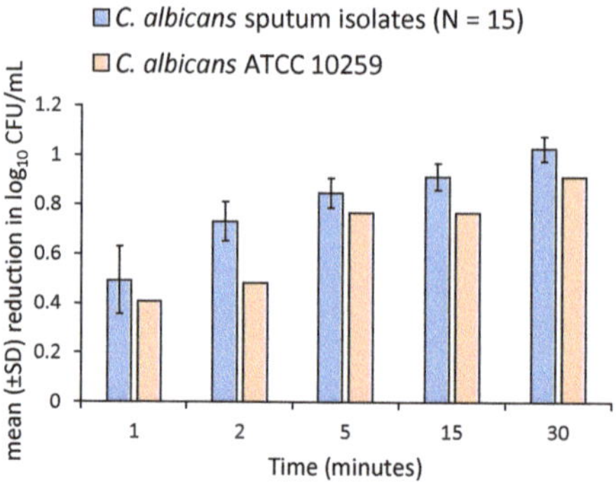

Figure 5. Bar chart of mean (±SD) reduction in $\log_{10}$ CFU/mL levels of 15 *C. albicans* sputum isolates as well as *C. albicans* ATCC 10259 from 1 to 30 min exposure to 2% of *L. angustifolia* essential oil. Within 1 min treatment mean cell reduction in sputum strains was 0.49 $\log_{10}$ CFU/mL (68%), while in 30 min their cell viability mean reduced 1.03 $\log_{10}$ CFU/mL (91%). In *C. albicans* ATCC 10259 cell count reduction ranged from 0.41 $\log_{10}$ CFU/mL (61%) to 0.91 $\log_{10}$ CFU/mL (88%) calculated in 1 min and 30 min, respectively.

3. Discussion

As natural aromatic plant products, EOs are mainly composed of terpens and terpenoids. LAEO is present with approximately 1.3% in the flowering aerial part of the plant and contains more than 100 compounds [20]. It was shown that the LAEO and its main constituents can kill or inhibit the growth of fungi [21]. The main components of this LAEO that exhibit fungistatic and fungicidal effects are monoterpene alcohols (60–65%), such as linalool (20–50% per fraction) and linalyl acetate (25–46% per fraction) [20]. The other components are cis-ocimene, terpinen4-ol, limonene, cineol, camphor, lavandulyl acetate, lavandulol and α-terpineol, β-caryophyllene, geraniol, α-pinene. Non-terpenic aliphatic components such as 3-octanon, 1-octen-3-ol, 1-octen-3-y-lacetate, 3-octanol are present as well [22]. They are determined based on the chemical compositions LAEO, but EOs components could differ among other *Lavandula* spp. [23]. A chemical analysis of *L. binaludensis* EO showed that γ-terpinen with participation of 71.6% is its major content [24]. Zuzarte et al. showed that the main component of *L. viridis* EO was 1,8-cineole with participation of 34.3–42.2% [25]. Out of 10 lavender EO components shown in Table 1, 8 are in range classified by the European Pharmacopoeia [19]. Two components of 1,8-cineole and camphor were in higher percentage (5.0% and 4.6%, respectively). These deviations in EOs' composition sometimes reflect non-uniformity in the bioactivity results and they could be consequences of climatic, geographical, seasonal and geological factors, physiological characteristics of plants and extraction procedures.

The action mechanism of essential oils remains somewhat controversial. An exceptional antifungal activity of EOs is probably based on their lypophilic nature which typically integrates into membrane structures thus increasing the cell permeability and leaching of intracellular membranes [26]. It has been noticed that among lavender EOs, the component α-pinene shows the best antifungal effect [25]. Other components such as linalool 1,8-cineol and linalyl acetate are also potent, contributing to the overall antifungal activity of [27,28]. In addition, linalool decreases *C. albicans* yeast cells rapidly, as the time kill essays show [21]. As an acyclic monoterpene alcohol, linalool differs from most of the monoterpenes that exhibit significant antifungal activity. It is reasonable to speculate that the presence of alcohol components primarily determines the antifungal activity and then whether the constituent has a cyclic or acyclic structure. Probably these terpene alcohol moieties disrupt the microbial cell membrane thus increasing the permeability to protons and larger ions [29].

In this investigation, LAEO exhibited considerable antifungal effects against *C. albicans* sputum isolates. Our results correlate with those of several previous studies in which MIC values for LAEO range between 0.5–2% [30,31]. However, LAEO MIC_{50} and MIC_{90} values were twice higher than the values obtained in the previous experiments. This may be a consequence of the denser inocula that we used, as well as the origin of the tested strains. D'Auria et al. showed that higher concentrations of LAEO are necessary for growth inhibition of *C. albicans* oropharingeal isolates [21]. Moreover, a small number of studies included the mentioned isolates in their research, meaning that we could thoroughly compare our results with them. Nevertheless, the authors who compared *Candida* susceptibility to EOs by vapor diffusion and microdilution reported that MIC values in the vapor phase were lower than in the liquid phase [32,33]. Mandras et al. obtained an LAEO MIC_{90} of 1% in the liquid phase; whereas in the vapor phase, the growth of all *C. albicans* strains tested at a 0.06% concentration was inhibited [34]. High volatility, poor water solubility and dispersibility of EOs and their components could be the main reasons for unequal inter-laboratory results.

The experiments performed in our study showed that LAEO MICs of the all-tested fluconazole-resistant *C. albicans* strains were not different from the MICs of fluconazole-susceptible strains implicating that *C. albicans* does not exhibit cross-resistance to the mentioned EO and fluconazole. This could be a consequence of different antifungal actions in which azoles interfere with cytochrome P-450-dependent enzyme lanosterol 14α-demethylase, which is responsible for the production of ergosterol and where several mechanisms of fluconazole resistance are possible: a decrease in the affinity of lanosterol 14α-demethylase, hyperproduction of the same enzyme, as well as a loss of sterol desaturase, leading to alteration in the ergosterol synthesis [8]. Moreover, EOs can disrupt the function of efflux pumps which, through an active transfer, scavenge the fluconazole across the plasma membrane allowing its easier entry inside the yeast cell [35]. Some authors implicate that EOs alter the synthesis of many cell wall polysaccharide components, such as β-glucan, chitin and mannan [17]. Surprisingly, our data have shown that strains in which the caspofungin MIC values were higher than 0.25 μg/mL did not require higher LAEO concentrations for their growth inhibition.

Grounded on the fact that MFC values were once or twice higher than MICs (Table 1), it is suggested that LAEO has more fungicidal than fungistatic activity during 24 h exposure [36]. Based on the MFC found for each *Candida* species, time-kill curves were constructed in order to evaluate the effect of MFC_{90} LAEO concentration during short time exposure. The kill curves obtained with 2% LAEO indicate that this concentration is primary fungistatic (< 3 $\log_{10}$ CFU/mL) against most *C. albicans* isolates for 30 min. Previous studies have shown that during a short time exposure, 2% concentration of EOs exhibit a predominantly fungistatic activity against *Candida*, but fungicidal activity may also be observed, depending on the strain, type of EO and the test conditions [21,37]. However, in our investigation, the aforementioned LAEO concentration led to 0.301 $\log_{10}$ CFU/mL (50%) yeast cells reduction within 5 min for all the tested strains. The fact that 99.9% of growth reduction was not achieved for most isolates during 30 min implies that the decreasing of the significant reduction effect observed in the first minutes of the exposure is probably in relation to fungal cell stress adaptation. This characteristic was also observed during exposure of EOs and their components to yeast cells where many stress responses and signaling pathways were induced, such as oxygen and free radical detoxification, heat shock proteins and chaperones, autophagy and vacuolar degradation mechanisms [38]. The same study showed that these responses are arising from early 15 min to late 90 min of exposure to certain EOs components.

Regarding the antifungal efficiency of the 2% LAEO against *C. albicans* strains, our EO was able to significantly reduce the number of yeast cells during a short time exposure. Unfortunately, only one study monitored the effect of LAEO against *Candida* during 30 min of exposure [21]. In the same study, 2% LAEO concentration killed 99% of the cells within 5 min and 100% of the cells were killed within 15 min. This study was performed on one strain only, so it is difficult to compare the results obtained then with the ones in our investigation. However, in our investigation the mean cell reduction calculated in 5 min and 15 min for sputum strains was approximately 86% and 88%, respectively. Similarly, cell reduction in *C. albicans* ATCC 10259 calculated within 5 min and 15 min was 83%.

4. Materials and Methods

4.1. Microorganisms

Fifteen species of *Candida albicans* isolated from sputum of pediatric patients with AML, diagnosed at the Institute for Mother and Child Health Care of Serbia "Dr Vukan Cupic", were included in this study. Initially, isolation was performed on Sabouraud dextrose agar (SDA) (HiMedia, Mumbai, India). Identification of *Candida* strains was performed after subcultivation on HiCromeTM Candida Differential Agar (HiMedia, Mumbai, India) according to the specific appearance of colonies defined by the manufacturer. For isolates that could not be identified by the aforementioned chromogenic medium, protein fingerprint analysis obtained by matrix-assisted laser desorption in ionization-time of flight mass spectrometry (MALDI-TOF MS) was used. After subcultivation on SDA at 37 °C for 24 h colony adhering to the toothpick was directly transferred onto a 96-spot polished steel target plate (Bruker Daltonics, Bremen, Germany) for spotting and allowed to air drying. The bacterial test standard was used as a positive control (Bruker Daltonics, Bremen, Germany). Immediately, 1 μL of 70% formic acid and alpha-cyano-4-hydroxycinnamic acid (HCCA) matrix (Bruker Daltonics, Bremen, Germany) were spotted onto the target plate. After drying at room temperature samples were analyzed using MALDI-TOF MS Biotyper Sirius one IVD System (Bruker Daltonics, Bremen, Germany) in automatic runs operated by flexControl, ver. 3.4.207.20 (Bruker Daltonics, Bremen, Germany). The yeast identification was achieved using MBT Compass software, ver. 4.1.100 (Bruker Daltonics, Bremen, Germany), based on the comparison of generated mass spectra with database (MBT Compass Library, Revision H, 3893 species/entries). The log score values > 1.7 indicated reliable species identification. American Type Culture Collection strain *C. albicans* ATCC 10259 as control strain suggested for EOs assays was included in the investigation as well. Complete mycological examinations were performed at the National Referential Medical Mycology Laboratory (NRMML), Faculty of Medicine, University of Belgrade (Serbia).

4.2. Essential Oil

L. angustifolia essential oil was obtained from Herba® (Belgrade, Serbia) lot N° 02560115 and stored at −18 °C until the chemical and mycological investigations were to be conducted.

4.3. Chemical Identification of Essential Oil

The gas chromatography with flame ionization detection (GC-FID) analyses were carried out with a HP-7890 Series A apparatus (Agilent Technologies Inc., Santa Clara, CA, USA) equipped with a split-splitless injector, a flame ionization detector (FID), and a HP-5 capillary column (25 m × 0.32 mm i.d., film thickness 0.52 mm). The oven temperature was programmed rising from 40 to 260 °C at 4 °C/min; injector temperature was 250 °C; detector temperature was 300 °C; carrier gas was H2 (1.0 mL/min). The relative contents expressed as percentages were obtained from electronic integration of the peak areas measured using FID.

The gas chromatography-mass spectrometry (GC/MS) analyses were performed under almost the same analytical conditions as the GC-FID analyses, with a HP G1800C Series II GCD analytical system (HewlettPackard, Palo Alto, CA, USA) equipped with a HP-5MS capillary column (30 m × 0.25 mm i.d., film thickness 0.25 mm). Helium (1.0 mL/min) was used as carrier gas, and the transfer-line temperature (MSD) was heated at 260 °C. Mass spectra were acquired in the EI mode (70 eV) over the m/z range 40–450 amu. Aliquot of 1 μL of sample diluted in EtOH (10 mL/mL) was injected in split mode (1:30). The identification of the constituents was performed by comparing their mass spectra and retention indices (RIs) with those obtained from authentic samples and/or 4 listed in the

NIST/Wiley mass-spectra libraries, using different types of searches (PBM/NIST/AMDIS) and available literature data [39,40].

4.4. Antifungal Agents

Fluconazole and caspofungin (Sigma–Aldrich, Taufkirchen, Germany) powders were dissolved in 100% dimethyl sulfoxide (DMSO) at final concentration 12.8 mg/mL and 0.8 mg/mL, respectively. Prepared stock solutions were stored at −70 °C until use.

4.5. Antifungal Susceptibility Testing

4.5.1. Preparation of Inoculums

Different colonies of 15 *Candida albicans* isolates and a control strain from overnight culture on SDA at 37 °C were suspended in 6 mL distilled water and homogenized with a gyratory vortex to the final density of the 0.5 McFarland standard. Working suspension was made by further 1:10 dilution of the standardized suspension resulting in 1–5 × 10^5 CFU/mL. Cell counts and viability of yeasts were confirmed in triplicates using a hemocytometer chamber by adding trypan blue.

4.5.2. Determination of LAEO, Fluconazole and Caspofungin MIC and MFC Values

LAEO stock solution was prepared in pure ethanol (1:2) and then diluted in RPMI 1640 broth with L-glutamine (Capricorn Scientific, Ebsdorfergrund, Germany), buffered to pH 7.0 with 0.165 M morpholinepropanesulfonic acid (MOPS; Sigma-Aldrich, Taufkirchen, Germany), supplemented with 2% glucose to obtain its final concentration of 8% (*vol./vol.*). Tween 80 in final concentration of 0.5% was added to enhance the EO solubility. Initially, 200 µL of freshly prepared LAEO stock solution was poured in the first rows of flat-bottomed 96-well microtiter plates. One hundred µL of RPMI 1640 with MOPS and 2% glucose were added in the remained wells. Nine serial two-fold dilutions were performed by the withdrawal of a 100 µL aliquot from the concentrated well into the succeeding one. Preparations of fluconazole and caspofungin work solutions, together with the susceptibility testing, were performed according to the EUCAST (European Committee on Antimicrobial Susceptibility Testing) document E.DEF 7.3.2. After adding of 100 µL appropriate working yeast suspension (1–5 × 10^5 CFU/mL), the final concentration of LAEO ranging between 4–0.008%, 64–0.125 µg/mL for fluconazole and 4–0.008 µg/mL for caspofungin were obtained. EO-free and antifungals-free grow control and sterility control wells were also included. The final solvent concentrations did not affect yeast growth as their concentration did not exceed 2% for ethanol and 1% for DMSO. All microtiter plates were incubated at 37 °C for 24 h in aerobic atmosphere.

After incubation, the plates were estimated to fungal growth. Minimal inhibitory concentration (MIC) of LAEO was defined as the lowest concentration, showing complete inhibition of visible growth. For fluconazole and caspofungin, MIC determinated the lowest concentration reducing $\geq$ 50% of the culture growth in comparison with the drug-free growth control reading by the microplate reader on 450 nm (Tecan Sunrise, Mannedorf, Switzerland). EUCAST clinical breakpoints are used to categorize the fluconazole susceptibility. Due to the significant inter-laboratory variation in caspofungin MIC ranges, a recommended epidemiological cut-off (ECV) value of 0.25 µg/mL was used to categorize the isolates into wild-type (WT) and non-WT [41]. MIC_{50} and MIC_{90} were defined as the lowest concentrations of each substance capable of inhibiting 50% and 90% of the tested isolates, respectively.

The minimal fungicidal concentration (MFC) of EO and tested drugs were estimated after MIC determination as the lowest concentration that kills 99.9% yeast cells after subculturing 10 µL of broth taken from all the wells without turbidity on SDA. MFC_{50} and MFC_{90} were defined as the lowest concentrations that kill 50% and 90% of tested isolates, respectively. Each experiment was performed in duplicates on three different dates. The results were reported as modal values.

4.6. Time-Kill Essay

In order to evaluate the further fungicidal effect of LAEO MFC_{90}, a time-kill assay on 15 *C. albicans* sputum isolates and *C. albicans* ATCC 10259 was performed in a liquid phase according to the method proposed by Klepser et al. with some modifications [42]. One hundred microliters of a working suspension 1–5 × 10^5 CFU/mL was poured onto flat-bottomed 96-well micro titration plates and 100 μL of 4% EO, previously prepared by dissolving in pure ethanol (1:2) and further diluting in RPMI 1640 broth with MOPS, 2% glucose and 0.5% Tween 80, were added (the final EO concentration was 2%). Comparatively, nontreated growth controls were prepared as well. After incubation at 0, 1, 2, 5, 15 and 30 min, 10 μL from each test samples were removed for three serial 1:10 dilutions (10 μL samples and 90 μL distilled water). Colonies were counted after 100 μL inoculation of the second and the third dilution on SDA and incubation at 37 °C for 48 h in aerobic atmosphere. The limit of detection was ≥ 10^2 CFU/mL. The experiment was performed in duplicates and average values are included for further observations. Time-kill curves were constructed by plotting logarithmic values of colony counts ($\log_{10}$ CFU/mL) as a function of time (in minutes). In order to determine fungicidal or fungistatic character of LAEO, a reduction in the growth ≥ 3 $\log_{10}$ CFU/mL of the starting inoculums was defined as a fungicidal activity of EO [43]. Approximate time intervals to achieve 50% (≥0.301 $\log_{10}$ CFU/mL), 90% (≥1 $\log_{10}$ CFU/mL) and 99.9% (≥3 $\log_{10}$ CFU/mL) of growth reduction were calculated with non-linear interpolation function using Excel (version 16; Microsoft Corp., Redmond, WA, USA) for Windows 10 (Microsoft Corp., Redmond, WA, USA). Ultimately, roughly counted, the overall reduction in *C. albicans* sputum strains for each time interval expressed as a $\log_{10}$ CFU/mL was calculated using the following Equation (1):

$$mean\ \log_{10}\mathrm{CFU/mL}\ reduction = -\log_{10}\left(1 - \frac{mean\ \%\ reduction}{100}\right) \quad (1)$$

4.7. Statistical Analysis

Normality distribution of the variables was verified using the Kolmogorov–Smirnov and Shapiro–Wilk tests. MIC values were compared by using the Mann–Whitney U test. For comparison of the time-kill test results Friedman test with post hoc pairwise comparison (Wilcoxon signed-rang test) with applied Bonferroni correction were used to determine the correlation between $\log_{10}$ CFU/mL values measured at different times. In all the cases, p values < 0.05 were considered as statistically significant. All statistical analyses were performed using Excel (version 16; Microsoft Corp., Redmond, WA, USA) for Windows 10 (Microsoft Corp., Redmond, WA, USA) and R Commander (R package ver. 2.7-1; R Core Team, Vienna, Austria).

5. Conclusions

As expected, the major components of LAEO were linalool and linalyl acetate. Our results demonstrated the efficacy of LAEO in growth reduction in *C. albicans* sputum isolates. The 2% solutions of LAEO possess potent and rapid antifungal activity that is primary fungistatic within 30 min. The same concentration is equally effective against fluconazole susceptible and fluconazole resistant, as well as against WT and non-WT caspofungin isolates. Although the prophylactic treatment of invasive candidiasis is based on systemic antifungal agents, further studies are necessary to assess the application of the essential oils as aromatherapy with high-risk patients.

Author Contributions: S.M. performed all the mycological experiments, analyzed the data and wrote a draft of the paper; J.A.S. designed all the experiments, reviewed and revised the draft manuscript; D.B. and D.P. performed the experiments and analyzed data related to chemical identification of essential oil. I.C.C., E.D. and A.D. oversaw the laboratory work, analyzed the data, reviewed and revised the draft manuscript. All authors have read and agreed to the published version of the manuscript.

Funding: This research was funded by the Ministry of Education, Science and Technological Development of the Republic of Serbia through the Grant Agreement with University of Belgrade—Faculty of Pharmacy No. 451-03-68/2022-14/200161.

Institutional Review Board Statement: Not applicable.

Informed Consent Statement: Not applicable.

Data Availability Statement: Not applicable.

Conflicts of Interest: The authors declare no conflict of interest. The funders had no role in the design of the study; in the collection, analyses, or interpretation of data; in the writing of the manuscript, or in the decision to publish the results.

References

1. Steinbach, W.J. Pediatric Invasive Candidiasis: Epidemiology and Diagnosis in Children. *J. Fungi* **2016**, *2*, 5. [CrossRef]
2. Pappas, P.G.; Lionakis, M.S.; Arendrup, M.C.; Ostrosky-Zeichner, L.; Kullberg, B.J. Invasive candidiasis. *Nat. Rev. Dis. Primers* **2018**, *4*, 18026. [CrossRef]
3. Lockhart, S. Current Epidemiology of *Candida* Infection. *Clin. Microbiol. Newsl.* **2014**, *36*, 131–136. [CrossRef]
4. Palazzi, D.L.; Arrieta, A.; Castagnola, E.; Halasa, N.; Hubbard, S.; Brozovich, A.A.; Fisher, B.T.; Steinbach, W.J. *Candida* speciation, antifungal treatment and adverse events in pediatric invasive candidiasis: Results from 441 infections in a prospective, multinational study. *Pediatr. Infect. Dis. J.* **2014**, *33*, 1294–1306. [CrossRef]
5. Hope, W.W.; Castagnola, E.; Groll, A.H.; Roilides, E.; Akova, M.; Arendrup, M.C.; Arikan-Akdagli, S.; Bassetti, M.; Bille, J.; Cornely, O.A.; et al. ESCMID guideline for the diagnosis and management of *Candida* diseases 2012: Prevention and management of invasive infections in neonates and children caused by *Candida* spp. *Clin. Microbiol. Infect.* **2012**, *18* (Suppl. S7), 38–52. [CrossRef]
6. Mukherjee, P.K.; Sheehan, D.; Puzniak, L.; Schlamm, H.; Ghannoum, M.A. Echinocandins: Are they all the same? *J. Chemother.* **2011**, *23*, 319–325. [CrossRef]
7. Arendrup, M.C.; Perlin, D.S. Echinocandin resistance: An emerging clinical problem? *Curr. Opin. Infect. Dis.* **2014**, *27*, 484–492.
8. Whaley, S.G.; Berkow, E.L.; Rybak, J.M.; Nishimoto, A.T.; Barker, K.S.; Rogers, P.D. Azole Antifungal Resistance in *Candida albicans* and Emerging Non-albicans *Candida* Species. *Front. Microbiol.* **2017**, *7*, 2173. [CrossRef] [PubMed]
9. Uzun, E.; Sariyar, G.; Adsersen, A.; Karakoc, B.; Ötük, G.; Oktayoglu, E.; Pirildar, S. Traditional medicine in Sakarya province (Turkey) and antimicrobial activities of selected species. *J. Ethnopharmacol.* **2004**, *95*, 287–296. [CrossRef]
10. Chassagne, F.; Samarakoon, T.; Porras, G.; Lyles, J.T.; Dettweiler, M.; Marquez, L.; Salam, A.M.; Shabih, S.; Farrokhi, D.R.; Quave, C.L. A Systematic Review of Plants with Antibacterial Activities: A Taxonomic and Phylogenetic Perspective. *Front. Pharmacol.* **2021**, *11*, 586548. [CrossRef] [PubMed]
11. Armijos, C.; Matailo, A.; Bec, N.; Salinas, M. Chemical composition and selective BuChE inhibitory activity of the essential oils from aromatic plants used to prepare the traditional Ecuadorian beverage horchata lojana. *J. Ethnopharmacol.* **2020**, *263*, 113162. [CrossRef] [PubMed]
12. Maia, M.F.; Moore, S.J. Plant-based insect repellents: A review of their efficacy, development and testing. *Malar. J.* **2011**, *10* (Suppl. S1), S11. [CrossRef]
13. Carvalhinho, S.; Costa, A.M.; Coelho, A.C.; Martins, E.; Sampaio, A. Susceptibilities of *Candida albicans* mouth isolates to antifungal agents, essentials oils and mouth rinses. *Mycopathologia* **2012**, *174*, 69–76. [CrossRef] [PubMed]
14. Noumi, E.; Snoussi, M.; Hajlaoui, H.; Trabelsi, N.; Ksouri, R.; Valentin, E.; Bakhrouf, A. Chemical composition, antioxidant and antifungal potential of *Melaleuca alternifolia* (tea tree) and *Eucalyptus globulus* essential oils against oral *Candida* species. *J. Med. Plants Res.* **2011**, *5*, 4147–4156.
15. Marcos-Arias, C.; Eraso, E.; Madariaga, L.; Quindós, G. In vitro activities of natural products against oral *Candida* isolates from denture wearers. *BMC Complement. Altern. Med.* **2011**, *11*, 119. [CrossRef] [PubMed]
16. Gucwa, K.; Milewski, S.; Dymerski, T.; Szweda, P. Investigation of the Antifungal Activity and Mode of Action of *Thymus vulgaris, Citrus limonum, Pelargonium graveolens, Cinnamomum cassia, Ocimum basilicum*, and *Eugenia caryophyllus* Essential Oils. *Molecules* **2018**, *23*, 1116. [CrossRef]
17. Rajkowska, K.; Kunicka-Styczyńska, A.; Maroszyńska, M.; Dąbrowska, M. The effect of thyme and tea tree oils on morphology and metabolism of *Candida albicans*. *Acta Biochim. Pol.* **2014**, *61*, 305–310. [CrossRef]
18. Mertas, A.; Garbusińska, A.; Szliszka, E.; Jureczko, A.; Kowalska, M.; Król, W. The influence of tea tree oil (*Melaleuca alternifolia*) on fluconazole activity against fluconazole-resistant *Candida albicans* strains. *Biomed. Res. Int.* **2015**, *2015*, 590470. [CrossRef] [PubMed]
19. Council of Europe. *European Pharmacopoeia*, 10th ed.; Council of Europe: Strasbourg, France, 2019.
20. Fleming, T. *PDR for Herbal Medicines*, 2nd ed.; Medical Economics Company: Montvale, NJ, USA, 2000.
21. D'Auria, F.D.; Tecca, M.; Strippoli, V.; Salvatore, G.; Battinelli, L.; Mazzanti, G. Antifungal activity of *Lavandula angustifolia* essential oil against *Candida albicans* yeast and mycelial form. *Med. Mycol.* **2005**, *43*, 391–396. [CrossRef] [PubMed]

22. European Medicines Agency. Assessment Report on Lavandula Angustifolia Miller, Aetheroleum and Lavandula Angustifolia Miller, Flos. Available online: https://www.ema.europa.eu/en/documents/herbal-report/final-assessment-report-lavandula-angustifolia-miller-aetheroleum-lavandula-angustifolia-miller-flos_en.pdf (accessed on 18 June 2022).
23. Karpiński, T.M. Essential Oils of *Lamiaceae* Family Plants as Antifungals. *Biomolecules* **2020**, *10*, 103. [CrossRef]
24. Minooeianhaghighi, M.H.; Sepehrian, L.; Shokri, H. Antifungal effects of *Lavandula binaludensis* and *Cuminum cyminum* essential oils against *Candida albicans* strains isolated from patients with recurrent vulvovaginal candidiasis. *J. Mycol. Med.* **2017**, *27*, 65–71. [CrossRef] [PubMed]
25. Zuzarte, M.; Gonçalves, M.J.; Cavaleiro, C.; Canhoto, J.; Vale-Silva, L.; Silva, M.J.; Pinto, E.; Salgueiro, L. Chemical composition and antifungal activity of the essential oils of *Lavandula viridis* L'Her. *J. Med. Microbiol.* **2011**, *60*, 612–618. [CrossRef] [PubMed]
26. D'agostino, M.; Tesse, N.; Frippiat, J.P.; Machouart, M.; Debourgogne, A. Essential Oils and Their Natural Active Compounds Presenting Antifungal Properties. *Molecules* **2019**, *24*, 3713. [CrossRef] [PubMed]
27. Mandras, N.; Roana, J.; Scalas, D.; Del Re, S.; Cavallo, L.; Ghisetti, V.; Tullio, V. The Inhibition of Non-albicans *Candida* Species and Uncommon Yeast Pathogens by Selected Essential Oils and Their Major Compounds. *Molecules* **2021**, *26*, 4937. [CrossRef]
28. Singulani, J.L.; Pedroso, R.S.; Ribeiro, A.B.; Nicolella, H.D.; Freitas, K.S.; Damasceno, J.L.; Vieira, T.M.; Crotti, A.E.; Tavares, D.C.; Martins, C.H.; et al. Geraniol and linalool anticandidal activity, genotoxic potential and embryotoxic effect on zebrafish. *Future Microbiol.* **2018**, *13*, 1637–1646. [CrossRef]
29. Maurya, A.; Prasad, J.; Das, S.; Dwivedy, A.K. Essential Oils and Their Application in Food Safety. *Front. Sustain. Food Syst.* **2021**, *5*, 653420. [CrossRef]
30. Bogdan, M.A.; Bungau, S.; Tit, D.M.; Zaha, D.C.; Nechifor, A.C.; Behl, T.; Chambre, D.; Lupitu, A.I.; Copolovici, L.; Copolovici, D.M. Chemical Profile, Antioxidant Capacity, and Antimicrobial Activity of Essential Oils Extracted from Three Different Varieties (Moldoveanca 4, Vis Magic 10, and Alba 7) of *Lavandula angustifolia*. *Molecules* **2021**, *26*, 4381. [CrossRef]
31. Perić, M.; Rajković, K.; Milić Lemić, A.; Živković, R.; Arsić Arsenijević, V. Development and validation of mathematical models for testing antifungal activity of different essential oils against *Candida* species. *Arch. Oral Biol.* **2019**, *98*, 258–264. [CrossRef]
32. Santomauro, F.; Donato, R.; Sacco, C.; Pini, G.; Flamini, G.; Bilia, A.R. Vapour and Liquid-Phase *Artemisia annua* Essential Oil Activities against Several Clinical Strains of *Candida*. *Planta Med.* **2016**, *82*, 1016–1020. [CrossRef]
33. Tyagi, A.K.; Malik, A. Liquid and vapour-phase antifungal activities of selected essential oils against *Candida albicans*: Microscopic observations and chemical characterization of *Cymbopogon citratus*. *BMC Complement. Altern. Med.* **2010**, *10*, 65. [CrossRef]
34. Mandras, N.; Nostro, A.; Roana, J.; Scalas, D.; Banche, G.; Ghisetti, V.; Del Re, S.; Fucale, G.; Cuffini, A.M.; Tullio, V. Liquid and vapour-phase antifungal activities of essential oils against *Candida albicans* and non-albicans *Candida*. *BMC Complement. Altern. Med.* **2016**, *16*, 330. [CrossRef] [PubMed]
35. Ahmad, A.; Khan, A.; Manzoor, N. Reversal of efflux mediated antifungal resistance underlies synergistic activity of two monoterpenes with fluconazole. *Eur. J. Pharm. Sci.* **2013**, *48*, 80–86. [CrossRef]
36. Siddiqui, Z.N.; Farooq, F.; Musthafa, T.N.M.; Ahmad, A.; Khan, A.U. Synthesis, characterization and antimicrobial evaluation of novel halopyrazole derivatives. *J. Saudi Chem. Soc.* **2013**, *17*, 237–243. [CrossRef]
37. D'Auria, F.D.; Laino, L.; Strippoli, V.; Tecca, M.; Salvatore, G.; Battinelli, L.; Mazzanti, G. In vitro activity of tea tree oil against *Candida albicans* mycelial conversion and other pathogenic fungi. *J. Chemother.* **2001**, *13*, 377–383. [CrossRef] [PubMed]
38. Zhang, Y.; Muend, S.; Rao, R. Dysregulation of ion homeostasis by antifungal agents. *Front Microbiol.* **2012**, *3*, 133. [CrossRef]
39. Adams, R.P. *Identification of Essential Oil Components by Gas Chromatography/Mass Spectrometry*, 4th ed.; Allured Publishing Corporation: Carol Stream, IL, USA, 2007.
40. Hochmuth, D.H. *MassFinder 3: Software for GC/MS Interpretation and Presentation, Mass Spectral Library Administration*; Hochmuth Scientific Consulting: Hamburg, Germany, 2006.
41. Espinel-Ingroff, A.; Alvarez-Fernandez, M.; Cantón, E.; Carver, P.L.; Chen, S.C.; Eschenauer, G.; Getsinger, D.L.; Gonzalez, G.M.; Govender, N.P.; Grancini, A.; et al. A multicenter study of epidemiological cutoff values and detection of resistance in *Candida* spp. to anidulafungin, caspofungin and micafungin using the Sensititre YeastOne colorimetric method. *Antimicrob. Agents Chemother.* **2015**, *59*, 6725–6732. [CrossRef]
42. Klepser, M.E.; Ernst, E.J.; Lewis, R.E.; Ernst, M.E.; Pfaller, M.A. Influence of test conditions on antifungal timekill curve results: Proposal for standardized methods. *Antimicrob. Agents Chemother.* **1998**, *42*, 1207–1212. [CrossRef]
43. Scorneaux, B.; Angulo, D.; Borroto-Esoda, K.; Ghannoum, M.; Peel, M.; Wring, S. SCY-078 Is Fungicidal against *Candida* Species in Time-Kill Studies. *Antimicrob. Agents Chemother.* **2017**, *61*, e01961-16. [CrossRef]

molecules

MDPI

Review

Seaweeds in the Oncology Arena: Anti-Cancer Potential of Fucoidan as a Drug—A Review

Jun-O Jin [1,2,*,†], Dhananjay Yadav [3,†], Kajal Madhwani [4,†], Nidhi Puranik [5], Vishal Chavda [6,7,*] and Minseok Song [3,*]

1 Shanghai Public Health Clinical Center & Institutes of Biomedical Sciences, Shanghai Medical College, Fudan University, Shanghai 201508, China
2 Department of Microbiology, University of Ulsan College of Medicine, Seoul 05505, Korea
3 Department of Life Sciences, Yeungnam University, Gyeongsan 38541, Korea
4 Department of Microbiology, BETS Science College, Sabar Kantha 385535, India
5 Department of Biochemistry & Genetics, Barkatullah University, Bhopal 462026, India
6 Department of Pathology, Stanford School of Medicine, Stanford University Medical Center, Stanford, CA 94305, USA
7 Department of Medicine, Multispeciality, Trauma and ICCU Center, Sardar Hospital, Ahmedabad 382350, India
* Correspondence: junojin@amc.seoul.kr (J.-O.J.); chavdavishal2@gmail.com (V.C.); minseok@yu.ac.kr (M.S.)
† These authors contributed equally to this work.

Citation: Jin, J.-O.; Yadav, D.; Madhwani, K.; Puranik, N.; Chavda, V.; Song, M. Seaweeds in the Oncology Arena: Anti-Cancer Potential of Fucoidan as a Drug—A Review. *Molecules* **2022**, *27*, 6032. https://doi.org/10.3390/molecules27186032

Academic Editors: Raffaele Pezzani and Sara Vitalini

Received: 27 July 2022
Accepted: 8 September 2022
Published: 16 September 2022

Abstract: Marine natural products are a discerning arena to search for the future generation of medications to treat a spectrum of ailments. Meanwhile, cancer is becoming more ubiquitous over the world, and the likelihood of dying from it is rising. Surgery, radiation, and chemotherapy are the mainstays of cancer treatment worldwide, but their extensive side effects limit their curative effect. The quest for low-toxicity marine drugs to prevent and treat cancer is one of the current research priorities of researchers. Fucoidan, an algal sulfated polysaccharide, is a potent therapeutic lead candidate against cancer, signifying that far more research is needed. Fucoidan is a versatile, nontoxic marine-origin heteropolysaccharide that has received much attention due to its beneficial biological properties and safety. Fucoidan has been demonstrated to exhibit a variety of conventional bioactivities, such as antiviral, antioxidant, and immune-modulatory characteristics, and anticancer activity against a wide range of malignancies has also recently been discovered. Fucoidan inhibits tumorigenesis by prompting cell cycle arrest and apoptosis, blocking metastasis and angiogenesis, and modulating physiological signaling molecules. This review compiles the molecular and cellular aspects, immunomodulatory and anticancer actions of fucoidan as a natural marine anticancer agent. Specific fucoidan and membranaceous polysaccharides from *Ecklonia cava*, *Laminaria japonica*, *Fucus vesiculosus*, *Astragalus*, *Ascophyllum nodosum*, *Codium fragile* serving as potential anticancer marine drugs are discussed in this review.

Keywords: seaweeds; marine algae; marine drugs; fucoidan; sulfated polysaccharide; cancer; adjuvant; prebiotics

1. Introduction

Modern pharmaceutical science is based on natural products (NPs) derived from cells and tissues of microorganisms, plants, and animals of both aquatic and terrestrial ecosystems. NPs are secondary or specialized metabolites employed as traditional medicines and are now regarded as pillars of conventional pharmacology. Secondary metabolites are critical prerequisites for competitiveness, stress tolerance, anti-predator adaption, antimicrobial immunity, anti-radiation, and marine anti-biofouling activities [1–3]. In order to overcome stress circumstances produced by fluctuating or variable environmental factors, for example, light intensity, temperature, moisture, and mechanical wounds, NPs are deemed important [4]. Moreover, NPs are critical modulators of complex interspecies interactions that have an impact on the survival and function of organisms [5]. As a result of the survival and environmental adaptation processes, very complex and diversified compounds that are extraordinarily potential therapeutics that synthetic small molecules cannot match are encouraged to be developed.

Although oceans cover more than three-quarters of the Earth's surface and marine natural products (MNPs) have a pronounced cytotoxic activity, most bioactive secondary metabolites have been acquired from terrestrial organisms [6]. The adaptation to distinct environmental conditions has resulted in diversification and sophisticated development of marine creatures from their terrestrial counterparts. Marine flora is taxonomically diverse, chemically distinct, biologically efficient, and highly specialized. Most sessile marine invertebrates have been found to produce active natural chemical products for a variety of ecological functions, such as defense against predators, parasites, illnesses, and competition, making aquatic ecosystems an important potential source of bioactive chemical compounds [7]. Hundreds of novel MNPs and over 36,000 marine compounds are identified each year from micro and macro-organisms such as fungus, bacteria, microalgae, macroalgae, sponges, corals, and tunicates. Many of these compounds interact with receptors and enzymes, highlighting their pharmacologic activity and constantly attracting enormous attention in the biomedical area [8]. Several marine compounds have demonstrated the ability to intervene in complex inhibitory actions and regulate several physiological functions such as cell signaling, permeability, angiogenesis, and apoptosis [9]. In general, the marine environment has created several highly effective marine-derived compounds with potential anticancer effects confirmed in-vitro, in-vivo, and in clinical trials.

Cancer remains a major cause of morbidity and mortality in industrial nations despite decades of basic and clinical research on promising new medicines. Cancer development is a dynamic and long-term process involving many complicated components and follows a step-by-step progression that eventually leads to metastasis. Initiation, promotion, and progression are the three crucial steps in creating various forms of human cancer. After cardiac disorders, cancer is the second biggest cause of death in industrialized countries, accounting for one out of every four deaths. Cancer is a multifaceted disease caused primarily by acquired genetic changes that provide tumor cells an edge in terms of survival or proliferation [10]. It is a complex multi-step process involving multiple factors such as smoking, occupational exposure, environmental pollution, unreasonable diet, genetics, viral infection, etc. [11]. Self-sufficiency in growth signals, evading apoptosis, resistance to anti-growth signals, unbounded replicative potential, genome instability, unceasing angiogenesis, tissue invasion, and metastasis are the hallmarks of cancer. Research must therefore continue to improve existing therapies and create new cures. Chemotherapy, radiation therapy, surgery, and combinations of these therapies have all been used to treat cancer. Regrettably, this therapeutics has various negative effects and could be replaced by MNPs [12]. In MNPs, particularly marine drugs, have been at the forefront of the issue as a chemotherapeutic substitute due to several advantages such as abundancy, inertness, high structural flexibility, renewable nature, biodegradability, biocompatibility, bioavailability, non-toxicity, easy processing and extractability [13,14]. Epidemiological studies have recently revealed that marine chemicals significantly impact this proclivity. Some natural extracts have been discovered that target specific signaling pathways to prevent or slow down the carcinogenesis process at various stages and exhibit properties such as specificity, easy stimulation of cancer cell apoptosis and minimal cytotoxicity [15]. Hence, the search for minimal toxicity natural substances is the demand of current research priorities. The anticancer therapeutic aspect of fucoidan as a natural marine drug is discussed in this review.

1.1. Anti-Cancer Marine Drugs

Since cancer is one of the world's deadliest diseases, the discovery of new anticancer drugs is extremely challenging. Recent improvements in marine bioprospecting have resulted in the identification of a large number of marine chemicals with anticancer potential [16,17]. However, whereas most marine compounds are obtained from shallow fauna, the latest anticancer compounds identified from deep sea species demonstrate how modern technology and deep sea sampling enable unequaled access to a previously untapped reservoir of biochemical diversity [18]. Only a few innovative synthetic compounds have become commercial treatments due to toxicity. Five anticancer drugs based on MNPs have previously been licensed for human use [19]. Most research into the anticancer capabilities of chemicals originating from marine invertebrates has focused on sponges, corals, and algae [20,21]. Complex ecosystems like coral reefs are characterized by severe competition for space and feeding pressure, which contain a high amount of bioactive metabolites. These metabolites have harmful or deterring effects and could be potent drugs against cancer. Various anticancer drugs are derivatives of these marine compounds that have been approved for clinical use, including cytarabine, vidarabine, nelarabine (prodrug of ara-G), fludarabine phosphate (prodrug of ara-A), trabectedin, eribulin mesylate, brentuximab vedotin, polatuzumab vedotin, enfortumab vedotin [22].

Marine plants are becoming more popular because of their ability to produce natural goods with outstanding health advantages. Marine algae is a rich source of vitamins, minerals, unsaturated fatty acid, polysaccharides and well-known bioactive compounds, including sulfated polysaccharides, phlorotannins and glycoproteins. Division Phaeophyta (known as brown seaweeds) includes the most studied species of marine plants, namely *Ecklonia, Laminaria, Undaria, Fucus, Astragalus, Codium, Asconodulum* and *Himanthalia* [23]. Seaweeds are abundant in active chemicals, like multi-ring sulphurouscyclics, macrolides and trace elements, as well as in polysaccharides, terpenoids, proteins, polyphenols, and sterols [24]. Brown algae, rich in sulfated polysaccharides and secondary plant metabolites, including fucoidans, phlorotannins, and fucoxanthin, can be used to find new medicinal agents [25–28]. These compounds have been proven to have anticancer capabilities in addition to antioxidant and immunomodulatory activities and favorable effects on chemotherapeutic side effects such as cardiovascular disease [29,30]. Antiproliferative and pro-apoptotic actions of fucoxanthin and phlorotannins have been documented [31,32]. However, the specific mechanisms have not been dismantled yet [33]. The latest research implies that the pure sulfated polysaccharides can be used as an anticancer fucoidan, which could aid in developing natural anticancer medications. Numerous studies have been undertaken to better understand the potential mechanism of fucoidan as a cancer-preventive drug in vitro and in experimental animal models and humans. There has been good documentation on the therapeutic potential of natural biodiversity along with its natural bioactive agents like as marine polysaccharides, especially fucoidan is present and this activity will enable us to develop new generations of cancer therapies [34].

1.2. Fucoidan

Fucoidan is a natural heteropolysaccharide that was isolated in 1913 by Kylin from sea brown algae, and he dubbed it "fucoidin" [35]. According to IUPAC standards, it is now known as "fucoidan"; however, it is also known as fucan, fucosan, or sulfated fucan [36]. The evolving term fucoidan covers a wide range of sulfated polysaccharides (FCSP) with a backbone of fucose (fucans), heterogeneous compositions (i.e., fucogalactans, xylofucoglucuromannan) and diverse origins [37]. Fucoidan has long been recognized as a therapeutic nutritional supplement in Asia due to its medicinal properties, including anticancer properties. Fucoidan has been a well-explored anticancer marine drug, with the first research findings appearing in the 1980s [38].

Fucoidan has been found to have several intriguing pharmacological properties, including antithrombotic, antitumoral, antiviral, and anti-inflammatory action (dos Santos Marlise A.) Animal research exploiting its numerous pharmacological characteristics has revealed anticancer and antimetastatic benefits [39]. Fucoidan delays cancer growth, exerts a cytotoxic effect, and manifests synergistic impact with anticancer chemotherapeutic drugs [40]. Over the last three years, published research on fucoidans has progressed, and a wide range of fucoidan extract products is now accessible. Fucoidans have recently been assessed, and their potential application as oncology treatments has been expanded [41]. The mechanism through which fucoidans can have a direct or indirect anticancer impact is becoming clearer. Anticancer activities of fucoidans include immune-modulatory and anti-inflammatory characteristics. Fucoidans' capacity to bind to Toll-like receptors and intervene with the action of vascular endothelial growth factors (VEGF) and matrix metalloproteinases (MMPs) could explain their anti-neoplastic properties [42].

During the development of anticancer medications, it is critical to understand the mechanisms by which certain chemicals exert their effects. Angiogenesis suppression, cell cycle arrest, apoptosis and/or necrosis induction, and immunological activation are all tactics used by anticancer medications to limit tumor growth [43]. The mode of action of algae bioactive is largely determined by their composition and chemical characteristics. However, various brown algae species have different fucoidan structures and contents.

Fucoidan Sources and Structure

Fucoidan is a water-soluble, sulfated, 1,2- or 1, 3- or 1,4-α-l-fucos polymer, found in brown marine algae as structural polysaccharides, accounting for 5–20% of dry algal weight [44,45]. Fucoidan is structurally diverse, containing fucose, uronic acid, galactose, xylose, arabinose, mannose, and glucose residues and various degrees of branching sulfate concentration, polydispersity, and irregular patterns of monomers [34,46]. The major sources of fucoidans are sea cucumber and brown algae. Brown seaweeds are a big selection of marine plants, including *Sargassum thunbergi, Ascophyllum nodosum, Fucus vesiculosus, Laminaria japonica, Fucus evanescens,* and *Laminaria cichorioides,* and so on, that are commonly found in diverse cold water sea areas (Saadaoui, Imen). Fucoidan is an anionic polysaccharide with a high concentration of l-fucose and sulfate ester groups. It is mostly derived from brown seaweed, but other algae species, such as *Fucus vesiculosus, Ascophyllum nodosum, Laminaria japonica,* and *Macrocystis pyrifera,* can be employed as sources, resulting in a variety of polymeric compositions [36,45].

Fucoidan molecular weight (MW) has been recorded in a wide range of sizes, ranging from 5–7 kDa to several hundred kDa. Low molecular weight fractions (LMWF), in particular, are thought to be more biocompatible [47]. The sugar content, glycosidic bonds, branching, and degree of sulfation and acetylation of fucoidan vary depending on the algal species [48]. *Laminaria japonica, Saccharina latissima, Sargassum polycystum,* and *Saccharina longicruris* predominantly contain galactose, while *Ascophyllum nodosum, Fucus serratus, Fucus vesiculosus* comprise glucose, mannose, xylose, and uronic acids as other monosaccharides [49–51]. The backbone structures of *Fucus vesiculosus, Fucus serratus,* and *Ascophyllum nodosum* are dominated by linear (1→3) and (1→4) glycosidic linkages, while *Chorda filum* has a complex backbone structure due to the addition of a few branched fucose residues to the main chains, and further *Analipus japonicus* and *Undaria pinnatifida* have highly complex and heavily branched backbone structures [51,52]. Exceptionally invertebrate fucans rarely have regular, repetitive patterns regarding sugar units, glycosidic linkages, branching points, sulfation, or acetylation [37]. However, sulfate content, degree of sulfation, and molecular weight are typically cited as factors influencing the bioactivities of fucoidan. Thus, due to structural variability and lack of regularity in fucoidan molecules, it is frequently difficult to formulate distinct conclusions about the chemical structures of fucoidans.

Fucoidan is a mucilaginous component from the brown algal surface. Different kinds of fucoidan can be obtained, and extraction procedures vary depending upon its source. In general, the extraction of fucoidan using dilute acid or alkalis takes a long time and a big volume of reagents. Chemical, physical, and/or enzymatic treatments and other purifying and fractionation techniques can be used to extract and purify fucoidans from seaweeds [27,53]. Moreover, microwave or ultrasound is utilized to create vibrations in cellular water molecules, breaking apart cells and enhancing the efficiency of the traditional water extraction process. Enzymatic extraction is a highly efficient and specific approach involving cell wall dissolution and cell content removal [26]. According to in vivo studies, anticancer activity exhibited by fucoidan is dependent on source, dosage, frequency, and mode of administration. As per current research, the metabolic rate of fucoidan is subjective to different methods of administration and concentration, which has typical effects on tumors [37].

2. Pharmacokinetics of Fucoidan

Several experimental activities have been carried out to address the in-vivo bioavailability so-called ADME studies of fucoidan, i.e., absorption, distribution, metabolism, and excretion. These fundamental pharmacokinetic factors aid in the development of dosing regimens. Experiments in lab rodents may reveal the absorption and bioavailability of fucoidan. In intestinal absorption studies on rats, *Fucus vesiculosus* fucoidan (737 kDa) was utilized. After treatment, a maximum concentration of fucoidan in serum was reached after 4 h, with residual absorbed fucoidan accumulating in the kidney. In addition, absorption of fucoidan from *C. okamuranus* in rats has also confirmed the accumulation of fucoidan in organs [54]. Furthermore, comparative bioavailability tests between LMWF and (moderate molecular weight fucoidan) MMWF have shown that LMWF has a better absorption rate and bioavailability; therefore, its biological potential is higher. Furthermore, in rats, following topical administration of fucoidan (MW 750 kDa) from *Fucus vesiculosus* demonstrated fine skin-penetrating characteristics. In comparison to an intravenous supply of the same dose, the topical use of the 100 mg/kg body weight resulted in a long half-life of the substance. Hence, fucoidan is taken up by endocytosis and has been detectable in human serum and urine during topical or oral fucoidan therapy [55]. Moreover, recent pharmacokinetic research of *Fucus vesiculosus* [56] confirmed that the MW of fucoidan plays a crucial role in its absorption and deposition, which could be linked to its prolonged half-life following topical application [57]. In addition, the pharmacokinetics of fucoidan may also be affected by the type of pharmaceutical formulation. According to Kimura et al., encapsulation of fucoidan in nanoparticles can boost at least partial cytotoxicity due to higher permeability. However, fucoidan's pharmacokinetics with respect to toxicity may still be favorable; data on its biodistribution in humans is currently lacking. Few clinical trials are presently ongoing focusing on the bio-distribution and tolerance of fucoidan. The bio-distribution, safety, and dosimetry of a tagged fucoidan are investigated in healthy persons or volunteers (ClinicalTrials.gov, Identifier: NCT03422055). Patients with stage III-IV non-small cell lung cancer (NSCLC) are being evaluated in a placebo-controlled trial. Fucoidan is added to their chemotherapy treatment to see how it affects their malignancy and quality of life (ClinicalTrials.gov, Identifier: NCT03130829). The outcomes of these clinical trials are crucial in understanding ADME and fucoidan toxicity in humans. Although more research on absorption across the intestinal tract is needed, it's fascinating to see scientists working on different aspects of fucoidan absorption. As a result of the massive amount of knowledge that is gradually becoming available, future developments of fucoidan as a drug will rely on well-informed decisions [58].

2.1. Fucoidan and Cancer

The anticancer mechanisms of fucoidan are multifaceted. Fucoidan has a broad range of impacts on cellular functions, including cell cycle regulation, RNA metabolism, protein metabolism, carbohydrate metabolism, bioenergetics, mitochondrial maintenance, and DNA repair pathways. Induction/inhibition of reactive oxygen species (ROS), mitochondrial instability, and caspase and poly (ADP-ribose) polymerase (PARP) cleavage are all aspects of it [59,60]. Increased radiation susceptibility [61], prolonged cell cycle arrest, improved immunological clearance, enhanced apoptosis, decreased DNA damage, and metastatic suppression seems to be direct and indirect anti-neoplastic actions of fucoidans. Fucoidans are independent checkpoint modulators that can be used with modern checkpoint inhibitors to treat cancer [62].

2.1.1. Anticarcinogenic Mechanism of Fucoidan

According to previous studies, the anticancer mechanism of fucoidan largely covers four components. They are (i) cell cycle arrest, (ii) induction of apoptosis, (iii) anti-angiogenesis and (iv) anti-inflammatory activities.

2.1.2. Role of Fucoidan in Cell Cycle Arrest and Apoptosis

Fucoidan inhibits cancer cell proliferation by regulating the cell cycle and blocking uncontrolled mitosis. Alekseyenko et al., 2007, performed Lewis lung cancer transplantation in C57 mice and found that tumor mass and lung metastasization were significantly reduced in fucoidan-treated mice, showing that fucoidan successfully suppressed tumor cell metastasis and proliferation in vivo [63]. Fucoidan (22.5 mg/mL) therapy-induced cell cycle arrest in the G2/M phase and inhibited cell growth in HCC cell lines [64]. Fucoidan boosted the number of G1/S cells in HUT-102 and NSCLC-N6 cell lines, while fucoidan promoted apoptosis and cell cycle arrest in HTLV-1 infected T cells and MCF-7 cells [65,66].

Programmed cell death, such as apoptosis, necroptosis and autophagy, is normally the result of an imposed form of cell cycle arrest. Apoptosis is essential for the survival of multicellular organisms. In general, apoptosis is required for homeostasis and is frequently deregulated in human diseases such as cancer [67]. In contrast to necrosis, apoptotic cells experience chromatin condensation, DNA fragmentation, cleavage of specific proteins, cellular DNA damage, morphological alterations such as blebbing and the budding of apoptotic bodies [68], as well as distinct energy-dependent biochemical mechanisms [69]. Fucoidan promotes apoptosis in cancer cells by aggressively activating apoptotic signals and their underlying downstream pathways. Eun et al., 2010 cocultured human colon cancer cells HT-29 and HCT116 with fucoidan derived from *Fucus vesiculosus* [70]. Fucoidan caused caspase-3, -7, -8, -9 activation, chromatin condensation, and PARP cleavage. According to the findings, fucoidan can effectively cause apoptosis via caspase-8 and -9-dependent pathways. In human lymphoma HS-Sultan cell lines, fucoidan suppresses tumor growth and promotes apoptosis [71]. Similarly, in MCF-7 cells, fucoidan extract enhanced mitochondrial depolarization by upregulating proapoptotic proteins Bax and Bad and downregulating antiapoptotic proteins Bcl-2 and Bcl-xl expression [72]. Furthermore, fucoidan treatment causes PARP cleavage and caspase-3/7 activation in MCF-7 cells, which are hallmarks of apoptosis [73]. Interestingly, Miyamoto et al., 2009 reported that MCF-7 cells require caspase-7, whereas activation of caspase-3 is not necessary for fucoidan-induced apoptosis [74].

Fucoidan can also show an antitumor effect through simultaneous cell cycle inhibition and induction of cancer cell apoptosis. In the human hepatoma SMMC-7721 cells, fucoidan therapy caused noteworthy growth inhibition and ROS-mediated apoptosis involving several distinctive features such as chromatin condensation or marginalization, vacuolization, lower glutathione consumption (GSH), mitochondrial swelling, and depolarization of the mitochondrial membrane potential [75]. Fucoidan inhibits PI3K, suppressing ERK and activates MAPK, limiting cancer cell proliferation and decreasing Bcl-2 to Bax ratio, inducing caspase-dependent apoptosis in BEL-7402 and LM3 cell lines [76]. Fucoidan inhibited cell growth in hepatocellular, cholangiocarcinoma, and gallbladder carcinoma cell lines by inducing apoptosis and inhibiting cell cycle progression in a dose- and time-dependent manner [64].

2.1.3. Fucoidan and Angiogenesis

Cancer cells rely on an adequate supply of oxygen and nutrients for cell proliferation and metastatic dissemination, highlighting angiogenesis and lymph angiogenesis are critical processes for tumorigenesis. Antiangiogenic therapy has proven to be an effective way to slow tumor growth; consequently, angiogenesis inhibitors are a popular target for cancer treatment [77]. Surplus sulfated fucoidan promotes antitumor and antiangiogenic actions in cancer cells. Fucoidan suppresses vascular endothelial growth factor (VEGF) expression, restricts angiogenesis, cuts off the neoplastic cells' nutrients and oxygen, lowers its volume, and impedes invasion and metastasis. Using blocking VEGF165 from attaching to its cell surface receptor vascular endothelial growth factor receptor-2 (VEGFR-2), Koyanagi et al., 2003 observed that both natural and oversulfated fucoidans dramatically reduced the mitogenic and chemotactic activities of VEGF165 on HUVEC. They also claimed that increasing the number of sulfate groups in the fucoidan molecule increases its antiangiogenic and antitumor properties [78]. Tse-hung et al., 2015 supplied fucoidan to Lewis lung cancer cell implanted mice, resulting in reduced serum and lung tissue VEGF levels when compared to controls [79]. Accordingly, in mice implanted with Sarcoma-180 cells, both native and oversulfated fucoidans suppressed neovascularization [78]. Moreover, the experimental angiogenesis test showed that the expression of the angiogenesis factor VEGF-A was dramatically reduced by 100 mg/mL fucoidan [80]. Fucoidan-mediated in vivo inhibition of neovascularization is also reported [78]. These findings clearly state that fucoidan's antitumor activity is attributable to its antiangiogenic effect.

Fucoidans exhibit entirely diverse effects on angiogenesis depending on their molecular weight and sulfate content. Natural fucoidan with a high molecular weight (30 kDa) has antiangiogenic properties, inhibiting proliferation, migration, endothelic tube formation, and vascular network formation, whereas low (4–9 kDa) and intermediate (15–20 kDa) molecular weight fucoidan stimulate angiogenesis and enhance HUVEC migration, respectively. Contrastingly, over-sulfated fucoidan seems to have a more potent antiangiogenic property than natural or desulphated fucoidan [81].

2.1.4. Fucoidan and Immunomodulation

Fucoidan remodels the immune system to improve the sustainability of cancer-specific T cells and NK cells and results in the elimination of cancer cells. Crude and modified fucoidans have been proven to have immunopotentiating effects in tumor-bearing mice, resulting in antitumor efficacy [27,53]. Several studies have confirmed crude fucoidan's profound cell survival and natural killer activity in vitro and in vivo [82]. Farzaneh et al., 2015 found that fucoidan limits the development of promyelocytic leukemia cells in vitro and in vivo by boosting the cytotoxic activity of NK cells and delaying tumor formation [40]. In line with this, fucoidan reduced cell proliferation in Huh7 human liver cancer cells more effectively by inhibiting the chemotaxin CXCL12 and CXCR4 pathway, an effective target for cancer therapy [54]. Fucoidan has a strong anti-inflammatory activity; however, the underlying mechanisms exerted by fucoidan have not been fully elucidated. Fucoidan has been shown to impede lymphocyte adhesion and invasion, enzyme inhibition, and induce apoptosis at various phases of the inflammatory process. The downregulation of MAPK and NF-B signaling pathways, as well as selectins and the concomitant decrease in the generation of proinflammatory cytokines, is the most often debated putative mode of action for fucoidan [83,84].

2.1.5. Fucoidan and Metastasis

Tumor cell proliferation and survival are frequently linked with deregulated intracellular signal transduction and persistent activation of cellular pathways [38]. Metastasis is a leading cause of cancer-related fatalities, accounting for up to 90% of all cases due to its systemic and pervasive nature and higher drug resistance. Metastasis is the multi-stage process through which cancer cells spread to different body areas through circulation, lymphatic system, or direct extension. The antimetastatic property of fucoidans is evidenced in both in-vitro and in-vivo studies. Fucoidan had moderate anticancer and antimetastatic effects following single and recurrent administration at a dose of 10 mg/kg [63]. Fucoidans from *L. saccharina*, *L. digitata*, *Fucus serratus*, *Fucus distichus*, and *Fucus vesiculosus* effectively inhibited MDA-MB-231 breast carcinoma cell adhesion, underlining anti-metastasis outcomes [84]. During metastasis, malignant cells break away from the primary tumor and invade nearby tissues by destroying the extracellular matrix (ECM) with MMPs. In a mice model of lung cancer, increased levels of fucoidan exhibited diminished MMP-2 activity and altered expression of adhesive and migratory proteins, such as integrin, VEGF-1 and -2, P-selectin and neuropilin-1 [85,86]. In the 4T1 xenograft model, fucoidan greatly lowers tumor volume and the number of metastatic lung nodules by minimizing cell migration and invasion while suppressing cell proliferation, colony formation, and expression of epithelial to mesenchymal transition (EMT) biomarkers in vitro [59,60]. An important function in the modulation of EMT in cancer cells is the molecular network of transforming growth factor-β (TGF-β) receptors (TGFRs). Fucoidan lowers the amount of TGF-RI and TGF-RII proteins, including Smad5/8 and Smad2/3 phosphorylation and Smad4 expression, affecting downstream molecular signaling. Fucoidan inhibits ECM degradation and reduces invasiveness in hepatoma HCC cells by downregulating the TGFR and SMAD signals [87]. Fucoidan also can efficiently reverse TGFRs, elicit EMT, stimulate the expression of epithelial markers, depress the expression of interstitial markers, and diminish the expression of transcriptional repressors, including Twist, Snail, and Slug. MAPK/ERK/PI3K/Akt/mTOR pathways regulate cell proliferation, differentiation, and death, and thereby the number of malignancies exploits all of them to be invasive or migratory. Antimetastatic effect of fucoidan is owing to its ability to block MAPK cascade, inactivate ERK1/2 pathway, inhibit phosphorylation of PI3K-Akt, and suppress mTOR and its downstream 4E-BP1 and p70S6K, and downregulate the expression of MMP-2 in time- and concentration-dependent manner. Remarkably, the highest concentration at which fucoidan has the most inhibitory effect is 200 mg/mL [88]. Fucoidan preferentially suppresses lung cancer by promoting Smurf2-dependent ubiquitination of TGF receptors [89], downregulation of MMPs, VEGF, and activation of tumor apoptotic signaling molecules. Besides this, fucoidan therapy also suppresses NF-kB and activator protein-1 (AP-1) transcription factors that are deregulated in cancer, inhibiting tumorigenesis, promotion, and metastasis [90,91]. AP-1 binds to the promoters of several cellular genes and starts their immediate gene expression as these genes are involved in the transcriptional regulation process. Fucoidan suppresses the AP-1 activation by inhibiting JunD expression in an HTLV-1-infected T-cell line, thereby inhibiting HTLV-1-infected T-cell proliferation [66]. Fucoidan inhibits Ik-Ba phosphorylation and raises total p65 in the cytoplasm while decreasing it in the nucleus, modulating inflammation, immunity, differentiation, cell growth, tumorigenesis and apoptosis [88]. Fucoidan inhibits the transcription of c-fos, c-jun, and AP-1 as well as decreases phosphorylation of ERK and JNK via inhibiting EGF-induced EGFR phosphorylation [92], enhancing antimetastatic activities (Figure 1).

Furthermore, post-transcriptional mechanisms have been linked to TGF signaling via microRNAs (miRs) and have been involved in regulating EMT. Apparently, dose-dependent overexpression of microRNA-29b (miR-29b) in human HCC cells was demonstrated by blocking its downstream DNA methyltransferase 3B target (DNMT3B) and enhancing tumor metastasis suppressor gene 1 (MTSS1) expression [87]. Fucoidans are promising candidates being used as functional foods and pharmaceuticals for cancer prevention, chemotherapeutic agonists, and nanotechnology-based targeted therapy due to their extensive bioactivities in cancer, notably in metastasis.

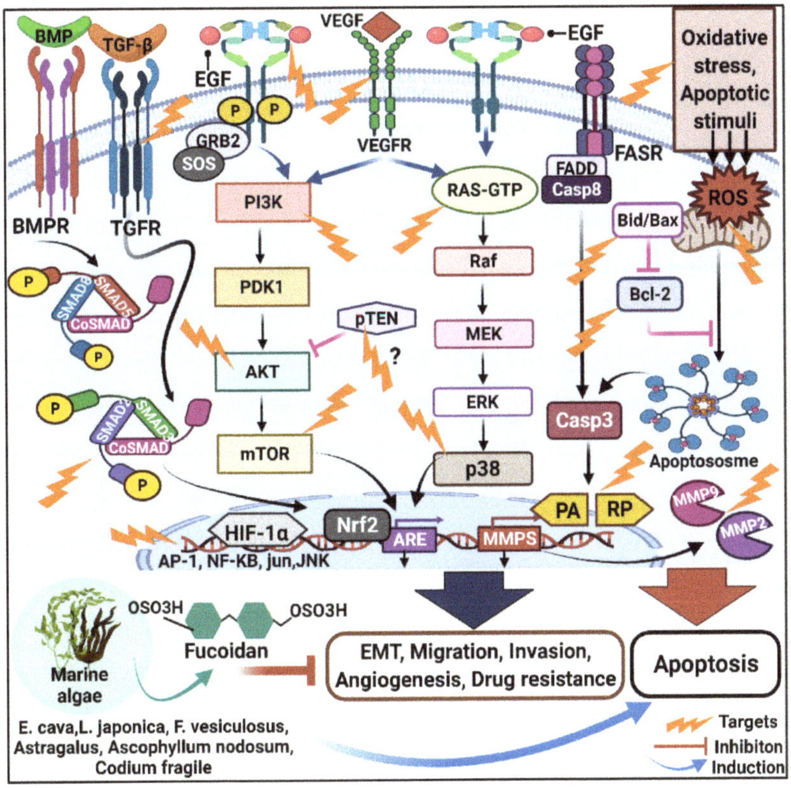

Figure 1. Schematic representation of following signaling pathways and molecules targeted by fucoidan to exert anticancer activities. Fucoidan from marine algae modulates TGF-β, PI3K/Akt/MAPK and extrinsic and intrinsic apoptosis pathways to enhance cytotoxic effects. The aforementioned pathways are deregulated in cancer, resulting in Epithelial-mesenchymal transition (EMT), migration, adhesion, drug resistance and metastasis. Fucoidan enhances antimetastatic effects, induces apoptosis, and serves as a potential marine anticancer drug.

2.1.6. Fucoidan and Gut Flora

Due to their commercial abundance, brown seaweeds are frequently used as food additives. Fermentation of high molecular Fucoidan extracted from Ascophyllum nodosum and Laminaria japonica lowers inflammatory response while promoting Ruminococcaceae and Lactobacillus, respectively [93]. Meanwhile, dietary fucoidan progressively restores intestinal villi by upregulating the expression of tight junction proteins such as ZO-1, Occludin, Claudin-1, and Claudin-8 via p38 MAPK and ERK1/2 activation. Eventually, fucoidan supplementation improves intestinal barrier function by enhancing intestinal microbiota diversity and causing changes in microbial composition with a higher Bacteroidetes/Firmicutes phylum ratio. Fucoidan induces the enrichment of short-chain fatty acids producer Prevotella at the genus level [94]. A study revealed that fucoidan could be used as a gut flora modulator for breast cancer prevention. More detailed research is needed to better understand the association between cancer and intestinal flora and the anticancer mechanism of fucoidan [93]. As a result, the improvement of gut microbiota, as well as antiobesogenic, anti-diabetic, anti-microbial, and anticancer bioactivities offered by Fucoidan, underlines the need for more research to analyze the structure-dependent fermentation of fucoidan to assign a prebiotic effect. Tables 1–3 illustrates preclinical and clinical studies of fucoidan and its anticancer action mechanism.

Table 1. Case studies of fucoidan in alternative medicine and complementary therapy: human clinical trials.

Fucoidan Source	Cancer Type	Results	Risk of Clinically Significant Interactions	Reference
U. pinnatifida-	Breast cancer	No significant changes	Absent	[95]
Low-molecular-weight fucoidan (LMF)	Metastatic colorectal cancer (mcrc)	Improved disease control rate	Absent	[96]
HiQ-fucoidan from *Laminaria japonica*	Lung cancer	Survival rates increased by approx. 50%	Reduced the occurrence of general fatigue	[59]
Cladosiphon okamuranus	Colorectal cancer	Endure prolonged chemotherapy without fatigue	-Suppressed general fatigue -No suppression of diarrhea and neurotoxicity	[97]
Mozuku, *Cladosiphon novae-caledoniae* Kylin	Advanced cancer	Decreased level of proinflammatory cytokines	Insignificant quality of life score	[98]

Table 2. In-vivo effect of fucoidan on tumor growth in tumor-bearing mice.

Fucoidan Source	Cancer Cell Type	Action Mechanism	Reference
Cladosiphon okamuranus (LMWF) IMWF and HMWF Oral administration	Colon 26	tumor growth↓ ↑survival time ↑NK cells in the spleen	[99]
Fucus vesiculosus	4T1 Lewis lung cancer cells B16	-Inhibition of angiogenesis -Induction of apoptosis -Prevention of metastasis	[59,78,79]
Fucus evanescence	Lewis lung cancer cells	Antitumor and antimetastatic activities	[63]
Sargassum plagiophyllum	Diethylnitrosamine-induced hepatocellular	Inhibition of carcinogen metabolism	[100]
Cladosiphon okamuranus Tokida	Sarcoma 180 (S-180)-xenograft	↑cytotoxicity via NO production by fucoidan-stimulated macrophages	[101]
Undaria pinnatifida	A20	Cytolytic activity by Th1 and NK cell activation	[102]
Ascophyllum nodosum	MOPC-315 plasma cell tumor	Anti-angiogenesis	[103]
Sargassum mcclurei	colon cancer DLD-1 cells	Anti-tumorigenesis	[104]
Ecklonia cava	SKOV3 tumor xenograft	-↑ROS-mediated apoptosis -antitumoral	[105]
	CT-26 carcinoma xenograft	↑NK cell-mediated anticancer immunity	[106]
Hydroclathrus clathratus	Sarcoma 180 xenograft	Suppressed tumor growth	[107]
Sargassum stenophyllum	B16F10 cells	Antiangiogenic and antitumoral	[108]
Stoechospermum marginatum	Ehrlich ascites tumor (EAT) cells	Angio-suppressive and antiproliferative activities	[109]
Sargassum fusiforme	A549	Immunomodulatory activity	[110]
	SPC-A-1	Anti-angiogenesis	[111]

Table 3. In-vitro effect of fucoidan on different cancer cell types.

Fucoidan Source	Cell Type	Action Mechanism	Action Characteristic	Reference
Sargassum pallidum	HepG2, A549, and MGC-803	Antitumor activity	-Antioxidant	[112]
Sargassum tortile	P-388	Increases cytotoxicity	-	[113]
Sargassum micracanthum	Human head and neck squamous cell carcinoma (HNSCC)	Anticancer efficacy	-	[114]
Undaria pinnatifida	A549 SMMC-7721 NB4, KG1a, HL60, and K562	-Induces apoptosis -Inhibit cell proliferation	Down-regulation of p38, PI3K/Akt, and the activation of the ERK1/2 MAPK pathway -NK-cell ↑ -Livin, XIAP mRNA ↓ Caspase-3,8,9 ↑ Bax-to-Bcl-2 ratio↑ Cytochrome c ↑ -ROS↑	[40,115]
Fucus vesiculosus	HT-29 MCF-7 MDA-MB-231 Lewis lung A549 H1975 Huh-7 SNU-761 SNU-3085 HL-60 NB4 THP-1 SUDHL-4 OCI-LY8 NU-DUL-1 TMD8 U293 DB	-Inhibit cell proliferation -Induce cell apoptosis -Inhibit metastasis	-IRS-1/PI3K/AKT↓ -Ras/Raf/ERK↓ -Caspase-7,8,9 activation, cytochrome c, Bax ↑ Bcl-2↓ -Smad2/3,Smad4↓ -NF-κB↓ -Inhibit VEGF,MMPs -Caspase-3↑ PARP cleavage -ERK1/2, MEK1/2, JNK ↑	[59,70,116,117]
Sargassum macrocarpum	MDA-MB-231, A549, and HCT116	Induces ROS-mediated apoptosis	Inhibits STAT3 Signaling	[118]
Sargassum muticum	MCF-7 and MDA-MB-231	Induce apoptosis, antioxidant, and antiangiogenesis effects	-	[119]
Sargassum angustifolium	HeLa and MCF-7	Cytotoxic activity	-	[120]
Sargassum cinereum	Caco-2 and HCT-15	Anticancer and apoptotic effect	Enhances ROS production	[121,122]
Sargassum filipendula	HeL	Induces apoptosis	Down-regulates Bcl-2	[123]
Ecklonia cava	CT-26	Induces apoptosis	Bcl-2/Bax signal pathway	[68]
Eisenia bicyclis	SK-MEL-28, DLD-1	Inhibited the colony formation	-	[124,125]
Hizikia fusiformis	PC3	Induces ROS-dependent apoptosis	Elevated expression of Fas, FasL, Bax and tBid, and decreased expression of Bcl-2 -reduced c-Flip expression and activated caspase-8, -9 and -3, leading to an increment of poly (ADP-ribose) polymerase (PARP) cleavage	[126]
Hydroclathrus clathratus	HL-60, MCF-7	Antiproliferative activity	Induced sub-G1 arrest	[127]
Saccharina	DLD-1 T-47D	Inhibit cell proliferation	Inhibit the binding of EGF receptor with EGF	[38]
Sargassum	DLD-1 Huh6 Huh7 SK-Hep1 HepG2	Inhibit cell proliferation	-Colony formation inhibition -TGF-β R1, 2↓ Phospho-Smad2/3↓ Smad 4 protein↓	[38]

Table 3. *Cont.*

Fucoidan Source	Cell Type	Action Mechanism	Action Characteristic	Reference
Cladosiphon	MCF-7	Induce cell apoptosis	PARP cleavage Caspase-7,8,9 ↑ Cytochrome C, Bax, Bid↑	[38]
Bifurcaria bifurcata	NSCLC-N6	Inhibit cell proliferation	The growth arrest is irreversible	[38]
Turbinaria conoides	A549	-Inhibit cell proliferation -Induce cell apoptosis	G0/G1 phase arrest	[38]
Sargassum latifolium	leukemia (1301 cells)	Chemopreventive activity	Antioxidant capacity	[128]
Sargassum glaucescens	MCF-7 HT-29	Induce apoptosis	Fragmented the DNA of cancer cells	[129]

Upwards arrow shows an increase (↑), a downwards arrow a decrease (↓).

3. Fucoidan from Different Sources of Marine Sea Weeds

3.1. *Ecklonia cava*

Ecklonia cava (*E. cava*) is a brown alga widely distributed in China and Korea. *E. Cava* comprises abundant bio-active sulfated polysaccharide complexes. According to recent marine investigations, *E. cava* possess noteworthy antioxidant activity and is consequently competent in reducing the development of apoptotic bodies. Sulfated polysaccharide isolated from *E. cava* substantially promotes apoptosis via regulating pro-apoptotic caspases-7 and -8, Bax, and Bcl-xL [130]. According to Ahn et al., 2015 protamex extract (PCP) of *E. cava*, which is rich in fucose and sulfate components, strongly correlates with inhibitory growth sequels rather than antiproliferative effects involving enhanced apoptotic sub-G1 DNA contents, partial DNA fragmentation, hypo diploid sub-G0/1-peak as well as dosage dependent activated PARP and caspase 9 by dramatically modulating the Bcl-2/Bax signal pathway, indicating anticancer impact on colorectal carcinoma cells [68]. Additionally, U937 leukemic cells showed similar growth inhibitory results with sulfated polysaccharides of *E. cava* [131,132]. Furthermore, numerous researchers have discovered that fucoidan extracted from *E. cava* induces death in MCF-7 and HCT-15 cells via the activation of caspases and the regulation of MAPK, including ERK, p38 kinase, and the Akt pathway, as well as alterations in Bcl-2 and Bax [71,74,133]. Yet, it remains to be seen whether PCP is linked to MAPK signaling pathways such as ERK, p38 kinase, and the Akt pathway in CT 26 cells [68]. However, the efficiency of fucoidan in inducing apoptosis varied between different types of colon cancer cells.

Several preclinical trials have recently reported that *E. cava* fucoidan (ECF) has significant immunomodulatory effects. In view of this, ECF-mediated potent immunostimulative effects and cytotoxic effects on B16 and CT-26 carcinoma growth were reported in the mice model. However, the immunostimulatory effects of ECF on DC activation and anticancer immunity have been thoroughly investigated. Accordingly, ECF suppresses tumor growth by boosting DC activation and anti-inflammatory cytokine production. However, ECF promotes immune activation but has little effect on cancer cells since it is not an Ag-specific immune response. Because of this, cancer is treated differently from a bacterial infection in the body. ECF requires cancer antigens to stimulate an Ag-specific immunosuppressive response that targets tumors directly and promotes apoptosis to limit tumor growth. As a matter of fact, currently, cytotoxic T cell (CTL) activation is extensively exploited in cancer immunotherapy. Cancer antigen-specified CTL activation causes identification and apoptosis of antigen-presenting cells (APCs) through production of inflammatory cytokine and cytotoxic proteins such as perforins and granzyme B. CD8+ DCs activate antigen-specific CTLs in mice when antigens are presented on major histocompatibility complex (MHC) class I, which interacts with TCRs and CD8 on T cells [134,135]. Apart from antigenic presentation, CD8+ DCs produce a lot of co-stimulators and proinflammatory cytokines, which aid CD8 T cells in being differentiated into CTLs [136]. Antigen presentation, cytokine production, and co-stimulation are all signs of DC maturation. Moreover, in vitro activation of CTLs mediated by bone marrow-derived dendritic cells (BMDCs) is challenging to explain in mice, as BMDCs show distinct phenotypes and functions than in vivo DCs. In a mouse model, the combination of ECF and ovalbumin (OVA) administration resulted in ECF-mediated splenic CD8+ and CD8-DCs activation, enhanced OVA-specific T cell proliferation and IFN-γ production. These findings show that ECF, as an immunostimulatory molecule, significantly enhances antigen-specific immune responses in rodents [134]. Thus, ECF may be effective as an adjuvant to improve immune response against human malignancies. In a previous study, we compared the effects of four fucoidans isolated from *Ecklonia cava*, *Macrocystis pyrifera*, *Undaria pinnatifida*, and *Fucus vesiculosus* on human peripheral blood dendritic cells (PBDCs). In contrast to the other three fucoidans, we found that ECF significantly upregulated co-stimulatory molecules, MHC class I and II, and the production of proinflammatory cytokines in monocyte-derived DCs (MODCs) and PBDCs. Based on these findings, the study proposed that ECF may have the ability to increase human immunological activation [137].

3.2. *Laminaria japonica*

Laminaria japonica (*L. japonica*) is uncommon edible seaweed in Asia. *L. japonica* is a type of edible seaweed that is widely consumed in Asia. Laminarin is a polysaccharide composed of (13)-beta-D-glucan and (16)-beta-linkage that is isolated from *L. japonica* [138]. Laminaria has been shown to have antioxidant, antibacterial, anti-inflammatory and hypolipidemic properties. It has also been used as herbal medicine to treat cancer, gastrointestinal issues, diabetes, and arsenic poisoning. Preliminary trials suggested that *L. japonica* polysaccharides serve as a potent immunomodulator playing a pivotal role in combating cancer. In this context, polysaccharides from *L. japonica* can stimulate immune cells of both innate and adaptive immune systems, increasing the release of cytokines such as IL-10, IL-6, and IL-1 α, rendering them promising anticancer immunotherapy. It stimulates macrophages that directly release anticancer inflammatory molecules like NO and TNF- α while promoting bone marrow dendritic cells (BMDCs) maturation to activate naive T lymphocytes, thereby resulting in tumor cytotoxicity. Meng et al., 2014 confirmed declining phagocytosis, decreased lysosomal count, upregulated expression of BMDC's major membrane components, including CD80, CD83, CD86, CD40, and MHC II, and increased BMDC-released production of IL-12 and TNF- α, all of which have been shown to significantly increase BMDC maturation by *L. japonica* polysaccharides [139]. Therefore, investigation of the immunological significance of laminarian indicated considerable BMDC maturation through increased expression levels of co-stimulatory molecules, IFN-γ and TNF-α cytokines production, and consequent cytotoxic T lymphocyte activation. As a result, it is reasonable to conclude that *L. japonica* polysaccharides have a high potential to boost BMDC maturation and can be employed as an immunological stimulant as well as a therapeutic adjuvant in immune-compromised individuals [139]. Furthermore, Lin et al., 2016 revealed that optimal fermentation products of *L. japonica* enhance anti-inflammatory effects by lowering ROS production, prostaglandin E2, NF-kB (p65) phosphorylation, and macrophagic NO production [140]. Fucoidan is a more powerful free radical scavenging antioxidant than laminarin due to the presence of the sulfate groups and the anionic charge [14]. As a result of these findings, laminarin has the potential to be employed as a novel immune stimulatory molecule for cancer immunotherapy.

3.3. *Fucus vesiculosus*

Fucoidan from *Fucus vesiculosus* (*F. vesiculosus*) induced apoptosis in HCT-15 cells by activating ERKs, p38, and inhibiting the PI3K/Akt signaling pathway [133]. Another study reported that *F. vesiculosus* fucoidans deplete cytosolic and nuclear NF-KB localization to suppress the invasion and migration of human lung cancer cells [88]. According to Ulf et al., 2015 treatment with *F. vesiculosus* resulted in a strong inhibition of viability in various pancreatic cancer cell lines. The fucoidan extract arrested the cell cycle at the S/G2 phase in cancerous proliferating cells due to the up-regulation of cell cycle inhibitors such as p21, p27, p57 and TP53INP1 while unaffecting non-malignant resting cells [141]. Hence, as a side effect of *F. vesiculosus* therapy, apoptosis was induced by PARP-cleavage in tumor cells rather than mitochondrial pathways. Intriguingly, in combination with autophagy inhibitors, enhanced cytotoxicity was reported. Overall, *F. vesiculosus* is a powerful cancer cell growth inhibitor; however, it neither causes mitochondrial inflammation nor causes caspase-dependent apoptosis, or necroptosis in normal cells, underlining the non-toxicity and antioxidant properties of fucoidan [141].

Researchers have been led to study the synergistic effects of fucoidan with the existing anticancer medicines. Doxorubicin (DOX) is renowned for its powerful dose-dependent anticancer efficacy, but a slew of significant side effects constrains it. Acute cardiotoxicity is a major side effect; hence, doxorubicin-loaded fucoidan nanoparticles are currently being investigated as a viable and safer option. Fucoidan, derived from *F. vesiculosus*, reduces doxorubicin-induced acute cardiotoxicity through modulating JAK2/STAT3-mediated apoptosis and autophagy [142]. By virtue of unique accumulation and preferential location in tumors, doxorubicin-loaded fucoidan nanoparticles demonstrated stronger anticancer activity in mice [143] or lower inhibitory concentration in cancer cells [144] compared to the free DOX model. Besides this, *F. vesiculosus* combined with paclitaxel had a potential antagonistic effect in breast cancer models, according to Burney et al., 2018 [145]. Moreover, recent in vivo human ovarian cancer murine model experiments illustrated no change in paclitaxel activity when given in combination with *F. vesiculosus*, contrary to prior in-vitro investigations. Additional studies are reasonable to delimitate mechanisms contributing to a discrepancy in the in vivo activity when given in combination with paclitaxel. As a preliminary stage, pharmacokinetic clinical research is presently underway to appraise the impact of fucoidan in conjunction with chemotherapy for solid tumor patients [145]. In vitro and in-vivo studies imply that *F. vesiculosus* could be a valuable marine drug that warrants additional research and development for clinical use.

3.4. *Astragalus membranaceus*

Astragalus membranaceus, a popular immunomodulatory herb used in China, has a long history in traditional Chinese medicine. Currently, it is commonly deployed in mixed herbal decoctions as an immunomodulating agent. *Astragalus polysaccharides* have been investigated extensively, especially for their immuno-potentiating features, such as induction of DCs, efficient vaccine adjuvant, and immuno-regulated anticancer activities. *A. membranaceus* polysaccharide raises IL-2, IL-12, and TNF-secretion while lowering IL-10 levels in the blood, conferring anticancer activity in vivo, improving host immune responses, and appears to be safe and effective for antitumor therapy [146]. According to cell viability detection data, *Astragalus polysaccharides* can reduce the proliferation of human lung cancer cell lines A549 and NCI-H358 by suppressing NF-KB transcription activity and down-regulating expression of p65, p50, CyclinD1, and Bcl-xL proteins. Similarly, *Astragalus polysaccharide* has shown up-regulation of p53 and PTEN expression through regulating p53/MDM2 feedback loops, signifying one of the alleged antiproliferative mechanisms [147]. APS activates macrophages and enhances the level of nitric oxide (NO) and tumor necrosis factor-α (TNF-α), both of which serve as pro-apoptotic inducers in MCF- 7 cells [148]. Notably, *Astragalus polysaccharides* have cytotoxic and growth-inhibitory effects against breast cancer by regulating novel molecular therapeutic targets such as EGFR and ANXA1 [149]. In addition, clinical trials have shown that Astragalus polysaccharides can improve the well-being of advanced non-small cell lung cancer (NSCLC) patients by increasing the efficiency of platinum-based chemotherapy [150]. Astragalus polysaccharides, an AKT signaling pathway blocker, can be used in conjunction with Apatinib to treat gastric cancer [151]. Intranasal administration of Astragalus polysaccharide upregulated CCR7 expression, resulting in expansion of $CD11C^+$ DCs population and activation NK and T cells in the mesenteric lymph nodes along with growth-inhibitory activity against pulmonary metastatic melanoma in mice. According to in vivo studies, combining *Astragalus polysaccharide* and anti-PD-L1 antibody, immune checkpoint blockade exhibits anti-invasive and antimetastatic effects. These findings suggest the application of *Astragalus polysaccharide* as a topical mucosal adjuvant to boost the anticancer efficacy of immune checkpoint inhibitors [152]. Antioxidant, anti-inflammatory, and proapoptotic *Astragalus polysaccharides* improve the cardiotoxicity of doxorubicin by modulating PI3k/Akt and p38MAPK pathway for high antitumor and antiglomerulonephritis efficacy [153]. Furthermore, *Astragalus polysaccharide* modulates the M1/M2 macrophage pool, enabling DC maturation and synergistically augmenting the anticancer effect of a conventional chemotherapeutic agent, cisplatin. These exploratory investigations set the groundwork for additional research into the curative potential of marine drugs as surrogate immunotherapy and/or the therapeutic viability of using them to treat cancer [154]. *Astragalus polysaccharide* also effectively induced erythroid differentiation in K562 cells by regulating the LMO2, Klf1, Klf3, Runx1, EphB4, and Sp1 genes, promoting γ-globin mRNA expression and fetal hemoglobin synthesis, thereby proving its potential clinical utility for leukemia treatment [155]. Furthermore, in patients with chronic myelogenous leukemia, *Astragalus polysaccharide* improved the immunological activity of plasmacytoid dendritic cells (pDCs) and accelerated their development and maturation [156]. In comparison to 5-fluorouracil (5-FU), Astragalus polysaccharides have high anticancer and immuno-enhancement actions in vivo, involving macrophagic pinocytosis, NK cell killing activity, tumor suppression, and T cell subsets expansion [157]. In line with this, *Astragalusx polysaccharide* alleviates immunosuppression of 5-FU by up-regulating the expression of IL-2, IFN-γ and TNF-α, ensuring adequacy as an immune adjuvant for chemotherapy [158]. Thus, *Astragalus polysaccharides* can be employed for cancer management in combination with conventional chemotherapies to enhance anticancer effects.

3.5. *Ascophyllum nodosum*

Ascophyllum nodosum comprises ascophyllan, fucoidan and alginate. Fucose and xylose are nearly equimolecular in ascophyllan. At the same time, fucoidan had a significantly greater ratio of fucose to xylose, and the concentrations of these monosaccharides in the alginate were extremely low. Regarding in vitro studies, ascophyllan exerts antiproliferative effects by induction of DNA fragmentation, cleavage of PARP, activation of caspases-9 and -3, altered the mitochondrial membrane permeability and typical apoptotic nuclear morphological changes while manifests immune-modulation by stimulating the secretion of TNF-α and granulocyte colony-stimulating factor (G-CSF) [159]. In mice models, intraperitoneal administration of ascophyllan exhibits antimetastatic effects via involving the immune system, specifically enhanced NK cell activity, and inhibiting ERK-mediated MMP-9 activity [160]. Ascophyllan stimulates the ERK, p38, and JNK signaling pathways and acts as an immunostimulant by inducing human monocyte-derived DCs (MDDCs), PBDCs, and blood dendritic cells antigen-BDCA1 and BDCA3 expression [161]. Ascophyllan therapy triggered IFN- γ production and CD69 overexpression and improved NK cell cytotoxicity, but cell proliferation remained unaffected. These findings show that ascophyllan stimulates NK cell activation in vitro and in vivo, and its effect is greater than fucoidan [162]. Nevertheless, several in vitro and in vivo studies imply that ascophyllan exhibit anti-inflammatory effects on stimulated macrophages primarily by reducing NO and ROS generation and activating NK cells in a concentration-dependent manner [163,164]. In general, ascophyllan exerts potential anticancer effects by enhancing immune-enhancing events involving DC activation, Th1 immune response and antibody generation. Hence, ascophyllan can be used as an adjuvant to develop therapeutic and preventive tumor vaccines [165].

3.6. *Codium fragile*

Codium fragile (*C. fragile*) polysaccharides are linear homopolymers comprising ß-1.4-linked D-mannose residues. According to research, the anticancer properties of *C. fragile* sulfated polysaccharides are based on the antioxidant and immunomodulatory effects. The antioxidant effects of *C. fragile* sulfated polysaccharides against hydrogen peroxide-mediated oxidative stress in vitro and in vivo models have been extensively reported. In Vero cells, *C. fragile* polysaccharide has shown cytoprotective benefits by reducing intracellular ROS, enhancing cell survival, and regulating apoptosis dose-dependently. In H2O2-stimulated zebrafish, *C. fragile* polysaccharides enhanced survival and restored heartbeat while diminishing ROS, cell mortality, and lipid peroxidation [166]. *C. fragile* polysaccharide improves anticancer immunity by stimulating human monocyte-derived dendritic cells (MDDCs) to secrete more proinflammatory cytokines and overexpress CD80, CD83, and CD86, as well as MHC class I and II. *C. fragile* polysaccharide also induced BDCA1+ and BDCA3+ subsets of PBDCs, promoting syngeneic helper and cytotoxic T cell activation and proliferation [152]. The polysaccharide from C. fragile induces bone marrow-derived dendritic cells (BMDCs) to activate NK cells and infiltrate T lymphocytes that produce IFN and TNF cytokines. *C. fragile* polysaccharide stimulated NK cells, resulting in elevation of cytotoxic mediators such as IFN-γ, perforin, IL-12, granzyme B, and CD69 overexpression, ensuing anticancer immune responses [167]. Furthermore, in the mouse model, combining *C. fragile* polysaccharide and cancer self-antigen therapy effectively reduced B16 tumor growth. C. fragile polysaccharide administration strengthened anti-PD-L1 antibody-mediated anticancer immunity in CT-26 carcinoma-bearing BALB/c mice. These data suggest that *C. fragile* polysaccharide, through increasing immune activation, could be employed as an adjuvant in cancer treatment [167] (Figure 2).

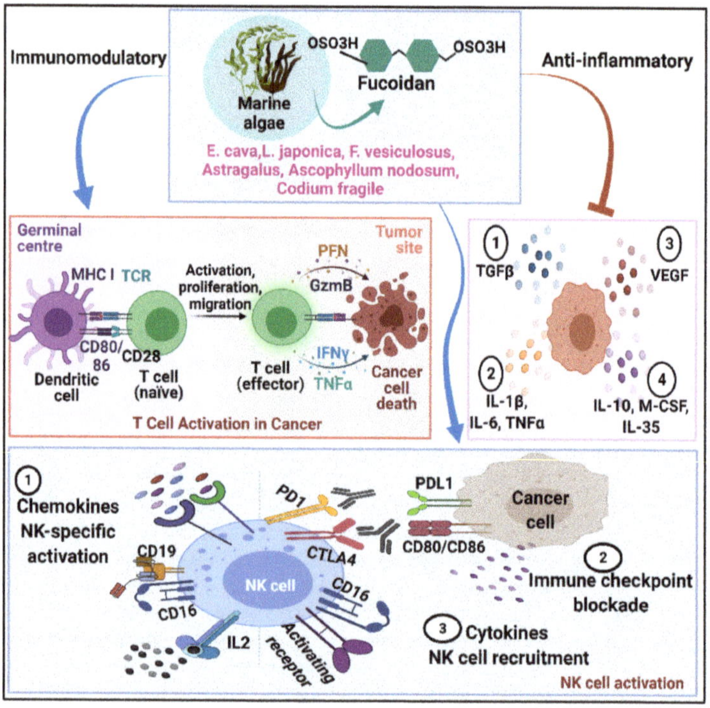

Figure 2. Depicts immunomodulatory features of fucoidan serve as a potential approach to managing cancer metastasis. Fucoidan from marine algae enhances immune cell proliferation, activation and migration. Fucoidan significantly activates NK cells and recruits them into the tumor microenvironment. It also stimulates dendritic cells to activate T cells and modulates immune responses against tumor cells. Fucoidan illustrates anti-inflammatory effects by suppressing the secretion of proinflammatory mediators at the tumor site.

4. Challenges and Future Prospects

The battle for life often leads a species to evolve its own distinctive weapons, such as speed, power, or even chemical toxins. These toxins have biologically powerful activity and distinct mechanisms of action, making them potent compounds for further therapeutic study [168,169]. Marine natural products have recently been praised for their exceptional bioactivity in extremely diluted circumstances [170,171]. Consequently, to find and develop new drugs further, it is reasonable to use marine natural products as a hit or lead compound [2,172]. One such strategy was the use of heparanase, a known anticancer drug target for developing anticancer therapy. Heparan sulfate, an essential part of the extracellular matrix, can only be broken down by heparanase, a mammalian enzyme. As a result, the extracellular matrix is remodeled, and growth factors and cytokines bound to heparan sulfate are released. Angiogenesis, immune cell migration, inflammation, wound healing, and metastasis is a few examples of physiological and pathological processes that are subsequently promoted [173]. Numerous studies have shown conclusively that heparanase plays a critical role in the advancement of cancer by degrading HSPGs, which in turn promotes neovascularization processes that support the invasion of metastatic cells [174,175]. Substances with a high level of heparanase inhibition and a low level of extracellular matrix-bound growth factor release or potentiation may hold hope for their potential as antiangiogenic and antimetastatic therapies [176]. Heparin and its derivatives, as well as other sulfated polysaccharides like fucoidan, have heparanase inhibitory activity, which enables them to block cancer cell invasion or prevent metastasis in in vitro experiments [177,178]. Although marine natural products offer a high potential for drug discovery, there are a few challenges associated with them. Despite extensive studies on fucoidan's anticancer potential and targeting capabilities, the use of fucoidan as a cancer treatment molecule or a drug delivery system has hit certain roadblocks. Firstly, pure fucoidan is indeed a costly and limited biomolecule as it is a natural polymer obtained from brown seaweed. It is difficult to obtain a sufficient quantity of these chemicals for future research [179]. Also, extraction and purification operations are sophisticated and involve several painstaking steps. *Fucus vesiculosus*, *Macrocystis pyrifera*, *Laminaria japonica* and *Undaria pinnatifida* have all been commercially marketed fucoidans. Due to the apparent expensive cost, fucoidan applications for research may have been limited, and its industrial application may have been held back. Secondly, another challenge is structural complexity. The monosaccharide content, molecular weight, and sulfation degree of fucoidan from various sources differ [180]. Due to its intricate structure, the manipulation or synthesis of marine natural products and the development of a less expensive synthetic mimic of fucoidan have been challenged [181,182]. Biological activity and its selectivity is the third obstacle. Since marine natural products are originated from marine species rather than mammals, their administration may cause undesirable biological processes in humans [183]. Fourthly, safety analysis and biodegradability testing are underway. Fucoidan has been eventually a food ingredient consumed for a long period by Asians. Even though fucoidan has been used as a food and dietary supplement for decades, studies have been undertaken to determine the safety and estimate hepatotoxicity and nephrotoxicity of pure extract and LMW fucoidan. The safety profile of fucoidan has been evaluated in vivo acute toxicity trials (4 weeks), and no significant adverse effects have been observed. Furthermore, in murine models, fucoidan from *F. vesiculosus*, *Laminaria japonica*, and *Undaria pinnatifida* exhibited no toxicological manifestations [184–186]. However, fucoidan is considered to be eliminated intact in urine since it is not a biodegradable polymer. Researchers must develop a consensus on the various families of medications that can be derived from marine algae, their mechanisms of action, and the possibilities of offering these phytochemical substances as dietary supplements to cancer patients [187]. Fifthly, the safety of fucoidan, whether utilized as nanoparticles or as a coating material for various nanosystems, has yet to be determined. In order to confirm its use as a safe and effective anticancer agent, more studies should be conducted to test the long-term safety of fucoidan nanoparticles.

Nowadays, studies are urged to support the potential evaluation of fucoidan in prospective clinical trials in combination with chemotherapy to improve cancer patient outcomes. Several preclinical investigations have already been conducted to evaluate the potential hepatic metabolism-mediated chemo-drug interactions with fucoidan. Fucoidan from *F. vesiculosus* demonstrated considerable synergistic influence when combined with paclitaxel and tamoxifen [145], whereas it had additive efficacy with topotecan and no antagonistic activity with letrozole [188]. Hence, *F. vesiculosus* fucoidan may have minimal drug-supplement interactions with either the CYP450 or the COMT hepatic metabolic pathways. Notably, fucoidan does not support tumor growth, and its immunomodulatory effect may inhibit tumor growth; however, this evidence needs to be validated in future clinical trials. Future research is needed to corroborate the findings of fucoidans in combination with chemotherapy [188,189].

Fucoidan is a potential biopolymer with vivid applications in cancer treatment. The polymer contains inherent anticancer qualities, as well as excellent drug delivery characteristics, and active targeting capabilities. An excellent application for fucoidan is cancer-targeted therapy. Fucoidan can be conjugated to cancer-specific target molecules or chemo drugs or encased in liposomes that release their payload exclusively in neoplastic cells. However, before clinical trials, the inclusive structure and other chemical features of fucoidan must be determined. Most studies on polysaccharide nanoparticles in the literature employ quite a common carrier material such as Chitosan (CS), polyethyleneimine (PEI), polyamidoamine (PAMAM), and poly-l-lysine (PLL), epirubicin (EPI)-carrageenan oligosaccharide (CAO)-gold (Au) nanoparticles for nano delivery. Furthermore, well-designed clinical trials are required to examine the effectiveness and safety of marine drugs in cancer patients.

5. Conclusions

Marine drugs are key in cancer care and can be regarded as future frontiers in pharmaceutical research. The anticancer mechanism shown by marine drugs commonly involves regulation of signal transduction, cell cycle arrest, cell apoptosis, and inhibition of migration and neo-angiogenesis, and also stimulates the immune responses and antioxidant system to prevent cancer. Maintenance of immune homeostasis and regulation are significant in cancer treatment, and according to recent research, immunomodulatory marine drugs have a high potential to reduce the negative effects of immunosuppressive chemotherapies. The pathophysiological pathways which play an imperative role in the therapeutic and preventative activities of marine medications against oncogenesis are VEGF/VEGFR2, TLR4/ROS/ER, TGFR/Smad7/Smurf2, PI3K/AKT/mTOR, p38/MAPK/ERK/JNK, TGFR/Smad/Snail, β-catenin/wnt, CXCL12/ CXCR4, and PBK/ topk. Marine polysaccharides primarily produce antitumor effects by direct cytotoxicity, immune stimulation, and synergistic effects with conventional anticancer medicines. Since fucoidan is a natural polymer, it requires extensive processing before medicinal use, and the structure of the extracted polymer is highly reliant on the source. The considerations above may limit its use on a large basis. As a result, more research is needed to establish strong structure-activity connections. Further exploration is therefore recommended to develop a solid structural activity correlation that paves the way for a polymer synthesis with expected activity and possible safety. Conclusively, fucoidan is an invaluable multifunctional therapeutic drug against cancer.

Author Contributions: Conceptualization, J.-O.J., D.Y. and V.C.; writing—original draft preparation, D.Y., K.M. and V.C.; writing—review and editing, N.P., D.Y., J.-O.J., V.C. and M.S.; supervision, J.-O.J. and M.S. All authors have read and agreed to the published version of the manuscript.

Funding: This work was supported by the National Research Foundation (NRF) of Korea (NRF-2021R1I1A3055750) and the National Institute of Biological Resources (NIBR202219101). JOJ was supported by the Basic Research Program through the NRF funded by the MSIT (NRF-2020R1A4A1016029).

Conflicts of Interest: The authors declare no conflict of interest.

References

1. Debbab, A.; Aly, A.H.; Lin, W.H.; Proksch, P. Bioactive compounds from marine bacteria and fungi. *Microb. Biotechnol.* **2010**, *3*, 544–563. [CrossRef]
2. Gerwick, W.H.; Moore, B.S. Lessons from the past and charting the future of marine natural products drug discovery and chemical biology. *Chem. Biol.* **2012**, *19*, 85–98. [CrossRef]
3. Reen, F.J.; Gutiérrez-Barranquero, J.A.; Dobson, A.D.W.; Adams, C.; Gara, F. Emerging concepts promising new horizons for marine biodiscovery and synthetic biology. *Mar. Drugs* **2015**, *13*, 2924–2954. [CrossRef]
4. Jacobo-Velázquez, D.A.; González-Agüero, M.; Cisneros-Zevallos, L. Cross-talk between signaling pathways: The link between plant secondary metabolite production and wounding stress response. *Sci. Rep.* **2015**, *5*, 8608. [CrossRef]
5. Parmentier, E.; Michel, L. Boundary lines in symbiosis forms. *Symbiosis* **2013**, *60*, 1–5. [CrossRef]
6. Hill, R.T.; Fenical, W. Pharmaceuticals from marine natural products: Surge or ebb? *Curr. Opin. Biotechnol.* **2010**, *21*, 777–779. [CrossRef]
7. Jimenez, P.C.; Wilke, D.V.; Branco, P.C.; Bauermeister, A.; Rezende-Teixeira, P.; Gaudêncio, S.P.; Costa-Lotufo, L.V. Enriching cancer pharmacology with drugs of marine origin. *Br. J. Pharmacol.* **2020**, *177*, 3–27. [CrossRef]
8. Molinski, T.F.; Dalisay, D.S.; Lievens, S.L.; Saludes, J.P. Drug development from marine natural products. *Nat. Rev. Drug Discov.* **2009**, *8*, 69–85. [CrossRef]
9. Fuentes, B. Antidiabetic drugs for stroke prevention in patients with type-2 diabetes. The neurologist's point of view. *Med. Clínica* **2018**, *150*, 275–281. [CrossRef]
10. Friedman, A. Cancer as multifaceted disease. *Math. Model. Nat. Phenom.* **2012**, *7*, 3–28. [CrossRef]
11. Stein, C.J.; Colditz, G.A. Modifiable risk factors for cancer. *Br. J. Cancer* **2004**, *90*, 299–303. [CrossRef]
12. Iwamoto, T. Clinical application of drug delivery systems in cancer chemotherapy: Review of the efficacy and side effects of approved drugs. *Biol. Pharm. Bull.* **2013**, *36*, 715–718. [CrossRef]
13. Bilal, M.; Iqbal, H.M.N. Marine seaweed polysaccharides-based engineered cues for the modern biomedical sector. *Mar. Drugs* **2020**, *18*, 7. [CrossRef]
14. Moroney, N.C.; O'Grady, M.N.; Lordan, S.; Stanton, C.; Kerry, J.P. Seaweed polysaccharides (laminarin and fucoidan) as functional ingredients in pork meat: An evaluation of anti-oxidative potential, thermal stability and bioaccessibility. *Mar. Drugs* **2015**, *13*, 2447. [CrossRef]
15. Senthilkumar, R.; Chen, B.-A.; Cai, X.-H.; Fu, R. Anticancer and multidrug-resistance reversing potential of traditional medicinal plants and their bioactive compounds in leukemia cell lines. *Chin. J. Nat. Med.* **2014**, *12*, 881–894. [CrossRef]
16. Schwartsmann, G.; da Rocha, A.B.; Berlinck, R.G.S.; Jimeno, J. Marine organisms as a source of new anticancer agents. *Lancet Oncol.* **2001**, *2*, 221–225. [CrossRef]
17. Blunt, J.W.; Carroll, A.R.; Copp, B.R.; Davis, R.A.; Keyzers, R.A.; Prinsep, M.R. Marine natural products. *Nat. Prod. Rep.* **2018**, *35*, 8–53. [CrossRef]
18. Li, K.; Chung-Davidson, Y.-W.; Bussy, U.; Li, W. Recent advances and applications of experimental technologies in marine natural product research. *Mar. Drugs* **2015**, *13*, 2694–2713. [CrossRef]
19. Ruiz-Torres, V.; Encinar, J.A.; Herranz-López, M.; Pérez-Sánchez, A.; Galiano, V.; Barrajón-Catalán, E.; Micol, V. An updated review on marine anticancer compounds: The use of virtual screening for the discovery of small-molecule cancer drugs. *Molecules* **2017**, *22*, 1037. [CrossRef]
20. Anjum, K.; Abbas, S.Q.; Shah, S.A.A.; Akhter, N.; Batool, S.; Hassan, S.S.U. Marine sponges as a drug treasure. *Biomol. Ther.* **2016**, *24*, 347–362. [CrossRef]
21. Cooper, E.L.; Hirabayashi, K.; Strychar, K.B.; Sammarco, P.W. Corals and their potential applications to integrative medicine. *Evid. Based Complementary Altern. Med.* **2014**, *2014*, 184959. [CrossRef]
22. Barreca, M.; Spanò, V.; Montalbano, A.; Cueto, M.; Díaz Marrero, A.R.; Deniz, I.; Erdoğan, A.; Bilela, L.L.; Moulin, C.; Taffin-De-Givenchy, E.; et al. Marine anticancer agents: An overview with a particular focus on their chemical classes. *Mar. Drugs* **2020**, *18*, 619. [CrossRef]
23. Bae, M.J.; Karadeniz, F.; Ahn, B.N.; Kong, C.S. Evaluation of effective mmp inhibitors from eight different brown algae in human fibrosarcoma ht1080 cells. *Prev. Nutr. Food Sci.* **2015**, *20*, 153–161. [CrossRef]
24. Kusaykin, M.; Bakunina, I.; Sovo, V.; Ermakova, S.; Kuznetsova, T.; Besednova, N.; Zaporozhets, T.; Zvyagintseva, T. Structure, biological activity, and enzymatic transformation of fucoidans from the brown seaweeds. *Biotechnol. J.* **2008**, *3*, 904–915. [CrossRef]
25. Kumar, S.R.; Hosokawa, M.; Miyashita, K. Fucoxanthin: A marine carotenoid exerting anti-cancer effects by affecting multiple mechanisms. *Mar. Drugs* **2013**, *11*, 5130–5147. [CrossRef]
26. Jin, J.-O.; Chauhan, P.S.; Arukha, A.P.; Chavda, V.; Dubey, A.; Yadav, D. The therapeutic potential of the anticancer activity of fucoidan: Current advances and hurdles. *Mar. Drugs* **2021**, *19*, 265. [CrossRef]
27. Ale, M.T.; Maruyama, H.; Tamauchi, H.; Mikkelsen, J.D.; Meyer, A.S. Fucoidan from *Sargassum* sp. And *Fucus vesiculosus* reduces cell viability of lung carcinoma and melanoma cells in vitro and activates natural killer cells in mice in vivo. *Int. J. Biol. Macromol.* **2011**, *49*, 331–336. [CrossRef]
28. Glombitza, K.W.; Hauperich, S.; Keusgen, M. Phlorotannins from the brown algae cystophora torulosa and *Sargassum spinuligerum*. *Nat. Toxins* **1997**, *5*, 58–63. [CrossRef]

29. Park, H.S.; Kim, G.Y.; Nam, T.J.; Deuk Kim, N.; Hyun Choi, Y. Antiproliferative activity of fucoidan was associated with the induction of apoptosis and autophagy in ags human gastric cancer cells. *J. Food Sci.* **2011**, *76*, T77–T83. [CrossRef]
30. Wang, S.K.; Li, Y.; White, W.L.; Lu, J. Extracts from new zealand *Undaria pinnatifida* containing fucoxanthin as potential functional biomaterials against cancer in vitro. *J. Funct. Biomater.* **2014**, *5*, 29–42. [CrossRef]
31. Hye, S.K.; Hae, Y.C.; Ji, Y.K.; Byeng, W.S.; Hyun, A.J.; Jae, S.C. Inhibitory phlorotannins from the edible brown alga ecklonia stolonifera on total reactive oxygen species (ros) generation. *Arch. Pharmacal Res.* **2004**, *27*, 194–198. [CrossRef]
32. Peng, J.; Yuan, J.-P.; Wu, C.-F.; Wang, J.-H. Fucoxanthin, a Marine Carotenoid Present in Brown Seaweeds and Diatoms: Metabolism and Bioactivities Relevant to Human Health. *Mar. Drugs* **2011**, *9*, 1806–1828. [CrossRef]
33. Chauhan, P.S.; Yadav, D.; Jin, J.O. Therapeutic potential of algal nanoparticles: A brief review. *Comb. Chem. High Throughput Screen* **2021**, *25*, 2443–2451. [CrossRef]
34. Jiao, G.; Yu, G.; Zhang, J.; Ewart, H.S. Chemical structures and bioactivities of sulfated polysaccharides from marine algae. *Mar. Drugs* **2011**, *9*, 196–223. [CrossRef]
35. Kylin, H. Zur biochemie der meeresalgen. *Hoppe-Seyler. Z. Physiol. Chem.* **1913**, *83*, 171–197. [CrossRef]
36. Berteau, O.; Mulloy, B. Sulfated fucans, fresh perspectives: Structures, functions, and biological properties of sulfated fucans and an overview of enzymes active toward this class of polysaccharide. *Glycobiology* **2003**, *13*, 29R–40R. [CrossRef]
37. Kopplin, G.; Rokstad, A.M.; Mélida, H.; Bulone, V.; Skjåk-Bræk, G.; Aachmann, F.L. Structural characterization of fucoidan from *Laminaria hyperborea*: Assessment of coagulation and inflammatory properties and their structure–function relationship. *ACS Appl. Bio Mater.* **2018**, *1*, 1880–1892. [CrossRef]
38. Lin, Y.; Qi, X.; Liu, H.; Xue, K.; Xu, S.; Tian, Z. The anti-cancer effects of fucoidan: A review of both in vivo and in vitro investigations. *Cancer Cell Int.* **2020**, *20*, 154. [CrossRef]
39. Bittkau, K.S.; Dörschmann, P.; Blümel, M.; Tasdemir, D.; Roider, J.; Klettner, A.; Alban, S. Comparison of the effects of fucoidans on the cell viability of tumor and non-tumor cell lines. *Mar. Drugs* **2019**, *17*, 441. [CrossRef]
40. Atashrazm, F.; Lowenthal, R.M.; Woods, G.M.; Holloway, A.F.; Dickinson, J.L. Fucoidan and cancer: A multifunctional molecule with anti-tumor potential. *Mar. Drugs* **2015**, *13*, 2327–2346. [CrossRef]
41. Sanjeewa, K.K.A.; Lee, J.-S.; Kim, W.-S.; Jeon, Y.-J. The potential of brown-algae polysaccharides for the development of anticancer agents: An update on anticancer effects reported for fucoidan and laminaran. *Carbohydr. Polym.* **2017**, *177*, 451–459. [CrossRef]
42. Dutot, M.; Grassin-Delyle, S.; Salvator, H.; Brollo, M.; Rat, P.; Fagon, R.; Naline, E.; Devillier, P. A marine-sourced fucoidan solution inhibits toll-like-receptor-3-induced cytokine release by human bronchial epithelial cells. *Int. J. Biol. Macromol.* **2019**, *130*, 429–436. [CrossRef]
43. Saadaoui, I.; Rasheed, R.; Abdulrahman, N.; Bounnit, T.; Cherif, M.; Al Jabri, H.; Mraiche, F. Algae-derived bioactive compounds with anti-lung cancer potential. *Mar. Drugs* **2020**, *18*, 197. [CrossRef]
44. Jankowski, V.; Vanholder, R.; Van Der Giet, M.; Henning, L.; Tölle, M.; Schönfelder, G.; Krakow, A.; Karadogan, S.; Gustavsson, N.; Gobom, J.; et al. Detection of angiotensin ii in supernatants of stimulated mononuclear leukocytes by matrix-assisted laser desorption ionization time-of-flight/time- of-flight mass analysis. *Hypertension* **2005**, *46*, 591–597. [CrossRef]
45. Li, B.; Lu, F.; Wei, X.; Zhao, R. Fucoidan: Structure and bioactivity. *Molecules* **2008**, *13*, 1671–1695. [CrossRef]
46. García-Ríos, V.; Ríos-Leal, E.; Robledo, D.; Freile-Pelegrin, Y. Polysaccharides composition from tropical brown seaweeds. *Phycol. Res.* **2012**, *60*, 305–315. [CrossRef]
47. Balboa, E.M.; Conde, E.; Moure, A.; Falqué, E.; Domínguez, H. In vitro antioxidant properties of crude extracts and compounds from brown algae. *Food Chem.* **2013**, *138*, 1764–1785. [CrossRef]
48. Ustyuzhanina, N.E.; Bilan, M.I.; Ushakova, N.A.; Usov, A.I.; Kiselevskiy, M.V.; Nifantiev, N.E. Fucoidans: Pro- or antiangiogenic agents? *Glycobiology* **2014**, *24*, 1265–1274. [CrossRef]
49. Zvyagintseva, T.N.; Shevchenko, N.M.; Chizhov, A.O.; Krupnova, T.N.; Sundukova, E.V.; Isakov, V.V. Water-soluble polysaccharides of some far-eastern brown seaweeds. Distribution, structure, and their dependence on the developmental conditions. *J. Exp. Mar. Biol. Ecol.* **2003**, *294*, 1–13. [CrossRef]
50. Rioux, L.-E.; Turgeon, S.L.; Beaulieu, M. Structural characterization of laminaran and galactofucan extracted from the brown seaweed *Saccharina longicruris*. *Phytochemistry* **2010**, *71*, 1586–1595. [CrossRef]
51. Bilan, M.I.; Grachev, A.A.; Shashkov, A.S.; Nifantiev, N.E.; Usov, A.I. Structure of a fucoidan from the brown seaweed *Fucus serratus* L. *Carbohydr. Res.* **2006**, *341*, 238–245. [CrossRef]
52. Chevolot, L.; Mulloy, B.; Ratiskol, J.; Foucault, A.; Colliec-Jouault, S. A disaccharide repeat unit is the major structure in fucoidans from two species of brown algae. *Carbohydr. Res.* **2001**, *330*, 529–535. [CrossRef]
53. Ale, M.T.; Meyer, A.S. Fucoidans from brown seaweeds: An update on structures, extraction techniques and use of enzymes as tools for structural elucidation. *RSC Adv.* **2013**, *3*, 8131–8141. [CrossRef]
54. Nagamine, T.; Hayakawa, K.; Kusakabe, T.; Takada, H.; Nakazato, K.; Hisanaga, E.; Iha, M. Inhibitory effect of fucoidan on huh7 hepatoma cells through downregulation of cxcl12. *Nutr. Cancer* **2009**, *61*, 340–347. [CrossRef]
55. Tokita, Y.; Nakajima, K.; Mochida, H.; Iha, M.; Nagamine, T. Development of a fucoidan-specific antibody and measurement of fucoidan in serum and urine by sandwich elisa. *Biosci. Biotechnol. Biochem.* **2010**, *74*, 350–357. [CrossRef]
56. Pozharitskaya, O.N.; Shikov, A.N.; Faustova, N.M.; Obluchinskaya, E.D.; Kosman, V.M.; Vuorela, H.; Makarov, V.G. Pharmacokinetic and tissue distribution of fucoidan from *Fucus vesiculosus* after oral administration to rats. *Mar. Drugs* **2018**, *16*, 132. [CrossRef]

57. Pozharitskaya, O.N.; Shikov, A.N.; Obluchinskaya, E.D.; Vuorela, H. The pharmacokinetics of fucoidan after topical application to rats. *Mar. Drugs* **2019**, *17*, 687. [CrossRef]
58. Luthuli, S.; Wu, S.; Cheng, Y.; Zheng, X.; Wu, M.; Tong, H. Therapeutic effects of fucoidan: A review on recent studies. *Mar. Drugs* **2019**, *17*, 487. [CrossRef]
59. Hsu, H.-Y.; Lin, T.-Y.; Hwang, P.-A.; Tseng, L.-M.; Chen, R.-H.; Tsao, S.-M.; Hsu, J. Fucoidan induces changes in the epithelial to mesenchymal transition and decreases metastasis by enhancing ubiquitin-dependent tgfβ receptor degradation in breast cancer. *Carcinogenesis* **2013**, *34*, 874–884. [CrossRef]
60. Hsu, H.-Y.; Lin, T.-Y.; Hu, C.-H.; Shu, D.T.F.; Lu, M.-K. Fucoidan upregulates tlr4/chop-mediated caspase-3 and parp activation to enhance cisplatin-induced cytotoxicity in human lung cancer cells. *Cancer Lett.* **2018**, *432*, 112–120. [CrossRef]
61. Malyarenko, O.S.; Zdobnova, E.V.; Silchenko, A.S.; Kusaykin, M.I.; Ermakova, S.P. Radiosensitizing effect of the fucoidan from brown alga fucus evanescens and its derivative in human cancer cells. *Carbohydr. Polym.* **2019**, *205*, 465–471. [CrossRef]
62. Fitton, J.H.; Stringer, D.N.; Park, A.Y.; Karpiniec, S.S. Therapies from fucoidan: New developments. *Mar. Drugs* **2019**, *17*, 571. [CrossRef]
63. Alekseyenko, T.V.; Zhanayeva, S.Y.; Venediktova, A.A.; Zvyagintseva, T.N.; Kuznetsova, T.A.; Besednova, N.N.; Korolenko, T.A. Antitumor and antimetastatic activity of fucoidan, a sulfated polysaccharide isolated from the okhotsk sea fucus evanescens brown alga. *Bull. Exp. Biol. Med.* **2007**, *143*, 730–732. [CrossRef]
64. Fukahori, S.; Yano, H.; Akiba, J.; Ogasawara, S.; Momosaki, S.; Sanada, S.; Kuratomi, K.; Ishizaki, Y.; Moriya, F.; Yagi, M.; et al. Fucoidan, a major component of brown seaweed, prohibits the growth of human cancer cell lines in vitro. *Mol. Med. Rep.* **2008**, *1*, 537–542. [CrossRef]
65. Banafa, A.M.; Roshan, S.; Liu, Y.-Y.; Chen, H.-J.; Chen, M.-J.; Yang, G.-X.; He, G.-Y. Fucoidan induces g1 phase arrest and apoptosis through caspases-dependent pathway and ros induction in human breast cancer mcf-7 cells. *J. Huazhong Univ. Sci. Technol. [Med. Sci.]* **2013**, *33*, 717–724. [CrossRef]
66. Haneji, K.; Matsuda, T.; Tomita, M.; Kawakami, H.; Ohshiro, K.; Uchihara, J.-N.; Masuda, M.; Takasu, N.; Tanaka, Y.; Ohta, T.; et al. Fucoidan extracted from *Cladosiphon okamuranus* tokida induces apoptosis of human t-cell leukemia virus type 1-infected t-cell lines and primary adult t-cell leukemia cells. *Nutr. Cancer* **2005**, *52*, 189–201. [CrossRef]
67. Burz, C.; Berindan-Neagoe, I.; Balacescu, O.; Irimie, A. Apoptosis in cancer: Key molecular signaling pathways and therapy targets. *Acta Oncol.* **2009**, *48*, 811–821. [CrossRef]
68. Ahn, G.; Lee, W.; Kim, K.N.; Lee, J.H.; Heo, S.J.; Kang, N.; Lee, S.H.; Ahn, C.B.; Jeon, Y.J. A sulfated polysaccharide of *Ecklonia cava* inhibits the growth of colon cancer cells by inducing apoptosis. *EXCLI J.* **2015**, *14*, 294–306. [CrossRef]
69. Johnstone, R.W.; Ruefli, A.A.; Lowe, S.W. Apoptosis: A link between cancer genetics and chemotherapy. *Cell* **2002**, *108*, 153–164. [CrossRef]
70. Kim, E.J.; Park, S.Y.; Lee, J.-Y.; Park, J.H.Y. Fucoidan present in brown algae induces apoptosis of human colon cancer cells. *BMC Gastroenterol.* **2010**, *10*, 96. [CrossRef]
71. Aisa, Y.; Miyakawa, Y.; Nakazato, T.; Shibata, H.; Saito, K.; Ikeda, Y.; Kizaki, M. Fucoidan induces apoptosis of human hs-sultan cells accompanied by activation of caspase-3 and down-regulation of erk pathways. *Am. J. Hematol.* **2005**, *78*, 7–14. [CrossRef] [PubMed]
72. Zhang, Z.; Alonzo, R.; Athitsos, V. Experiments with Computer Vision Methods for Hand Detection. In Proceedings of the 4th International Conference on PErvasive Technologies Related to Assistive Environments, Crete, Greece, 25–27 May 2011; Association for Computing Machinery: New York, NY, USA, 2011; p. 21.
73. Teruya, T.; Konishi, T.; Uechi, S.; Tamaki, H.; Tako, M. Anti-proliferative activity of oversulfated fucoidan from commercially cultured *Cladosiphon okamuranus* tokida in u937 cells. *Int. J. Biol. Macromol.* **2007**, *41*, 221–226. [CrossRef] [PubMed]
74. Yamasaki-Miyamoto, Y.; Yamasaki, M.; Tachibana, H.; Yamada, K. Fucoidan induces apoptosis through activation of caspase-8 on human breast cancer mcf-7 cells. *J. Agric. Food Chem.* **2009**, *57*, 8677–8682. [CrossRef]
75. Yang, L.; Wang, P.; Wang, H.; Li, Q.; Teng, H.; Liu, Z.; Yang, W.; Hou, L.; Zou, X. Fucoidan derived from *Undaria pinnatifida* induces apoptosis in human hepatocellular carcinoma smmc-7721 cells via the ros-mediated mitochondrial pathway. *Mar. Drugs* **2013**, *11*, 1961–1976. [CrossRef]
76. Duan, Y.; Li, J.; Jing, X.; Ding, X.; Yu, Y.; Zhao, Q. Fucoidan induces apoptosis and inhibits proliferation of hepatocellular carcinoma via the p38 mapk/erk and pi3k/akt signal pathways. *Cancer Manag. Res.* **2020**, *12*, 1713–1723. [CrossRef]
77. Nishida, N.; Yano, H.; Nishida, T.; Kamura, T.; Kojiro, M. Angiogenesis in cancer. *Vasc. Health Risk Manag.* **2006**, *2*, 213–219. [CrossRef]
78. Koyanagi, S.; Tanigawa, N.; Nakagawa, H.; Soeda, S.; Shimeno, H. Oversulfation of fucoidan enhances its anti-angiogenic and antitumor activities. *Biochem. Pharmacol.* **2003**, *65*, 173–179. [CrossRef]
79. Huang, T.-H.; Chiu, Y.-H.; Chan, Y.-L.; Chiu, Y.-H.; Wang, H.; Huang, K.-C.; Li, T.-L.; Hsu, K.-H.; Wu, C.-J. Prophylactic administration of fucoidan represses cancer metastasis by inhibiting vascular endothelial growth factor (vegf) and matrix metalloproteinases (mmps) in lewis tumor-bearing mice. *Mar. Drugs* **2015**, *13*, 1882–1900. [CrossRef]
80. Liu, F.; Wang, J.; Chang, A.K.; Liu, B.; Yang, L.; Li, Q.; Wang, P.; Zou, X. Fucoidan extract derived from *Undaria pinnatifida* inhibits angiogenesis by human umbilical vein endothelial cells. *Phytomedicine* **2012**, *19*, 797–803. [CrossRef]
81. Wang, Y.Q.; Miao, Z.H. Marine-derived angiogenesis inhibitors for cancer therapy. *Mar. Drugs* **2013**, *11*, 903–933. [CrossRef]

82. Ale, M.T.; Maruyama, H.; Tamauchi, H.; Mikkelsen, J.D.; Meyer, A.S. Fucose-containing sulfated polysaccharides from brown seaweeds inhibit proliferation of melanoma cells and induce apoptosis by activation of caspase-3 in vitro. *Mar. Drugs* **2011**, *9*, 2605–2621. [CrossRef] [PubMed]
83. Zhang, X.W.; Liu, Q.; Thorlacius, H. Inhibition of selectin function and leukocyte rolling protects against dextran sodium sulfate-induced murine colitis. *Scand. J. Gastroenterol.* **2001**, *36*, 270–275. [CrossRef] [PubMed]
84. Cumashi, A.; Ushakova, N.A.; Preobrazhenskaya, M.E.; D'Incecco, A.; Piccoli, A.; Totani, L.; Tinari, N.; Morozevich, G.E.; Berman, A.E.; Bilan, M.I.; et al. A comparative study of the anti-inflammatory, anticoagulant, antiangiogenic, and antiadhesive activities of nine different fucoidans from brown seaweeds. *Glycobiology* **2007**, *17*, 541–552. [CrossRef] [PubMed]
85. Bachelet, L.; Bertholon, I.; Lavigne, D.; Vassy, R.; Jandrot-Perrus, M.; Chaubet, F.; Letourneur, D. Affinity of low molecular weight fucoidan for p-selectin triggers its binding to activated human platelets. *Biochim. Biophys. Acta (BBA) Gen. Subj.* **2009**, *1790*, 141–146. [CrossRef] [PubMed]
86. Rouzet, F.; Bachelet-Violette, L.; Alsac, J.-M.; Suzuki, M.; Meulemans, A.; Louedec, L.; Petiet, A.; Jandrot-Perrus, M.; Chaubet, F.; Michel, J.-B.; et al. Radiolabeled fucoidan as a p-selectin targeting agent for in vivo imaging of platelet-rich thrombus and endothelial activation. *J. Nucl. Med.* **2011**, *52*, 1433–1440. [CrossRef] [PubMed]
87. Yan, M.-D.; Yao, C.-J.; Chow, J.-M.; Chang, C.-L.; Hwang, P.-A.; Chuang, S.-E.; Whang-Peng, J.; Lai, G.-M. Fucoidan elevates microrna-29b to regulate dnmt3b-mtss1 axis and inhibit emt in human hepatocellular carcinoma cells. *Mar. Drugs* **2015**, *13*, 6099–6116. [CrossRef]
88. Lee, H.; Kim, J.S.; Kim, E. Fucoidan from seaweed *Fucus vesiculosus* inhibits migration and invasion of human lung cancer cell via pi3k-akt-mtor pathways. *PLoS ONE* **2012**, *7*, e50624. [CrossRef]
89. Kang, Y.; Wang, Z.-J.; Xie, D.; Sun, X.; Yang, W.; Zhao, X.; Xu, N. Characterization and potential antitumor activity of polysaccharide from gracilariopsis lemaneiformis. *Mar. Drugs* **2017**, *15*, 100. [CrossRef]
90. Busch, S.; Renaud, S.J.; Schleussner, E.; Graham, C.H.; Markert, U.R. Mtor mediates human trophoblast invasion through regulation of matrix-remodeling enzymes and is associated with serine phosphorylation of stat3. *Exp. Cell Res.* **2009**, *315*, 1724–1733. [CrossRef]
91. Wu, W.-S.; Wu, J.-R.; Hu, C.-T. Signal cross talks for sustained mapk activation and cell migration: The potential role of reactive oxygen species. *Cancer Metastasis Rev.* **2008**, *27*, 303–314. [CrossRef]
92. Lee, N.Y.; Ermakova, S.P.; Choi, H.K.; Kusaykin, M.I.; Shevchenko, N.M.; Zvyagintseva, T.N.; Choi, H.S. Fucoidan from laminaria cichorioides inhibits ap-1 transactivation and cell transformation in the mouse epidermal jb6 cells. *Mol. Carcinog.* **2008**, *47*, 629–637. [CrossRef] [PubMed]
93. Kang, Y.M.; Lee, B.J.; Kim, J.I.; Nam, B.H.; Cha, J.Y.; Kim, Y.M.; Ahn, C.B.; Choi, J.S.; Choi, I.S.; Je, J.Y. Antioxidant effects of fermented sea tangle (*Laminaria japonica*) by lactobacillus brevis bj20 in individuals with high level of γ-gt: A randomized, double-blind, and placebo-controlled clinical study. *Food Chem. Toxicol.* **2012**, *50*, 1166–1169. [CrossRef] [PubMed]
94. Cherry, P.; Yadav, S.; Strain, C.R.; Allsopp, P.J.; McSorley, E.M.; Ross, R.P.; Stanton, C. Prebiotics from seaweeds: An ocean of opportunity? *Mar. Drugs* **2019**, *17*, 327. [CrossRef] [PubMed]
95. Tocaciu, S.; Oliver, L.J.; Lowenthal, R.M.; Peterson, G.M.; Patel, R.; Shastri, M.; McGuinness, G.; Olesen, I.; Fitton, J.H. The effect of *Undaria pinnatifida* fucoidan on the pharmacokinetics of letrozole and tamoxifen in patients with breast cancer. *Integr. Cancer Ther.* **2018**, *17*, 99–105. [CrossRef]
96. Tsai, H.-L.; Tai, C.-J.; Huang, C.-W.; Chang, F.-R.; Wang, J.-Y. Efficacy of low-molecular-weight fucoidan as a supplemental therapy in metastatic colorectal cancer patients: A double-blind randomized controlled trial. *Mar. Drugs* **2017**, *15*, 122. [CrossRef]
97. Ikeguchi, M.; Yamamoto, M.; Arai, Y.; Maeta, Y.; Ashida, K.; Katano, K.; Miki, Y.; Kimura, T. Fucoidan reduces the toxicities of chemotherapy for patients with unresectable advanced or recurrent colorectal cancer. *Oncol. Lett.* **2011**, *2*, 319–322. [CrossRef]
98. Tobiume, K.; Matsuzawa, A.; Takahashi, T.; Nishitoh, H.; Morita, K.-I.; Takeda, K.; Minowa, O.; Miyazono, K.; Noda, T.; Ichijo, H. Ask1 is required for sustained activations of jnk/p38 map kinases and apoptosis. *EMBO Rep.* **2001**, *2*, 222–228. [CrossRef]
99. Azuma, K.; Ishihara, T.; Nakamoto, H.; Amaha, T.; Osaki, T.; Tsuka, T.; Imagawa, T.; Minami, S.; Takashima, O.; Ifuku, S. Effects of oral administration of fucoidan extracted from *Cladosiphon okamuranus* on tumor growth and survival time in a tumor-bearing mouse model. *Mar. Drugs* **2012**, *10*, 2337–2348. [CrossRef]
100. Suresh, V.; Anbazhagan, C.; Thangam, R.; Senthilkumar, D.; Senthilkumar, N.; Kannan, S.; Rengasamy, R.; Palani, P. Stabilization of mitochondrial and microsomal function of fucoidan from *Sargassum plagiophyllum* in diethylnitrosamine induced hepatocarcinogenesis. *Carbohydr. Polym.* **2013**, *92*, 1377–1385.
101. Takeda, K.; Tomimori, K.; Kimura, R.; Ishikawa, C.; Nowling, T.K.; Mori, N. Anti-tumor activity of fucoidan is mediated by nitric oxide released from macrophages. *Int. J. Oncol.* **2012**, *40*, 251–260.
102. Maruyama, H.; Tamauchi, H.; Iizuka, M.; Nakano, T. The role of nk cells in antitumor activity of dietary fucoidan from *Undaria pinnatifida* sporophylls (mekabu). *Planta Med.* **2006**, *72*, 1415–1417. [CrossRef] [PubMed]
103. Bogen, B.; Fauskanger, M.; Haabeth, O.A.; Tveita, A. Cd4+ t cells indirectly kill tumor cells via induction of cytotoxic macrophages in mouse models. *Cancer Immunol. Immunother.* **2019**, *68*, 1865–1873. [CrossRef] [PubMed]
104. Thinh, P.D.; Menshova, R.V.; Ermakova, S.P.; Anastyuk, S.D.; Ly, B.M.; Zvyagintseva, T.N. Structural characteristics and anticancer activity of fucoidan from the brown alga *Sargassum mcclurei*. *Mar. Drugs* **2013**, *11*, 1456–1476. [CrossRef] [PubMed]
105. Ahn, J.-H.; Yang, Y.-I.; Lee, K.-T.; Choi, J.-H. Dieckol, isolated from the edible brown algae *Ecklonia cava*, induces apoptosis of ovarian cancer cells and inhibits tumor xenograft growth. *J. Cancer Res. Clin. Oncol.* **2015**, *141*, 255–268. [CrossRef]

106. Zhang, W.; An, E.-K.; Park, H.-B.; Hwang, J.; Dhananjay, Y.; Kim, S.-J.; Eom, H.-Y.; Oda, T.; Kwak, M.; Lee, P.C.-W. *Ecklonia cava* fucoidan has potential to stimulate natural killer cells in vivo. *Int. J. Biol. Macromol.* **2021**, *185*, 111–121. [CrossRef]
107. Wang, H.; Chiu, L.C.; Ooi, V.E.; Ang, P.O., Jr. A potent antitumor polysaccharide from the edible brown seaweed *Hydroclathrus clathratus*. *Bot. Mar.* **2010**, *53*, 265–274. [CrossRef]
108. Dias, P.F.; Siqueira, J.M.; Vendruscolo, L.F.; Neiva, T.D.J.; Gagliardi, A.N.R.; Maraschin, M.; Ribeiro-do-Valle, R.M. Antiangiogenic and antitumoral properties of a polysaccharide isolated from the seaweed *Sargassum stenophyllum*. *Cancer Chemother. Pharmacol.* **2005**, *56*, 436–446. [CrossRef]
109. Vinayak, R.; Puttananjaiah, S.; Chatterji, A.; Salimath, B. Anti-proliferative and angio-suppressive effect of stoechospermum marginatum (c. Agardh) kutzing extract using various experimental models. *Nutr. Res. Pract.* **2014**, *8*, 377–385. [CrossRef]
110. Chen, X.; Nie, W.; Yu, G.; Li, Y.; Hu, Y.; Lu, J.; Jin, L. Antitumor and immunomodulatory activity of polysaccharides from *Sargassum fusiforme*. *Food Chem. Toxicol.* **2012**, *50*, 695–700. [CrossRef]
111. Chen, H.; Zhang, L.; Long, X.; Li, P.; Chen, S.; Kuang, W.; Guo, J. *Sargassum fusiforme* polysaccharides inhibit vegf-a-related angiogenesis and proliferation of lung cancer in vitro and in vivo. *Biomed. Pharmacother.* **2017**, *85*, 22–27. [CrossRef]
112. Ye, H.; Wang, K.; Zhou, C.; Liu, J.; Zeng, X. Purification, antitumor and antioxidant activities in vitro of polysaccharides from the brown seaweed *Sargassum pallidum*. *Food Chem.* **2008**, *111*, 428–432. [CrossRef] [PubMed]
113. Numata, A.; Kanbara, S.; Takahashi, C.; Fujiki, R.; Yoneda, M.; Usami, Y.; Fujita, E. A cytotoxic principle of the brown alga *Sargassum tortile* and structures of chromenes. *Phytochemistry* **1992**, *31*, 1209–1213. [CrossRef]
114. Ahn, M.H.; Shin, J.A.; Yang, S.O.; Choi, W.S.; Jang, S.; Kang, S.C.; Cho, S.D. Metabolite profiling of a *Sargassum micracanthum* methanol extract with in vitro efficacy against human head and neck squamous cell carcinoma aggressiveness. *Arch. Oral Biol.* **2022**, *137*, 105386. [CrossRef] [PubMed]
115. Boo, H.J.; Hyun, J.H.; Kim, S.C.; Kang, J.I.; Kim, M.K.; Kim, S.Y.; Cho, H.; Yoo, E.S.; Kang, H.K. Fucoidan from *Undaria pinnatifida* induces apoptosis in a549 human lung carcinoma cells. *Phytother. Res.* **2011**, *25*, 1082–1086. [CrossRef] [PubMed]
116. Rui, X.; Pan, H.F.; Shao, S.L.; Xu, X.M. Anti-tumor and anti-angiogenic effects of fucoidan on prostate cancer: Possible jak-stat3 pathway. *BMC Complement. Altern. Med.* **2017**, *17*, 378. [CrossRef] [PubMed]
117. Teng, H.; Yang, Y.; Wei, H.; Liu, Z.; Liu, Z.; Ma, Y.; Gao, Z.; Hou, L.; Zou, X. Fucoidan suppresses hypoxia-induced lymphangiogenesis and lymphatic metastasis in mouse hepatocarcinoma. *Mar. Drugs* **2015**, *13*, 3514–3530. [CrossRef]
118. Choi, Y.K.; Kim, J.; Lee, K.M.; Choi, Y.J.; Ye, B.R.; Kim, M.S.; Ko, S.G.; Lee, S.H.; Kang, D.H.; Heo, S.J. Tuberatolide b suppresses cancer progression by promoting ros-mediated inhibition of stat3 signaling. *Mar. Drugs* **2017**, *15*, 55. [CrossRef]
119. Namvar, F.; Mohamad, R.; Baharara, J.; Zafar-Balanejad, S.; Fargahi, F.; Rahman, H.S. Antioxidant, antiproliferative, and antiangiogenesis effects of polyphenol-rich seaweed (*Sargassum muticum*). *Biomed. Res. Int.* **2013**, *2013*, 604787. [CrossRef]
120. Vaseghi, G.; Sharifi, M.; Dana, N.; Ghasemi, A.; Yegdaneh, A. Cytotoxicity of *Sargassum angustifolium* partitions against breast and cervical cancer cell lines. *Adv. Biomed. Res.* **2018**, *7*, 43. [CrossRef]
121. Narayani, S.S.; Saravanan, S.; Ravindran, J.; Ramasamy, M.S.; Chitra, J. In vitro anticancer activity of fucoidan extracted from *Sargassum cinereum* against caco-2 cells. *Int. J. Biol. Macromol.* **2019**, *138*, 618–628. [CrossRef]
122. Somasundaram, S.N.; Shanmugam, S.; Subramanian, B.; Jaganathan, R. Cytotoxic effect of fucoidan extracted from *Sargassum cinereum* on colon cancer cell line hct-15. *Int. J. Biol. Macromol.* **2016**, *91*, 1215–1223. [CrossRef] [PubMed]
123. Silva Costa, L.; Silva Telles, C.B.; Medeiros Oliveira, R.; Duarte Barreto Nobre, L.T.; Dantas-Santos, N.; Barros Gomes Camara, R.; Santana Santos Pereira Costa, M.; Almeida-Lima, J.; Melo-Silveira, R.F.; Lopes Albuquerque, I.R.; et al. Heterofucan from *Sargassum filipendula* induces apoptosis in hela cells. *Mar. Drugs* **2011**, *9*, 603–614. [CrossRef] [PubMed]
124. Menshova, R.V.; Ermakova, S.P.; Anastyuk, S.D.; Isakov, V.V.; Dubrovskaya, Y.V.; Kusaykin, M.I.; Um, B.-H.; Zvyagintseva, T.N. Structure, enzymatic transformation and anticancer activity of branched high molecular weight laminaran from brown alga *Eisenia bicyclis*. *Carbohydr. Polym.* **2014**, *99*, 101–109. [CrossRef] [PubMed]
125. Ermakova, S.; Men'shova, R.; Vishchuk, O.; Kim, S.-M.; Um, B.-H.; Isakov, V.; Zvyagintseva, T. Water-soluble polysaccharides from the brown alga *Eisenia bicyclis*: Structural characteristics and antitumor activity. *Algal Res.* **2013**, *2*, 51–58. [CrossRef]
126. Choi, E.O.; Lee, H.; Park, C.; Kim, G.-Y.; Cha, H.-J.; Kim, S.; Kim, H.-S.; Jeon, Y.-J.; Hwang, H.J.; Choi, Y.H. Ethanol extracts of hizikia fusiforme induce apoptosis in human prostate cancer pc3 cells via modulating a ros-dependent pathway. *Asian Pac. J. Trop. Biomed.* **2020**, *10*, 78. [CrossRef]
127. Wang, H.; Ooi, E.V.; Ang, P.O., Jr. Antiviral polysaccharides isolated from hong kong brown seaweed *Hydroclathrus clathratus*. *Sci. China C Life Sci.* **2007**, *50*, 611–618. [CrossRef] [PubMed]
128. Gamal-Eldeen, A.M.; Ahmed, E.F.; Abo-Zeid, M.A. In vitro cancer chemopreventive properties of polysaccharide extract from the brown alga, *Sargassum latifolium*. *Food Chem. Toxicol.* **2009**, *47*, 1378–1384. [CrossRef]
129. Taheri, A.; Ghaffari, M.; Bavi, Z.; Sohili, F. Cytotoxic effect of the extract of seaweed *Sargassum glaucescens* against breast (mcf-7) and colorectal (ht-29) cancer cell lines. *KAUMS J. (FEYZ)* **2018**, *22*, 292–301.
130. Fedorov, S.N.; Ermakova, S.P.; Zvyagintseva, T.N.; Stonik, V.A. Anticancer and cancer preventive properties of marine polysaccharides: Some results and prospects. *Mar. Drugs* **2013**, *11*, 4876–4901. [CrossRef]
131. Athukorala, Y.; Kim, K.-N.; Jeon, Y.-J. Antiproliferative and antioxidant properties of an enzymatic hydrolysate from brown alga, *Ecklonia cava*. *Food Chem. Toxicol.* **2006**, *44*, 1065–1074. [CrossRef]

132. Athukorala, Y.; Ahn, G.N.; Jee, Y.H.; Kim, G.Y.; Kim, S.H.; Ha, J.H.; Kang, J.S.; Lee, K.W.; Jeon, Y.J. Antiproliferative activity of sulfated polysaccharide isolated from an enzymatic digest of *Ecklonia cava* on the u-937 cell line. *J. Appl. Phycol.* **2009**, *21*, 307–314. [CrossRef]
133. Hyun, J.H.; Kim, S.C.; Kang, J.I.; Kim, M.K.; Boo, H.J.; Kwon, J.M.; Koh, Y.S.; Hyun, J.W.; Park, D.B.; Yoo, E.S.; et al. Apoptosis inducing activity of fucoidan in hct-15 colon carcinoma cells. *Biol. Pharm. Bull.* **2009**, *32*, 1760–1764. [CrossRef] [PubMed]
134. Park, H.B.; Hwang, J.; Lim, S.M.; Zhang, W.; Jin, J.O. Dendritic cell-mediated cancer immunotherapy with *Ecklonia cava* fucoidan. *Int. J. Biol. Macromol.* **2020**, *159*, 941–947. [CrossRef] [PubMed]
135. Cohn, L.; Delamarre, L. Dendritic cell-targeted vaccines. *Front. Immunol.* **2014**, *5*, 255. [CrossRef]
136. Banchereau, J.; Palucka, A.K. Dendritic cells as therapeutic vaccines against cancer. *Nat. Rev. Immunol.* **2005**, *5*, 296–306. [CrossRef]
137. Zhang, W.; Park, H.B.; Yadav, D.; Hwang, J.; An, E.K.; Eom, H.Y.; Kim, S.J.; Kwak, M.; Lee, P.C.; Jin, J.O. Comparison of human peripheral blood dendritic cell activation by four fucoidans. *Int. J. Biol. Macromol.* **2021**, *174*, 477–484. [CrossRef]
138. Song, K.; Xu, L.; Zhang, W.; Cai, Y.; Jang, B.; Oh, J.; Jin, J.-O. Laminarin promotes anti-cancer immunity by the maturation of dendritic cells. *Oncotarget* **2017**, *8*, 38554–38567. [CrossRef]
139. Meng, J.; Cao, Y.; Meng, Y.; Luo, H.; Gao, X.; Shan, F. Maturation of mouse bone marrow dendritic cells (bmdcs) induced by *Laminaria japonica* polysaccharides (ljp). *Int. J. Biol. Macromol.* **2014**, *69*, 388–392. [CrossRef]
140. Lin, H.T.V.; Lu, W.J.; Tsai, G.J.; Chou, C.T.; Hsiao, H.I.; Hwang, P.A. Enhanced anti-inflammatory activity of brown seaweed *Laminaria japonica* by fermentation using bacillus subtilis. *Process. Biochem.* **2016**, *51*, 1945–1953. [CrossRef]
141. Geisen, U.; Zenthoefer, M.; Peipp, M.; Kerber, J.; Plenge, J.; Managò, A.; Fuhrmann, M.; Geyer, R.; Hennig, S.; Adam, D. Molecular mechanisms by which a *Fucus vesiculosus* extract mediates cell cycle inhibition and cell death in pancreatic cancer cells. *Mar. Drugs* **2015**, *13*, 4470–4491. [CrossRef]
142. Zhang, J.; Sun, Z.; Lin, N.; Lu, W.; Huang, X.; Weng, J.; Sun, S.; Zhang, C.; Yang, Q.; Zhou, G.; et al. Fucoidan from *Fucus vesiculosus* attenuates doxorubicin-induced acute cardiotoxicity by regulating jak2/stat3-mediated apoptosis and autophagy. *Biomed. Pharmacother.* **2020**, *130*, 110534. [CrossRef] [PubMed]
143. Pawar, V.K.; Singh, Y.; Sharma, K.; Shrivastav, A.; Sharma, A.; Singh, A.; Meher, J.G.; Singh, P.; Raval, K.; Kumar, A.; et al. Improved chemotherapy against breast cancer through immunotherapeutic activity of fucoidan decorated electrostatically assembled nanoparticles bearing doxorubicin. *Int. J. Biol. Macromol.* **2019**, *122*, 1100–1114. [CrossRef] [PubMed]
144. Lee, K.W.; Jeong, D.; Na, K. Doxorubicin loading fucoidan acetate nanoparticles for immune and chemotherapy in cancer treatment. *Carbohydr. Polym.* **2013**, *94*, 850–856. [CrossRef] [PubMed]
145. Burney, M.; Mathew, L.; Gaikwad, A.; Nugent, E.K.; Gonzalez, A.O.; Smith, J.A. Evaluation fucoidan extracts from *Undaria pinnatifida* and *Fucus vesiculosus* in combination with anticancer drugs in human cancer orthotopic mouse models. *Integr. Cancer Ther.* **2018**, *17*, 755–761. [CrossRef]
146. Yang, B.; Xiao, B.; Sun, T. Antitumor and immunomodulatory activity of *Astragalus membranaceus* polysaccharides in h22 tumor-bearing mice. *Int. J. Biol. Macromol.* **2013**, *62*, 287–290. [CrossRef]
147. Zong, A.; Cao, H.; Wang, F. Anticancer polysaccharides from natural resources: A review of recent research. *Carbohydr. Polym.* **2012**, *90*, 1395–1410. [CrossRef]
148. Li, W.; Song, K.; Wang, S.; Zhang, C.; Zhuang, M.; Wang, Y.; Liu, T. Anti-tumor potential of *Astragalus polysaccharides* on breast cancer cell line mediated by macrophage activation. *Mater. Sci. Eng. C* **2019**, *98*, 685–695. [CrossRef]
149. Li, W.; Hu, X.; Li, Y.; Song, K. Cytotoxicity and growth-inhibiting activity of *Astragalus polysaccharides* against breast cancer via the regulation of egfr and anxa1. *J. Nat. Med.* **2021**, *75*, 854–870. [CrossRef]
150. Wu, C.-Y.; Ke, Y.; Zeng, Y.-F.; Zhang, Y.-W.; Yu, H.-J. Anticancer activity of *Astragalus polysaccharide* in human non-small cell lung cancer cells. *Cancer Cell Int.* **2017**, *17*, 115. [CrossRef]
151. Wu, J.; Yu, J.; Wang, J.; Zhang, C.; Shang, K.; Yao, X.; Cao, B. *Astragalus polysaccharide* enhanced antitumor effects of apatinib in gastric cancer ags cells by inhibiting akt signalling pathway. *Biomed. Pharmacother.* **2018**, *100*, 176–183. [CrossRef]
152. Zhang, W.; Hwang, J.; Park, H.-B.; Lim, S.-M.; Go, S.; Kim, J.; Choi, I.; You, S.; Jin, J.-O. Human peripheral blood dendritic cell and t cell activation by *Codium fragile* polysaccharide. *Mar. Drugs* **2020**, *18*, 535. [CrossRef] [PubMed]
153. Cao, Y.; Ruan, Y.; Shen, T.; Huang, X.; Li, M.; Yu, W.; Zhu, Y.; Man, Y.; Wang, S.; Li, J. *Astragalus polysaccharide* suppresses doxorubicin-induced cardiotoxicity by regulating the pi3k/akt and p38mapk pathways. *Oxidative Med. Cell. Longev.* **2014**, *2014*, 674219. [CrossRef] [PubMed]
154. Bamodu, O.A.; Kuo, K.-T.; Wang, C.-H.; Huang, W.-C.; Wu, A.T.H.; Tsai, J.-T.; Lee, K.-Y.; Yeh, C.-T.; Wang, L.-S. *Astragalus polysaccharides* (pg2) enhances the m1 polarization of macrophages, functional maturation of dendritic cells, and t cell-mediated anticancer immune responses in patients with lung cancer. *Nutrients* **2019**, *11*, 2264. [CrossRef] [PubMed]
155. Yang, M.; Qian, X.-H.; Zhao, D.-H.; Fu, S.-Z. Effects of *Astragalus polysaccharide* on the erythroid lineage and microarray analysis in k562 cells. *J. Ethnopharmacol.* **2010**, *127*, 242–250. [CrossRef] [PubMed]
156. Liu, L.-M.; Zhang, L.-S. effect of *Astragalus polysaccharide* on the function and maturation of plasmacytoid dendritic cells from chronic myelogenous leukemia before and after treatment. *Zhonghua Xue Ye Xue Za Zhi = Zhonghua Xueyexue Zazhi* **2010**, *31*, 740–743.
157. Liu, A.-J.; Yu, J.; Ji, H.-Y.; Zhang, H.-C.; Zhang, Y.; Liu, H.-P. Extraction of a novel cold-water-soluble polysaccharide from *Astragalus membranaceus* and its antitumor and immunological activities. *Molecules* **2018**, *23*, 62. [CrossRef]

158. Li, W.; Hu, X.; Wang, S.; Jiao, Z.; Sun, T.; Liu, T.; Song, K. Characterization and anti-tumor bioactivity of *Astragalus polysaccharides* by immunomodulation. *Int. J. Biol. Macromol.* **2020**, *145*, 985–997. [CrossRef]
159. Nakayasu, S.; Soegima, R.; Yamaguchi, K.; Oda, T. Biological activities of fucose-containing polysaccharide ascophyllan isolated from the brown alga *Ascophyllum nodosum*. *Biosci. Biotechnol. Biochem.* **2009**, *73*, 961–964. [CrossRef]
160. Abu, R.; Jiang, Z.; Ueno, M.; Isaka, S.; Nakazono, S.; Okimura, T.; Cho, K.; Yamaguchi, K.; Kim, D.; Oda, T. Anti-metastatic effects of the sulfated polysaccharide ascophyllan isolated from *Ascophyllum nodosum* on b16 melanoma. *Biochem. Biophys. Res. Commun.* **2015**, *458*, 727–732. [CrossRef]
161. Zhang, W.; Kwak, M.; Park, H.B.; Okimura, T.; Oda, T.; Lee, P.C.; Jin, J.-O. Activation of human dendritic cells by ascophyllan purified from *Ascophyllum nodosum*. *Mar. Drugs* **2019**, *17*, 66. [CrossRef]
162. Zhang, W.; Okimura, T.; Oda, T.; Jin, J.O. Ascophyllan induces activation of natural killer cells in mice in vivo and in vitro. *Mar. Drugs* **2019**, *17*, 197. [CrossRef] [PubMed]
163. Nakano, K.; Kim, D.; Jiang, Z.; Ueno, M.; Okimura, T.; Yamaguchi, K.; Oda, T. Immunostimulatory activities of the sulfated polysaccharide ascophyllan from *Ascophyllum nodosum* in in vivo and in vitro systems. *Biosci. Biotechnol. Biochem.* **2012**, *76*, 1573–1576. [CrossRef] [PubMed]
164. Liang, Y.; Zha, S.; Tentaku, M.; Okimura, T.; Jiang, Z.; Ueno, M.; Hirasaka, K.; Yamaguchi, K.; Oda, T. Suppressive effects of sulfated polysaccharide ascophyllan isolated from *Ascophyllum nodosum* on the production of no and ros in lps-stimulated raw264.7 cells. *Biosci. Biotechnol. Biochem.* **2021**, *85*, 882–889. [CrossRef] [PubMed]
165. Zhang, W.; Okimura, T.; Xu, L.; Zhang, L.; Oda, T.; Kwak, M.; Yu, Q.; Jin, J.-O. Ascophyllan functions as an adjuvant to promote anti-cancer effect by dendritic cell activation. *Oncotarget* **2016**, *7*, 19284–19298. [CrossRef]
166. Wang, L.; Oh, J.Y.; Je, J.G.; Jayawardena, T.U.; Kim, Y.-S.; Ko, J.Y.; Fu, X.; Jeon, Y.-J. Protective effects of sulfated polysaccharides isolated from the enzymatic digest of *Codium fragile* against hydrogen peroxide-induced oxidative stress in in vitro and in vivo models. *Algal Res.* **2020**, *48*, 101891. [CrossRef]
167. Park, H.B.; Hwang, J.; Zhang, W.; Go, S.; Kim, J.; Choi, I.; You, S.; Jin, J.-O. Polysaccharide from *Codium fragile* induces anti-cancer immunity by activating natural killer cells. *Mar. Drugs* **2020**, *18*, 626. [CrossRef]
168. Harvey, A.L. Toxins and drug discovery. *Toxicon* **2014**, *92*, 193–200. [CrossRef]
169. De Souza, J.M.; Goncalves, B.D.C.; Gomez, M.V.; Vieira, L.B.; Ribeiro, F.M. Animal toxins as therapeutic tools to treat neurodegenerative diseases. *Front. Pharmacol.* **2018**, *9*, 145. [CrossRef]
170. Kiuru, P.; D'Auria, M.V.; Muller, C.D.; Tammela, P.; Vuorela, H.; Yli-Kauhaluoma, J. Exploring marine resources for bioactive compounds. *Planta Med.* **2014**, *80*, 1234–1246. [CrossRef]
171. Mayer, A.M.S.; Guerrero, A.J.; Rodríguez, A.D.; Taglialatela-Scafati, O.; Nakamura, F.; Fusetani, N. Marine pharmacology in 2014–2015: Marine compounds with antibacterial, antidiabetic, antifungal, anti-inflammatory, antiprotozoal, antituberculosis, antiviral, and anthelmintic activities; affecting the immune and nervous systems, and othermiscellaneousme. *Mar. Drugs* **2020**, *18*, 5. [CrossRef]
172. Huang, L.-G.; Li, J.-P.; Pang, X.-M.; Chen, C.-Y.; Xiang, H.-Y.; Feng, L.-B.; Su, S.-Y.; Li, S.-H.; Zhang, L.; Liu, J.-L. Microrna-29c correlates with neuroprotection induced by fns by targeting both birc2 and bak1 in rat brain after stroke. *CNS Neurosci. Ther.* **2015**, *21*, 496–503. [CrossRef] [PubMed]
173. Jayatilleke, K.M.; Hulett, M.D. Heparanase and the hallmarks of cancer. *J. Transl. Med.* **2020**, *18*, 453. [CrossRef] [PubMed]
174. Rivara, S.; Milazzo, F.M.; Giannini, G. Heparanase: A rainbow pharmacological target associated to multiple pathologies including rare diseases. *Future Med. Chem.* **2016**, *8*, 647–680. [CrossRef] [PubMed]
175. Hammond, E.; Khurana, A.; Shridhar, V.; Dredge, K. The role of heparanase and sulfatases in the modification of heparan sulfate proteoglycans within the tumor microenvironment and opportunities for novel cancer therapeutics. *Front. Oncol.* **2014**, *4*, 195. [CrossRef] [PubMed]
176. Zhao, H.; Liu, H.; Chen, Y.; Xin, X.; Li, J.; Hou, Y.; Zhang, Z.; Zhang, X.; Xie, C.; Geng, M.; et al. Oligomannurarate sulfate, a novel heparanase inhibitor simultaneously targeting basic fibroblast growth factor, combats tumor angiogenesis and metastasis. *Cancer Res.* **2006**, *66*, 8779–8787. [CrossRef]
177. Parish, C.R.; Coombe, D.R.; Jakobsen, K.B.; Bennett, F.A.; Underwood, P.A. Evidence that sulphated polysaccharides inhibit tumour metastasis by blocking tumour-cell-derived heparanases. *Int. J. Cancer* **1987**, *40*, 511–518. [CrossRef]
178. Soeda, S.; Ishida, S.; Shimeno, H.; Nagamatsu, A. Inhibitory effect of oversulfated fucoidan on invasion through reconstituted basement membrane by murine Lewis lung carcinoma. *Jpn. J. Cancer Res.* **1994**, *85*, 1144–1150.
179. Montaser, R.; Luesch, H. Marine natural products: A new wave of drugs? *Future Med. Chem.* **2011**, *3*, 1475–1489. [CrossRef]
180. Bilan, M.I.; Usov, A.I. Structural analysis of fucoidans. *Nat. Prod. Commun.* **2008**, *3*, 1639–1648. [CrossRef]
181. Choudhary, A.; Naughton, L.M.; Montánchez, I.; Dobson, A.D.W.; Rai, D.K. Current status and future prospects of marine natural products (mnps) as antimicrobials. *Mar. Drugs* **2017**, *15*, 272. [CrossRef]
182. Lear, M.J.; Hirai, K.; Ogawa, K.; Yamashita, S.; Hirama, M. A convergent total synthesis of the kedarcidin chromophore: 20-years in the making. *J. Antibiot.* **2019**, *72*, 350–363. [CrossRef] [PubMed]
183. Lindequist, U. Marine-derived pharmaceuticals–challenges and opportunities. *Biomol. Ther.* **2016**, *24*, 561–571. [CrossRef] [PubMed]

184. Hwang, P.-A.; Lin, H.-T.V.; Lin, H.-Y.; Lo, S.-K. Dietary supplementation with low-molecular-weight fucoidan enhances innate and adaptive immune responses and protects against mycoplasma pneumoniae antigen stimulation. *Mar. Drugs* **2019**, *17*, 175. [CrossRef] [PubMed]
185. Kim, K.-J.; Lee, O.-H.; Lee, H.-H.; Lee, B.-Y. A 4-week repeated oral dose toxicity study of fucoidan from the sporophyll of *Undaria pinnatifida* in sprague–dawley rats. *Toxicology* **2010**, *267*, 154–158. [CrossRef]
186. Etman, S.M.; Elnaggar, Y.S.R.; Abdallah, O.Y. Fucoidan, a natural biopolymer in cancer combating: From edible algae to nanocarrier tailoring. *Int. J. Biol. Macromol.* **2020**, *147*, 799–808. [CrossRef]
187. Kiruba, N.J.M.; Pradeep, M.A.; Thatheyus, A.J. Discovering promising anti-cancer drug candidates from marine algae. *Sci. Int.* **2018**, *6*, 44–50. [CrossRef]
188. Mathew, L.; Burney, M.; Gaikwad, A.; Nyshadham, P.; Nugent, E.K.; Gonzalez, A.; Smith, J.A. Preclinical evaluation of safety of fucoidan extracts from *Undaria pinnatifida* and *Fucus vesiculosus* for use in cancer treatment. *Integr. Cancer Ther.* **2017**, *16*, 572–584. [CrossRef]
189. Bovet, L.; Samer, C.; Daali, Y. Preclinical evaluation of safety of fucoidan extracts from *Undaria pinnatifida* and *Fucus vesiculosus* for use in cancer treatment. *Integr. Cancer Ther.* **2019**, *18*, 1–2. [CrossRef]

molecules MDPI

Article

Nutrients and Main Secondary Metabolites Characterizing Extracts and Essential Oil from Fruits of *Ammodaucus leucotrichus* Coss. & Dur. (Western Sahara)

Mohamed Lamin Abdi Bellau [1], Matteo Andrea Chiurato [2], Annalisa Maietti [3], Giancarlo Fantin [3,†], Paola Tedeschi [3], Nicola Marchetti [3], Massimo Tacchini [2], Gianni Sacchetti [2,*] and Alessandra Guerrini [2]

1 Pharmacist of the Sahrawi Refugee Camps, 37000 Tindouf, Algeria
2 Pharmaceutical Biology Lab., Research Unit 7 of Terra&Acqua Tech Technopole Lab., Department of Life Sciences and Biotechnology, University of Ferrara, Malborghetto di Boara, 44123 Ferrara, Italy
3 Department of Chemistry, Pharmaceutical and Agricultural Sciences, University of Ferrara, 44121 Ferrara, Italy
* Correspondence: scg@unife.it; Tel.: +39-053-229-3774
† The present paper has been written in memory of Giancarlo Fantin, dear friend and colleague.

Abstract: The ethnobotany of the Sahrawi people considers various species of plants and crude drugs as food, cooking spices and traditional health remedies. From among these, the fruits of *Ammodaucus leucotrichus* Coss. & Dur. (Apiaceae), known as Saharan cumin, were chosen for our research. The present paper reports a proximate composition and mineral element analysis of various samples of *A. leucotrichus* fruits, collected during the balsamic period (full fruiting) from plants grown in Bir Lehlu (Western Sahara) and purchased in a local market (Tindouf). These analyses pointed out interesting nutritional values of the crude drug. Decoction and alcoholic extract, analyzed by HPLC-DAD, evidenced ammolactone-A and R-perillaldehyde as the two main isolated constituents, particularly in the ethanolic extracts (ammolactone-A, market sample: 51.71 ± 0.39 mg/g dry extract; wild sample: 111.60 ± 1.80 mg/g dry extract; R-perillaldehyde, market sample: 145.95 ± 0.35 mg/g dry extract; wild sample: 221.40 ± 0.30 mg/g dry extract). The essential oils, obtained through hydrodistillation, were characterized by GC-MS and evidenced R-perillaldehyde (market sample: 53.21 ± 1.52%; wild sample: 74.01 ± 1.75%) and limonene (market sample: 35.15 ± 1.68%; wild sample: 19.90 ± 1.86%) as the most abundant compounds. The R configuration of perillaldehyde was ascertained and a complete description of the ^{1}H and ^{13}C NMR spectra of ammolactone-A was performed.

Keywords: *Ammodaucus leucotrichus*; proximate analysis; mineral element analysis; decoction; alcoholic extract; essential oil; secondary metabolites; ammolactone-A; R-perillaldehyde; chemical characterization

Citation: Abdi Bellau, M.L.; Chiurato, M.A.; Maietti, A.; Fantin, G.; Tedeschi, P.; Marchetti, N.; Tacchini, M.; Sacchetti, G.; Guerrini, A. Nutrients and Main Secondary Metabolites Characterizing Extracts and Essential Oil from Fruits of *Ammodaucus leucotrichus* Coss. & Dur. (Western Sahara). *Molecules* **2022**, *27*, 5013. https://doi.org/10.3390/molecules27155013

Academic Editors: Sara Vitalini and Raffaele Pezzani

Received: 14 July 2022
Accepted: 4 August 2022
Published: 6 August 2022

1. Introduction

The majority of the Sahrawi population has been living as refugees in the Hammada Desert in the southwestern part of Tindouf province (Algeria) for nearly 50 years. The refugees remain highly dependent on humanitarian aid for their survival, which has been reduced in recent years, causing a decrease in the distribution of fresh products, animal protein, etc., and impacting the population's nutritional intake. The 2016 nutrition surveys revealed an improvement in some indicators, such as a decrease in global acute malnutrition (GAM) (2010), but a significant increase in others, such as the prevalence of anemia in pregnant (60%) and lactating women (72%), and in children under five years of age (28.4%). Reduced humanitarian funding is accompanied by the limited availability of food from local agricultural production. Although several international projects have encouraged agriculture aimed at diversifying food needs in refugee camps, the results are still marginal [1,2].

In addition to malnutrition and anemia, there are several diseases related to refugee status such as dental disease, and gastrointestinal or respiratory infections, with a high seasonal and environmental component, and a significant increase in non-communicable ailments, such as cardiovascular and thyroid ailments, hypertension, asthma and diabetes, which mainly affect the adult population which does not enjoy healthy ageing. In this context, it should be remembered that Saharawi traditional medicine uses local medicinal plants for the treatment of various diseases, such as gastrointestinal, airways respiratory tract disorders and diabetes [3].

Our research work fits into this complex scenario. Starting from the study of the local plants widely used in the Saharawi tradition for food and the treatment of diseases, and also through interviews with female healers, our research focused on *A. leucotrichus* Coss. & Dur. (Apiaceae), an endemic plant in North Africa, especially in the southern Algerian Sahara and Tassili regions [4], whose fruits are widely used in traditional medicine and in culinary recipes. For this reason, this crude drug, or its infusion and/or decoction as traditional preparations, could be a potential source of nutritional elements and useful metabolites for the treatment of disorders related to unbalanced nutrition.

A. leucotrichus, known in Algeria as "Kammûn es-sofi" and "hairy cumin", is an aromatic small annual species, glabrous and having a characteristic smell reminiscent of *Cuminum cyminum* L. The fruit is a diachene 8–10 mm long and is covered with very dense and fuzzy hairs. In the literature, the essential oil obtained from *A. leucotrichus* fruits has mainly been studied for its phytochemical composition and antioxidant, antimicrobial and anti-inflammatory activities [5]. The main component, perillaldehyde, appears to have the S absolute configuration [6]. Among the main secondary metabolites are documented ammolactone-A [7] and flavonoid derivatives of which luteolin-O-(malonylglucoside) is the most abundant [3]. The main macroelements are K (2283 mg/100 g dw) and Ca (1555 mg/100 g dw), while Fe (22 mg/100 g dw) shows the most relevant result among the microelements. The fruits also contain a good amount of lipids (11 g/100 g dw), characterized by the presence of linoleic acid (C18:2n6), oleic acid (C18:1n9), α-linolenic acid (C18:3n3) and palmitic acid (C16:0) as the predominant fatty acids [3].

This research work has defined the absolute configuration of perillaldehyde and performed the complete description of the ^{1}H and ^{13}C NMR spectra of ammolactone-A, filling in the limited literature data. For the first time, perillaldehyde and ammolactone-A were detected and quantified in decoction and alcoholic extracts of *A. leucotrichus* fruits and prepared using wild-collected samples and market-purchased ones. Finally, a comparison with literature data was performed for nutritional parameters of the fruits and essential oils obtained from them by hydrodistillation.

2. Results and Discussion

2.1. Proximate Analysis and Mineral Composition

The proximate analysis, carried out on the fruits of *A. leucotrichus* (Table 1), showed quite similar residual moisture content in both samples, around 10–12%, consistent with the typical moisture content of dried aromatic plants [8]. The protein contents were also comparable (10.77 ± 0.07 g/100 g and 9.14 ± 0.62 g/100 g) for both samples but lower than in the literature data, while the lipids were almost three times more in the sample harvested in Bir Lehlu, which was similar to previously published data, than in those purchased in the Tindouf market (11.30 ± 0.14 g/100 g vs 4.05 ± 0.01 g/100 g); in both cases, these values are clearly lower than those of cumin, a spice used for culinary preparations and also belonging to the Apiaceae family [9], of which *A. leucotrichus* can be considered as a succedaneum.

Table 1. Proximate analysis of *A. leucotrichus* samples compared to literature data for *A. leucotrichus* and *C. cyminum*.

Proximate Analysis	*A. leucotrichus* Market	*A. leucotrichus* Wild	*A. leucotrichus* [1]	*Cuminum cyminum* [2]
Humidity (g/100 g)	12.65 ± 0.07	10.59 ± 0.33	/	6.44
Proteins (g/100 g)	10.77 ± 0.07	9.14 ± 0.62	13.1 ± 0.9	19.00
Lipids (g/100 g)	4.05 ± 0.01	11.30 ± 0.14	11.1 ± 0.3	29.17
Total ash (g/100 g)	14.43 ± 0.02	10.00 ± 0.01	10.8 ± 0.2	7.36
Total fiber (g/100 g)	74.51 ± 2.04	72.00 ± 3.55	/	51.3
Insoluble fiber (g/100 g)	66.68 ± 1.66	63.85 ± 4.01	/	46.4
Soluble fiber (g/100 g)	7.83 ± 0.38	8.15 ± 0.45	/	4.89

[1] Literature data [3]. [2] Literature data [9,10].

As far as fiber analysis is concerned, *A. leucotrichus* not only showed a high content of this nutrient for both samples (74.51 ± 2.04 g/100 g of total fiber for the market sample, 72.00 ± 3.55 g/100 g for wild-collected sample) but, above all, the percentage of insoluble fiber was found to be more concentrated than the soluble one, especially for the market sample (66.68 ± 1.66 g/100 g). This result is very interesting because high-fiber diets could be associated with the prevention and treatment of certain diseases such as diabetes, also common among the Saharawi population; consequently, this plant could be considered a potential nutraceutical ingredient for the prevention and treatment of this disease [11].

Concerning the analysis of minerals (Table 2), both samples showed a high concentration of K and Ca content quite similar to cumin, but lower than data previously published for *A. leucotrichus* fruits. The health benefits of potassium could be relevant for blood pressure, bone density and risk of kidney stones. Calcium is an important macroelement involved in the regulation of muscle contraction, blood coagulation, the transmission of nerve impulses, the regulation of cell permeability, the activity of numerous enzymes (promoting the release of insulin by pancreatic cells) and the growth and fortification of teeth and bones [12,13]. The samples highlighted a high iron content (132.0 ± 4.4 mg/100 g and 86 ± 11 mg/100 g, respectively, for market and wild-collected samples), and good zinc (1.64 ± 0.02 mg/100 g and 2.98 ± 0.29 mg/100 g, respectively, for market and wild-collected samples) and manganese (3.84 ± 0.04 mg/100 g and 2.93 ± 0.01 mg/100 g, respectively, for market and wild-collected samples) values. With particular reference to iron, the data of our samples were more interesting than those previously published for *A. leucotrichus* and *C. cyminum*.

Table 2. Mineral determination of *A. leucotrichus* samples compared to literature data for *A. leucotrichus* and *C. cyminum*.

Minerals	*A. leucotrichus* Market	*A. leucotrichus* Wild	*A. leucotrichus* [1]	*Cuminum cyminum* [2]
Macroelements				
Na (mg/100 g)	110 ± 19	156 ± 4	160 ± 6	168
Mg (mg/100 g)	254 ± 36	234 ± 47	236.6 ± 0.4	337
K (mg/100 g)	1949 ± 99	1636 ± 88	2283 ± 0.4	1790
Ca (mg/100 g)	737 ± 10	691 ± 48	1555 ± 2	917
Microelements				
Fe (mg/100 g)	132.0 ± 4.4	86 ± 1.1	22 ± 2	14.0
Zn (mg/100 g)	1.64 ± 0.02	2.98 ± 0.29	1.72 ± 0.04	3.0
Cu (mg/100 g)	0.68 ± 0.03	0.64 ± 0.01	0.39 ± 0.03	0.87
Mn (mg/100 g)	3.84 ± 0.04	2.93 ± 0.01	7.6 ± 0.4	3.46

[1] Literature data [3]. [2] Literature data [9,10].

These results have importance especially when compared with the nutritional values of cumin; in fact, the high content of such minerals could help in reducing problems related to their deficiency, such as growth delay and anemia, which have a significant incidence in the Saharawi population.

Regarding the fatty acids profile (Table 3), both samples showed a high percentage of oleic acid (86.07% and 86.7%, respectively, for market and wild-collected samples) and a lower percentage of linoleic acid (8.33% and 9.27%), with the oleic-linoleic ratio > 7. This profile recalls that of olive oil, a lipidic food with a composition most similar to endogenous nutritional needs. Several researchers have suggested that its consumption is associated with a reduced risk for several chronic illnesses, such as diabetes, hypertension, obesity and cardiovascular diseases. Furthermore, the fatty acids profile of *A. leucotrichus* presented a percentage of palmitic acid lower than olive oil (7–17%). Foods with a high content of saturated fatty acids, such as palmitic acid, are considered to be among the factors preventing cardiovascular diseases [14,15]. The fatty acids profile of our samples was different from previously published data, especially with reference to the percentage of palmitic acid, linoleic and α-linolenic acid; on the other hand, it seemed to be very similar in composition to cumin.

Table 3. Fatty acids (relative percentage) of *A. leucotrichus* samples compared to literature data for *A. leucotrichus* and *C. cyminum*.

Fatty Acid	*A. leucotrichus* Market	*A. Leucotrichus* Wild	*A. leucotrichus* [1]	*Cuminum cyminum* [2]
C14:0 (myristic)	0.06	<0.05	0.87	0.10
C16:0 (palmitic)	3.83	3.33	21.2	6.13
C16:1 (palmitoleic)	0.40	0.33	1.06	1.13
C18:0 (stearic)	1.10	<0.05	0.10	1.83
C18:1n9c (oleic)	86.07	86.7	53.8	73.17
C18:2n6c (linoleic)	8.33	9.27	1.99	16.68
C18:3n3 (α-linolenic)	0.22	0.38	10.6	0.97

[1] Literature data [3]. [2] Literature data [10].

2.2. Chemical Characterization of *A. leucotrichus* Essential Oil, Extracts and Their Main Components

The composition of the essential oils of *A. leucotrichus* is shown in Table 4; the main compounds identified belong to the class of oxygenated monoterpenes, including R-perillaldehyde as a major component (53.31% for market sample, 74.01% for wild sample), followed by hydrocarbon monoterpenes, with limonene as the main compound (35.15% for market sample, 19.90% for wild sample).

Table 4. Essential oil composition of *A. leucotrichus* samples.

No.	Component [1]	*A. leucotrichus* Market Area% [2]	*A. leucotrichus* Wild Area% [2]	AI Exp [3]	AI Lit [4]
1	α-pinene	0.77 ± 0.02	1.28 ± 0.02	928	932
2	camphene	3.33 ± 0.12	0.09 ± 0.01	943	946
3	β-pinene	0.33 ± 0.04	0.90 ± 0.04	972	974
4	myrcene	-	0.21 ± 0.02	986	988
5	p-mentha-1(7),8-diene	0.16 ± 0.01	-	1001	1003
6	3-carene	2.1 ± 0.03	1.15 ± 0.01	1005	1008
7	**limonene**	**35.15 ± 1.68**	**19.90 ± 1.86**	1024	1024
8	**R-perillaldehyde**	**53.21 ± 1.52**	**74.01 ± 1.75**	1272	1269
9	perillyl alcohol	2.41 ± 0.07	1.01 ± 0.11	1296	1294
10	methylperillate	0.71 ± 0.02	0.91 ± 0.02	1395	1392
11	*cis*-β-caryophyllene	0.22 ± 0.02	-	1400	1408
12	germacrene D	0.35 ± 0.01	-	1483	1485
13	δ-amorphene	-	0.10 ± 0.02	1514	1511
	Total identified	98.74	99.56		

[1] Components are listed in order of elution and their nomenclature is in accordance with the NIST (National Institute of Standards and Technology) library. [2] Relative peak areas, calculated by GC-FID. [3] AI exp: linear retention indices calculated on Varian VF-5 ms column. [4] AI lit: linear retention indices [16]. The main compounds are in bold.

A comparison of the percentages of the main compounds (limonene and R-perillaldehyde) revealed in *A. leucotrichus* essential oil in previous papers is summarized in Table 5 and shows that our results are very similar to those of the literature in terms of quality, but from a quantitative point of view there are remarkable differences, probably due to the different harvesting period, latitude, climate and soil type. In fact, the content of perillaldehyde fluctuates between a maximum of 84.4% and a minimum of 53.2%, while that of limonene is between 35.15% and 1.7%.

Table 5. Main components of *A. leucotrichus* essential oils in literature and their extraction yields.

Compound	*A. leucotrichus* Market (Area%)	*A. leucotrichus* Wild (Area%)	(Area%) [A]	(Area%) [B]	(Area%) [C]	(Area%) [D]	(Area%) [E]
limonene	35.15	19.90	28.8	1.7	26.8	8.2	6.9–29.2
perillaldehyde	53.21	74.01	56.4	84.4	63.6	87.9	60.1–37.5
Yield	2.00%	3.80%	0.7%	2.58%	2.76%	1.6%	2.0–2.1%

A = [17]; B = [18]; C = [19]; D = [20]; E = [21].

Interestingly, the composition of the essential oil of *A. leucotrichus* is very similar to that of the essential oil of another Iranian plant, namely *Dracocephalum surmandinum*, where the two main components of the volatile fraction are characterized by perillaldehyde (54.3%) and limonene (30.1%) [22], and of the essential oil of leaf *Perilla frutescens* (L.) Britt., the best-known source of S-(−)-perillaldehyde [23].

The yield of essential oils (Table 4) was respectively 2.00 ± 0.02% for the market sample and 3.80 ± 0.06% for the wild-collected sample, representing the best value with respect to the literature data.

The isolation of the molecules, perillaldehyde and ammolactone-A, from the ethanol extract, allowed them to be chemically characterized. The GC-MS of perillaldehyde showed a purity of 96% and an experimental mass fragmentation comparable to literature data [16] (m/z): 151 (15), 150 (15), 135 (50), 122 (45), 107 (70), 93 (60), 91 (75), 79 (100), 68 (55), 67 (100), 77 (45), 53 (30). As reported by Chebrouk et al. (2019) [6], this molecule in *A. leucotrichus* had the absolute configuration of S-(−)-perillaldehyde. The injection of isolated perillaldehyde and commercial S-(−)-perillaldehyde (Sigma Aldrich, Burlington, MA, USA) in GC-FID, equipped with chiral column, highlighted that the isolated perillaldehyde is not superimposable with the commercial one and corresponded to enantiomer R (Figure 1), as we recently reported [24]. The findings are in contrast to previous literature data in which the absolute configuration S has been supposed without experimental evidence. In addition, the optical rotation was: $[\alpha]_D^{20}$ = +115 (c 10, ethanol), opposite to the commercial S-(−)-perillaldehyde.

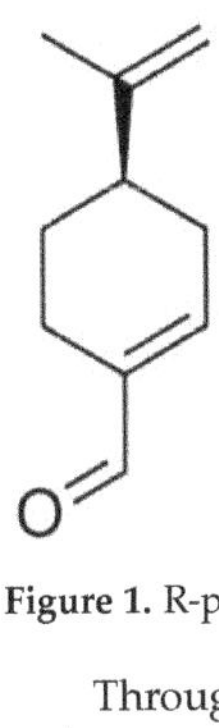

Figure 1. R-perillaldehyde.

Through GC-MS analysis we established that ammolactone-A had a purity of 97% and an experimental mass fragmentation comparable to literature data [7] (m/z): 332 (1), 293 (3), 248 (20), 233 (5), 230 (25), 205 (20), 190 (30), 175 (35), 169 (10), 167 (20), 159 (50), 157 (50), 145 (45), 133 (35), 132 (75), 120 (45), 107 (35), 105 (35), 85 (25), 81 (65), 79 (65), 57

(100). The same authors described only partially the NMR spectra of this molecule. For this reason, we performed the complete description of its ^{1}H and ^{13}C NMR spectra. ^{1}H NMR (400MHz, $CDCl_3$): 5.52 (1H, m, H-3), 5.42 (1H, ddd, J = 11.2 Hz; J = 9.4 Hz; J = 0. 8 Hz, H-8), 4.57 (1H, dd, J = 11,6 Hz, J = 9.1 Hz, H-6), 3.05 (1H, ddd, J = 11.2 Hz; J = 9.3 Hz; J = 9.2 Hz, H-7), 2.70 (1H, dq, J = 9.3 Hz; J = 7.8 Hz, H-11), 2.62 (1H, brs, OH-10), 2.60 (1H, dd, J = 11.4 Hz; J = 5.6 Hz, H-5), 2.41 (1H, m, H-1), 2.37 (1H, sex, J = 7.0 Hz, H-17), 2.18 (1H, m, H-2), 2.09 (1H, m, H-2′), 2.07 (1H, dd, J = 14.7 Hz, J = 9.6 Hz, H-9), 1.87 (3H, dt, J = 2.8 Hz; J = 1.5 Hz, CH_3 C-4), 1.75 (1 H, dd, J =14.7 Hz; J = 1.8 Hz; J = 0.8 Hz, H-9′), 1.68 (1H, dquint, J = 13.6 Hz, J = 7.4 Hz, H-18), 1.48 (1H, dquint, J = 13.6 Hz; J = 7.4 Hz, H-18′), 1.31 (3H, d, J = 7.8 Hz, CH_3 C-13), 1.22 (3H, s, CH_3 C10), 1.14 (3H, d, J = 7.0 Hz, CH_3 C-17), 0.92 (3H, t, J = 7.4 Hz, CH_3 C-18).

^{13}C NMR (400 MHz, $CDCl_3$): 179.4 (C12), 177.4 (C16), 147.2 (C4), 125.4 (C3), 80.8 (C6), 71.6 (C10), 67.3 (C8), 55.7 (C17), 50.2 (C5), 45.6 (C7), 43.3 (C9), 41.3 (C1), 36.3 (C11), 32.4 (C2), 31.3 (C14), 26.8 (C18), 18.9 (C15), 16.6 (C20), 13.6 (C13), 11.8 (C19).

All the proton resonances were associated with those of the directly attached carbon atoms through the DEPT and 2D NMR HMQC experiments. The proton multiplets were arranged in sequence through the COSY experiment, yielding the three spin systems evidenced in bold in Figure 2. The key HMBC correlations, marked with the arrows in Figure 2, detect the heteronuclear correlations of the quaternary carbons C(4), C(10), C(12) and C(16) of ammolactone-A.

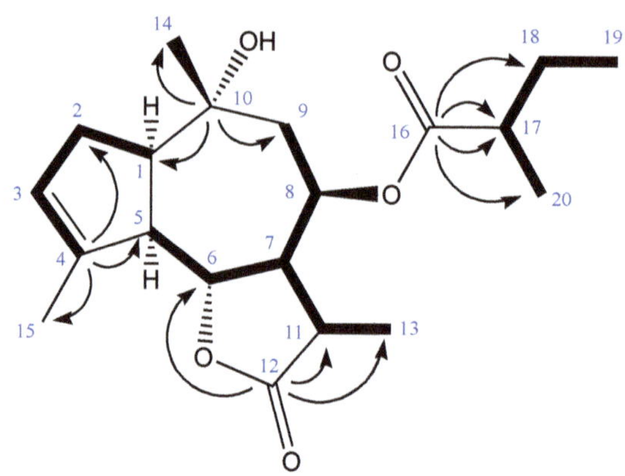

Figure 2. Proton connectivity network by COSY (bold lines) and key HMBC cross-peaks of ammolactone-A. Further HMBC cross-peaks are omitted for clarity.

As shown in Table 6, the ethanol extract from the wild-collected sample evidenced a higher amount of both ammolactone-A and R-perillaldehyde, the two main compounds detected. The decoction had a lower content of these two metabolites. It should be emphasized that there are no references in the literature concerning the characterization of these preparations.

Table 6. HPLC-DAD quantitative analysis of fruits extracts of *A. leucotrichus*.

Component	*A. leucotrichus* Market Decoction [1]	*A. leucotrichus* Market Ethanolic Extract [1]	*A. leucotrichus* Wild Decoction [1]	*A. leucotrichus* Wild Ethanolic Extract [1]
ammolactone-A	5.68 ± 0.12	51.71 ± 0.39	3.32 ± 0.22	111.6 ± 1.8
R-perillaldehyde	24.32 ± 0.93	145.95 ± 0.35	35.88 ± 0.60	221.4 ± 0.3

[1] mg/g dry extract ± (sd).

3. Materials and Methods

3.1. Plant Material

The fruits of *Ammodaucus leucotrichus* Coss. & Dur. were collected from a wild population of plants at Bir Lehlu (coordinates: 26°20′58″ N 09°34′32″ W, Western Sahara) and purchased in Tindouf local market. The samples authentication was performed by Dr. Mohamed Lamin Abdi Bellau and Prof. Alessandra Guerrini through the IUCN Centre For Mediterranean Cooperation (2005) [25]. Wild plant material was dried at room temperature for 15 days. The voucher specimen (code no. AMM.022.001) is stored in the Herbarium of the University of Ferrara (Italy). The present research is compliant with the Nagoya protocol.

3.2. Proximate Analysis

To determine the moisture, the samples were finely ground with a knife mill (Grindomix GM200, Retsch, Dusseldorf, Germany), then dried in an oven at 110 °C until a constant weight. Moisture was expressed as g/100 g.

Proteins (total nitrogen compounds) were determined on 1 g of dry matter, according to the Kjeldahl method [26]. The protein content was determined by means of a conversion factor equal to 6.25.

Total mineral contents (total ashes) were determined on 1 g of dry matter, in a muffle furnace at 570 °C overnight. Total ashes were expressed as g/100 g.

To determine the total lipidic content, about 3 g of dried crude drug was put in a thimble and introduced into the Soxhlet extraction unit (VelpScientifica, Usmate, Milan, Italy) with 50 mL of diethyl ether (CarloErba, Rodano, Milan, Italy). The lipids were dissolved in 3 mL of hexane and transesterified with 1.5 mL of 5% sodium hydroxide in methanol to obtain free fatty acid methyl esters. Sample volume of 1 µL was injected into the gas chromatography-mass spectrometry (Varian Saturn 2100 MS/MS ion trap mass spectrometer coupled to a Varian 3900 gas-chromatograph). The separation was achieved with a capillary column Zebron ZB-WAX (Phenomenex) (60 m, 0.25 mm i.d., 25 µm film thickness) supplied with helium carrier gas at 1 mL/min constant flow. The injector temperature was 260 °C and the oven temperature program was the following: start 100 °C for 2 min, ramp to 200 °C at 10 °C/min and hold for 58 min.

Dietary fiber (total, soluble and insoluble) was determined on 1 g of duplicate dried samples as per the instruction protocol developed for the Megazyme Total Dietary Fiber Assay Kit (Megazyme International, Co., Wicklow, Ireland) and based on the combined enzymatic and gravimetric method of Lee et al. (1992) [27] and Prosky et al. (1988) [28].

3.3. Mineral Composition Determination

Na, K, Mg, Zn, Ca, Fe, Cu and Mn were determined according to previously described methods [29]. One gram of dry matter was mineralized in CEM's digestion vessels (PTFE mod. SV140, FKV) with HNO_3-H_2O_2 in an oven (DK6 heating digester, VELP) with a temperature program and coupled with a module for steam extraction (EM 5, FKV).

A Perkin–Elmer (Perkin-Elmer, Inc.,Waltham, MA, USA) (mod. 1100B) atomic absorption spectrometer (AAS) was used for the analysis. The spectrometer was equipped with a deuterium background corrector and single-element Intensitron (Perkin–Elmer, Inc.,Waltham, MA, USA). Standard solutions of each element were prepared by diluting reference standard solutions for AAS (BDH certified atomic absorption reference solutions). The samples were checked against reference standards and measured for their absorbance after instrument calibration. An average of five readings of absorbance was taken for all samples.

3.4. Hydrodistillation of *A. leucotrichus* Fruits

The pale blue essential oils, with a characteristic odor, were obtained performing a standardized hydrodistillation process, optimized for *A. leucotrichus* crude drug, that used 20 g of dried fruits in 500 mL of water in a Clevenger-type apparatus for 3 h. Three separate distillations were performed for each sample and successively the obtained essential oils were pooled, dried over anhydrous sodium sulfate (Na_2SO_4) and stored at 4 °C in amber glass vials until analysis [30].

3.5. Preparation of Ethanolic Extract and Decoction

An aliquot of 4.0 g of shredded fruits of *A. leucotrichus* was extracted with 200 mL of ethanol by ultrasound assisted maceration (Ultrasonik 104X, Ney Dental International, MED-WOW, Nicosia, Cyprus), total volume 10.4 L, internal dimensions: 146 × 292 × 241 mm, frequency analysis: 48 kHz) for 1 h at 25 °C. Three distinct extractions were performed for each sample. The combined extracts were then filtered and concentrated with a rotary evaporator (RV 10 digital, IKA®-Werke GmbH & Co. KG, Staufen im Breisgau, Germany).

To 2.5 g of shredded fruits of *A. leucotrichus* was added 50 mL of distilled water. The mixture was placed on a heating plate under magnetic stirring and once it reached 100 °C it was left in such conditions for 15 min [31]. Three distinct extractions were performed for each sample. The combined decoctions were then filtered and lyophilized.

3.6. Isolation of Chemical Constituents and Quantitative Chemical Characterization of Extracts and Essential Oil

The isolation of unknown compounds was performed in a silica gel chromatographic column (silica gel 60 220–440 mesh, particle size: 0.035–0.070 mm, Sigma-Aldrich, Milan, Italy). An aliquot of 500 mg of ethanolic extract was suspended in 2 mL of mobile phase: hexane: ethyl acetate (8:2). Precoated silica gel plates (silica gel 60 F_{254}; thickness 0.25 mm; Merck, Milan, Italy) with the same above mobile phase were used to control the fraction separations; after development, the plate was sprayed with phosphomolybdic acid solution (20% phosphomolybdic acid in EtOH) [32] and heated to 120 °C. The isolated molecules showed an intense blue color on a yellow background. The solvents of collected fractions were evaporated to dryness with a rotary evaporator (RV 10 digital, IKA®-Werke GmbH & Co. KG, Staufen im Breisgau, Germany). The isolation by silica gel column gave 40 mg of perillaldehyde and 70 mg of ammolactone-A. White crystals of ammolactone-A were recrystallized from hot cyclohexane.

The GC-MS was used to analyze the essential oil, the identity and the purity of the separated molecule. The GC-MS analysis was performed with a Varian 3800 chromatograph (Varian, Palo Alto, CA, USA) equipped with a Varian Factor Four VF-5 ms column (5%-phenyl-95%-dimethylpolysiloxane, internal diameter: 0.25 mm, length: 30 m) interconnected with a Varian mass spectrometer SATURN MS-4000 (Varian,, Palo Alto, CA, USA), with electronic impact ionization, ion trap analyzer and software provided with the NIST database for the identification of components. The experimental conditions used were the following: helium carrier gas (1 mL/min), a split ratio of 1:50, ionization energy (EI) 70 eV, emission current of 10 μA, scan rate of 1 scan/sec, mass range 40–400 Da. For the analysis, the oven initial temperature of 70 °C was increased to 230 °C with a rate of 4 °C/min and maintained at 230 °C for 10 min; finally, it was brought from 230 to 280 °C with an increase of 5 °C/min. The total time of acquisition of the chromatogram was 70 min. The arithmetic index of perillaldehyde was determined by adding a C_8–C_{32} n-alkanes (Sigma-Aldrich) mixture to the essential oil before injecting into the GC–MS equipment, following the same conditions reported above [33].

The pure perillaldehyde was compared with the commercial S-(−)-perillaldehyde (Sigma Aldrich) through GC analyses, performed with a Thermo Focus-gas chromatograph equipped with a flame ionization detector and a chiral Megadex 5 column (25 m × 0.25 mm), with the following temperature program: 80 to 200 °C, rate 2 °C min^{-1}. Optical rotation was measured at 20 ± 2 °C in ethanol, 10% concentration; $[\alpha]_D^{20}$D value is given in 10^{-1}deg cm^2 g^{-1}.

The ^1H, ^{13}C, DEPT NMR and the 2D NMR experiments (COSY, HMQC, HMBC) spectra for the characterization of ammolactone-A were recorded with a Varian Mercury Plus 400, operating at 400 MHz (^1H) and 100 MHz (^{13}C), respectively. The chemical shifts were referenced to the residual solvent signal ($CDCl_3$: δ_H 7.26, δ_C 77.16). The 2D NMR experiments (COSY, HMQC, HMBC) were processed using the MestReNova (Santiago de Compostela, Spain) Version 6.0.2-5475 software.

HPLC analyses of ethanolic extract and decoction were performed using JASCO modular HPLC system (Tokyo, Japan, model PU 2089) coupled to a diode array apparatus (MD 2010 Plus) linked to an injection valve with a 20 μL sampler. A Zorbax-Eclipse plus-C18 (250 × 4.6 mm, 5 μm) was used at a flow rate of 1.0 mL/min. The mobile phase consisted of solvent solution A (water/phosphoric acid pH = 2.9) and B (acetonitrile). The adopted gradient system consisted of the following steps: isocratic condition of 75:25 *v*/*v* (A/B) until 10 min; gradual changing to 20:80 *v*/*v* up to 30 min; progressive raise to 0:100 *v*/*v* up to 35 min and isocratic mode up to 45 min; back to starting point (75:25 *v*/*v*) in 10 min. The injection volume was 40 μL. The chromatograms were observed at 190 nm. For perillaldehyde the calibration range was 500–10 μg/mL, the correlation coefficient (r^2) 0.993, the limit of quantification (LOQ) 7 μg/mL, the limit of detection (LOD) 2 μg/mL; for ammolactone-A the calibration range was 500–2.5 μg/mL, the correlation coefficient (r^2) 0.999, the limit of quantification (LOQ) 8 μg/mL, the limit of detection (LOD) 3 μg/mL.

4. Conclusions

The crude drug of *A. leucotrichus* is mainly known in the literature for its essential oil and its antimicrobial and anti-inflammatory activities, mainly related to the presence of perillaldehyde: in this work, we established the absolute configuration of the molecule (R), which differs from that found in *P. frutescens* (S). We also provided the NMR characterization of ammolactone-A, which is present, together with perillaldehyde, in the alcohol extract and decoction we prepared from two samples, one collected in the wild and the other purchased in a market. These two secondary metabolites were quantified in these two preparations. The essential oil obtained by hydrodistillation had R-perillaldehyde and limonene as the main components.

The samples showed a high fiber content, which could be useful for the prevention and treatment of certain diseases such as diabetes, which is common among the Saharawi population. The high content of Ca, K and Fe could help in reducing problems related to deficiency in these minerals, such as growth delay and anemia, which also have a strong incidence in the Saharawi population.

Author Contributions: The contributions of authors are as follows: conceptualization, A.G. and A.M.; methodology, A.G. and A.M.; formal analysis, M.L.A.B., M.T., P.T., G.F. and A.G.; data curation, G.F., N.M., M.A.C. and A.G.; writing—original draft preparation, M.A.C., A.G. and G.S.; writing—review and editing, A.G. and G.S.; supervision, G.S.; funding acquisition, A.G. All authors have read and agreed to the published version of the manuscript.

Funding: This research was funded by a FIR2016 (Guerrini) grant of the University of Ferrara, FFABR 2017-MIUR (Guerrini).

Institutional Review Board Statement: Not applicable.

Informed Consent Statement: Not applicable.

Data Availability Statement: Data is contained within the article.

Acknowledgments: The authors are grateful to the Sahrawi Ministry of Public Health and Omar Mih, representative of the Polisario Front in Italy.

Conflicts of Interest: The authors declare no conflict of interest.

Sample Availability: Samples of the crude drug, extracts and compounds are available from the authors.

References

1. Eyt, F. *Analyse et Opportunités Pour une Approche Résilience Commune, Camps de Réfugiés de Tindouf*; World Food Programme: Rome, Italy, 2019.
2. Ministerio De Salud Pública Saharaui. *Plan Estratégico de Salud 2016–2020*; Ministerio De Salud Pública Saharaui: Agadir, Morocco, 2015.
3. Ziani, B.E.C.; Rached, W.; Bachari, K.; Alves, M.J.; Calhelha, R.C.; Barros, L.; Ferreira, I.C.F.R. Detailed chemical composition and functional properties of Ammodaucus leucotrichus Coss. & Dur. and Moringa oleifera Lamarck. *J. Funct. Foods* **2019**, *53*, 237–247.
4. Louail, Z.; Kameli, A.; Benabdelkader, T.; Bouti, K.; Hamza, K.; Krimat, S. Antimicrobial and antioxidant activity of essential oil of Ammodaucus leucotrichus Coss. *Dur. seeds. J. Mater. Environ. Sci.* **2016**, *7*, 2328–2334.
5. Idm'hand, E.; Msanda, F.; Cherifi, K. Medicinal uses, phytochemistry and pharmacology of Ammodaucus leucotrichus. *Clin. Phytosci.* **2020**, *6*, 6. [CrossRef]
6. Chebrouk, F.; Madani, K.; Cherfaoui, B.; Boukenna, L.; Válega, M.; Mendes, R.F.; Paz, F.A.A.; Bachari, K.; Talhi, O.; Silva, A.M.S. Hemi-Synthesis of Chiral Imine, Benzimidazole and Benzodiazepines from Essential Oil of *Ammodaucus leucotrichus* subsp. *leucotrichus*. *Molecules* **2019**, *24*, 975. [CrossRef] [PubMed]
7. Muckensturm, B.; Diyani, F.; Le Nouen, D.; Fkih-Tetouani, S.; Reduron, J.P. Ammolactone, a guaianolide from a medicinal plant, Ammodaucus leucotrichus. *Phytochemistry* **1997**, *44*, 907–910. [CrossRef]
8. European Directorate for the Quality of Medicines & HealthCare. Herbal drugs and herbal drug preparations. In *European Pharmacopoeia*, 10th ed.; Deutscher Apotheker Verlag: Stuttgart, Germany, 2019.
9. Pradeep, K.U.; Geervani, P.; Eggum, B.O. Common Indian spices: Nutrient composition, consumption and contribution to dietary value. *Plant Food Hum. Nutr.* **1993**, *44*, 137–148. [CrossRef]
10. Food Data Central. U.S. Department of Agriculture. Agriculture Research Service. Available online: https://fdc.nal.usda.gov/fdc-app.html#/food-details/170923/nutrients (accessed on 11 July 2022).
11. Barber, T.M.; Kabisch, S.; Pfeiffer, A.F.H.; Weickert, M.O. The Health Benefits of Dietary Fibre. *Nutrients* **2020**, *12*, 3209. [CrossRef]
12. Chatterjee, R.; Yeh, H.C.; Edelman, D.; Brancati, F. Potassium and risk of Type 2 diabetes. *Expert Rev. Endocrinol. Metab.* **2011**, *6*, 665–672. [CrossRef]
13. Chung, M.; Balk, E.M.; Brendel, M.; Ip, S.; Lau, J.; Lee, J.; Lichtenstein, A.; Patel, K.; Raman, G.; Tatsioni, A.; et al. Vitamin D and calcium: A systematic review of health outcomes. *Evid. Rep. Technol. Assess.* **2009**, *183*, 1–420.
14. Jimenez-Lopez, C.; Carpena, M.; Lourenço-Lopes, C.; Gallardo-Gomez, M.; Lorenzo, J.M.; Barba, F.J.; Prieto, M.A.; Simal-Gandara, J. Bioactive Compounds and Quality of Extra Virgin Olive Oil. *Foods* **2020**, *9*, 1014. [CrossRef]
15. Briggs, M.A.; Petersen, K.S.; Kris-Etherton, P.M. Saturated Fatty Acids and Cardiovascular Disease: Replacements for Saturated Fat to Reduce Cardiovascular Risk. *Healthcare* **2017**, *5*, 29. [CrossRef] [PubMed]
16. Adams, R.P. *Identification of Essential Oil Components by Gas Chromatography/Mass Spectrometry*, 4th ed.; Allured Busines Media: Carol Stream, IL, USA, 2007.
17. Gherraf, N.; Zellagui, A.; Kabouche, A.; Lahouel, M.; Salhi, R.; Rhouati, S. Chemical constituents and antimicrobial activity of essential oils of Ammodaucus leucotricus. *Arab. J. Chem.* **2017**, *10*, S2476–S2478. [CrossRef]
18. Abu Zarga, M.H.; Al-Jaber, H.I.; Baba Amer, Z.Y.; Sakhrib, L.; Al-Qudah, M.A.; Al-Humaidi, J.Y.G.; Abaza, I.F.; Afifi, F.U. Chemical Composition, Antimicrobial and Antitumor Activities of Essential Oil of Ammodaucus leucotrichus Growing in Algeria. *J. Biol. Act. Prod. Nat.* **2013**, *3*, 224–231.
19. Velasco-Negueruela, A.; Pèrez-Alonso, M.J.; Pèrez de Paz, P.L.; Pàla-Paùl, J.; Sanz, J. Analysis by gas chromatography-mass spectrometry of the volatiles from the fruits of Ammodaucus leucotrichus subsp. leucotrichus and subsp. nanocarpus grown in North Africa and the Canary Islands, respectively. *J. Chromat. A* **2006**, *1108*, 273–275. [CrossRef] [PubMed]
20. El-Haci, I.A.; Bekhechi, C.; Atik-Bekkara, F.; Mazari, W.; Gherib, M.; Bighelli, A.; Casanova, J.; Tomi, F. Antimicrobial Activity of Ammodaucus leucotrichus Fruit Oil from Algerian Sahara. *Nat. Prod. Commun.* **2014**, *9*, 711–712. [CrossRef] [PubMed]
21. Dahmane, D.; Dob, T.; Krimat, S.; Nouasri, A.; Metidji, H.; Ksouri, A. Chemical composition, antioxidant and antibacterial activities of the essential oils of medicinal plant Ammodaucus leucotrichus from Algeria. *J. Essent. Oil Res.* **2017**, *29*, 48–55. [CrossRef]
22. Sonboli, A.; Esmaeili, M.A.; Gholipour, A.; Kanani, M.R. Composition, cytotoxicity and antioxidant activity of the essential oil of Dracocephalum surmandinum from Iran. *Nat. Prod. Commun.* **2010**, *5*, 341–344. [CrossRef] [PubMed]
23. Zhou, F.; Dai, O.; Peng, C.; Xiong, L.; Ao, H.; Liu, F.; Zhou, Q.M. Pro-Angiogenic Effects of Essential Oil from Perilla frutescens and Its Main Component (Perillaldehyde) on Zebrafish Embryos and Human Umbilical Vein Endothelial Cells. *Drug Des. Dev. Ther.* **2021**, *15*, 4985–4999. [CrossRef]
24. Catanzaro, E.; Turrini, E.; Kerreb, T.; Sioen, S.; Baeyens, A.; Guerrini, A.; Abdi Bellau, M.L.; Sacchetti, G.; Paganetto, G.; Krysko, V.D.; et al. Perillaldehyde is a new ferroptosis inducer with a relevant clinical potential for acute myeloid leukemia therapy. *Biomed. Pharmacother.* 2022; submitted for publication.
25. IUCN Centre For Mediterranean Cooperation. *A Guide to Medicinal Plants in North Africa*; IUCN Centre For Mediterranean Cooperation: Malaga, Spain, 2005.
26. International Dairy Federation. Determination of nitrogen content. In *Kjeldahl Method, Norma FIL-IDF, No 20B*; International Dairy Federation: Brussels, Belgium, 1993.

27. Lee, S.; Prosky, L.; DeVries, J. Determination of total, soluble, and insoluble dietary fiber in foods: Enzymatic-gravimetric method, MES-TRIS buffer: Collaborative study. *J. Assoc. Off. Anal. Chem.* **1992**, *75*, 395–416. [CrossRef]
28. Prosky, L.; Asp, N.G.; Scheweizer, T.F.; DeVries, J.W.; Furda, I. Determination of insoluble and soluble, and total dietary fibre in foods and food products: Interlaboratory study. *J. Assoc. Off. Anal. Chem.* **1988**, *71*, 1017–1023.
29. Ertl, K.; Goessler, W. Grains, whole flour, white flour, and some final goods: An elemental comparison. *Eur. Food Res. Technol.* **2018**, *244*, 2065–2075. [CrossRef]
30. Tacchini, M.; Echeverria Guevara, M.P.; Grandini, A.; Maresca, I.; Radice, M.; Angiolella, L.; Guerrini, A. Ocimum campechianum Mill. from Amazonian Ecuador: Chemical Composition and Biological Activities of Extracts and Their Main Constituents (Eugenol and Rosmarinic Acid). *Molecules* **2020**, *26*, 84. [CrossRef] [PubMed]
31. Makhloufi, A.; Moussaoui, A.; Lazouni, H.A. Antibacterial activities of essential oil and crude extracts from Matricaria pubescens (Desf.) growing wild in Bechar, South west of Algeria. *J. Med. Plant Res.* **2012**, *6*, 3124–3128. [CrossRef]
32. Wagner, H.; Bladt, S. *Plant Drug Analysis: A Thin Layer Chromatography Atlas*, 2nd ed; Springer: Berlin, Germany, 1996; p. 364.
33. Scalvenzi, L.; Radice, M.; Toma, L.; Severini, F.; Boccolini, D.; Bella, A.; Guerrini, A.; Tacchini, M.; Sacchetti, G.; Chiurato, M.A.; et al. Larvicidal activity of *Ocimum campechianum*, *Ocotea quixos* and *Piper aduncum* essential oils against *Aedes aegypti*. *Parasite* **2019**, *26*, 23. [CrossRef] [PubMed]

molecules

MDPI

Article

Curcumin Modifies the Activity of Plasmatic Antioxidant Enzymes and the Hippocampal Oxidative Profile in Rats upon Acute and Chronic Exposure to Ozone

Abraham Alberto Ramírez-Mendoza [1], Mario Alberto Ramírez-Herrera [1], Cesar Ricardo Cortez-Álvarez [2], Sendar Daniel Nery-Flores [3], Aldo Rafael Tejeda-Martínez [4], Marina María de Jesús Romero-Prado [1] and María Luisa Mendoza-Magaña [1,*]

1 Laboratorio de Neurofisiología, Departamento de Fisiología, Centro Universitario de Ciencias de la Salud, Universidad de Guadalajara, Guadalajara 44340, Mexico; abrahamalberto.ramirez@alumnos.udg.mx (A.A.R.-M.); amario@cucs.udg.mx (M.A.R.-H.); marina.rprado@academicos.udg.mx (M.M.d.J.R.-P.)

2 Departamento de Farmacobiología, Centro Universitario de Ciencias Exactas e Ingenierías, Universidad de Guadalajara, Guadalajara 44430, Mexico; cesar.cortez@academicos.udg.mx

3 Departamento de Investigación en Alimentos, Facultad de Ciencias Químicas, Universidad Autónoma de Coahuila, Saltillo 25280, Mexico; sendar.nery@uadec.edu.mx

4 Laboratorio de Neurobiología Molecular, División de Neurociencias del Centro de Investigación Biomédica de Occidente, IMSS, Guadalajara 44340, Mexico; tejedamartinez.ar@gmail.com

* Correspondence: luisa.mendoza@academicos.udg.mx; Tel./Fax: +52-33-10585313

Citation: Ramírez-Mendoza, A.A.; Ramírez-Herrera, M.A.; Cortez-Álvarez, C.R.; Nery-Flores, S.D.; Tejeda-Martínez, A.R.; Romero-Prado, M.M.d.J.; Mendoza-Magaña, M.L. Curcumin Modifies the Activity of Plasmatic Antioxidant Enzymes and the Hippocampal Oxidative Profile in Rats upon Acute and Chronic Exposure to Ozone. *Molecules* **2022**, 27, 4531. https://doi.org/10.3390/molecules27144531

Academic Editors: Raffaele Pezzani and Sara Vitalini

Received: 14 June 2022
Accepted: 13 July 2022
Published: 15 July 2022

Publisher's Note: MDPI stays neutral with regard to jurisdictional claims in published maps and institutional affiliations.

Abstract: Ozone (O_3) is an oxidating tropospheric pollutant. When O_3 interacts with biological substrates, reactive oxygen and nitrogen species (RONS) are formed. Severe oxidative damage exhausts the endogenous antioxidant system, which leads to the decreased activity of antioxidant enzymes such as catalase (CAT), glutathione peroxidase (GPx), and superoxide dismutase (SOD). Curcumin (CUR) is a natural polyphenol with well-documented antioxidant and anti-inflammatory properties. The aim of this work is to evaluate the effects of curcumin on CAT, GPx, and SOD activity and the inhibition of oxidative damage after the acute and chronic exposure to O_3. Fifty male Wistar rats were divided into five experimental groups: the intact control, CUR-fed control, exposed-to-O_3 control, CUR-fed (preventive), and CUR-fed (therapeutic) groups. These two last groups received a CUR-supplemented diet while exposed to O_3. These experiments were performed during acute- and chronic-exposure phases. In the preventive and therapeutic groups, the activity of plasma CAT, GPx, and SOD was increased during both exposure phases, with slight differences; concomitantly, lipid peroxidation and protein carbonylation were inhibited. For this reason, we propose that CUR could be used to enhance the activity of the antioxidant system and to diminish the oxidative damage caused by exposure to O_3.

Keywords: curcumin; antioxidant properties; ozone

1. Introduction

Globally, air pollution affects millions of people to such an extent that it has been associated with chronic degenerative diseases [1]. Among air pollutants, O_3 stands out due to its high oxidizing power. O_3 is produced via the photochemical reactions of volatile organic compounds and nitrogen oxides (NOx) [2]. When O_3 interacts with the biologic epithelia, reactive oxygen and nitrogen species (RONS) are quickly produced. However, when these oxidative species are generated in low concentrations, they are efficiently inactivated by endogenous antioxidant enzymes. Furthermore, during acute exposure to O_3, the endogenous antioxidant system is able to revert the oxidative damage once the exposure has ended [3–5]. Conversely, if RONS reach high concentrations and the exposure period becomes chronic, the antioxidant system is easily overcome, and oxidative damage is disseminated in tissues and organs. During these events, chained oxidative reactions occur through a variety of oxidant metabolites that affect most biomolecules [5,6]. Furthermore, the expression and activity of pro-oxidant enzymes (cyclo-oxygenase 2, lipo-oxygenase 5, and inducible nitroxide synthase) are increased by RONS through the activation of the nuclear factor kappa-light-chain-enhancer of activated B cells (NF-κB) [7]. Previous work by our group has demonstrated that NF-κB overactivation increases the concentration of inflammatory cytokines [8]. Similarly, the overactivation of NF-κB increases the expression of Kelch-like ECH-associated protein 1 (KEAP1), which binds to the nuclear factor erythroid 2-related factor 2 (Nrf2), leading to its proteosomal degradation. This fact could have consequences in decreasing the expression and activity of endogenous antioxidant enzymes [9]. On the other hand, the activation of Nrf2 increases the expression of antioxidant enzymes [10,11]. Among antioxidant enzymes, catalase (CAT), superoxide dismutase (SOD), and glutathione peroxidase (GPx) have the capacity to reduce RONS activity by blocking or delaying oxidative damage and protecting cellular components, such as proteins, lipids, carbohydrates, and DNA [10,12]. Despite the oxidative harmful effects caused by O_3, there are few pharmacological approaches used to revert the damage process. The effects of several natural and synthetic molecules with antioxidant properties that might reduce the oxidative damage caused by RONS have been documented in the literature [13]. However, there are serious concerns regarding the secondary effects and adverse interactions that arise during their long-term consumption. For example, when vitamin C is administrated in the long term, it causes gastric disorders and ulcers. Vitamin E causes hemorrhagic events and mortality during long-term administration. Chronic melatonin administration has been related to daytime sleepiness, headaches, dizziness, and hypothermia [14–16]. Particularly, the effect of tibolone against the oxidative damage caused by ozone in the CNS has been evaluated and has demonstrated beneficial effects; however, the long-term administration of tibolone should be avoided due to an increased risk of different kind of cancers [17,18].

Curcumin (CUR) is a phenolic molecule extracted from *Curcuma longa* Lin rhizome, a plant that belongs to the Zingiberaceae family [19]. CUR exhibits potent antioxidant activity against RONS [20], and it can be administered for long periods of time without exerting side effects [21]. A variety of beneficial pharmacological effects of CUR, such as anticancer, antiparasitic, anti-inflammatory, and signaling pathway modulation effects, among others. These effects have been explained on the basis of multiple molecular targets on which CUR acts [19,22–25]. Thus, we are interested in studying this molecule and its activity in different damage models. CUR has been reported to increase the activity of antioxidant enzymes and to provide protection from oxidative damage [26]; however, this effect has not been studied in persistent oxidative damage during acute or chronic exposure to O_3. In previously published articles, we demonstrated that curcumin (CUR) decreased NF-κB activation and reduced astrocytosis, microgliosis, and apoptosis in the hippocampus during acute and chronic O_3 exposure in preventive and therapeutic approaches [8,27]. In fact, the hippocampus is crucial for short-term memory and learning, being one of the most important structures of the central nervous system, but it is also the most sensitive to oxidative damage, as it is directly connected to the olfactory mucosa, which is the first access site for exogenous pollutants [28]. The aim of this work is to evaluate effects of curcumin on the activity of plasma antioxidant enzymes and the oxidative profile in the hippocampi of rats exposed to acute and chronic doses of O_3.

2. Results

2.1. *Curcumin Modified Plasma Antioxidant Enzymes Activities during Ozone Exposure*

2.1.1. CAT Activity

CAT activity was determined by means of a spectrophotometric method that detects the presence of H_2O_2 in biological samples. Thus, a higher level of H_2O_2 means lower CAT activity. In Figure 1a shows that the plasma of rats exposed to O_3 in the acute phase (AO, 0.3702 ± 0.057) had the lowest activity, which was significant when compared to the acute-intact group (0.7270 ± 0.040). In contrast, the dietary administration of CUR significantly improved the CAT activity in the AP (0.787 ± 0.031) and AT (0.772 ± 0.047) groups ($p < 0.01$). When compared against the control groups, AI and AC, AP and AT showed similar levels of activity.

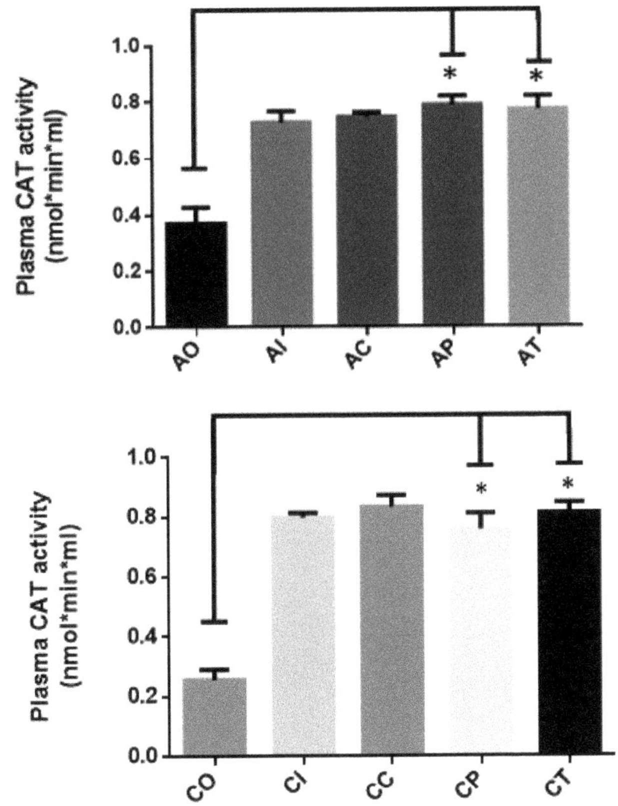

a

b

Figure 1. The catalase activity in rats exposed to ozone during the acute and chronic phases: (**a**) the preventive and therapeutic administration of dietary curcumin (5.6 mg/Kg) resulted in significantly increased activity (* $p < 0.01$) when evaluated at the end of the acute phase when compared to the control group exposed to ozone; acute O_3 (AO), acute intact (AI), acute CUR (AC), acute preventive (AP), acute therapeutic (AT); (**b**) the effect of curcumin, when evaluated at the end of the chronic phase, maintained the same profile as observed in the acute phase, with a significance of $p < 0.01$ (*); chronic O_3 (CO), chronic intact (CI), chronic CUR (CC), chronic preventive (CP), chronic therapeutic (CT).

In the chronic phase (Figure 1b), the CAT activity was similar to the acute phase. Again, dietary CUR caused a significant increase in CAT activity in the CP (0.757 ± 0.053) and CT (0.8115 ± 0.035) groups ($p < 0.01$) when compared to CO (0.255 ± 0.034), as observed in Figure 1b. When these groups were compared to the unexposed CI and CC groups, no significant differences were found.

2.1.2. SOD Activity

The SOD assay implies the use of the tetrazolium salt WST-1, which reacts with the superoxide anion generated by the xanthine oxidase. Thus, the dismutation of the superoxide anion into H_2O_2 and O_2 catalyzed by SOD is related to WST-1 reducing into WST-1 formazan. The resulting data are expressed in terms of the activity percentage rate.

Figure 2a depicts that the effects of dietary CUR on SOD activity significantly increased in the AP group (77.74 ± 3.62), as the percentage of the activation rate increased significantly ($p < 0.01$) versus the OA group (48.34 ± 5.72). The AI group (40.83 ± 2.20, $p < 0.01$) also showed lower SOD activity compared to the AC group; thus, the increased SOD activity in the AC group could be due to a CUR effect. Additionally, the effect of dietary CUR in the AT group also increased SOD activity versus the AI group ($p < 0.01$).

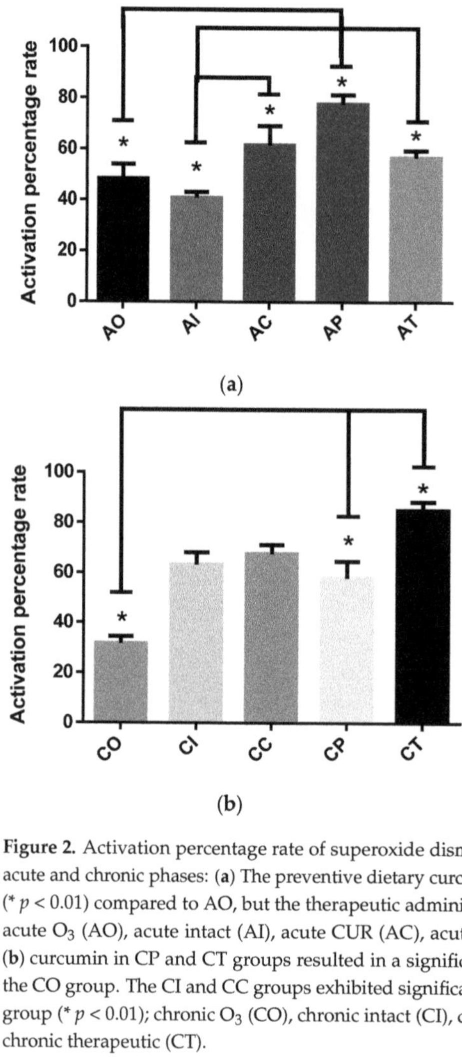

Figure 2. Activation percentage rate of superoxide dismutase in rats exposed to ozone during the acute and chronic phases: (**a**) The preventive dietary curcumin administration increased SOD activity (* $p < 0.01$) compared to AO, but the therapeutic administration failed to increase the SOD activity; acute O_3 (AO), acute intact (AI), acute CUR (AC), acute preventive (AP), acute therapeutic (AT); (**b**) curcumin in CP and CT groups resulted in a significant SOD increase (* $p < 0.01$) compared to the CO group. The CI and CC groups exhibited significant increases in SOD activity versus the CO group (* $p < 0.01$); chronic O_3 (CO), chronic intact (CI), chronic CUR (CC), chronic preventive (CP), chronic therapeutic (CT).

After the chronic phase was completed (Figure 2b), dietary CUR significantly increased SOD activity in the CP (58.02 ± 6.64) and CT (85.0 ± 3.41) groups ($p < 0.01$) when compared to the CO group (31.67 ± 2.68). Furthermore, the CT group showed better SOD activity when compared to CP ($p < 0.05$). The groups CI (63.33 ± 4.83) and CC (67.50 ± 3.70) also showed increased SOD activity ($p < 0.01$) compared to the group exposed to O_3 (CO). Additionally, the SOD activity was significantly increased in the CI group (panel b) when versus the AI group (panel a).

2.1.3. GPx Activity

The GPx assay is mainly based on the oxidation of GSH to generate GSSG, inducing the reduction of cumene hydroperoxide. The glutathione reductase reduces GSSG to produce GSH, thereby consuming NADPH. The decrease in NADPH is proportional to the GPx activity. In the acute phase (Figure 3a), dietary CUR significantly increased the GPx activity in the AP (36.63 ± 1.55, $p < 0.01$) and AT (31.90 ± 2.77, $p < 0.05$) groups compared to the AO group (20.37 ± 3.08). The AI and AC groups showed low GPx activity, similar to the AO group. In the chronic phase (Figure 3b), the CP group (23.14 ± 3.21) and CT group (34.97 ± 2.27) exhibited a significant increase in GPx activity caused by CUR ($p < 0.01$) compared to the CO group (10.11 ± 2.31). The CI group (19.72 ± 1.88) also showed a significant increase in GPx activity when compared to the CO group ($p < 0.01$).

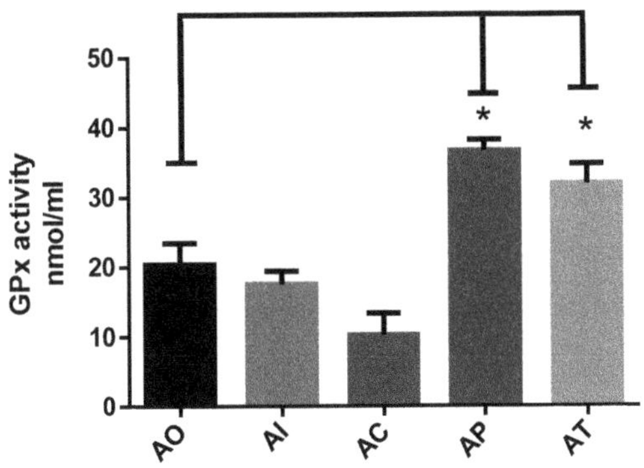

(a)

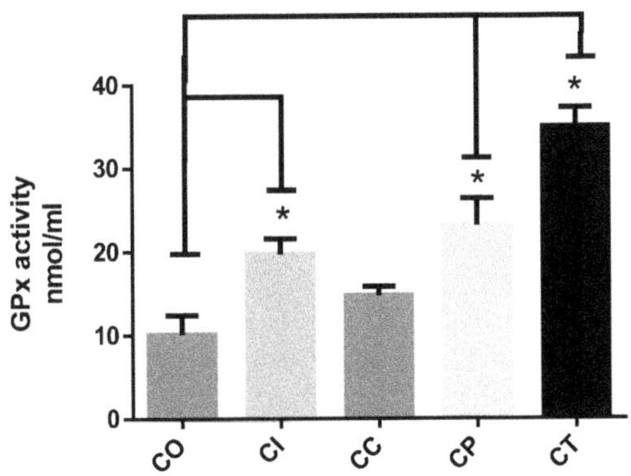

(b)

Figure 3. The glutathione peroxidase (GPx) activity in rats exposed to ozone during the acute and chronic phases: (**a**) the preventive administration of dietary curcumin in the preventive approach at the end of the acute phase presented increased activity of GPx (* $p < 0.05$) when compared to the AO group. Similarly, the AT group showed increased GPx activity ($p < 0.01$) compared to the acute ozone-exposed control group OA; acute O_3 (AO), acute intact (AI), acute CUR (AC), acute preventive (AP), acute therapeutic (AT); (**b**) the effect of curcumin in the CP and CT groups (* $p < 0.01$) resulted significantly increased GPx activity (* $p < 0.05$) compared to the chronic ozone-exposed control CO group; chronic O_3 (CO), chronic intact (CI), chronic CUR (CC), chronic preventive (CP), chronic therapeutic (CT).

2.1.4. Inhibition of MDA and 4-HNE Formation by CUR

The oxidative damage to hippocampal lipids was significantly inhibited by CUR in the acute phase (Figure 4a). Comparing the percentages in the AO group (18.07% ± 8.89) versus the percentages of the AP (95.02% ± 0.4638) and AT (97.35% ± 0.4279) groups, the difference was statistically significant ($p < 0.0001$). There were no significant differences among the AP and AT groups versus the AI and AC groups. In the chronic phase (Figure 4b), the percentage of oxidative damage to lipids depicted a similar profile, in which the CP group (90.40% ± 1.765) and the CT group (98.23% ± 0.222) showed significant inhibition caused by CUR ($p < 0.0001$) compared to CO (23.92% ± 5.847).

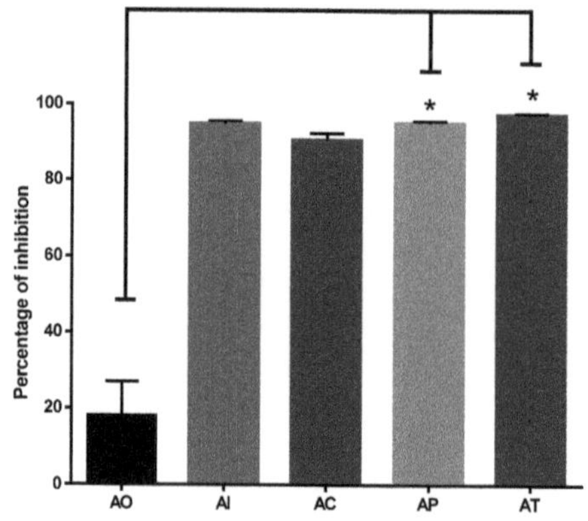

(**a**)

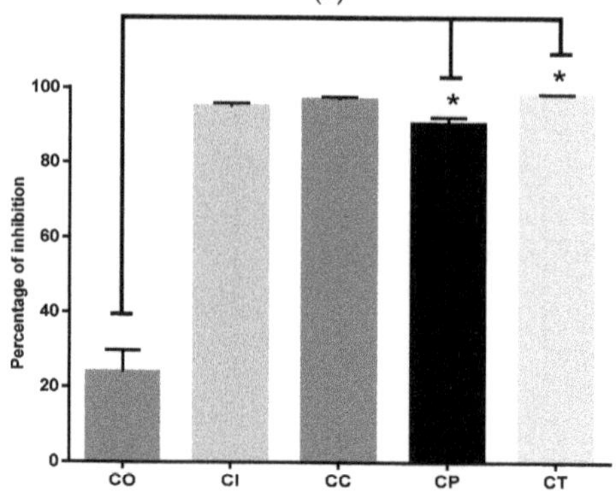

(**b**)

Figure 4. Inhibition of oxidative damage to lipids by curcumin in acute and chronic phases after exposure to ozone: (**a**) in the acute exposure to ozone, the administration of dietary curcumin in preventive and therapeutic modes significantly inhibited the oxidative damage to lipids (* $p < 0.01$) compared to the control group exposed to ozone OA; acute O_3 (AO), acute intact (AI), acute CUR (AC), acute preventive (AP), acute therapeutic (AT); (**b**) the effect of curcumin in preventive and therapeutic administration significantly inhibited oxidative damage to lipids (* $p < 0.01$), compared to the CO group; chronic O_3 (CO), chronic intact (CI), chronic CUR (CC), chronic preventive (CP), chronic therapeutic (CT).

2.1.5. Curcumin Inhibited Oxidative Damage to Hippocampal Proteins

Dietary administration of CUR protected hippocampal proteins during exposure to O_3. In the acute phase (Figure 5a), the AO group only inhibited protein oxidation by 18.11% ± 9.91; meanwhile, CUR conferred protection to proteins by inhibiting oxidative damage by 99.59% ± 0.168 ($p < 0.0001$) in the AP group. Similar protection was observed in the AT group (97.17% ± 0.137). When the chronic phase was complete (Figure 5b), the CO group showed antioxidant protection of 18.17% ± 4.87. In contrast, dietary CUR protected against oxidative protein damage in the CP, with a percentage of 99.50 ± 0.038 ($p < 0.0001$), and for the CT group, the percentage was 96.48 ± 0.34 ($p < 0.0001$). The percentage of protein oxidative damage in the CI group was 95.91 ± 1.83, and for the CC, group the percentage was 98.82 ± 0.08.

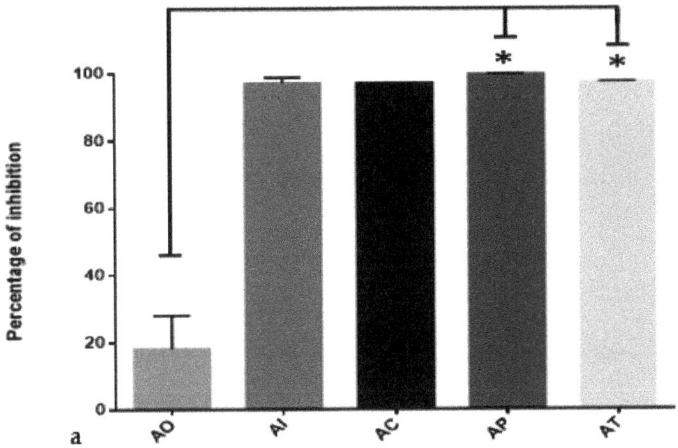

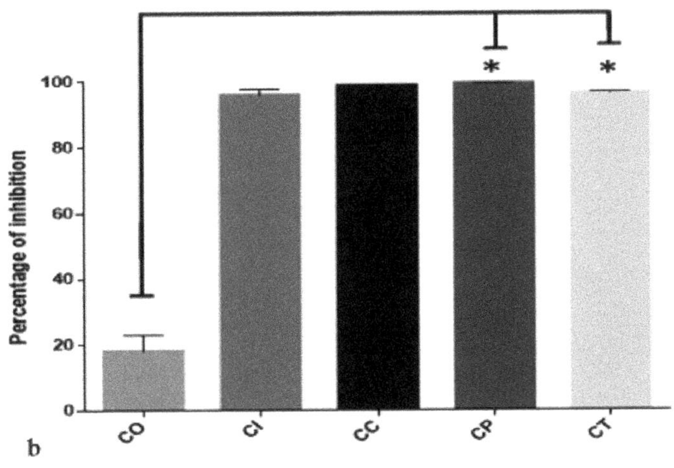

Figure 5. Inhibition of oxidative damage to hippocampal proteins exerted by curcumin against the ozone exposure in the acute and chronic phases: (**a**) the administration of dietary curcumin to the AP and AT groups significantly inhibited (* $p < 0.0001$) protein oxidation of hippocampal tissue in rats exposed to ozone when compared to the OA group, where oxidative damage was at the maximum level; acute O_3 (AO), acute intact (AI), acute CUR (AC), acute preventive (AP), acute therapeutic (AT); (**b**) this image depicts the effect of curcumin in preventive and therapeutic administration during chronic exposure to ozone; in both cases, the oxidative damage was markedly inhibited (* $p < 0.0001$) compared to damage determined in the chronic ozone-exposure control group (CO); chronic O_3 (CO), chronic intact (CI), chronic CUR (CC), chronic preventive (CP), chronic therapeutic (CT).

3. Discussion

In this work, the preventive and therapeutic dietary administration of CUR was analyzed, as it exerted differentiated modulatory effects on the activity of the plasma antioxidant enzymes in rats during acute and chronic exposure to O_3. However, the of CAT, SOD, and GPx activity was increased by CUR, with slight differences between preventive and therapeutic administration. Some of the differences that were identified among the measured enzymes may respond to dynamic regulations within the antioxidant system depending on whether or not it is under the influence of curcumin. Contrarily, CUR inhibited the oxidative damage caused by O_3, regardless of the exposure time or the CUR administration mode. In this case, the direct antioxidant activity of CUR as a scavenger for RONS should be highlighted.

The variability found in diverse reports analyzing the effects of CUR or its derivatives with respect to the behavior of endogenous antioxidant enzymes may be due to factors such as administration mode (preventive, therapeutic, continue, intermittent, or single dose), dose level (high, medium, or low), the age and gender of the animals used in experimental models, injury mode (single, continuous, intermittent, acute, chronic), the nature of the injury agent (physical, chemical, mechanic), and the experimental design, among other factors [29–34].

In our experiments, we obtained dramatic results for the CAT, SOD, and GPx activity. Some differences are important to clarify with respect to the age of the animals and the activation sequence. The basal activity of CAT remained unaltered in the acute versus the chronic phase in the intact group as well as in the CUR-fed group. Contrarily, the group exposed to O_3 exhibited a significant decrease in CAT activity during both exposure phases (49% acute phase, and 68% chronic phase). The SOD activity in the intact group during the acute phase was not different from the activity determined for the O_3-exposed group; for its part, CUR improved SOD activity in the AC group. Furthermore, SOD activity was not affected by O_3 during the acute phase, as it could had been protected by GPx, which showed normal activity in this exposure phase; concomitantly, the CAT activity was depleted at the same time, and this could indicate that it was consumed at this point. Similar behavior has been previously reported [3]. In contrast, the protective effect of CUR administered in the preventive or therapeutic mode induced a recovery of the CAT, SOD, and GPx activity at levels similar to those in the intact condition. These results are in accordance with previous reports [35].

As a plausible explanation, the decreased CAT, SOD, and GPx activity could be due to the low activation of Nrf2 [36] as consequence of increased activation of NFkB by RONS. Previous studies have documented that the activation of NFkB leads to an overexpression of KEAP1, which binds Nrf2, inhibiting its translocation and increasing its proteosomal degradation [9,37]. This could be a possible explanation for the decline in the activity of the antioxidant endogenous enzymes. In spite of this mechanism having been previously reported, it remains to be further documented in similar studies. Furthermore, RONS may cause oxidative damage to CAT at high levels, reducing its activity. In our experiments, the exposure to O_3 caused a significant decrease in CAT activity in the AO and CO groups; meanwhile, the AI and AC groups showed high basal activity, similar to the preventive and therapeutic groups that recovered their CAT activity due to the effects of CUR. Furthermore, the SOD and GPx activity was not affected in the acute phase, which could be due to the predominant formation of H_2O_2 caused by O_3 and its neutralization by CAT, whose activity was exhausted during acute exposure. On the contrary, the activity of both enzymes was decreased by O_3 in the chronic phase. This effect could be the result of the activation of NFkB, leading to the increased expression of KEAP1, which, in turn, binds to Nrf2, leading to its further proteosomal degradation and the consequent decreased expression of CAT, SOD, and GPx [38].

In the preventive mode, CUR could first act as an antioxidant by neutralizing RONS and then be involved in priming the activation of Nrf2 through the reduction of serine residues, leading to its nuclear translocation [9]. Furthermore, CUR can avoid the activation of NFkB through the inhibition of the I kappa B kinase [39,40]; this would decrease NFkB dissociation from IkBa and would consequently also decrease the expression of KEAP1, leading to reduced Nrf2 proteosomal degradation [41]. Thus, when the oxidant insult by O_3 begins, both transcriptional factors are modulated by CUR. In the therapeutic mode, CUR exerts action focused on reverting the activation of NFkB and inhibiting the degradation of Nrf2 by reducing the expression of KEAP1, favoring its activation and reactivating the expression and activity of CAT, SOD, and GPx [8,9].

The SOD activity in the O_3 control group in the acute phase did not exhibit significant activity modification. This could mean that SOD activity may have been restored every day after the 4 h of daily exposure. However, after O_3 exposure for 60 days (chronic phase), the SOD activity was no longer restored and caused a 50% decrease. Similar findings have been previously reported [4,42]. On the other hand, the intact control in the chronic phase presented increased SOD activity (64.47%) compared to the acute phase; this could indicate the maturation of the naturally occurring endogenous antioxidant system, as previously shown [43]. In contrast, the beneficial effect of CUR was evidenced in the therapeutic group in both phases by increasing the enzyme activity. These findings agree with previous reports [44,45]. Furthermore, this is the first time that the effect of CUR supporting the activity of SOD against the damage caused by acute or chronic exposure to ozone in a preventive mode has been reported.

In the acute phase (15 days), the GPx activity was not significantly altered by O_3 exposure, which could be due to a repairing process during the 20 h interval without exposure. However, GPx activity was significantly increased when the animals were exposed to ozone and treated with CUR in the preventive and therapeutic approaches. Furthermore, the chronic exposure to O_3 caused a significant decrease in GPx activity, meaning that the 60-day exposure to O_3 impeded the success of the repair process. This decrease was reverted by CUR in the therapeutic mode and was avoided in the preventive mode.

The GPx activity in the chronic phase of the O_3 control group showed a 50% decrease in activity in the chronic phase compared to in the acute phase, as described in vitro by [3].

Using this model of oxidative damage at an O_3 dose of 0.7 ppm for four hours (h) every day, the CO group showed that it can increase oxidative damage in the cell membrane by increasing MDA, 4-HNE, and protein carbonylation [8,42] in such a way that it allows the over-activation of NF-κB, which induces the overexpression of KEAP1. In this context, when KEAP1 sequesters Nrf2 in such a way that it decreases the activity of antioxidant enzymes [9], this allows the onset of oxidative stress. When Nrf2 stops translocating to the nucleus, the activity of antioxidant enzymes decreases, as demonstrated in this work and as previously documented [3].

To assess the estimation of oxidative damage to lipids, MDA and 4HNE metabolites are commonly determined, and these are generated during RONS interaction with cells. The implemented methods report the use of mesylate to estimate both metabolites simultaneously [46]. The results demonstrating the oxidative damage to lipids obtained in the AP, CP, AT, and CT groups showed that CUR maintained similar percentages of oxidative inhibition similar to the intact group in both phases. These findings indicate that CUR performed its direct antioxidant effect as an RONS scavenger and indirectly mediated Nrf2 activation and NFκB inactivation. This effect is very similar to that reported in the work carried out by Mendoza-Magaña et al., 2021, and Guerrero-Hue et al., 2019 [27,47]. Furthermore, the results of the protein carbonylation assay also showed that CUR was able to inhibit this oxidative damage process in the experimental groups that were exposed to ozone and fed with CUR in the preventive and therapeutic modes. These findings have been reported [48,49].

Supported by the findings and analysis performed in this report and other studies carried out previously, we propose testing the protective effects that CUR may exert in human populations inhabiting in cities with continuously high ozone levels throughout the year and evaluating its effects on the endogenous antioxidant system, which would improve the inhibition of oxidative damage, cognitive performance, and academic profile in volunteers of different ages. Considering that CUR may prevent and revert oxidative damage due to its scavenger properties and because of the enhancement of the antioxidant enzymes activity evaluated in the present work, its beneficial effects should continue to be evaluated in experimental models and clinical trials. We suggest that it might be used to prevent pollutant-related diseases, such as Parkinson's and Alzheimer's disease, as well as a complimentary therapy for these oxidative damage-related disorders.

4. Materials and Methods

4.1. Animals

This study was performed using 50 male Wistar rats (*Rattus norvegicus*) that were 21-days-old and that had an average body weight of 130 g. They were kept in light and dark cycles of 12 X 12 h and at 50–60% relative humidity with free access to food and water. The experiments were carried out according to the National Institutes of Health Guide for the Care and Use of Laboratory Animals (NIH Publications No. 8023, revised 1978), which are as well established by the Ethics Committee of the Health Sciences Center (CUCS, Universidad de Guadalajara), under the approval number, CI-00512, and by the Ministry of Health of Jalisco State 76/UG-JAL/2011.

4.2. Diet

A turmeric ethanolic extract was obtained using a Soxhlet extraction apparatus. The infrared spectrum profile was compared to a CUR standard diluted in ethanol (Sigma Aldrich, St. Louis, MO, USA), showing and confirming the identity of the extracted molecule (Figure 1). The concentration of CUR in the extract was determined by UV spectrometry and adjusted to 12 mg/mL. This solution was employed to impregnate the food pellets (Prolab®RMH Laboratory Animal diet, 2500 Rodent 5P14), and ethanol was eliminated by ventilated evaporation at 58 °C for 4 h. These procedures were performed in the dark to avoid CUR photodegradation. This food provided an approximate daily CUR dose of 5.6 mg/Kg. The diet was dynamically adjusted by increasing the grams of food served according to the increases in the animal´s body weight.

4.3. Experimental Design

Animals were randomly distributed into five experimental groups of 10 rats each. Every experimental group was adapted to the handling procedures and accommodation in the exposure acrylic chamber for four hours over a 7-day period prior to initializing the experiment. The adaptation was carried out to minimize the effect of human contact. The design was established by considering two periods of O_3 exposure: an acute phase (A, 15 days) and a chronic phase (C, 60 days), which led to the formation of two subgroups. Thus, the final experimental groups were named as follows: acute intact group (AI, $n = 5$) and chronic intact group (CI, $n = 5$) were exposed to O_3-free air and food without CUR; the groups fed with CUR supplementation and exposed to O_3-free air in both phases were denoted as AC ($n = 5$) and CC ($n = 5$). The O_3-exposed groups were exposed to 0.7 ppm for 4 h and were denoted as AO ($n = 5$) and CO ($n = 5$). Time of exposure to O_3 and/or the time of the supplemented diet with CUR defined the following groups as the therapeutic groups or the preventive groups for each exposure period. The preventive groups received food with CUR supplementation for the first 7 days, and afterwards, O_3 exposure began and continued until both exposure phases were completed (AP, $n = 5$ and CP, $n = 5$). The therapeutic groups were exposed to 0.7 ppm of O_3 for the first 7 days, and afterwards the diet with CUR supplementation was served until the end of each exposure time, defining the groups as therapeutic acute (AT, $n = 5$) and therapeutic chronic (CT, $n = 5$).

The Table 1 summarizes the experimental groups and their nomenclature.

Table 1. Experimental groups and design.

	Acute Phase
AO	Acute O_3 (exposure to 0.7 ppm of O_3 for 4 h for 15 days).
AI	Acute intact (exposed to O_3 free air for 4 h and without CUR supplementation for 15 days).
AC	Acute CUR (diet supplemented with CUR and exposed to O_3-free air for 15 days).
AP	Acute preventive (diet supplemented with CUR provided 7 days prior to exposure to O_3 for 15 days and continued CUR supplementation).
AT	Acute therapeutic (exposure to O_3 7 days prior to the administration of the diet supplemented with CUR for 15 days and continued exposure to O_3 until day 15).
	Chronic phase
CO	Chronic O_3 (exposure to 0.7 ppm of O_3 for 4 h for 60 days).
CI	Chronic intact (exposure to O_3-free air for 4 h and without CUR supplementation for 60 days).
CC	Chronic CUR (exposure to O_3-free air for 4 h with diet supplemented with CUR administered for 60 days).
CP	Chronic preventive (diet supplemented with CUR provided 7 days prior to exposure to O_3 for 60 days; CUR-supplemented feeding continued until day 60).
CT	Chronic therapeutic (exposure to O_3 for 7 days prior to the administration of the diet supplemented with CUR for 60 days; exposure to O_3 continued until day 60).

4.4. Ozone Exposure

Experimental were groups exposed to O_3 (0.7 ppm), as this dose was previously reported to cause antioxidant defense depletion [50], or to O_3 free air and were placed inside a hermetic acrylic chamber (65 × 25 × 45 cm L/H/D) coupled to a premix chamber (40 × 24 × 45 cm) daily for 4 h; the exposure chamber was built using a similar design as previously reported [8,51,52]. The premix chamber received O_3 generated by a Certizon C100 apparatus (Sander Elektroapparatebau GmbH, Uetze, Alemania), which was fed with medical-grade oxygen. The O_3 generated was mixed with O_3-free air to adjust the flux to the mentioned concentration. The concentration of O_3 was monitored with a semiconductor (ES-600, Ozone Solutions Inc., Hull, ID, USA) to yield an adequate atmosphere with constant flow of 1.6 to 1.2 L/min. As part of biosecurity actions, the released O_3 from the chamber was inactivated with neutralizing filters made of a sodium nitrate, potassium carbonate, glycerol, methanol, and water solution before being released into the air.

4.5. Plasma Sample Obtention

Once the experimental groups concluded their exposure phase, the animals were euthanasia via intraperitoneal injection with a lethal dose of sodium pentobarbital (90 mg/Kg). Blood was extracted via intracardiac punction using heparinized syringes; plasma samples were separated by centrifugation at 2500 rpm at 4 °C, and 10 μL/mL of phosphate-buffered solution containing a homemade antiprotease cocktail (bestatin, leupentin, aprotinin, PMSF, EDTA and EGTA) was added to each sample (Sigma Chemical, St. Louis, MO, USA). Samples were frozen at −80 °C and stored until use.

4.6. Hippocampus Processing

Immediately after blood collection, the animals were decapitated, and the head was placed on an ice-cold surface while the hippocampus was dissected. The hippocampus has been reported to be a brain structure that is highly sensitive to oxidative damage at a dose of 0.7 ppm of O_3 [50]. Two hippocampi were weighted separately to prepare two 10% homogenate samples in PBS. A 10 μL/mL amount of 0.5 M butylhydroxytoluene (BTH) was added to samples used for malondialdehyde and 4 hydroxynonenal (MDA/4-HNE) determination. The other received 10 μL/mL of an antiprotease cocktail with 0.2 mM mercaptoethanol (Sigma Chemical, St. Louis, MO, USA) for protein carbonylation analysis. Tissue samples were frozen at −80 °C and stored until use [8].

4.7. Estimation of Antioxidant Enzyme Activity

The SOD, CAT, GPX activity in plasma samples from the experimental groups obtained after completing the acute and chronic exposure times was determined using a commercial kit according to the manufacturer's instructions (Abcam, Cambridge, UK). Plasma samples were stored at −80 °C until analysis; subsequently, the homogeneity of the protein content was determined by the Lowry method. This assay is based on the Biuret reaction and has additional steps to increase the sensibility. In the biuret reaction, copper interacts with four nitrogen atoms of the peptides to form a cuprous complex. The Lowry method adds phosphomolybdic/phosphotungstic acid, which is also known as the Folin–Ciocalteu reagent. This reagent interacts with cuprous ions and the side chains of tyrosine, tryptophan, and cysteine to produce a blue-green color that can be detected between 650 and 750 nm [52]. Once analyzed, all samples are diluted to the same concentration as before the enzyme activity assays to avoid variation due to different protein contents.

The analysis of CAT activity is based on a two-step procedure. The samples containing CAT were incubated in the presence of a known concentration of H_2O_2; the unconverted H_2O_2 reacts with the OxiRed probe, producing a product measurable at 570 nm. Therefore, the catalase activity contained in the sample is reversely proportional to the optical density obtained. The procedure was performed according to the kit's manufacturer's instructions (ab83464).

The SOD activity is measurable through the superoxide anions produced by xanthine oxidase activity. Superoxide anions are dismutated by SOD, generating hydrogen peroxide and oxygen. Superoxide anions act on the tetrazolium salt WST-1 to produce formazan dye, which is soluble in water and measurable at 450 nm. Thus, greater SOD activity is revealed in the sample by the lower amount of the formazan dye produced. The procedure was carried out according to the manufacturer's instructions. Furthermore, the results are expressed as the activation percentage rate, as indicated by the manufacturer (ab65354).

The GPx activity is based on the reduction of cumene hydroperoxide while it oxidizes glutathione (GSH) to glutathione disulfide (GSSH). GSSH is reduced to GSH due to the consumption of NADPH by glutathione reductase (GR). Thus, the decrease in NADPH at 340 nm is proportional to the GPx activity. The procedure was performed according to manufacturer's instructions (ab102530).

All plasma samples were tested in duplicate for the mentioned assays. Determinations were performed in a microplate reader (Multiskan Go., Thermo Scientific, Waltham, MA, USA).

4.8. Inhibition of MDA and 4-HNE Formation by CUR

The method to determine the concentration of oxidized lipid metabolites in hippocampus homogenate samples was carried out according to manufacturer´s instructions (Cat. # FR12, Oxford Biomedical Res., Oxford, MI, USA). In brief, samples were centrifuged at 3000× *g* for five min at 4 °C, and 250 μL of the sample was added in each centrifuged tube. Afterwards, 812.5 μL of 11-Methyl-2-phenylindole was added and mixed and incubated at 45 °C for 40 min. Subsequently, 187.5 μL of mesylate was added and quenched in an ice bath. The samples were incubated at 45 °C over 45 min. Finally, the samples were maintained at 4 °C, and 200 μL of the supernatant was placed in a 96-well microtiter plate reader and analyzed at λ 595 nm absorbance in triplicate. The calibration curve was prepared using 650 μL of chromogen solution with progressive concentrations with malonaldehyde bis (dimethyl acetal) at a concentration of 0.315 and with 10 nmol/mL. To assess the estimation of oxidative damage to lipids, MDA and 4HNE metabolites are commonly determined, and these are generated during RONS interaction with cells. The implemented methods report the use of mesylate to estimate both metabolites simultaneously [46]. Results are expressed as the percentage inhibition of oxidative damage by considering a 100% the level obtained in the AO and CO groups.

4.9. Protein Carbonylation

The protein carbonylation assay reveals the level of oxidative damage to proteins induced by O_3 exposure. The OxyBlot kit was used according to manufacturer's instructions (Cat. # S7150, Merck Millipore Corp., Billerica, MA, USA, EE. UU.).

The protein concentration in the samples was determined by the Bradford method, and 4 μg of protein/μL was used. Duplicate samples were denaturalized with 5 μL of 12% sodium lauryl sulphate added to each sample. Half samples from each group were incubated with the derivatizing reagent 2,4-Dinitrophenylhydrazine (DNPH). The remaining samples were incubated with a non-derivatizing control reagent. Afterwards, the reaction was halted, and samples were electrophoresed in 10% of SDS–polyacrylamide gels using a mini-PROTEAN chamber (Bio-Rad, Hercules, CA, USA) at 100 V. Consecutively, proteins were electrotransferred to PVDF membranes using a blot module (Bio-Rad, Hercules, CA, USA) at 25 V for 12 h at 4 °C. At the end, the PVDF membranes with the transferred proteins were blocked with 5% of skimmed milk in PBS for 12 h at 4 °C. Following these steps, samples were incubated with rabbit IgG anti-DNPH (1:150). Peroxidase labelled IgG anti-rabbit (1:500) was used as a secondary antibody, and incubation was performed for 1 h at room temperature. After washing, the membranes were exposed to HRP quimioluminiscent Immobilon reagent (Millipore Corp., Billerica, MA, USA, EE. UU.) to visualize the oxidized protein bands. Analysis of the blot images was performed using Image Studio Lite Ver 5.2®, and the integrated optical density (IOD) data were generated per sample and per experimental group to perform the statistical analysis. The results were expressed in the percentage of inhibition of oxidative damage to proteins.

5. Conclusions

The dietary use of CUR showed high efficiency as a regulator of antioxidant enzymes and a protector of the oxidative damage caused by O_3. These findings have started to show that CUR can be useful against the environmental contingencies that usually occur in cities with high atmospheric pollution.

Author Contributions: Conceptualization, M.L.M.-M. and M.A.R.-H.; data curation, A.A.R.-M.; formal analysis, A.A.R.-M., A.R.T.-M. and M.A.R.-H.; funding acquisition, M.L.M.-M. and M.A.R.-H.; investigation, A.A.R.-M. and M.L.M.-M.; methodology, A.A.R.-M. and S.D.N.-F.; project administration, C.R.C.-Á. and M.M.d.J.R.-P.; resources, M.L.M.-M. and M.A.R.-H.; software, A.R.T.-M. and A.A.R.-M.; supervision, M.A.R.-H. and M.L.M.-M.; validation, C.R.C.-Á. and A.A.R.-M.; visualization, M.L.M.-M.; writing—review and editing, M.A.R.-H. All authors have read and agreed to the published version of the manuscript.

Funding: This research received financial support by the PROSNI program granted to María Luisa Mendoza-Magaña and the project granted to Mario Alberto Ramírez-Herrera through the Programa de Fortalecimiento de la Investigación y el Posgrado 2020.

Institutional Review Board Statement: The study was conducted according to the guidelines of the Ethics and Biosecurity Committees of the Centro Universitario de Ciencias de la Salud (CI-00512) and Secretaría de Salud de Jalisco (15-UG-JAL/2012).

Informed Consent Statement: Not applicable.

Data Availability Statement: Data supporting reported results can be found in the Research Coordination Department of the Centro Universitario de Ciencias de la Salud and with the Corresponding Author.

Acknowledgments: The authors thank Michele Brennan-Bourdon for her careful language review. Mario Alberto Ramírez-Herrera and María Luisa Mendoza-Magaña also contributed to this research work. Aldo Rafael Tejeda-Martínez and Sendar Daniel Nery-Flores are thanked for being present during the experimental process. Cesar Ricardo Cortez-Álvarez and Marina María de Jesús Romero-Prado are thanked for their supervision and attention to detail in this work.

Conflicts of Interest: The authors declare no conflict of interest.

Sample Availability: Samples of the compounds are not available.

References

1. Hooper, L.G.; Kaufman, J.D. Ambient Air Pollution and Clinical Implications for Susceptible Populations. *Ann. Am. Thorac. Soc.* **2018**, *15*, S64–S68. [CrossRef] [PubMed]
2. Gong, C.; Liao, H.; Zhang, L.; Yue, X.; Dang, R.; Yang, Y. Persistent ozone pollution episodes in North China exacerbated by regional transport. *Environ. Pollut.* **2020**, *265*, 115056. [CrossRef] [PubMed]
3. Lee, Y.K.; Mok Kim, S.; Han, S. Ozone-induced inactivation of antioxidant enzymes. *Biochimie* **2003**, *85*, 947–952. [CrossRef] [PubMed]
4. Rivas-Arancibia, S.; Vazquez-Sandoval, R.; Gonzalez-Kladiano, D.; Schneider-Rivas, S.; Lechuga-Guerrero, A. Effects of ozone exposure in rats on memory and levels of brain and pulmonary superoxide dismutase. *Environ. Res.* **1998**, *76*, 33–39. [CrossRef]
5. Rivas-Arancibia, S.; Hernández-Zimbrón, L.F.; Rodríguez-Martínez, E.; Borgonio-Pérez, G.; Velumani, V.; Durán-Bedolla, J. Chronic exposure to low doses of ozone produces a state of oxidative stress and blood-brain barrier damage in the hippocampus of rat. *Adv. Biosci. Biotechnol.* **2013**, *4*, 24. [CrossRef]
6. Tsikas, D. Assessment of lipid peroxidation by measuring malondialdehyde (MDA) and relatives in biological samples: Analytical and biological challenges. *Anal. Biochem.* **2017**, *524*, 13–30. [CrossRef]
7. Morgan, M.J.; Liu, Z.G. Crosstalk of reactive oxygen species and NF-κB signaling. *Cell Res.* **2011**, *21*, 103–115. [CrossRef]
8. Nery-Flores, S.D.; Mendoza-Magaña, M.L.; Ramírez-Herrera, M.A.; Ramírez-Vázquez, J.d.J.; Romero-Prado, M.M.d.J.; Cortez-Álvarez, C.R.; Ramírez-Mendoza, A.A. Curcumin Exerted Neuroprotection against Ozone-Induced Oxidative Damage and Decreased NF-κB Activation in Rat Hippocampus and Serum Levels of Inflammatory Cytokines. *Oxidative Med. Cell. Longev.* **2018**, *2018*, 9620684. [CrossRef]
9. Ren, L.; Zhan, P.; Wang, Q.; Wang, C.; Liu, Y.; Yu, Z.; Zhang, S. Curcumin upregulates the Nrf2 system by repressing inflammatory signaling-mediated Keap1 expression in insulin-resistant conditions. *Biochem. Biophys. Res. Commun.* **2019**, *514*, 691–698. [CrossRef]
10. Cuadrado, A.; Rojo, A.I.; Wells, G.; Hayes, J.D.; Cousin, S.P.; Rumsey, W.L.; Attucks, O.C.; Franklin, S.; Levonen, A.-L.; Kensler, T.W.; et al. Therapeutic targeting of the NRF2 and KEAP1 partnership in chronic diseases. *Nat. Rev. Drug Discov.* **2019**, *18*, 295–317. [CrossRef]
11. Holze, C.; Michaudel, C.; Mackowiak, C.; Haas, D.A.; Benda, C.; Hubel, P.; Pennemann, F.L.; Schnepf, D.; Wettmarshausen, J.; Braun, M.; et al. Oxeiptosis, a ROS-induced caspase-independent apoptosis-like cell-death pathway. *Nat. Immunol.* **2018**, *19*, 130–140. [CrossRef] [PubMed]
12. Apostolova, N.; Victor, V.M. Molecular strategies for targeting antioxidants to mitochondria: Therapeutic implications. *Antioxid. Redox Signal.* **2015**, *22*, 686–729. [CrossRef] [PubMed]
13. Poljsak, B. Strategies for reducing or preventing the generation of oxidative stress. *Oxidative Med. Cell. Longev.* **2011**, *2011*, 194586. [CrossRef] [PubMed]
14. Lykkesfeldt, J.; Michels, A.J.; Frei, B. Vitamin C. *Adv. Nutr.* **2014**, *5*, 16–18. [CrossRef]
15. Miller, E.R.; Pastor-Barriuso, R.; Dalal, D.; Riemersma, R.A.; Appel, L.J.; Guallar, E. Meta-Analysis: High-Dosage Vitamin E Supplementation May Increase All-Cause Mortality. *Ann. Intern. Med.* **2005**, *142*, 37–46. [CrossRef]
16. Besag, F.M.C.; Vasey, M.J.; Lao, K.S.J.; Wong, I.C.K. Adverse Events Associated with Melatonin for the Treatment of Primary or Secondary Sleep Disorders: A Systematic Review. *CNS Drugs* **2019**, *33*, 1167–1186. [CrossRef]
17. Lambrinoudaki, I. Progestogens in postmenopausal hormone therapy and the risk of breast cancer. *Maturitas* **2014**, *77*, 311–317. [CrossRef]
18. Løkkegaard, E.C.L.; Mørch, L.S. Tibolone and risk of gynecological hormone sensitive cancer. *Int. J. Cancer* **2018**, *142*, 2435–2440. [CrossRef]
19. Aggarwal, B.B.; Sung, B. Pharmacological basis for the role of curcumin in chronic diseases: An age-old spice with modern targets. *Trends Pharmacol. Sci.* **2009**, *30*, 85–94. [CrossRef]
20. Menon, V.P.; Sudheer, A.R. Antioxidant and anti-inflammatory properties of curcumin. *Adv. Exp. Med. Biol.* **2007**, *595*, 105–125. [CrossRef]
21. Sharifi-Rad, J.; Rayess, Y.E.; Rizk, A.A.; Sadaka, C.; Zgheib, R.; Zam, W.; Sestito, S.; Rapposelli, S.; Neffe-Skocińska, K.; Zielińska, D.; et al. Turmeric and Its Major Compound Curcumin on Health: Bioactive Effects and Safety Profiles for Food, Pharmaceutical, Biotechnological and Medicinal Applications. *Front. Pharmacol.* **2020**, *11*, 01021. [CrossRef] [PubMed]
22. Pérez-Arriaga, L.; Mendoza-Magaña, M.L.; Cortés-Zárate, R.; Corona-Rivera, A.; Bobadilla-Morales, L.; Troyo-Sanromán, R.; Ramírez-Herrera, M.A. Cytotoxic effect of curcumin on Giardia lamblia trophozoites. *Acta Trop.* **2006**, *98*, 152–161. [CrossRef] [PubMed]
23. Goel, A.; Jhurani, S.; Aggarwal, B.B. Multi-targeted therapy by curcumin: How spicy is it? *Mol. Nutr. Food Res.* **2008**, *52*, 1010–1030. [CrossRef] [PubMed]

24. Gupta, S.C.; Patchva, S.; Aggarwal, B.B. Therapeutic roles of curcumin: Lessons learned from clinical trials. *AAPS J.* **2012**, *15*, 195–218. [CrossRef] [PubMed]
25. Gupta, S.C.; Kismali, G.; Aggarwal, B.B. Curcumin, a component of turmeric: From farm to pharmacy. *BioFactors* **2013**, *39*, 2–13. [CrossRef]
26. Lin, X.; Bai, D.; Wei, Z.; Zhang, Y.; Huang, Y.; Deng, H.; Huang, X. Curcumin attenuates oxidative stress in RAW264.7 cells by increasing the activity of antioxidant enzymes and activating the Nrf2-Keap1 pathway. *PLoS ONE* **2019**, *14*, e0216711. [CrossRef]
27. Mendoza-Magaña, M.L.; Espinoza-Gutiérrez, H.A.; Nery-Flores, S.D.; Ramírez-Mendoza, A.A.; Cortez-Álvarez, C.R.; Bonnet-Lemus, R.d.M.; Ramírez-Herrera, M.A. Curcumin Decreases Hippocampal Neurodegeneration and Nitro-Oxidative Damage to Plasma Proteins and Lipids Caused by Short-Term Exposure to Ozone. *Molecules* **2021**, *26*, 4075. [CrossRef]
28. Dorado-Martinez, C.; Paredes-Carbajal, C.; Mascher, D.; Borgonio-Perez, G.; Rivas-Arancibia, S. Effects of different ozone doses on memory, motor activity and lipid peroxidation levels, in rats. *Int. J. Neurosci.* **2001**, *108*, 149–161. [CrossRef]
29. Dai, C.; Xiao, X.; Zhang, Y.; Xiang, B.; Hoyer, D.; Shen, J.; Velkov, T.; Tang, S. Curcumin Attenuates Colistin-Induced Peripheral Neurotoxicity in Mice. *ACS Infect. Dis.* **2020**, *6*, 715–724. [CrossRef]
30. El-Bahr, S.M. Effect of curcumin on hepatic antioxidant enzymes activities and gene expressions in rats intoxicated with aflatoxin B1. *Phytother. Res. PTR* **2015**, *29*, 134–140. [CrossRef]
31. Giergiel, M.; Lopucki, M.; Stachowicz, N.; Kankofer, M. The influence of age and gender on antioxidant enzyme activities in humans and laboratory animals. *Aging Clin. Exp. Res.* **2012**, *24*, 561–569. [CrossRef] [PubMed]
32. Guo, J.; Cao, X.; Hu, X.; Li, S.; Wang, J. The anti-apoptotic, antioxidant and anti-inflammatory effects of curcumin on acrylamide-induced neurotoxicity in rats. *BMC Pharmacol. Toxicol.* **2020**, *21*, 62. [CrossRef] [PubMed]
33. Kaplán, P.; Tatarková, Z.; Lichardusová, L.; Kmeťová Sivoňová, M.; Tomašcová, A.; Račay, P.; Lehotský, J. Age-Associated Changes in Antioxidants and Redox Proteins of Rat Heart. *Physiol. Res.* **2019**, *68*, 883–892. [CrossRef] [PubMed]
34. Samarghandian, S.; Azimi-Nezhad, M.; Farkhondeh, T.; Samini, F. Anti-oxidative effects of curcumin on immobilization-induced oxidative stress in rat brain, liver and kidney. *Biomed Pharm.* **2017**, *87*, 223–229. [CrossRef]
35. Bahadır, A.; Ceyhan, A.; Öz Gergin, Ö.; Yalçın, B.; Ülger, M.; Özyazgan, T.M.; Yay, A. Protective effects of curcumin and beta-carotene on cisplatin-induced cardiotoxicity: An experimental rat model. *Anatol. J. Cardiol.* **2018**, *19*, 213–221. [CrossRef]
36. Shaw, P.; Chattopadhyay, A. Nrf2-ARE signaling in cellular protection: Mechanism of action and the regulatory mechanisms. *J. Cell. Physiol.* **2020**, *235*, 3119–3130. [CrossRef]
37. Nery-Flores, S.D.; Ramírez-Vázquez, J.J.; Mendoza-Magaña, M.L.; Ramírez-Herrera, M.A.; Cortez-Álvarez, C.R.; Romero-Prado, M. Experimental exposure to ozone induces activation and translocation of NFkB and is reverted by curcumin. *Toxicol. Lett.* **2016**, *259*, S75. [CrossRef]
38. Ulasov, A.V.; Rosenkranz, A.A.; Georgiev, G.P.; Sobolev, A.S. Nrf2/Keap1/ARE signaling: Towards specific regulation. *Life Sci.* **2022**, *291*, 120111. [CrossRef]
39. Fu, J.; Shi, Q.; Song, X.; Xia, X.; Su, C.; Liu, Z.; Song, E.; Song, Y. Tetrachlorobenzoquinone exhibits neurotoxicity by inducing inflammatory responses through ROS-mediated IKK/IκB/NF-κB signaling. *Environ. Toxicol. Pharmacol.* **2016**, *41*, 241–250. [CrossRef]
40. Jobin, C.; Bradham, C.A.; Russo, M.P.; Juma, B.; Narula, A.S.; Brenner, D.A.; Sartor, R.B. Curcumin Blocks Cytokine-Mediated NF-κB Activation and Proinflammatory Gene Expression by Inhibiting Inhibitory Factor I-κB Kinase Activity. *J. Immunol.* **1999**, *163*, 3474–3483.
41. Shin, J.W.; Chun, K.S.; Kim, D.H.; Kim, S.J.; Kim, S.H.; Cho, N.C.; Na, H.K.; Surh, Y.J. Curcumin induces stabilization of Nrf2 protein through Keap1 cysteine modification. *Biochem. Pharmacol.* **2020**, *173*, 113820. [CrossRef] [PubMed]
42. Martinez-Campos, C.; Lara-Padilla, E.; Bobadilla-Lugo, R.A.; Kross, R.D.; Villanueva, C. Effects of Exercise on Oxidative Stress in Rats Induced by Ozone. *Sci. World J.* **2012**, *2012*, 135921. [CrossRef] [PubMed]
43. Ehrenbrink, G.; Hakenhaar, F.S.; Salomon, T.B.; Petrucci, A.P.; Sandri, M.R.; Benfato, M.S. Antioxidant enzymes activities and protein damage in rat brain of both sexes. *Exp. Gerontol.* **2006**, *41*, 368–371. [CrossRef] [PubMed]
44. Sankar, P.; Telang, A.G.; Kalaivanan, R.; Karunakaran, V.; Suresh, S.; Kesavan, M. Oral nanoparticulate curcumin combating arsenic-induced oxidative damage in kidney and brain of rats. *Toxicol. Ind. Health* **2016**, *32*, 410–421. [CrossRef]
45. Santana-Martínez, R.A.; Silva-Islas, C.A. The Therapeutic Effect of Curcumin in Quinolinic Acid-Induced Neurotoxicity in Rats is Associated with BDNF, ERK1/2, Nrf2, and Antioxidant Enzymes. *Antioxidants* **2019**, *8*, 388. [CrossRef]
46. Schleicher, J.; Dahmen, U. Computational Modeling of Oxidative Stress in Fatty Livers Elucidates the Underlying Mechanism of the Increased Susceptibility to Ischemia/Reperfusion Injury. *Comput. Struct. Biotechnol. J.* **2018**, *16*, 511–522. [CrossRef]
47. Guerrero-Hue, M.; García-Caballero, C.; Palomino-Antolín, A.; Rubio-Navarro, A.; Vázquez-Carballo, C.; Herencia, C.; Martín-Sanchez, D.; Farré-Alins, V.; Egea, J.; Cannata, P.; et al. Curcumin reduces renal damage associated with rhabdomyolysis by decreasing ferroptosis-mediated cell death. *FASEB J. Off. Publ. Fed. Am. Soc. Exp. Biol.* **2019**, *33*, 8961–8975. [CrossRef]
48. Dkhar, P.; Sharma, R. Attenuation of age-related increase of protein carbonylation in the liver of mice by melatonin and curcumin. *Mol. Cell. Biochem.* **2013**, *380*, 153–160. [CrossRef]
49. Deng, H.; Wan, M.; Li, H.; Chen, Q.; Li, R.; Liang, B.; Zhu, H. Curcumin protection against ultraviolet-induced photo-damage in Hacat cells by regulating nuclear factor erythroid 2-related factor 2. *Bioengineered* **2021**, *12*, 9993–10006. [CrossRef]

50. Nery-Flores, S.D.; Ramirez-Herrera, M.A.; Mendoza-Magana, M.L.; Romero-Prado, M.M.J.; Ramirez-Vazquez, J.J.; Banuelos-Pineda, J.; Espinoza-Gutierrez, H.A.; Ramirez-Mendoza, A.A.; Tostado, M.C. Dietary Curcumin Prevented Astrocytosis, Microgliosis, and Apoptosis Caused by Acute and Chronic Exposure to Ozone. *Molecules* **2019**, *24*, 2839. [CrossRef]
51. Barragán-Mejía, M.G.; Castilla-Serna, L.; Calderón-Guzmán, D.; Hernández-Islas, J.L.; Labra-Ruiz, N.A.; Rodríguez-Pérez, R.A.; Angel, D.S. Effect of nutritional status and ozone exposure on rat brain serotonin. *Arch. Med. Res.* **2002**, *33*, 15–19. [CrossRef]
52. Goldring, J.P.D. Measuring Protein Concentration with Absorbance, Lowry, Bradford Coomassie Blue, or the Smith Bicinchoninic Acid Assay Before Electrophoresis. *Methods Mol. Biol.* **2019**, *1855*, 31–39. [CrossRef] [PubMed]

molecules

MDPI

Article

Optimisation of Vitamin B12 Extraction from Green Edible Seaweed (*Ulva lactuca*) by Applying the Central Composite Design

Deny Susanti [1,2,*], Fatin Shazwani Ruslan [1], Muhammad Idham Shukor [1], Normawaty Mohammad Nor [3], Nurul Iman Aminudin [1,2], Muhamad Taher [4,5,*] and Junaidi Khotib [6,*]

1 Department of Chemistry, Kulliyyah of Science, International Islamic University Malaysia, Kuantan 25200, Malaysia; ftnshazwanii@gmail.com (F.S.R.); idhamshukor97@gmail.com (M.I.S.); nuruliman@iium.edu.my (N.I.A.)
2 Sustainable Chemistry Research Group (SusChemRG), Kulliyyah of Science, International Islamic University Malaysia, Kuantan 25200, Malaysia
3 Department of Marine Science, Kulliyyah of Science, International Islamic University Malaysia, Kuantan 25200, Malaysia; normahwaty@iium.edu.my
4 Department of Pharmaceutical Technology, Kulliyyah of Pharmacy, International Islamic University Malaysia, Kuantan 25200, Malaysia
5 Pharmaceutics and Drug Translational Research Group, Kulliyyah of Pharmacy, International Islamic University Malaysia, Kuantan 25200, Malaysia
6 Department of Pharmacy Practice, Faculty of Pharmacy, Airlangga University, Surabaya 60115, Indonesia
* Correspondence: deny@iium.edu.my (D.S.); mtaher@iium.edu.my (M.T.); junaidi-k@ff.unair.ac.id (J.K.)

Abstract: Vitamin B12, only found naturally in animal-based foods, is essential for brain functions and various chemical reactions in the human body. Insufficient vitamin B12 leads to vitamin B12 deficiency, common among strict vegetarians due to their limited intake of animal-based foods. Nevertheless, extensive studies have demonstrated that macroalgae, specifically the *Ulva lactuca* species, are rich in vitamin B12 and could be further exploited in future dietary applications. In the current study, the ideal extraction method of vitamin B12 from dried *U. lactuca* was developed and optimised to achieve the maximum vitamin B12 yield. The effects of several extraction parameters, including the solvent-to-solvent, methanol:water (MeOH:H_2O), and solute-to-solvent ratios, and pH on the total vitamin B12 content were analysed through a two-level factorial and central composite design. The highest vitamin B12 content, particularly cyanocobalamin (CN-Cbl), was recovered through the ultrasonic-assisted extraction (UAE) of oven-dried *U. lactuca* at 3 g:60 mL of solute-to-solvent and 25:75% of MeOH to H_2O ratios at pH 4. The extraction of CN-Cbl from oven-dried *U. lactuca* that employed the UAE method has elevated CN-Cbl content recovery compared to other extraction methods.

Keywords: seaweed; extraction; vitamin B12; central composite design

Citation: Susanti, D.; Ruslan, F.S.; Shukor, M.I.; Nor, N.M.; Aminudin, N.I.; Taher, M.; Khotib, J. Optimisation of Vitamin B12 Extraction from Green Edible Seaweed (*Ulva lactuca*) by Applying the Central Composite Design. *Molecules* **2022**, *27*, 4459. https://doi.org/10.3390/molecules27144459

Academic Editors: Raffaele Pezzani, Sara Vitalini and Young Hae Choi

Received: 31 May 2022
Accepted: 9 July 2022
Published: 12 July 2022

1. Introduction

A healthy diet reduces the risk of several diseases, such as heart disease and malnutrition, improving the well-being of individuals. However, nutritional deficiency is often observed among the elderly who might have suffered from nutritional anaemia, including magnesium, folate, and vitamin B12 (cobalamin) deficiencies [1]. Recently, vitamin deficiencies, including that of vitamin B12, were reported to have affected individuals of all ages, including pregnant and lactating mothers worldwide [2]. Strict vegetarians are even more vulnerable to vitamin B12 deficiency, as they limit their consumption of meat products [2]. The frequency of vitamin B12 deficiency among vegetarians was approximately 62%, while 25 to 86%, 21 to 41%, and 11 to 90% were evaluated in pregnant women, children, adolescents, and elderly, respectively [3]. Therefore, the issue needs to be taken seriously to reduce the number of people suffering from vitamin B12 deficiency.

Vitamin B12 (Figure 1) represents all potentially active cobalamins. Cobalamins are a group of cobalt-containing compounds (corrinoids) with a lower axial ligand containing cobalt-coordinated nucleotide (5,6-dimethylbenzimidazole as a base). There are various forms of vitamin B12, namely cyanocobalamin (CN-Cbl), hydroxocobalamin (OH-Cbl), adenosylcobalamin (Adl-Cbl), and methylcobalamin (Me-Cbl). The various forms of vitamin B12 are differentiated based on the β-axial ligand occupancy of the corrin ring [4]. Anaerobic bacteria naturally synthesise vitamin B12. Animal-based foods also contain vitamin B12.

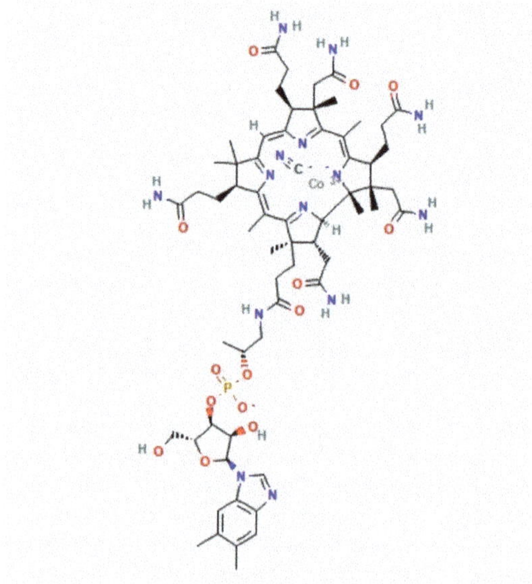

Figure 1. Vitamin B12 chemical structure [5].

Vegetarians and individuals with vitamin B12 deficiency are required to supplement their diet. For instance, vitamin B12 could be acquired through supplements or food fortified with vitamins, such as nutraceuticals extracted from natural resources such as seaweed, specifically *Ulva lactuca*. Seaweed-based foods exhibited potential in reducing the risks of several diseases, including nutritional deficiency [6], since they are low in calories and high in vitamins, minerals, bioactive metabolites, dietary fibres, and polysaccharides [7].

Wahlstrom et al. [8] stated that the *Ulva* species, particularly *U. lactuca*, is widely known for their edible properties due to their high nutritional value, since they contain high levels of polysaccharides, proteins, vitamins, and trace mineral contents. Furthermore, *U. lactuca* displayed antioxidant, antimicrobial, antiviral, and anti-inflammatory properties [9]. Therefore, the seaweed may exhibit potential as an alternative source of vitamins, especially for the elderly and strict vegetarians. In another report, the green edible seaweed was suggested as a source of vitamin B. A daily intake of 1.4 g/day of *U. lactuca* was sufficient to meet the daily requirements of vitamin B12 [10]. Therefore, in the current investigation, the potential of *U. lactuca* as a source of vitamin B12 was evaluated.

Currently, studies are conducted to identify the best drying processes, extraction methods, and optimised parameters in extracting a high quality and quantity of vitamin B12 from *U. lactuca*. The present investigation compared the extraction methods of vitamin B12 from several types of dried *U. lactuca* to obtain a standardised procedure, and the best extraction parameters to extract an excellent quality and quantity of vitamin B12. Three extraction methods, boiling, orbital shaking, and ultrasonic-assisted extraction (UAE), were employed to extract vitamin B12 from the oven-, sun-, air-, and freeze-dried *U. lactuca* under various extraction parameters, including (i) solvent-to-solvent and solute-to-solvent ratios and (ii) pH.

The parameters were analysed and scrutinised individually using a two-level factorial design during the screening procedure. The two-level factorial experiments functioned as excellent screening tools in estimating the overall effects of the main factors and the interactions between the factors [11]. Gottipati and Mishra [12] also highlighted that the factorial design helped examine pre-treatment variations. Subsequently, a central composite design (CCD) based on the response surface methodology (RSM) was employed for optimal parameters. The optimisation of the factors was crucial to develop a standardised extraction method that would produce the desired amount of vitamin B12. The range for each factor examined in the current investigation was varied according to previous reports, which were 25:75 to 75:25%, 3:60 to 3:90 g/mL, and pH 3 to 5, for the solvent-to-solvent and solute-to-solvent ratios and pH, respectively [13,14].

2. Results

2.1. *Screening the Extraction Methods and Drying Conditions*

Three extraction procedures, boiling, orbital shaking, and UAE, were employed to extract vitamin B12 from the oven-, sun-, air-, and freeze-dried *U. lactuca* samples. According to the HPLC qualitative analysis, the *U. lactuca* contained CN-Cbl, one of the vitamin B12 analogues. The retention time was recorded at 1.9 min, which corresponded to the retention time of standard CN-Cbl.

Each extraction procedure of the dried sample was denoted with an abbreviation in the other section.

The highest CN-Cbl content (Table S1, Supplementary Data) was observed in the sun-dried *U. lactuca* samples extracted following the orbital shaking procedure (0.0356 mg/mL). The oven-dried samples extracted through the UAE method comprised 0.0236 mg/mL, while the oven-dried samples extracted by the boiling method consisted of 0.0210 mg/mL CN-Cbl. Although the SDO method yielded the highest CN-Cbl content, the model from the two-level factorial was insignificant ($p > 0.05$). Resultantly, eliminating the factors improved the model during the optimisation process. In contrast, the ODU method exhibited a remarkably higher CN-Cbl content (Figure 2) than those extracted employing the boiling (ODB) technique. It has been demonstrated that the mean CN-Cbl content obtained from the UAE method was 43% more than its heat-extracted counterparts [15].

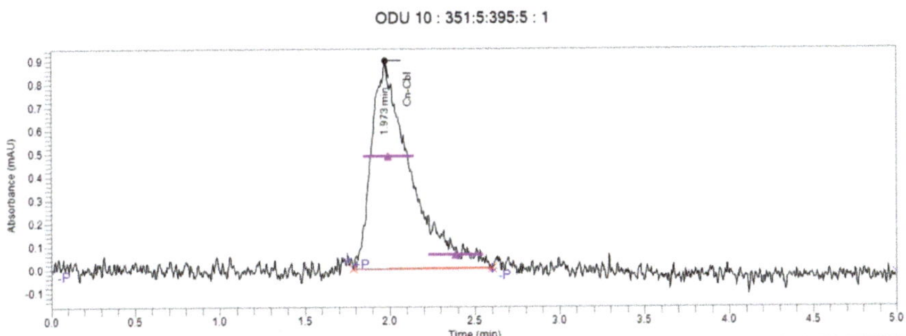

Peak #	RT (min)	Component Name	Channel	Area	Height	BL	Final Amount	Units
1	1.973	Cn-Cbl	351:5:395:5	14,231	889	BB	30.0780	ppm

Figure 2. HPLC chromatogram of optimised purified vitamin B12 by CCD, oven dried *U. lactuca* extracted by UAE (ODU) run 10.

The oven-dried *U. lactuca* samples exhibited a positive and significant effect ($p < 0.05$) on the CN-Cbl yield when extracted under the most suitable parameters and extraction method. Moreover, Dang et al. [16] emphasised that the optimal temperature for oven-drying was 40 °C when drying brown seaweed, *Hormosira banksia*, as rapid degradation of the components occurred when the drying temperature was over 40 °C. The temperature applied in the oven drying method in the present investigation was similar. However, the drying technique should still be cautiously screened and selected beforehand depending on the compound of interest which needs to be extracted. For instance, the oven-drying technique reduced the total flavonoid content yield by 30% and the power of ferric reducing antioxidants by 17% obtained from *Chlorella vulgaris* algae samples [17].

2.2. Screening for Significant Factors Using the Two-Level Factorial Design

Apart from determining the suitable extraction and drying methods for *U. lactuca* to obtain vitamin B12, screening for significant factors was also crucial. The ideal experimental design was determined by screening the concerned factors, the solvent-to-solvent ratio (A), pH (B), and solute-to-solvent ratio (C), utilising the two-level factorial design. Table S1 in Supplementary Data displays the two-level factorial designs with five centre points and the experimental layout representing 21 experimental runs with the concentration of CN-Cbl as the response. The concentration of CN-Cbl obtained from oven-, sun-, air-, and freeze-dried *U. lactuca*, extracted through the boiling, orbital shaking, and UAE methods, are also included. The significance of each model, individual model terms, and lack of fit were determined based on their respective *F*- and *p*-values. The results showed that the ODB, ADB, ODU, and ADU models were significant, with a p-value < 0.05, suggesting that the models were fitted adequately. However, referring to the ANOVA analysis presented in Tables S2–S5 (Supplementary Data), the model for ODU (Table S4) exhibited the highest *F*-value at 76.53, while the model for ODB was 22.26, ADB at 4.57, and ADU was 10.49. The highest CN-Cbl content was also recorded from the ODU model. Therefore, the ODU model was further optimised to interpret the relationship between the significant factors and the response by employing the CCD.

According to the ANOVA analysis (Table S4 in Supplementary Data), the main effects of factors A and B were significant ($p < 0.05$). The B factor was more significant than A, which exhibited the *F*-values of 49.69 and 10.53, respectively. Moreover, the two-level interactions of solvent-to-solvent ratio and pH (AB), solvent-to-solvent ratio and solute-to-solvent ratio (AC), pH and solute-to-solvent ratio (BC), and the overall interaction (ABC) exerted significant positive effects on the total CN-Cbl yield from the dried *U. lactuca* samples. The findings also suggested that the single interaction of the main effects (A and B) and their interactions (AB, AC, BC, and ABC) influenced the total CN-Cbl extracted.

The effects of an individual factor, C, on the yield of CN-Cbl was insignificant ($p > 0.05$). The observation might be due to the less remarkable statistical relevance in the extraction procedure possessed by the factor [18]. Scientifically, the solvent type and strength are the leading influential elements in the optimisation procedure of an extraction method. Moreover, the *F*-value of the curvature was 60.40 with a significant p-value. Hence, the optimisation procedure by CCD was required to approximate the response and explain the relationship between the significant factors and the response from the ODU samples.

2.3. Significant Factors That Affected the CN-Cbl Yield from the Oven-Dried U. lactuca Samples That Employed the UAE Technique

The two-level factorial design of ODU indicated that the solvent-to-solvent ratio and pH demonstrated significant effects over the concentration of CN-Cbl content obtained from the extraction process. Although a single solute-to-solvent ratio exhibited an insignificant effect on the yield, eliminating the effect was not required. The observation was because the solute-to-solvent ratio interacted with other factors in the two, and in their overall interactions. The diagnostic analysis also revealed that the two-level factorial computed an identical distribution of the normal probability of residuals and residuals-versus-predicted-value plots. On the other hand, Figure 3 displayed no outliers in the residual analysis. Consequently, the solute-to-solvent ratio was fixed at the minimum value of 3 g:60 mL in the subsequent CCD procedure. The minimum and maximum values for factor A were within 0:100–50:50%, and 3–5 for factor B. The values were determined based on the desirability ramp, as shown in Figure 4.

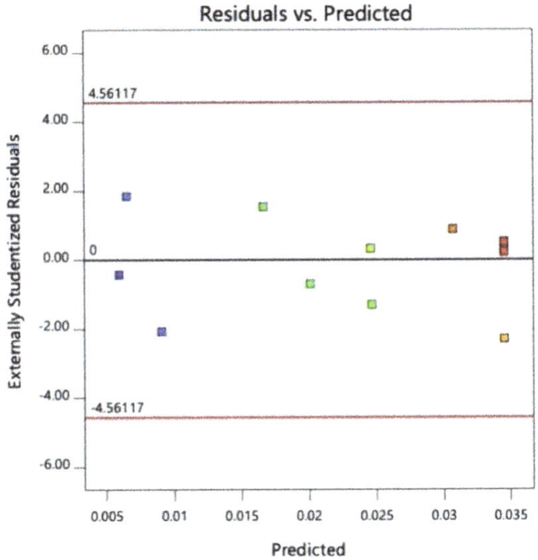

Figure 3. Residual analysis of ODU from 2-Level Factorial.

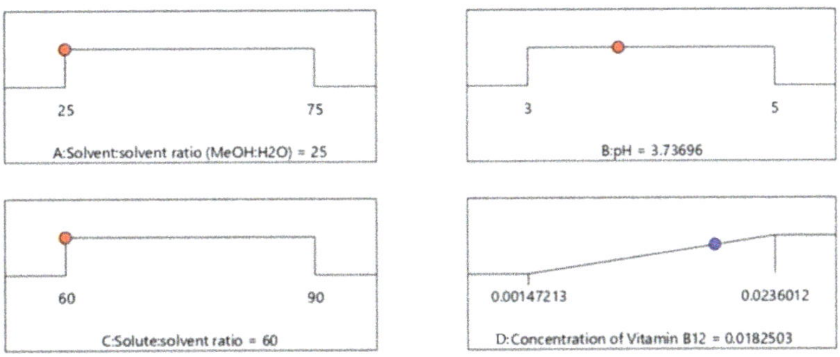

Figure 4. Desirability ramp for numerical optimization of independent variable, solvent:solvent ratio, pH and solute:solvent ration for the concentration of CN-Cbl of ODU sample. A: solvent:solvent ratio (MeOH:H_2O); B: pH; C: solvent:solvent ratio; D: concentration of CN-Cbl.

2.4. Optimising of the ODU CN-Cbl Yield through CCD

The CN-Cbl yield was further optimised using a second-order model, the CCD. It was necessary to approximate the response in this study since the curvature in the initial model was significant ($p < 0.05$). The CCD was applied by augmenting the factorial design with appropriate axial points for the analysis. The α-value of the CCD was set at 1, and 5 centre points were added, generating 13 runs. Table 1 displays the optimised design layout of the CCD for UAE oven-dried *U. lactuca* samples with the concentration of CN-Cbl as the response.

Table 1. Optimised design layout of CCD for the extraction of vitamin B12 from the *U. lactuca* oven-dried samples utilising the UAE technique.

Run	Solvent:Solvent Ratio (MeOH:H_2O)	pH	Conc. of Vitamin B12 (mg/mL)
1	0:100	3	0.0250
2	50:50	3	0.0084
3	0:100	5	0.0072
4	50:50	5	0.0055
5	0:100	4	0.0323
6	50:50	4	0.0189
7	25:75	3	0.0225
8	25:75	5	0.0192
9	25:75	4	0.0357
10	25:75	4	0.0355
11	25:75	4	0.0351
12	25:75	4	0.0358
13	25:75	4	0.0305

Table 2 presents the ANOVA analysis of the CCD. Only two significant factors, A and B, were optimised during the extraction of CN-Cbl. The solute to solvent ratio was kept constant at 3 g of the dried sample extracted in 60 mL of the solvent. Based on the ANOVA analysis, the generated quadratic model was desirable (*F*-value at 49.26) and demonstrated an insignificant lack of fit. Individual factors A and B remained significant, as well as the two-level interactions between AB, contributing to the concentration of CN-Cbl.

Table 2. The ANOVA analysis of vitamin B12 extracted from the oven-dried *U. lactuca* samples that employed the UAE method.

Source	Sum of Squares	df	Mean Square	*F*-Value	*p*-Value	
Model	0.0003	5	0.0003	49.26	<0.0001	Significant
A	0.0002	1	0.0002	27.14	0.0012	
B	0.0001	1	0.0001	15.64	0.0055	
AB	0.0001	1	0.0001	9.06	0.0196	
A^2	0.0002	1	0.0002	37.27	0.0005	
B^2	0.0005	1	0.0005	85.86	<0.0001	
Residual	0.0000	7	6.179×10^{-7}			
Lack of fit	0.0000	3	7.511×10^{-6}	1.45	0.3538	Not significant
Pure error	0.0000	4	5.180×10^{-6}			
Cor total	0.0016	12				

The solvent-to-solvent ratio squared (A^2) and pH squared (B^2) were included in the significant model terms after the model was fitted with the second order. According to the normal probability plots of CCD (Figure 5), all the residuals lay close enough to form a straight line, indicating that errors were distributed normally. The residuals in the plot of residuals versus predicted value were structureless, thus demonstrating that the investigated model satisfied the assumptions. Therefore, the fitted quadratic model was correct and desirable. The R^2 value of the quadratic model was 0.9724, indicating that 97.24 % of the variation in the responses was explained. The predicted R^2 value, 0.8343, was in reasonable agreement with the adjusted R^2 value of 0.9526, indicating the adequacy of the quadratic model, as the difference was less than 0.2. The adequacy precision ratio of 16.9051 was also desirable as the ratio was exceeded 4. Consequently, the generated quadratic model was fitted to navigate the design space.

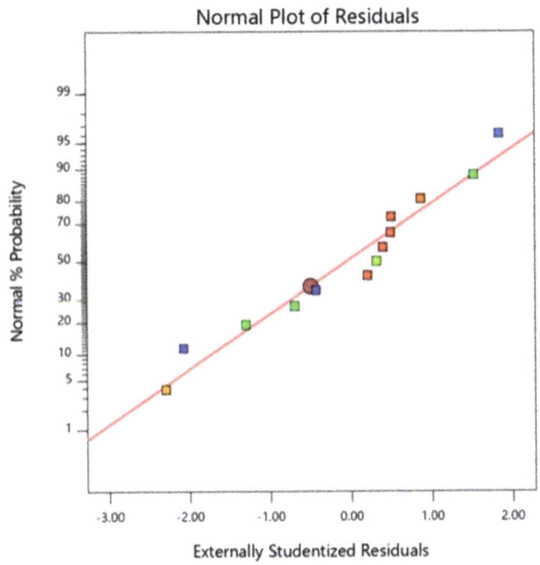

Figure 5. Normal probability plot of ODU sample extracted by UAE method using Central Composite Design.

Equations (1) and (2) are the final empirical models in terms of the coded and actual factors, respectively, for the concentration of CN-Cbl (mg/mL).

$$\text{Concentration of CN-Cbl} = 0.0346 - 0.0053\ A - 0.0040\ B + 0.0037\ AB - 0.0091\ A^2 - 0.0139\ B^2, \quad (1)$$

$$\text{Concentration of CN-Cbl} = (-0.0160010 - 0.000080)A + (0.103123)B + 0.000150(A \times B) - 0.000015(A^2) - 0.013860(B^2) \quad (2)$$

where A is the solvent-to-solvent ratio, B is the pH, and C is the solute-to-solvent ratio.

2.5. Analysis of the Optimum Region by Employing CCD

Optimum parameters in extraction procedures play crucial roles in benefiting the yield of an extract. In the CCD, the optimum regions of the parameters were analysed based on the contour plots of the factors. The contour plot of the AB interaction is displayed in Figure 6a, with the solute-to-solvent ratio kept constant at 3 g:60 mL. According to the results, the optimum point, the highest CN-Cbl content, was located at the centre of the contour. The observations indicated that the CN-Cbl content increased when the extraction was performed at 25:75% MeOH:H_2O and pH 4.

The 3D surface plot of the AB interaction, illustrated in Figure 6b, manifested a curve was with a maximum point at the solvent-to-solvent ratio approximately within 20:80–30:70% and pH 4–4.5. The lowest concentration of CN-Cbl (0.0055 mg/mL) was obtained when dried *U. lactuca* was extracted at 50:50% solvent to solvent ratio and pH 5, followed by 0.0072 mg/mL at 0:100% solvent to solvent ratio and pH 5, and 0.0084 mg/mL at 50:50% solvent to solvent ratio and pH 3. Therefore, 25:75% solvent to solvent ratio and pH 4 were selected as the optimum parameters for the new extraction protocol.

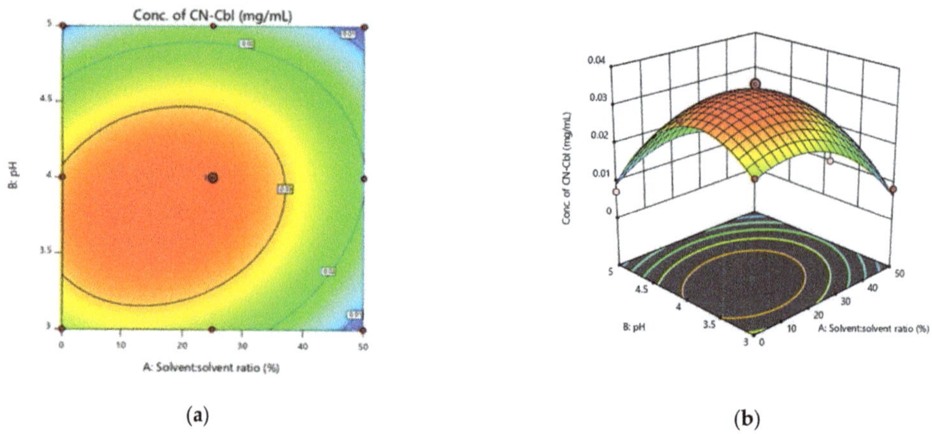

Figure 6. Surface response of the concentration of CN-Cbl recovered as a function of (**a**) AB interaction in contour plot and (**b**) AB interaction in the 3D surface plot.

2.6. Model Conformation and Validation Test

Experimental runs listed in Table S6 (Supplementary Data) were employed to verify the adequacy of the generated model. The concentration of CN-Cbl content in the oven-dried *U. lactuca* for each run was at a 95% prediction interval based on the model developed in the CCD. All the tested runs exhibited error percentages less than 15% between the actual and predicted mean for the concentration of CN-Cbl content. The overall data mean was also within the 95% prediction interval low and 95% prediction interval high range. Therefore, the developed model was reasonably well fitted and precisely presented the effects of the extraction parameters on the concentration of CN-Cbl content.

2.7. Microbiological Assay of Vitamin B12

The purified samples were subjected to microbiological assays that utilised *E. coli* and *L. leichmanii* to confirm the presence of vitamin B12 in the oven-dried *U. lactuca* samples extracted by the UAE method. A zone of *E. coli* and *L. leichmanii* growth were observed surrounding the wells containing CN-Cbl standard and the purified sample. The observations confirmed the presence of vitamin B12, specifically CN-Cbl. There was no growth surrounding the wells of the control, which contained distilled water.

2.8. Liquid Chromatography/Electronspray Ionisation-Mass Spectroscopy/Mass Spectroscopy (LC/ESI-MS/MS) Analysis of Vitamin B12

The purified extract containing CN-Cbl compound extracted from oven-dried *U. lactuca* by utilising the UAE method was further analysed and characterised through LC/ESI-MS/MS. Figure 7 demonstrates the presence of M + H CN-Cbl, which was 1355.5232 m/z.

A similar spectrum was observed in a previous study that analysed the CN-Cbl content extracted from *Spirulina platensis*. The spectrum displayed the presence of M + H CN-Cbl, which was 1356.58 m/z [14]. The current study results confirmed that extracting vitamin B12 from the oven-dried *U. lactuca* by employing the UAE method enhanced the amount of CN-Cbl recovered.

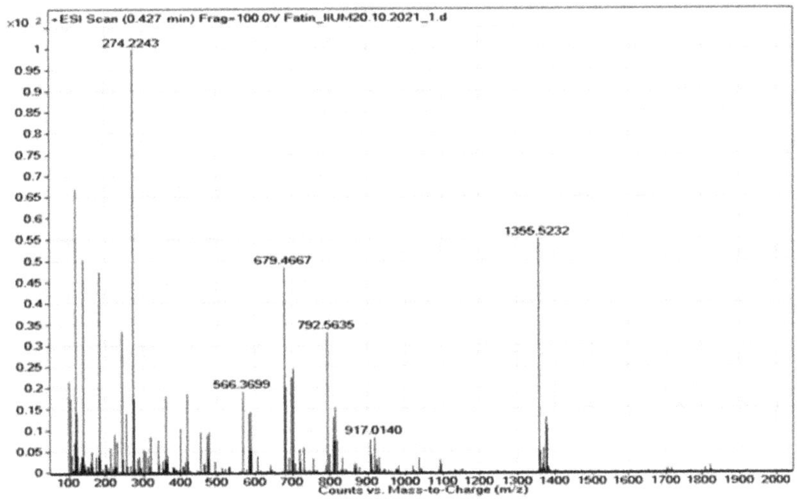

Figure 7. Mass spectrum of CN-Cbl in the purified oven-dried *U. lactuca* extract.

3. Discussion

Vitamin B12 molecules are complex, very reactive, and water-soluble. Therefore, the measurement and extraction of vitamin B12 from food, plant, and algae samples present an enormous analytical challenge. Commonly, the vitamin B12 content in algal materials is low, alongside the presence of other metabolites [19]. The focus of this research was to vary the extraction procedure and drying conditions of vitamin B12 from *U. lactuca* to maximise the yield of vitamin B12. The investigation also introduced novel extraction methods, and reduced the use of hazardous solvents during the extraction of vitamin B12. In addition, in the present study, the extraction solvent was standardised to ease future research. Previous studies discussed that certain organisms might contain the vitamin B12 analogues such as OH-Cbl, Adl-Cbl, and Me-Cbl. However, due to their unstable nature and extraction conditions, the components might be lost during the extraction procedure [14]. Therefore, methanol and water were used in the present study as the extraction solvent since vitamin B12 is a water-soluble vitamin and, owing to the low boiling point of methanol, the decomposition of thermolabile bioactive compounds was prevented [20].

Our data showed that the ODU method exhibited a remarkably higher CN-Cbl content than those extracted employing the boiling (ODB) technique. The same trend was observed in a study on CN-Cbl extraction from a solid dietary supplement that compared the final CN-Cbl contents from UAE and heat extraction methods [15]. Moreover, numerous studies have applied the novel extraction method to extract compounds, such as polyphenols, antioxidants, and prebiotics from macroalgae [16,21]. Our findings agreed that the UAE method that operated at low temperatures enabled the preservation of thermolabile compounds and prevented the structure from being entirely damaged [15,22]. Hence, obtaining *U. lactuca* following the UAE method might be adequate to liberate CN-Cbl prior to quantitative analysis. The procedure could be a standardised protocol in future extractions of CN-Cbl from macroalgae.

The different drying conditions for the extraction procedure of vitamin B12 from *U. lactuca* were examined. The oven-dried *U. lactuca* samples exhibited a positive and significant effect ($p < 0.05$) on the CN-Cbl yield when extracted under the most suitable parameters and extraction method. The findings agreed with the reports from Silva et al. [23], highlighting that oven drying stabilised the *Ulva* species, such as *U. rigida*, during the extraction of specific bioactive components within the optimised temperature. Resultantly, the UAE method was selected as the best procedure in extracting CN-Cbl from the oven-dried *U. lactuca*.

Based on the AB interaction in 3D surface and contour plot, lowering and increasing the solvent-to-solvent ratio and pH beyond the optimum point exhibited no improvement in the concentration of CN-Cbl extracted. The best solvent-to-solvent ratio utilised was 25:75% MeOH:H_2O. Kumudha and Sarada [14] highlighted that water was the best solvent to extract vitamins since it could extract several water-soluble molecules, including vitamin B12. Other solvents introduced in the extraction of vitamins include water with ethanol and dimethyl sulphoxide. The current study employed water with MeOH as the solvent since MeOH exhibited the highest efficiency as the eluting solvent for vitamin B12 [24]. According to Fang et al. [24], the suggested solvent-to-solvent ratio of 50% MeOH:50% H_2O was sufficient to meet the requirement. However, 25% MeOH:75% H_2O was evaluated as the optimum solvent-to-solvent ratio in this investigation, which might be attributed to the different samples used.

The highest CN-Cbl content was recorded at pH 4, indicating pH 4 was the optimum pH to yield the maximum CN-Cbl content from the dried *U. lactuca* samples. A declining trend in CN-Cbl content was observed when the pH was decreased to 3 and increased to 5. According to Chandra-Hioe et al. [15], the extraction of vitamin B12 required specific extraction conditions as it is only stable at lower pH, in the absence of ultraviolet, and the presence of other vitamin B varieties, such as thiamine, niacin, and ascorbic acid. Therefore, the decreasing CN-Cbl yield as the pH was increased to 5 might be because several vitamin B12 compounds were destroyed. Vitamin B12 is unstable under highly acidic and alkaline conditions [25]. Bajaj and Singhal [26] reported that the concentrations of vitamin B12 in an aqueous solution with pH 2 were significantly lower than in solutions with pH up to 6. The observations demonstrated that the degradation rate of vitamin B12 accelerated progressively under a very low and high pH.

Microbiological assay using *E. coli* and *L. leichmanii* led to the confirmation of the presence of vitamin B12 analogues of CN-Cbl compound in the purified sample extract. It was then further analysed and characterised through LC/ESI-MS/MS. Considering these findings, the extraction processes of *U. lactuca* extracts were mainly affected by the extraction method and several important parameters, and these might be a suitable candidate for the novel standardised and simplified extraction process of vitamin B12 from green edible macroalgae. In conclusion, our study demonstrated that *U. lactuca* extracts possessed vitamin B12 analogues, particularly CN-Cbl, which usually only found in animal-based products. The extract also showed some significant effects towards various parameters and extraction conditions. The high content of vitamin B12 analogues, particularly CN-Cbl present in *U. lactuca* extracts, may indicate the suitable conditions and parameters for a new standardised extraction protocol developed in this study.

4. Materials and Methods

4.1. Sample Collection and Preparation

The *U. lactuca* samples were collected fresh from Merambong Island, Johor, Malaysia, at the coordinates (Latitude 1°18′60.00″ N, Longitude: 103°36′59.99″ E), in January 2020 and January 2021, with water salinity and range temperature of 12 ppt and 25–31 °C, respectively. The species was identified and validated by algae expert, Assoc. Prof Dr, Normawaty Mohammad Nor (Department of Marine Science and Technology, IIUM) during the sample collection. Potential epiphytes were removed from the fresh algae and rinsed on the spot with seawater. Subsequently, the samples were oven-, sun-, air-, and freeze-dried to a constant weight. The dried samples were ground in a grinder for 5 min to obtain a fine and homogeneous powder pre-storage in different sealed bags at room temperature for further investigation.

The drying methods employed are presented as follows:

- Air drying: The *U. lactuca* samples were placed in a single layer on a clean flipchart paper and air-dried in the laboratory using fan (speed 300 rpm) to keep air circulating until the sample reached a constant weight (temperature 25 °C).
- Sun-drying: The *U. lactuca* samples were placed in a single layer on a clean aluminium foil and dried under direct sunlight until the sample reached a constant weight (temperature 32 ± 2 °C).
- Oven drying: The *U. lactuca* samples were placed on an aluminium foil in a single layer and dried in an oven drier (Sanyo, OSA, JP) until a constant weight was obtained. The oven temperature was set to 40 °C.
- Freeze drying: The *U. lactuca* samples were frozen in a freezer (Sanyo, OSA, JP) overnight at −80 °C and then being placed in a chamber that combines a chilled condenser and a vacuum pump to aid sublimation of water. The freeze-drier (Thermo Scientific, WLM, USA) was set at a cryo-temperature of −50 °C.

4.2. Extraction of Vitamin B12

Three grams each of the ground air-, oven-, sun-, and freeze-dried *U. lactuca* samples were added into 100 mL amber Schott bottles. Next, the samples were suspended in methanol (MeOH) (Qrec Sdn. Bhd., KL, MY) and ultra-pure water (Scope-T, CN) solvent mixture at three different ratios, which were 25: 75, 50:50, and 75:25 % of MeOH: H_2O. The solute-to-solvent ratios were also varied, from 3 g:60 mL, 3 g:75 mL, and 3 g:90 mL, respectively. The experimental design for the different extraction procedures and parameters applied were performed as mentioned in the Supplementary Data Table S1. The extraction of vitamin B12 from the dried *U. lactuca* samples was conducted in the dark since vitamin B12 is light sensitive [27]. The vitamin B12 extracts were centrifuged at 6000 revolutions per minute (rpm) for 10 min (Hettich Zentrifugen, BW, DE). The supernatant was evaporated in a vacuum to procure pure vitamin B12 extracts. The pH of the extracts was adjusted to within 3 to 5, based on the experimental design obtained from the two-level factorial and CCD designs by RSM. The extraction methods utilised are listed as follows:

- Boiling Extraction Method: The dried *U. lactuca* samples were boiled using hotplate (Thermolyne Thermo Fischer Scientific, Waltham, MA, USA) for 20 min at 65 °C [14]. Upon extraction, the extracts obtained were centrifuged and kept in below 15 °C chiller (Sanyo, OSA, JP) before being purified.
- Orbital Shaker Extraction Method: The dried *U. lactuca* samples were shaken on an orbital shaker (Thermo Fischer Scientific, WLM, USA) for 30 min [28] at 200 rpm and centrifuged upon extraction. The extracts were kept in below 15 °C chiller before being purified.
- Ultrasonic Assisted Extraction Method: The dried *U. lactuca* samples were suspended in the solvent mixture and extracted in an ultrasonic water bath (WiseClean, CH) with the power and frequency set to 665 Watt and 20 kHz, respectively, for 30 min. The extracts obtained were cooled and kept in a chiller below 15 °C before being purified.

4.3. Purification and Determination of Vitamin B12

The purification of the vitamin B12 extracts was completed by passing the samples containing vitamin B12 through an Amberlite XAD-2 column (Merck KGaA, Darmstadt HE, DE). First, the column was filled with pure MeOH (Merck & Co., Kenilworth, NJ, USA) and an Amberlite XAD-2 resins slurry to 15–16 cm bed height. Next, the MeOH was drained and replaced with deionised water to reach equilibrium. After 15 min, the column was drained, and the vitamin B12 extracts were loaded. Finally, the vitamin B12 was drained very slowly for 3 h and eluted with 80% MeOH to increase the purification efficiency [14].

The concentrated solution was spotted on a silica gel thin-layer chromatography (TLC) plate (Merck Millipore, Burlington, MA, USA) and developed with 7:4:5:1:1 of 1-butanol (Merck & Co., Kenilworth, NJ, USA), chloroform (Merck & Co., Kenilworth, NJ, USA), acetic acid (Merck & Co., Kenilworth, NJ, USA), ammonia (Merck & Co., Kenilworth, NJ, USA), and water (H_2O) solvent in the dark at room temperature. The TLC plate was first rinsed with a dichloromethane (Merck KGaA, Darmstadt HE, DE) and MeOH (Qrec Sdn. Bhd., KL, MY) solution at a 1:1 ratio before being spotted with the vitamin B12 extracts. The spots on the TLC plate were dried, collected, extracted with 80% MeOH, evaporated to dryness under reduced pressure, and dissolved in 50 mL of distilled water [29]. The concentrated solution was further characterised through high-performance liquid chromatography (HPLC) (Perkin Elmer, Waltham, MA, USA).

4.4. *High-Performance Liquid Chromatography (HPLC) Analysis*

4.4.1. Sample Solutions

Vials containing crude extract samples were prepared by diluting the extract with ultra-pure water in 1:2 mL crude extract to ultra-pure water ratio. The vials containing the extracts were vortexed to ensure homogeneity. The dissolved crude extract was filtered using Millipore® SLHN033NB Millex® HN Syringe Filter Nylon Membrane; 0.45 μm in diameter (Merck Millipore, Burlington, MA, USA).

4.4.2. Standard Solutions

Analytical standard stock solutions of vitamin B12 (Merck & Co., Kenilworth, NJ, USA) were purchased from Merck for quantification purposes. The standards consisted of CN-Cbl, OH-Cbl, Adl-Cbl, and Me-Cbl. The stock solutions were diluted with HPLC grade MeOH (Merck & Co., Kenilworth, NJ, USA). The stock solutions were employed to create three to five calibration points for the calibration curve. Each calibration curve for CN-Cbl, OH-Cbl, Adl-Cbl, and Me-Cbl was plotted, individually.

4.4.3. Chromatographic Conditions

The concentrated and purified samples were analysed through a reverse-phase HPLC using Luna® C18(2) column (Phenomenex, Torrance, CA, USA). The samples were eluted with MeOH (A) (HPLC grade (Merck & Co., Kenilworth, NJ, USA) and 0.1% acetic acid (B) HPLC grade (Merck & Co., Kenilworth, NJ, USA) as the mobile phase in an isocratic mode for seven minutes. The injection volume, oven temperature, and flow rate were set at 10 μL, 30 °C, and 0.9 mL/min, respectively. The chromatograms were obtained at a maximum wavelength of 351 nm with an ultraviolet (UV) Flexar PDA Plus detector (Perkin Elmer, Waltham, MA, USA). All data obtained were analysed using Perkin Elmer Chromera software version 4.1.0.6386.

4.5. *Statistical Analysis*

Response surface methodology (RSM) two-level factorial design and Central Composite Design (CCD) were employed to determine the correlation of the factors (MeOH:H_2O and solute:solvent ratios and pH). The model significance and interactional effects between the factors were observed based on the F-test and p-values, respectively. The effects were significantly different at $p < 0.05$ and a confidence level of 95%. The two-way analysis of variance (ANOVA) was employed to analyse the model and the effects of the factors. Initially, the two-level factorial design with centre points was applied to screen out the most significant factors. Subsequently, the CCD was used to develop and analyse the optimised extraction procedure model. The computational analysis by the software generated the correlation values.

4.6. *Microbiological Assays of Vitamin B12 Using Eschericia coli and Lactobacillus leichmanii*

E. coli, American Type Culture Collection (ATCC) 25922 (Kwik-StikTM, Saint Cloud, MN, USA), a Gram-negative strain, and *L. leichmanii*, (ATCC) 7830 (Kwik-StikTM, Saint Cloud, MN, USA), a Gram-positive strain, were grown in a maintenance medium at 37 °C, mixed with vitamin B12 assay agar, and pour plated. The CN-Cbl standards and purified samples (20 μL) were inoculated into wells bored in the solid agar medium. The plates and distilled water used as control were incubated at 37 °C for 24 h. The zone of growth was observed after 24 h.

4.7. Tandem Mass Spectrometry/Mass Spectroscopy (MS/MS) Analysis of Vitamin B12

The electron spray ionisation mass spectrum (ESI-MS) (Bruker, Billerica, MA, USA) was performed in positive mode utilising a linear gradient of MeOH (A) (Merck & Co., Kenilworth, NJ, USA) and 0.1 % acetic acid (B) HPLC grade (Merck & Co., Kenilworth, NJ, USA) as the mobile phase. The Positive-ion tandem mass spectrometry (MS/MS) experiments were performed in product mode on time-of-flight (TOF) mass spectrometer (Agilent 6545). The cone and desolvation gas flow were set at 28 and 1000 L/h.

The samples were introduced into the mass spectrometer through a direct flow injection system through a solvent delivery at a 0.3 mL/min flow rate. The capillary and cone voltages were set at 3.00 kV and 28.00 V, respectively. The source, desolvation, and column temperature were set at 120, 400, and 35 °C, respectively. The mass spectra of the samples were recorded and compared to the mass spectra of other samples reported by previous studies.

5. Conclusions

A non-conventional extraction method using ultrasonic extraction (UAE) has improved the extraction yield of vitamin B12, particularly CN-Cbl, from the green edible macroalgae *U. lactuca*. The UAE employed several important parameters, such as solvent-to-solvent, solute-to-solvent ratio, and pH. The highest CN-Cbl was obtained in the sun-dried *U. lactuca* assisted by the orbital shaking procedure (0.0356 mg/mL). The ANOVA analysis showed that low solute-to-solvent ratio, optimised solvent-to-solvent ratio, and enhanced pH increased the recovery of CN-Cbl. The highest CN-Cbl content was recovered when the extraction of oven-dried *U. lactuca* that utilised the UAE method was conducted at 3 g:60 mL solute to solvent ratio and 25:75% MeOH: H_2O at pH 4. The CN-Cbl content obtained from the oven-dried *U. lactuca* using the UAE technique was elevated compared to the boiling and orbital shaking extraction methods. The optimised parameters would help future research on how to obtain the highest yields of vitamin B12 that would be useful in the nutraceutical industry.

Supplementary Materials: The following supporting information can be downloaded at: https://www.mdpi.com/article/10.3390/molecules27144459/s1, Table S1: The concentrations of CN-Cbl obtained through the boiling, orbital shaking and ultrasonic assisted extraction methods; Table S2: Table of ANOVA for the [Cn-Cbl] extracted from ODB using 2-Level Factorial; Table S3: Table of ANOVA for the [Cn-Cbl] extracted from ADB using 2-Level Factorial; Table S4: Table of ANOVA for the [Cn-Cbl] extracted from ODU using 2-Level Factorial; Table S5: Table of ANOVA for the [Cn-Cbl] extracted from ADU using 2-Level Factorial; Table S6: Table of experimental run for the validation of CCD.

Author Contributions: D.S., N.M.N. and F.S.R.: Conceptualization, investigation, methodology, formal analysis, writing, and visualization. N.I.A. and M.I.S.: Conceptualization, methodology, validation, review, and editing. M.T. and J.K.: Validation, formal analysis, review, and editing. D.S. and N.M.N.: Conceptualization, resources, formal analysis, validation, review and editing, visualization, and overall supervision. All authors have read and agreed to the published version of the manuscript.

Funding: This research was funded by MINISTRY OF HIGHER EDUCATION (MOHE) MALAY SIA through the Fundamental Research Grant Scheme, grant number FRGS/1/2019/WAB01/UIAM/02/5.

Institutional Review Board Statement: Not applicable.

Informed Consent Statement: Not applicable.

Data Availability Statement: Data is contained within the article or Supplementary Material.

Acknowledgments: We would like to thank the Ministry of Education, Culture, Research and Technology of the Republic of Indonesia via Research Grant PDUPT for 2021–2022 for supporting this publication.

Conflicts of Interest: The authors declare no conflict of interest.

Sample Availability: The seaweed sample is available from the authors.

List of Abbreviations

ODB	Oven-dried: extracted by boiling method
SDB	Sun-dried sample, extracted by boiling method
ADB	Air-dried sample, extracted by boiling method
FDB	Freeze-dried, extracted by boiling method
ODO	Oven-dried, extracted by orbital shaking method
SDO	Sun-dried, extracted by orbital shaking method
ADO	Air-dried, extracted by orbital shaking method
FDO	Freeze-dried, extracted by orbital shaking method
ODU	Oven-dried, extracted by ultrasonic-assisted extraction method
SDU	Sun-dried, extracted by ultrasonic-assisted extraction method
ADU	Air-dried, extracted by ultrasonic-assisted extraction method
FDU	Freeze-dried, extracted by ultrasonic-assisted extraction method

References

1. Shahar, S.; Budin, S.B.; Bakar, M.A.; Umar, N.A.; Halim, J.M. Anaemia and Cognitive Function among Chinese Elderly in Old Folks Homes. *J. Sains Kesihat. Malaysia Malays. J. Health Sci.* **2005**, *3*, 1–15.
2. Del Bo, C.; Riso, P.; Gardana, C.; Brusamolino, A.; Battezzati, A.; Ciappellano, S. Effect of Two Different Sublingual Dosages of Vitamin B12 on Cobalamin Nutritional Status in Vegans and Vegetarians with a Marginal Deficiency: A Randomized. *Clin. Nutr.* **2019**, *38*, 575–583. [CrossRef] [PubMed]
3. Wong, C.W. Vitamin B12 Deficiency in the Elderly: Is It Worth Screening. *Hong Kong Med. J.* **2015**, *21*, 155–164. [CrossRef] [PubMed]
4. Langan, R.C.; Goodbred, A.J. Vitamin B12 Deficiency: Recognition and Management. *Am. Fam. Physician* **2017**, *96*, 384–389. [CrossRef] [PubMed]
5. National Center for Biotechnology Information. PubChem Compound Summary for CID 5311498, Cyanocobalamin. Available online: https://pubchem.ncbi.nlm.nih.gov/compound/Cyanocobalamin (accessed on 11 July 2022).
6. Jayasinghe, P.S.; Pahalawattaarachchi, V.; Ranaweera, K.K.D.S. Seaweed Based Jam as a Source of Nutrition. Available online: https://scholar.google.com/scholar?cluster=6002363050367060791&hl=en&as_sdt=0,5 (accessed on 10 May 2022).
7. Ganesan, A.R.; Tiwari, U.; Rajauria, G. Seaweed Nutraceuticals and Their Therapeutic Role in Disease Prevention. *Food Sci. Hum. Wellness* **2019**, *8*, 252–263. [CrossRef]
8. Wahlström, N.; Nylander, F.; Malmhäll-Bah, E.; Sjövold, K.; Edlund, U.; Westman, G.; Albers, E. Composition and Structure of Cell Wall Ulvans Recovered from Ulva Spp. along the Swedish West Coast. *Carbohydr. Polym.* **2020**, *233*, 115852. [CrossRef]
9. Yu-Qing, T.; Mahmood, K.; Shehzadi, R.; Ashraf, M.F. Ulva Lactuca and Its Polysaccharides: Food and Biomedical Aspects. *J. Biol.* **2016**, *6*, 140–151.
10. Macartain, P.; Gill, C.I.R.; Brooks, M.; Campbell, R.; Rowland, I.R. Nutritional Value of Edible Seaweeds. *Nutr. Rev.* **2007**, *65*, 535–543. [CrossRef]
11. Golshani, T.; Jorjani, E.; Chelgani, S.C.; Shafaei, S.Z.; Nafechi, Y.H. Modeling and Process Optimization for Microbial Desulfurization of Coal by Using a Two-Level Full Factorial Design. *Int. J. Min. Sci. Technol.* **2013**, *23*, 261–265. [CrossRef]
12. Gottipati, R.; Mishra, S. Process Optimization of Adsorption of Cr (VI) on Activated Carbons Prepared from Plant Precursors by a Two-Level Full Factorial Design. *Chem. Eng. J.* **2020**, *160*, 99–107. [CrossRef]
13. Jalilian, N.; Najafpour, G.D.; Khajouei, M. Enhanced Vitamin B12 Production Using Chlorella Vulgaris. *Int. J. Eng. Trans. A Basics* **2019**, *32*, 1–9. [CrossRef]
14. Kumudha, A.; Sarada, R. Effect of Different Extraction Methods on Vitamin B12 from Blue Green Algae, Spirulina Platensis. *Pharm. Anal. Acta* **2015**, *6*, 337. [CrossRef]
15. Chandra-Hioe, M.V.; Xu, H.; Arcot, J. The Efficiency of Ultrasonic-Assisted Extraction of Cyanocobalamin Is Greater than Heat Extraction. *Heliyon* **2020**, *6*, e03059. [CrossRef] [PubMed]
16. Dang, T.T.; Van Vuong, Q.; Schreider, M.J.; Bowyer, M.C.; Van Altena, I.A.; Scarlett, C.J. Optimisation of Ultrasound-Assisted Extraction Conditions for Phenolic Content and Antioxidant Activities of the Alga Hormosira Banksii Using Response Surface Methodology. *J. Appl. Phycol.* **2017**, *29*, 3161–3173. [CrossRef]
17. Madhubalaji, C.K.; Mudaliar, S.N.; Chauhan, V.S.; Sarada, R. Evaluation of Drying Methods on Nutritional Constituents and Antioxidant Activities of Chlorella Vulgaris Cultivated in an Outdoor Open Raceway Pond. *J. Appl. Phycol.* **2021**, *33*, 1419–1434. [CrossRef]
18. Azwanida, N. A Review on the Extraction Methods Use in Medicinal Plants, Principle, Strength and Limitation. *Med. Aromat. Plants* **2015**, *4*, 3–8. [CrossRef]

19. Watanabe, F.; Yabuta, Y.; Bito, T.; Teng, F. Vitamin B12-Containing Plant Food Sources for Vegetarians. *Nutrients* **2014**, *6*, 1861–1873. [CrossRef]
20. Ruslan, F.S.; Susanti, D.; Taher, M.; Mohammad, N.F. Optimization of Supercritical Fluid Extraction of Asiaticoside from Centella Asiatica Using Central Composite Design (CCD). *Sep. Sci. Technol.* **2021**, *56*, 2766–2774. [CrossRef]
21. Rodrigues, D.; Sousa, S.; Silva, A.; Amorim, M.; Pereira, L.; Rocha-Santos, T.A.P.; Gomes, A.M.P.; Duarte, A.C.; Freitas, A.C. Impact of Enzyme- and Ultrasound-Assisted Extraction Methods on Biological Properties of Red, Brown, and Green Seaweeds from the Central West Coast of Portugal. *J. Agric. Food Chem.* **2015**, *63*, 3177–3188. [CrossRef]
22. Ciko, A.M.; Jokić, S.; Šubarić, D.; Jerković, I. Overview on the Application of Modern Methods for the Extraction of Bioactive Compounds from Marine Macroalgae. *Mar. Drugs* **2018**, *16*, 348. [CrossRef]
23. Silva, A.F.R.; Abreu, H.; Silva, A.M.S.; Cardoso, S.M. Effect of Oven-Drying on the Recovery of Valuable Compounds from Ulva Rigida, Gracilaria Sp. and Fucus Vesiculosus. *Mar. Drugs* **2019**, *17*, 90. [CrossRef] [PubMed]
24. Fang, F.; Kang, X.J.; Liu, Z.Y.; Ma, Y.Q.; Gu, Z.Z. Use of Packed-Fiber Solid-Phase Extraction for Sample Clean-up and Preconcentration of Vitamin B12 before Determination. *Chinese Chem. Lett.* **2009**, *20*, 1491–1494. [CrossRef]
25. Lovander, M.D.; Lyon, J.D.; Parr, D.L.; Wang, J.; Parke, B.; Leddy, J. Critical Review—Electrochemical Properties of 13 Vitamins: A Critical Review and Assessment. *J. Electrochem. Soc.* **2018**, *165*, G18–G49. [CrossRef]
26. Bajaj, S.R.; Singhal, R.S. Degradation Kinetics of Vitamin B12 in Model Systems of Different PH and Extrapolation to Carrot and Lime Juices. *J. Food Eng.* **2020**, *272*, 109800. [CrossRef]
27. House, E.; Estate, W.I.; Jigani, B.; Rd, L.; Technologies, A.; Park, B.; Rd, J.L. No Title. *Single-Laboratory Valid. AOAC.* **2020**, *1*, 3–8. [CrossRef]
28. Ohmori, S.; Kataoka, M.; Koyama, H. Stability of Cyanocobalamin in Sugar-Coated Tablets. *Int. J. Pharm.* **2007**, *337*, 161–168. [CrossRef]
29. Watanabe, F.; Schwarz, J.; Takenaka, S.; Miyamoto, E.; Ohishi, N.; Nelle, E.; Hochstrasser, R.; Yabuta, Y. Characterization of Vitamin B12 Compounds in the Wild Edible Mushrooms Black Trumpet (Craterellus Cornucopioides) and Golden Chanterelle (Cantharellus Cibarius). *J. Nutr. Sci. Vitaminol.* **2012**, *58*, 438–441. [CrossRef]

MDPI
St. Alban-Anlage 66
4052 Basel
Switzerland
Tel. +41 61 683 77 34
Fax +41 61 302 89 18
www.mdpi.com

Molecules Editorial Office
E-mail: molecules@mdpi.com
www.mdpi.com/journal/molecules

www.ingramcontent.com/pod-product-compliance
Lightning Source LLC
Chambersburg PA
CBHW061808210926
51850CB00006B/647